Advances in Soil Science

SOIL EROSION AND CARBON DYNAMICS

Advances in Soil Science

Series Editor: B. A. Stewart

Published Titles

Interacting Processes in Soil Science
R. J. Wagenet, P. Baveye, and B. A. Stewart

Soil Management: Experimental Basis for Sustainability and Environmental Quality
R. Lal and B. A. Stewart

Soil Management and Greenhouse Effect
R. Lal, J. M. Kimble, E. Levine, and B. A. Stewart

Soils and Global Change
R. Lal, J. M. Kimble, E. Levine, and B. A. Stewart

Soil Structure: Its Development and Function
B. A. Stewart and K. H. Hartge

Structure and Organic Matter Storage in Agricultural Soils
M. R. Carter and B. A. Stewart

Methods for Assessment of Soil Degradation
R. Lal, W. H. Blum, C. Valentine, and B. A. Stewart

Soil Processes and the Carbon Cycle
R. Lal, J. M. Kimble, R. F. Follett, and B. A. Stewart

Global Climate Change: Cold Regions Ecosystems
R. Lal, J. M. Kimble, and B. A. Stewart

Assessment Methods for Soil Carbon
R. Lal, J. M. Kimble, R. F. Follett, and B. A. Stewart

Soil Erosion and Carbon Dynamics
E.J. Roose, R. Lal, C. Feller, B. Barthès, and B. A. Stewart

Advances in Soil Science

SOIL EROSION AND CARBON DYNAMICS

Edited by

**Eric J. Roose
Rattan Lal
Christian Feller
Bernard Barthès
Bobby A. Stewart**

Boca Raton   London   New York

A CRC title, part of the Taylor & Francis imprint, a member of the
Taylor & Francis Group, the academic division of T&F Informa plc.

On the cover: A typical landscape of red ferrallitic soils on the high plateau of central Madagascar near Antananarivo during the rainy season. The hilltop is covered by overgrazed grassland deeply eroded around the cattle trails which join the village and the springs in the valley. The hills lose carbon, nutrients, soil, and water, but might be nourishing the rice paddies below.

Published in 2006 by
CRC Press
Taylor & Francis Group
6000 Broken Sound Parkway NW, Suite 300
Boca Raton, FL 33487-2742

© 2006 by Taylor & Francis Group, LLC
CRC Press is an imprint of Taylor & Francis Group

No claim to original U.S. Government works
Printed in the United States of America on acid-free paper
10 9 8 7 6 5 4 3 2 1

International Standard Book Number-10: 1-56670-688-2 (Hardcover)
International Standard Book Number-13: 978-1-56670-688-9 (Hardcover)
Library of Congress Card Number 2005050888

This book contains information obtained from authentic and highly regarded sources. Reprinted material is quoted with permission, and sources are indicated. A wide variety of references are listed. Reasonable efforts have been made to publish reliable data and information, but the author and the publisher cannot assume responsibility for the validity of all materials or for the consequences of their use.

No part of this book may be reprinted, reproduced, transmitted, or utilized in any form by any electronic, mechanical, or other means, now known or hereafter invented, including photocopying, microfilming, and recording, or in any information storage or retrieval system, without written permission from the publishers.

For permission to photocopy or use material electronically from this work, please access www.copyright.com (http://www.copyright.com/) or contact the Copyright Clearance Center, Inc. (CCC) 222 Rosewood Drive, Danvers, MA 01923, 978-750-8400. CCC is a not-for-profit organization that provides licenses and registration for a variety of users. For organizations that have been granted a photocopy license by the CCC, a separate system of payment has been arranged.

Trademark Notice: Product or corporate names may be trademarks or registered trademarks, and are used only for identification and explanation without intent to infringe.

Library of Congress Cataloging-in-Publication Data

Soil erosion and carbon dynamics / edited by Eric J. Roose ... [et al.].
    p. cm. -- (Advances in soil science)
    Papers presented at a symposium held in Montpellier, France, September 23-28, 2002.
  Includes bibliographical references and index.
  ISBN 1-56670-688-2 (alk. paper)
  1. Soil erosion--Congresses. 2. Carbon cycle (Biogeochemistry)--Congresses. 3. Soils--Carbon content--Congresses. I. Roose, Eric. II. Advances in soil science (Boca Raton, Fla.)

S622.2S64 2005
631.4'5--dc22   2005050888

Visit the Taylor & Francis Web site at
http://www.taylorandfrancis.com

and the CRC Press Web site at
http://www.crcpress.com

Preface

Global climate change is one of the most serious environmental issues of the twenty-first century. Change in climate has already occurred, with a global increase in temperature of about 0.6°C during the twentieth century, and considering different scenarios, projected increase in temperature will continue to 4 to 6°C by the end of the twenty-first century. Drastic consequences may sometimes be positive, but are most often negative.

The carbon (C) cycle plays a significant role in the global climate change, both in the causes of and the solutions for its mitigation. The stakes are explicitly stated in the proceedings of several international conventions (Rio in 1992, Kyoto in 1997, Marrakech in 2001), and are relevant to the sustainable management of soil and water resources for agriculture.

An important point to consider is that about 2000 Gt C directly related to the atmospheric CO_2 is stored as a terrestrial sink in the continental biosphere (vegetation + soil). So the biosphere naturally acts as a C sink. The challenge is whether this sink can be increased by anthropic intervention involving recommended land use and soil/crop management.

The notion of C sequestration in agroecosystems is a generic concept including C storage (derived from CO_2) in the soil-plant system and fluxes (negative or positive) of others non-CO_2 greenhouse gases (GHGs, i.e., CH_4 and N_2O expressed on an equivalent CO_2-C basis) induced by ecosystem management. This book focuses more on CO_2-C and organic C fluxes in soils, sediments, and rivers than on the others GHGs.

There is a need to quantify these different C fluxes and how they are affected by management alternatives leading to an increase in soil-plant C storage, plant productivity, biodiversity, and protection against erosion.

While the magnitude and severity of soil erosion are well documented, fluxes of eroded C are rarely quantified, and a lot of uncertainties exist about the transfer of particulate organic C from the plot level to the large watershed level into the river, lake, or marine sediments. Thus, the original objective of this book is to collate quantitative data on eroded C fluxes from the scale of the agricultural plot to the large watershed.

Quantifying eroded C at the plot level is important for the true evaluation of the C sequestration due to an improved management system. As a matter of fact, in many studies, soil "C sequestration" at the plot scale is calculated by difference between C stocks under conventional (reference plot) management and the improved management, the difference being attributed to a storage of atmospheric CO_2 in the soil (via the plant). However, a part of the difference may also be due to transfers of particulate (and soluble) C by erosion and sedimentation. In many situations, erosion is not negligible on the reference plot, and the eroded C flux cannot be considered as a CO_2-flux. Therefore, "C sequestration" and the impact of so-called improved management may be grossly overestimated. Thus, the second objective of this book is to quantify, for different soil management practices, the magnitude of eroded C, and to compare this magnitude to the true value of C sequestration to determine if these are of the same order of magnitude.

There is little experimental data about the fate of eroded C (particulate and soluble) during its transfer along the slope and its deposition into lake or marine sediments. Water eroded C is generated by the breakdown of soil aggregates, a first process that exposes soil organic C to microbial processes and thus increase C mineralization and CO_2 emissions. However, little information is available on C turnover during transportation and after deposition. It is not known whether CO_2 fluxes increase during the transport of particulate and soluble C. And finally, if organic C buried with the sediments is easily mineralized or protected vis-à-vis that *in situ* or in the original soil. In other words, does eroded C contribute to C sequestration or to CO_2 emissions? Thus the third main objective of this book is to discuss the fate of the eroded C, whether it is a source or sink for atmospheric CO_2?

This volume is based on the first symposium of the international colloquium Land Uses, Erosion and Carbon Sequestration held in Montpellier, France, September 23 through 28, 2002. The three main objectives presented above also formed the basis and organizational/scientific structure of this symposium on carbon erosion. In addition to the introduction and conclusion sections, the book is divided into three main parts:

- Basic concepts and general approaches to the global C cycle, erosion, and eroded C
- Soil and C erosion from agroecosystems at the plot scale
- Solubilization and carbon transfers in rivers and deposition in sediments

This book contains twenty of the invited papers presented at the symposium on carbon erosion. The symposium was organized by a team of researchers from the Réseau Erosion from IRD (Institut de recherche pour le développement, Dr. E. Roose, Dr. G. De Noni, Dr. C. Prat, Montpellier, France), from the laboratory IRD-MOST (Matière organique des sols tropicaux, Dr. C. Feller, B. Barthès, Montpellier, France), from CIRAD (Centre de coopération internationale en recherche agronomique pour le développement, Dr. F. Ganry, Dr. G. Bourgeon, Montpellier, France), and from the Carbon Management and Sequestration Center of the Ohio State University (Dr. R. Lal, Columbus, Ohio, U.S.).

Collation and synthesis of this research information would not have been possible without scientific, financial, and personal support from IRD (Délégation à l'information et à la communication, M. N. Favier and T. Mourier, Paris, and research unit Seq-C, Dr. C. Feller), from CIRAD (Agronomie program, Dr. F. Ganry), from CNRS (Centre national de la recherche scientifique, Laboratoire des mécanismes de transfert en géologie, Dr. J. L. Probst, Toulouse, France), from INRA (Institut national de la recherche agronomique, Dr. M. Robert, Paris), from ESSC (European Society for Soil Conservation, Dr. J. L. Rubio, Valencia, Spain), from WASWC (World Association for Soil and Water Conservation, Dr. S. Sombapanit, Bangkok), from the European COST 623 (Dr. J. Baade, Jena, Germany), from CTA (European Technical Center for Agricultural and Rural Cooperation, Wageningen, The Netherlands), from the French Ministries of Environment, Culture (Délégation à la défense de la langue française), and Foreign Affairs (Direction de la coopération scientifique, universitaire et de recherche), from the Conseil Général de l'Hérault (France), and from the Agropolis association (Montpellier).

The editors thank all the authors for their outstanding efforts to document and present their research and their current understanding of soil erosion and carbon dynamics. Their efforts have advanced our knowledge with regard to the fate of eroded carbon and management strategies to minimize the loss of carbon by accelerated erosion.

The editors also thank the staff of CRC/Taylor and Francis for their help and support in the timely publication of this volume, and for making this information available to the scientific community. In addition to oversight by the editorial board, several manuscripts were reviewed by Drs. M. Meybeck, O. Planchon, and P. Seyler. Editorial changes were incorporated in each manuscript by the staff of the Carbon Management and Sequestration Center of the Ohio State University, Columbus, Ohio, U.S. All these efforts have advanced the frontiers of soil science and improved the understanding of the fate of eroded carbon at scales ranging from a plot to a watershed, and linking soil erosion to the global carbon cycle.

—The editorial board

About the Editors

Eric Roose, Ph.D., received an agronomist engineer degree from the Louvain University (Belgium), Ph.D. in Soil Science from Abidjan University (Ivory Coast), and Docteur ès-Science from the Orléans University (France). He is a soil scientist who specializes in soil conservation and restoration at the Institut de Recherche pour le Développement (IRD). Born in Brussels, he has lived in France and Ivory Coast and worked in many developing countries of western, central, and northern Africa and French West Indies since 1963. From 1964 to 1978, Dr. Roose was a soil scientist at the Adiopodoumé IRD Center, near Abidjan conducting long-term experiments on Ferralllitic and Ferrugineous tropical soils and studying dynamics under natural and cultivated ecosystems between the equatorial rain forest condition of southern Ivory Coast to the sudano-sahelian savannas of Central Burkina Faso. There he assessed erosion parameters of the USLE model applied to western Africa, soil and water conservation practices, soil carbon dynamics in relation to land uses, nutrients leaching in lysimeters, colloids and water balance, and earth worms and termite activities in relation to soil erosion. He developed three rainfall simulators and many devices to measure infiltration rate on steep slopes, soil surface features, erosion risks in Africa and in France. He coordinated a network of 600 French-speaking researchers and developers working on soil erosion and restoration in 50 countries and organized 16 international symposia and coedited 23 Bulletins du Réseau Erosion (> 10,000 pages on the Net). He has authored, edited, or coedited 360 scientific publications (3 books written and 17 coedited, 162 scientific papers, 110 symposia papers, 50 expert reports). In 2000, Dr. Roose joined the research unit Carbon Sequestration at the IRD center of Montpellier and organized the international Colloquium Land uses, Erosion, and Carbon Sequestration. He continues to research on carbon erosion/sequestration, carbon related to soil aggregation and erodibility, and traditional systems of soil and water conservation in the Mediterranean mountains. He has participated in several expert meetings organized by FAO, FIDA, World Bank, NGOs, and French Cooperation. Dr. Roose is a fellow of the International Union for Soil Science and the French Society of Soil Science, a member of the board of the European Society for Soil Conservation (ESSC) and of the International Soil Conservation Organization (ISCO), the vice president of the World Association for Soil and Water Conservation (WASWC), and the president of the French Erosion Network at the Agence Universitaire de la Francophonie (AUF). He is the recipient of the WASWC, AUF, and of the Académie française d'Agriculture de Paris awards.

Rattan Lal, Ph.D., is a professor of soil physics in the School of Natural Resources and Director of the Carbon Management and Sequestration Center, FAES/OARDC at The Ohio State University. Prior to joining Ohio State in 1987, he was a soil physicist for 18 years at the International Institute of Tropical Agriculture, Ibadan, Nigeria. In Africa, Professor Lal conducted long-term experiments on land use, watershed management, soil erosion processes as influenced by rainfall characteristics, soil properties, methods of deforestation, soil-tillage and crop-residue management, cropping systems including cover crops and agroforestry, and mixed/relay cropping methods. He also assessed the impact of soil erosion on crop yield and related erosion-induced changes in soil properties to crop growth and yield. Since joining The Ohio State University in 1987, he has continued research on erosion-induced changes in soil quality and developed a new project on soils and climate change. He has demonstrated that accelerated soil erosion is a major factor affecting emission of carbon from soil to the atmosphere. Soil-erosion control and adoption of conservation-effective measures can lead to carbon sequestration and mitigation of the greenhouse effect. Other research interests include soil compaction, conservation tillage, mine soil reclamation, water table management, and sustainable use of soil and water resources of the tropics for enhancing food security. Professor Lal is a fellow of the Soil Science Society of America, American Society of Agronomy, Third World Academy of Sciences, American Association for the

Advancement of Sciences, Soil and Water Conservation Society, and Indian Academy of Agricultural Sciences. He is the recipient of the International Soil Science Award, the Soil Science Applied Research Award, the Soil Science Research Award of the Soil Science Society of America, the International Agronomy Award, the Environment Quality Research Award of the American Society of Agronomy, the Hugh Hammond Bennett Award of the Soil and Water Conservation Society, and the Borlaug Award. He is the recipient of an honorary degree of Doctor of Science from Punjab Agricultural University, India and of the Norwegian University of Life Sciences, Aas, Norway. He is past president of the World Association of the Soil and Water Conservation and the International Soil Tillage Research Organization. He was a member of the U.S. National Committee on Soil Science of the National Academy of Sciences (1998 to 2002). He has served on the Panel on Sustainable Agriculture and the Environment in the Humid Tropics of the National Academy of Sciences. He has authored and coauthored about 1,100 research publications. He has also written nine and edited or coedited 43 books.

Christian Feller, Ph.D. in organic chemistry and "Docteur d'Etat" in soil science, is a soil scientist with the Institut de Recherche pour le Développement (IRD), with a specialization in soil organic matter (SOM) functions and dynamics. He has lived and worked in France (Paris, Nancy, Montpellier) and in different developing countries including Senegal, French West Indies, Brazil, and Madagascar. Since 1975, he has studied the nature and the different functions of SOM in tropical soils in relation to fertility and environment. He developed a laboratory methodology for assessment and characterization of soil functional organic compounds and was pioneer in proposing the use of the isotope ^{13}C in natural abundance as a powerful tool for the quantitative evaluation of the origin and dynamics of SOM. Since 1998, he has been involved in the evaluation of the role of agricultural systems in carbon sequestration in tropical soils taking into account the nature of the soils, the amount and quality of organic matter applied, the forms and dynamics of the sequestered soil carbon, and the importance of aggregation in the protection of soil organic carbon (SOC) against mineralization. In collaboration with IRD colleagues, he has demonstrated that different alternatives to conventional agricultural systems are efficient (between 0 to 1.5 tC/ha/year) in sequestering SOC in tropical soils, including agroforestry systems, no-tillage and cover crop systems, no burning of sugarcane residues, improved pastures, etc. Additional research interests include application of soil science to other disciplines such as road geotechnic, archaeology, traditional farmer's perception of soil fertility and the environment, and the history of soil science. Dr. Feller is a fellow of the Soil Science Society of America, American Society of Agronomy, International Union for Soil Science, French Society of Soil Science, and French Academy of Agriculture. He is past director of different Units of Research (UR) at IRD and past director of the Institut Fédératif de Recherche (IFR 124) "Ecosystems" in Montpellier. He has authored and coauthored about 149 research publications. He has also written two books, and edited or coedited four more.

Bernard Barthès is an agronomist engineer and soil scientist. He has worked at the French Institut de Recherche pour le Développement (IRD) since 1985, first in (French) Guiana, then in Congo-Brazzaville, and presently in Montpellier, France. In Guiana, he studied the three-dimensional organization of soils and how it influences the distribution of natural or cultivated vegetation. He observed that the spatial distribution of some tree species of rain forest as well as cassava tuber yields were strongly influenced by soil parameters such as the thickness of permeable horizons and waterlogging. In Congo, his work related to the hydrophysical properties of soils under cassava-based cropping systems. In Montpellier, he has been studying the influence of land use and cropping systems on soil organic matter, aggregation and erodibility, and the relationships between these soil parameters. He showed that topsoil aggregate stability is a simple and relevant indicator but also an important determinant of soil's susceptibility to runoff and water erosion, especially in Mediterranean and tropical regions. Additionally, he contributed to studies that demonstrated the interest of legume cover crops for sustainable soil management in maize-based cropping systems in Benin. From 2001 to 2004, Bernard Barthès was the director of the IRD-MOST laboratory (Matière Organique des Sols Tropicaux) in Montpellier. He is author or coauthor of about 30 articles and 30 other scientific papers. He is a member of the French association for soil sciences (AFES).

Bobby A. Stewart, Ph.D., is Distinguished Professor of Soil Science at West Texas A&M University, Canyon, Texas. He is also Director of the Dryland Agriculture Institute and former Director of the USDA Conservation and Production Research Laboratory at Bushland, Texas; past president of the Soil Science Society of America; and member of the 1990–1993 Committee on Long Range Soil and Water National Research Council, National Academy of Sciences. Dr. Stewart is very supportive of education and research on dryland agriculture. The Stewart Dryland Agriculture Scholarship was established in 1994 to honor Dr. Stewart on his retirement from the USDA Agriculture Research Service Conservation and Production Research Laboratory and to provide scholarships for undergraduate and graduate students with a demonstrated interest in dryland agriculture.

Contributors

Saadi Abdeljaoued
Department of Geology
University of Tunis
Tunis, El Manar, Tunisia

Jean Albergel
IRD
Montpellier, France

Alain Albrecht
IRD
UR SeqBio
Montpellier, France

Mourad Arabi
Institut National Recherche Forestiere
Ain D'Heb, Medea, Algeria

C. D. Arbelo
Department of Soil Science and Geology
Faculty of Biology
University of La Laguna Avda. Astrofísico
 Francisco Sánchez
La Laguna, Tenerife, Canary Islands, Spain

C. M. Armas
Department of Soil Science and Geology
Faculty of Biology
University of La Laguna Avda. Astrofísico
 Francisco Sánchez
La Laguna, Tenerife, Canary Islands, Spain

Anastase Azontonde
Laboratoire des Sciences du Sol, Eaux et
 Environnement
Cotonou, Benin

Bernard Barthès
IRD
UR SeqBio
Montpellier, France

Boris Bellanger
UPMC-INRA-CNRS
Biogéochimie des milieux continentaux
Université Pierre & Marie Curie
Paris, France

Abdellah Ben Mamou
Department of Geology
University of Tunis
Tunis, El Manar, Tunisia

J. Bernadou
IRD
Érosion et changements d'usage des terres
Vientiane, Laos

Martial Bernoux
IRD
UR SeqBio
Montpellier, France

A. Bilgo
Institut de l'Environnement et de Recherches
 Agricoles
Burkina Faso

Eric Blanchart
IRD
UR SeqBio
Montpellier, France

Jean-Loup Boeglin
Laboratoire des Mécanismes et Transferts en
 Géologie
Toulouse, France

Philippe Bonté
UMR
Laboratoire des Sciences du Climat et de
 l'Environnement
Gif-sur-Yvette, France

Anja Boye
World Agroforestry Centre
Kisumu, Kenya

Jean-Jacques Braun
Indian Institute of Science
Department of Metallurgy
Bangalore, India

Jean-Pierre Bricquet
IRD
Montpellier, France

Michel Brossard
IRD
UR SeqBio
Montpellier, France

Didier Brunet
IRD
UR SeqBio
Montpellier, France

Plinio Barbosa de Camargo
CENA
Piracicaba, Brazil

Carlos C. Cerri
CENA-USP
Piracicaba, Brazil

Carlos E. P. Cerri
CENA-USP
Piracicaba, Brazil

Vincent Chaplot
IRD
Érosion et changements d'usage des terres
Vientiane, Laos

R. Chikowo
Department of Soil Science and Agricultural Engineering
University of Zimbabwe
Harare, Zimbabwe

T. Chirwa
Zambia-ICRAF Agroforestry Project
Chipata, Zambia

C. Colas
TGM (DGO, UMR CNRS-EPOC 5805)
Université Bordeaux
Talence, France

Marcelo Corrêia Bernardes
CENA
Piracicaba, Brazil

A. Coynel
TGM
Université Bordeaux
Talence, France

Vincent Eschenbrenner
IRD
UR SeqBio
Montpellier, France

Henri Etcheber
Centre de Recherche sur les Environnements Sédimentaires et Océaniques
University Bordeaux
Bordeaux, France

Christian Feller
IRD
UR SeqBio
Antananarivo, Madagascar

Jacques Fournier
Ecole Inter-Etats des Techniciens de l'Hydraulique et de l'Equipement Rural
Burkina Faso

Cyril Girardin
UPMC-INRA-CNRS
Biogéochimie des milieux continentaux
Université Pierre & Marie Curie
Paris, France

Gerard Govers
Physical and Regional Geography Research Group
Leuven, Belgium

J. A. Guerra
Department of Soil Science and Geology
Faculty of Biology
University of La Laguna Avda. Astrofísico Francisco Sánchez
La Laguna, Tenerife, Canary Islands, Spain

Jean Luc Guyot
IRD-LMTG
Université P. Sabatier
Toulouse, France

Goswin Heckrath
Danish Institute for Agricultural Sciences
Department of Crop Physiology and Soil Science
Tjele, Denmark

Victor Hien
Institut de l'Environnement et de Recherches Agricoles
Burkina Faso

Sylvain Huon
UPMC-INRA-CNRS
Biogéochimie des milieux continentaux
Université Pierre & Marie Curie
Paris, France

Bounmanh Khamsouk
Vauclin, Martinique

Rattan Lal
Carbon Management and Sequestration Center
The Ohio State University
Columbus, Ohio

Andre Laraque
IRD-LMTG
Université P. Sabatier
Toulouse, France

Yves Le Bissonnais
INRA
Science du Sol
Montpellier, France

Maria Inês Lopes de Oliveira
EMBRAPA Cerrados
Planaltina, Brazil

P. L. Mafongoya
Zambia-ICRAF Agroforestry Project
Chipata, Zambia

Taoufik Mansouri
Laboratory of Remote Sensing and GIS
University of Gafsa
Gafsa, Tunisia

André Mariotti
UPMC-INRA-CNRS
Biogéochimie des milieux continentaux
Université Pierre & Marie Curie
Paris, France

Luiz Antonio Martinelli
CENA
Piracicaba, Brazil

Dominique Masse
Institut de Recherches pour le Développement (IRD)
Burkina Faso

Mohamed Mazour
Université de Tlemcen, Hydrologie
Tlemcen, Algeria

Nadjia Mededjel
INRF
Mansourah, Tlemcen, Algeria.

Michel Meybeck
UMR SISYPHE
Université Paris
Paris, France

J. L. Mora
Department of Soil Science and Geology
Faculty of Biology
University of La Laguna Avda. Astrofísico Francisco Sánchez
La Laguna, Tenerife, Canary Islands, Spain

Patricia Moreira-Turcq
IRD-LMTG
Université P. Sabatier
Toulouse, France

Boutkhil Morsli
INRF
Mansourah, Tlemcen, Algeria

Jefferson Mortatti
CENA
University São Paulo
Piracicaba, Brazil

Jules-Rémy Ndam-Ngoupayou
Département des Sciences de la Terre
University Yaoundé
Cameroon

G. Nyamadzawo
Department of Soil Science and Agricultural Engineering
University of Zimbabwe
Harare, Zimbabwe

P. Nyamugafata
Department of Soil Science and Agricultural Engineering
University of Zimbabwe
Harare, Zimbabwe

Brunot Nyeck
Département des Sciences de la Terre
University Yaoundé
Cameroon

Robert Oliver
Laboratoire MOST (CIRAD)
Montpellier, France

Jean-Claude Olivry
IRD-LMTG
Université P. Sabatier
Toulouse, France

Jean-Pierre Henry Balbaud Ometto
CENA
Piracicaba, Brazil

Didier Orange
IRD-UR ECU
Paris, France

Pascal Podwojewski
IRD UR
Erosion et changement d'usage des sols
Hanoi, Vietnam

Jean-Luc Probst
Laboratoire des Mécanismes et Transferts en
  Géologie
Toulouse, France

Tim A. Quine
University of Exeter
Department of Geography
Exeter, England

Michel Robert (deceased)
Ministère de l'Ecologie et du Développement
  Durable
Académie d'Agriculture
Paris, France

Antonio Rodríguez Rodríguez
Department of Soil Science and Geology
Faculty of Biology
University of La Laguna Avda. Astrofísico
  Francisco Sánchez
La Laguna, Tenerife, Canary Islands, Spain

Eric Roose
IRD
UR SeqBio
Montpellier, France

Anneke de Rouw
IRD UR
Erosion et changement d'usage des sols (ECU)
Vientiane, Laos

Georges Serpantié
Institut de Recherches pour le Développement
Antananarivo, Madagascar

Patrick Seyler
IRD-LMTG
Université Paul Sabatier
Toulouse, France

Stéphane Sogon
UMR
Laboratoire des Sciences du Climat et de
  l'Environnement
Gif-sur-Yvette, France

Christian Valentin
Erosion et changement d'usage des sols
Vientiane, Laos

Kristof Van Oost
University of Exeter
Department of Geography
Exeter, England

Fernando Velasquez
Centro de Ecologia de Boconó
Universidad de los Andes
Boconó, Venezuela

Cécile Villenave
Laboratoire d'Ecologie Microbienne
Université Lyon,
Villeurbanne, France

Patrick Zante
IRD
UR LISAH
Montpellier, France

Contents

Section 1
Basic Concepts

Chapter 1 Global Change and Carbon Cycle: The Position of Soils and Agriculture 3
Michel Robert

Chapter 2 Soil Carbon Sequestration 13
Martial Bernoux, Christian Feller, Carlos C. Cerri, Vincent Eschenbrenner, and Carlos E. P. Cerri

Chapter 3 Influence of Soil Erosion on Carbon Dynamics in the World 23
Rattan Lal

Chapter 4 Modeling Soil Erosion Induced Carbon Fluxes between Soil and Atmosphere on Agricultural Land Using SPEROS-C 37
Kristof Van Oost, Tim A. Quine, Gerard Govers, and Goswin Heckrath

Section 2
Erosion at the Plot Scale

Chapter 5 Soil Carbon Erosion and Its Selectivity at the Plot Scale in Tropical and Mediterranean Regions 55
Eric Roose and Bernard Barthès

Chapter 6 Organic Carbon in Forest Andosols of the Canary Islands and Effects of Deforestation on Carbon Losses by Water Erosion 73
Antonio Rodríguez Rodríguez, C. D. Arbelo, J. A. Guerra, J. L. Mora, and C. M. Armas

Chapter 7 Soil Carbon Dynamics and Losses by Erosion and Leaching in Banana Cropping Systems with Different Practices (Nitisol, Martinique, West Indies) 87
Eric Blanchart, Eric Roose, and Bounmanh Khamsouk

Chapter 8 Influence of Land Use, Soils, and Cultural Practices on Erosion, Eroded Carbon, and Soil Carbon Stocks at the Plot Scale in the Mediterranean Mountains of Northern Algeria 103
Boutkhil Morsli, Mohamed Mazour, Mourad Arabi, Nadjia Mededjel, and Eric Roose

Chapter 9 Carbon, Nitrogen, and Fine Particles Removed by Water Erosion on Crops, Fallows, and Mixed Plots in Sudanese Savannas (Burkina Faso) 125
A. Bilgo, Georges Serpantié, Dominique Masse, Jacques Fournier, and Victor Hien

Chapter 10 Effect of a Legume Cover Crop on Carbon Storage and Erosion in an Ultisol under Maize Cultivation in Southern Benin..143
Bernard Barthès, Anastase Azontonde, Eric Blanchart, Cyril Girardin, Cécile Villenave, Robert Oliver, and Christian Feller

Chapter 11 Organic Carbon Associated with Eroded Sediments from Micro-Plots under Natural Rainfall from Cultivated Pastures on a Clayey Ferralsol in the Cerrados (Brazil) 157
Didier Brunet, Michel Brossard, and Maria Inês Lopes de Oliveira

Chapter 12 Runoff, Soil, and Soil Organic Carbon Losses within a Small Sloping-Land Catchment of Laos under Shifting Cultivation 167
Vincent Chaplot, Yves Le Bissonnais, and J. Bernadou

Chapter 13 Soil Erodibility Control and Soil Carbon Losses under Short-Term Tree Fallows in Western Kenya 181
Anja Boye and Alain Albrecht

Chapter 14 Soil and Carbon Losses under Rainfall Simulation from Two Contrasting Soils under Maize-Improved Fallows Rotation in Eastern Zambia......................197
G. Nyamadzawo, P. Nyamugafata, R. Chikowo, T. Chirwa, and P. L. Mafongoya

Section 3
Carbon Transfer in Rivers

Chapter 15 Origins and Behaviors of Carbon Species in World Rivers 209
Michel Meybeck

Chapter 16 Carbon, Nitrogen, and Stable Carbon Isotope Composition and Land-Use Changes in Rivers of Brazil..239
Luiz Antonio Martinelli, Plinio Barbosa de Camargo, Marcelo Corrêia Bernardes, and Jean-Pierre Henry Balbaud Ometto

Chapter 17 Organic Carbon Transported by the Equatorial Rivers: Example of Congo-Zaire and Amazon Basins..255
Patrick Seyler, A. Coynel, Patricia Moreira-Turcq, Henri Etcheber, C. Colas, Didier Orange, Jean-Pierre Bricquet, André Laraque, Jean Luc Guyot, Jean-Claude Olivry, and Michel Meybeck

Chapter 18 Soil Carbon Stock and River Carbon Fluxes in Humid Tropical Environments: The Nyong River Basin (South Cameroon) 275
Jean-Loup Boeglin, Jean-Luc Probst, Jules-Rémy Ndam-Ngoupayou, Brunot Nyeck, Henri Etcheber, Jefferson Mortatti, and Jean-Jacques Braun

Chapter 19 Organic Carbon in the Sediments of Hill Dams in a Semiarid Mediterranean Area..289
Jean Albergel, Taoufik Mansouri, Patrick Zante, Abdellah Ben Mamou, and Saadi Abdeljaoued

Chapter 20 Monitoring Soil Organic Carbon Erosion with Isotopic Tracers: Two Case Studies on Cultivated Tropical Catchments with Steep Slopes (Laos, Venezuela) 301
Sylvain Huon, Boris Bellanger, Philippe Bonté, Stéphane Sogon, Pascal Podwojewski, Cyril Girardin, Christian Valentin, Anneke de Rouw, Fernando Velasquez, Jean-Pierre Bricquet, and André Mariotti

Section 4
Conclusions

Chapter 21 Erosion and Carbon Dynamics: Conclusions and Perspectives..........................331
Eric Roose, Michel Meybeck, Rattan Lal, and Christian Feller

Index 339

Section 1
Basic Concepts

CHAPTER 1

Global Change and Carbon Cycle: The Position of Soils and Agriculture

Michel Robert

CONTENTS

1.1 Introduction 3
1.2 Climate Change: Causes and Consequences 4
1.3 Carbon Cycle in the Continental Biosphere 5
    1.3.1 Past and Present Stocks and Fluxes 6
    1.3.2 Increasing C Sequestration 8
1.4 Main Challenges to Soil C Sequestration 9
    1.4.1 Soil Carbon Sequestration and International Conventions 9
        1.4.1.1 Climatic Change (UNFCC Convention) 9
        1.4.1.2 Convention on Biodiversity (UNCBD) 9
        1.4.1.3 Convention on Desertification (UNFCCD) 9
    1.4.2 Development of a Sustainable Agriculture 10
1.5 Conclusions 11
References 12

1.1 INTRODUCTION

Global climate change is the most serious environmental problem of the twenty-first century (IPCC, 2001). The change in climate has already begun with a global increase in temperature during the twentieth century, and considering different scenarios, projected increase in temperature will continue until the end of the twenty-first century. The consequences are important and may be sometimes positive, but are most often negative.

The carbon cycle plays a significant role in global climate change both in its causes and in its remediation. International conventions and the sustainable management of the land by agriculture can be important forces in using the carbon cycle for the remediation of global warming brought about by the greenhouse effect.

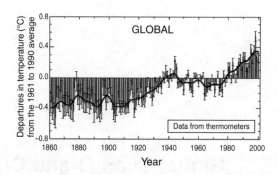

Figure 1.1 Variations of the earth surface temperature for the past 140 years. (From IPCC, 2001. Climatic change 2001. Synthesis report. A contribution of Working Groups I, II, and III to the third assessment report of the Intergovernmental panel on climate change. R. T. Watson and the Core Writing Team, Eds. Cambridge University Press, United Kingdom, 398 p.)

1.2 CLIMATE CHANGE: CAUSES AND CONSEQUENCES

IPCC (2001) reported an increase in temperature of +0.6°C during the twentieth century (Figure 1.1) and the projected increase in surface temperature, based on a large range of scenarios, ranges between 1.4 to 5.8°C (and even higher increase for the northern latitudes). If natural causes have played a great role in explaining the precedent historical variations (for example what is called in Europe "the short ice period"), the recent increase in temperature is now well correlated with anthropic causes and especially with the increase in concentration of different greenhouse gases: CO_2, CH_4, N_2O (Figure 1.2). The respective emissions of greenhouse gases by different countries

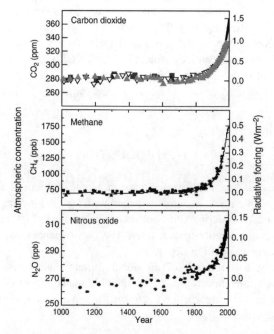

Figure 1.2 Variation of the atmospheric concentration of the main greenhouse gas. (From IPCC, 2001. Climatic change 2001. Synthesis report. A contribution of Working Groups I, II, and III to the third assessment report of the Intergovernmental panel on climate change. R. T. Watson and the Core Writing Team, Eds. Cambridge University Press, United Kingdom, 398 p.)

is also well known. For example, CO_2 emissions in 1995 were 1400 Mt for U.S. and more than 800 for China and Europe (15 countries). However, global emissions are increasing, even if the international convention on climate in Rio (1992) decided to stabilize them. Furthermore, 5.2% of emission reduction below the 1990 level would not have been sufficient even if the Kyoto protocol was ratified by all Annex I or developed countries.

What are the main consequences of increase in global emissions? An increase in temperature in boreal, subboreal, and alpine regions has a destabilization effect on both vegetation and soil (with very high organic matter content). In subhumid and semiarid areas, the increase in evapotranspiration may increase the water deficit and exacerbate the desertification process. Other effects on precipitation are difficult to predict and only general trends are known, with an increase of 5 to 10% over mid and high latitudes of the northern hemisphere and a decrease in the tropics. In addition, more variability in droughts and rains is also predicted along with attendant effects on runoff, erosion, and frequency of floods.

With regards to the effect on biomass production (and yield), the increase in CO_2 concentration can increase yield (10% for cereals, 25% for grasslands, and even 30% for trees), with drastic change in phenology. The CO_2 fertilizer effect will occur if water and nutrients are not limiting. The effect may be less in rainfed agriculture of drylands and in acid tropical or forest soils of low inherent fertility. Despite uncertainties, it is predicted that climate change will induce great perturbations and will accentuate the imbalance between north and south or rich and poor nations.

1.3 CARBON CYCLE IN THE CONTINENTAL BIOSPHERE

The global carbon cycle (Figure 1.3) shows that the world soils contain 1500 to 2000 Gigatonnes (Gt) of carbon depending on the soil depth (Table 1.1). In contrast, vegetation, mainly perennial, contains 600 to 700 Gt C.

Vegetation and soil can be either sources or sinks for atmospheric carbon. Prior to the industrial development in the nineteenth and twentieth centuries, deforestation and land cultivation were the main sources of greenhouse gas emission. Currently, they still contribute 1.7 to 2 Gt per year mainly through deforestation of 10 to 15 million ha in the Southern Hemisphere. The use of fossil carbon is now the main source of CO_2 and it is well known that the reduction of this source represents the only long-term solution, with all other solutions being either short term or complementary.

An important point to consider is that about 2 Gt of carbon as CO_2 is presumably being stored as a terrestrial sink in the continental biosphere (vegetation + soil). So the biosphere seems to

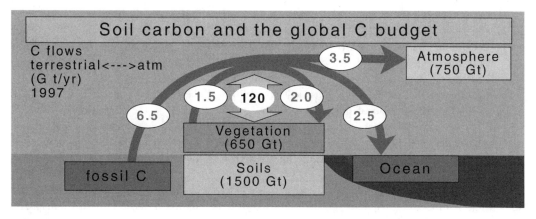

Figure 1.3 The terrestrial carbon cycle.

Table 1.1 Total Stocks of Soil Organic Carbon (SOC) (Pg C) and Mean C Content (kg C/m^2) by Major Agro-Ecological Zone (for upper 0.3 m and 1 m)

Agro-Ecological Zone	Spatially Weighted SOC Pools (Pg C)		Mean SOC Density (kg/m^2)	
	to 0.3 m depth	to 1 m depth	to 0.3 m depth	to 1 m depth
Tropics, warm humid	92–95	176–182	5.2–5.4	10.0–10.4
Tropics, warm seasonally dry	63–67	122–128	3.6–3.8	7.0–7.3
Tropics, cool	29–31	56–59	4.4–4.7	8.4–8.9
Arid	49–55	91–100	2.0–2.2	3.7–4.1
Subtropics with summer rains	33–36	64–68	4.5–4.7	8.6–9.1
Subtropics with winter rains	18–20	37–41	3.6–3.9	7.2–8.0
Temperate oceanic	20–22	40–44	5.8–6.4	11.7–12.9
Temperate continental	121–126	233–243	5.6–5.9	10.8–11.3
Boreal	203–210	478–435	9.8–10.2	23.1–24
Polar and Alpine (excl. land ice)	57–63	167–188	7.0–7.8	20.6–23.8

From Batjes 1996.

act naturally as a carbon sink and the challenge is to know if this sink can be increased by anthropic intervention.

Such intervention is included in the Kyoto protocol: Article 3.3 considers the role of reforestation and afforestation in the net changes in greenhouse gas (with quotas established for different countries); Article 3.4 addresses similar changes in agricultural soils, land use changes, and forestry (IPCC, 2000). At the Bonn meeting of the Conference of Parties of the Kyoto protocol (and also at Marrakech), C sequestration in agricultural soils was accepted without limitation of tonnage and area. However, a lot of questions still remain about the amount of C that could be sequestered and the verification of the changes in C pool in the soil.

1.3.1 Past and Present Stocks and Fluxes

Different factors determine the partitioning of the C stock. At the global scale, biomass production (Robert and Saugier, 2003) is an important factor that is also correlated with the rainfall amount. It influences soil organic matter (SOM) distribution in major ecological zones (Table 1.1; Batjes, 1996). High C contents are associated with humid forest soils. But low temperatures are responsible for the highest stocks in the boreal zone or in mountainous regions. Poor drainage and inundation are the main factors of peat formation. Soil type and properties (for example texture) also influence the antecedent soil carbon content.

Entropic factors such as land use, land cover, and agricultural practices govern actual C stock in soils. This is well illustrated by the C content of French soils (Arrouays et al., 2001) which ranges from more than 100 t/ha for some natural soils to 30 t/ha in arable or vineyard soils (Figure 1.4). These low values are explained by the significant loss of C that occurs during the first years of soil cultivation of grassland or forest soils, mainly due to tillage (Figure 1.5).

It is estimated that arable lands have lost about 40% of their carbon content in less than 50 years. A relevant illustration of this loss is by the soil C variation in the Corn Belt of the U.S. (Figure 1.6). The drastic effect of tillage is explained by the so-called de-protection of soil organic matter physically protected inside soil aggregates (Balesdent et al., 2000). The result is an increase in mineralization of organic matter and the flux of CO_2 into the atmosphere (Reicosky and Lindstrom, 1995).

Considering the possible influence of the increase in temperature on C mineralization of organic soils in the Northern Hemisphere and the influence of deforestation in the Southern Hemisphere, it is important to find the way to favor physical protection of soil organic matter, or to develop alternative management solutions.

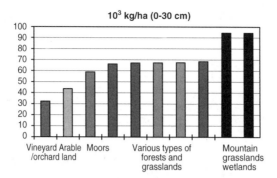

Figure 1.4 Variation of soil C content (t/ha for 30 cm depth) with soil cover and soil use. (From Arrouays, D. and W. Deslais, et al. 2001. The carbon content of top soil and its geographical distribution in France. *Soil Use and Management* 17:7–11. With permission.)

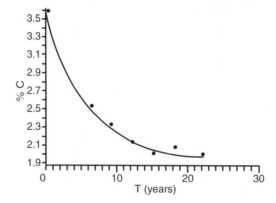

Figure 1.5 Carbon evolution in the Rothamsted Highfield grass-to-arable conversion experiment.

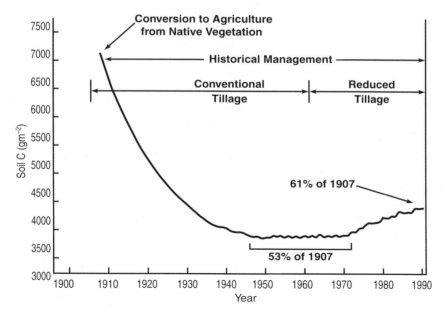

Figure 1.6 Simulated total soil carbon changes (0 to 20 cm depth) from 1907 to 1990 for the central U.S. cornbelt (Smith, 1999).

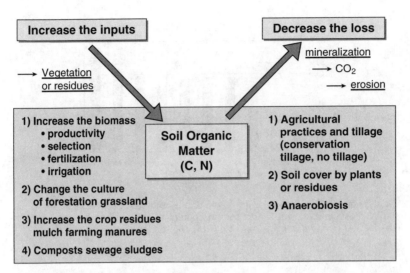

Figure 1.7 Management of soil organic matter in agriculture. (From Robert, M. 2001. Soil carbon sequestration for improved land management. *World Soil Resources Report 96*. FAO, Rome, 57 p. French and Spanish versions, 2002.)

1.3.2 Increasing C Sequestration

Some solutions will not be discussed in detail here even if they have a high potential of C sequestration: agroforestry, for example, offers a sustainable alternative to deforestation (Schroeder, 1994) with a potential of several t/ha/yr of C sequestration in both soils and trees; likewise, the better management of pastures (more than 3 billions ha estimated by FAO) represents an important potential C sink (more than 0.5 t/ha/yr).

For cultivated land with the lowest — often critical — organic matter content, two solutions are proposed:

- The first is to change cropland to forest or grassland: with increase of C stocks at the rate of more than 0.5 t C/ha/yr.
- The second is to change the agronomic practices (Figure 1.7; Robert, 2001).

Different scenarios are possible based on either increasing C inputs, decreasing C losses, or some combination of both. In order to increase C inputs, it is necessary to increase biomass (vegetation and or organic residues). To decrease loss, it is necessary to decrease or eliminate tillage practices and at the same time to cover the soil with mulch. In this way, losses of C by both mineralization and erosion are reduced.

Different historical agronomic systems are available:

- Since 1930, the U.S. has developed conservation tillage in which the soil is covered with a minimum of 30% of crop residue mulch.
- Conservation agriculture (First World meeting in 2001 in Madrid) has more strict principles: no soil tillage and permanent soil cover either by mulch or cover crops, which implies direct sowing.

The French institute CIRAD has developed very diverse agronomic systems adapted for different cropping systems (Capillon and Seguy, 2002).

1.4 MAIN CHALLENGES TO SOIL C SEQUESTRATION

Carbon sequestration in soil is important considering projected climatic change and the Kyoto protocol (Rosenzweig and Hillel, 2000), and links exist with other international conventions (on biodiversity or desertification). Organic matter is also a key component in soil and a key factor of soil quality and soil protection (Rees et al., 2001; Doran et al., 1996; Chenu and Robert, 2003). So the principal challenge for C sequestration is the sustainability of agriculture.

1.4.1 Soil Carbon Sequestration and International Conventions

1.4.1.1 Climatic Change (UNFCC Convention)

Important discussions still exist on the possibility to sequester C in soil. Several reports exist on this subject at the global scale (IPCC, 2000; Lal, 1997; Lal et al., 1998, 2000; Batjes, 1996; Robert, 2001). More recent reports have involved several researchers from Europe and France (EECP, 2003; INRA, 2002). Two main factors explain the differences of opinion among researchers. First, an assessment of the annual change in soil C stock is needed for each land use or change in management. Differences of sequestration potential exist in relation with climate, soil type, and so forth, and data from long-term field experiments are lacking. So only a limited range of values for soil C sequestration is available (for example from 0.1 to 1 t C/ha/yr for zero tillage). The second factor concerns the land area that can be converted to conservation tillage (millions or billions of ha?).

Different scenarios for Europe or France give interesting comparative values that are close to 1/1000 related to the global soil C stock of the reference area. It corresponds to 3 to 5% of the global CO_2 emission. At the global scale the potential is close to 2 Gt C per year (for a period of 20 to 50 years). So the soil contribution to prevent the greenhouse effect and the climatic change is not the long-term solution but an immediate solution that buys us time until more important remedial measures of reducing C emission can be identified.

It is also important that Article 3.4, the Clean Development Mechanism, is included to encourage, with financial support, developing countries to choose better adapted land management systems to sequester C in soil.

Furthermore, it is important to protect SOC stocks in northern latitudes against projected increase in temperatures. The question of deforestation is also sensitive to countries in the south. Considering the large size of the flux (1.7 Gt yr), it is important to use economic incentives to encourage adoption of alternative solutions (e.g., agroforestry and new farming systems). Furthermore, while this chapter deals only with CO_2, it is important to assess all greenhouse gases (N_2O and CH_4 fluxes).

1.4.1.2 Convention on Biodiversity (UNCBD)

While this convention is an important step, soil biodiversity is insufficiently known and taken into account. Soil organic matter is both a source of energy and nutrients for living organisms. Independent of the climate change, SOM represents one of the main factors that determine biological functioning and biodiversity (Soberon et al., 2000) (Figure 1.8). Other important factors in agriculture are tillage and pesticides.

1.4.1.3 Convention on Desertification (UNFCCD)

Soil organic matter is also a key factor that plays a role in exacerbating desertification (decrease in SOM) or in its prevention or remediation process (Figure 1.9; Hillel and Rosenzweig, 2002).

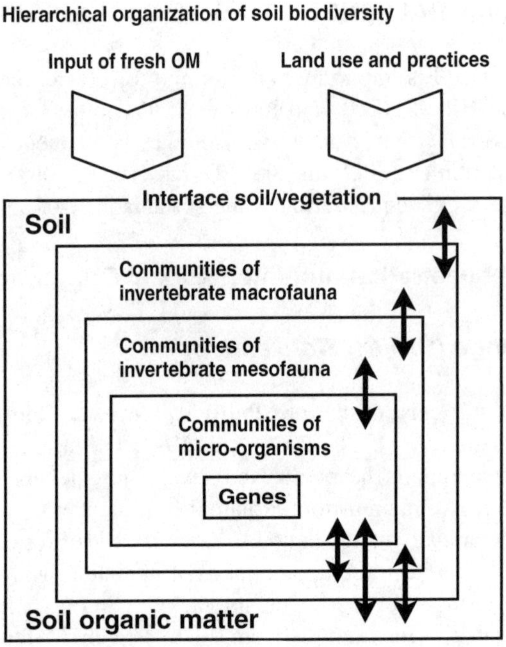

Figure 1.8 Role of soil organic matter in biodiversity. (From Robert, M. 2001. Soil carbon sequestration for improved land management. *World Soil Resources Report 96*. FAO, Rome, 57 p. French and Spanish versions, 2002.)

This action of organic matter is linked with the development of the vegetation and the soil cover and soil protection (see conservation agriculture).

1.4.2 Development of a Sustainable Agriculture

At the global scale, erosion (by wind or water) is the main process of soil degradation. The Global Assessment of Soil Degradation (GLASSOD; Oldeman, 1994) showed that erosion affects more than 1 billion ha. The main causal factors affecting erosion are deforestation, overgrazing, and agricultural management.

Different methods have been developed in the past to prevent erosion including terraces. In the U.S., the use of conservation tillage since 1940 has been effective to prevent wind erosion (another occurrence of the Dust Bowl). With complete soil coverage in South America and particularly in Brazil, conservation agriculture has been effective in reducing soil erosion by water even on steep slopes. Two factors are important for agricultural ecosystems: soil cover by mulch or vegetation and increase in SOM in different pools.

The increase in SOM improves soil quality and biological functioning with positive effects on structural stability (Tisdall and Oades, 1982). Soils with high SOM content have a better chemical fertility and strong mechanisms of recycling of the nutrients by the plants. Conservation agriculture is practiced on about 70 million ha mainly in the U.S., South America, Australia, and Canada. The area is increasing at the rate of 1 million ha each year. It has other benefits to farmers as well: increase in yields and savings in energy and labor (Madrid, 2001: conservation agriculture). However, some problems exist with the use of herbicides.

Europe is developing a strategy for soil protection with a priority for erosion control (+ desertification), organic matter management (+ biodiversity), and prevention of contamination. Soil protection is an eco-condition to receive ratification of the Common Agriculture Policy (within the

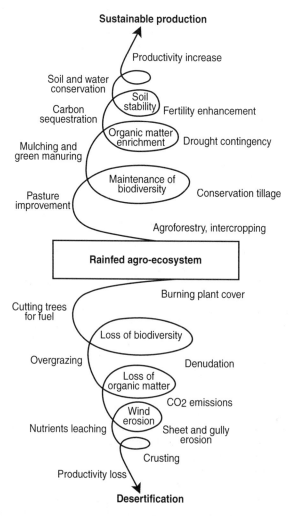

Figure 1.9 Role of soil and soil organic matter in the desertification process. (From Hillel, D. and C. Rosenzweig. 2002. Desertification in relation to climate variability and change. *Advance in Agronomy* 77:1–38. With permission.)

last reform) similar to what is occurring in the U.S. In 2004, the working groups on SOM and erosion will propose win-win recommendations for action that will imply new forms of agriculture in Europe, including adoption of reduced tillage that permits an increase in both soil cover and SOM.

1.5 CONCLUSIONS

Climatic change is a major problem of the twenty-first century. Despite numerous scenarios and uncertainties, preventive actions are necessary, and all sectors must take active part.

Soil C sequestration can have a significant role that is site specific and involves changes in forestry and agriculture through policy incentives. Recommended land use changes are also beneficial to soil protection (especially against erosion) and environmental quality, required conditions for sustainable agriculture. An increase in SOM can also contribute to the success of the three great conventions of the United Nations.

REFERENCES

Arrouays, D. and W. Deslais, et al. 2001. The carbon content of top soil and its geographical distribution in France. *Soil Use and Management* 17:7–11.
Balesdent, J., C. Chenu, and M. Balabane. 2000. Relationship of soil organic matter dynamics to physical protection and tillage. *Soil and Tillage Research* 53: 215–220.
Batjes, N. H. 1996. Total carbon and nitrogen in the soils of the world. *European Journal of Soil Science* 47: 151–163.
Capillon, A. and L. Seguy. 2002. Ecosystèmes cultivés et stockage du carbone. Cas des systèmes de culture en semis direct avec couverture végétale. *C. R. Acad. Agric. Fr.* 88, 5:63–70.
Chenu, C. and M. Robert. 2003. Importance of organic matter for soil properties and functions. Communication at the SCAPE meeting Alicante.
Doran, J. W., M. Sarrantonio, and M. A. Liebig. 1996. Soil health and sustainability. *Advances in Agronomy* 56:1–53.
ECCP European climate change programme. 2003.Working group sinks related to agricultural soils, final report, 83 p. Bruxelles.
Garcia-Torres L., et al. eds. 2001. Conservation agriculture: a worldwide challenge. I World Congress Madrid, vol. 1, keynote contribution, 387 p.
Hillel, D. and C. Rosenzweig. 2002. Desertification in relation to climate variability and change. *Advances in Agronomy* 77:1–38.
Houghton, R. A. 1995. Changes in the storage of terrestrial carbon since 1850. p. 45–65, in R. Lal, J. Kimble, E. Levine, and B. A. Stewart, eds., *Soils and Global Change*. CRC Press/Lewis Publishers.
INRA. 2002. Contribution à la lutte contre l'effet de serre. Stocker du carbone dans les sols agricoles de France? D. Arrouays, J. Balesdent, P. A. Jayet, J. F. Soussana, and P. Stengel, eds. *INRA*. Paris, 332 p.
IPCC. 2000. Land use, land-use change, and forestry. A special report of IPCC. R.T. Watson, I.R. Noble, B. Bolin, N.H. Ravindranath, D.J. Verardo, and D.J. Dokken, Eds. Cambridge University Press, United Kingdom, 377 p.
IPCC. 2001. Climatic change 2001. Synthesis report. A contribution of Working Groups I, II, and III to the third assessment report of the intergovernmental panel on climate change. R. T. Watson and the Core Writing Team, eds. Cambridge University Press, United Kingdom, 398 p.
Johnston, A. E. 1973. The effects of ley and arable cropping systems on the amounts of soil organic matter in the Rothamsted and Woburn Ley arable experiments. Rothamsted Experimental Station Annual Report for 1972, Part 2, 131–159.
Lal, R. 1997. Residue management, conservation tillage and soil restoration for mitigating greenhouse effect by CO_2 enrichment. *Soil and Tillage Research* 43:81–107.
Lal, R., J. M. Kimble, and R. F. Follett, eds. 1998. *Management of Carbon Sequestration in Soil*. CRC Press, 480 pp.
Lal, R., J. M. Kimble, and, B. A. Stewart, eds. 2000. *Global Climate Change and Tropical Ecosystems*. CRC Press/Lewis Publishers.
Oldeman, L. R. 1994. The global extent of soil degradation, pp. 99–117, in Greenland, D. J. and Szabolcs, I., eds., *Soil Resilience and Sustainable Land Use*. CAB International Wallingford.
Rees, R. M, B. C. Ball, C. D. Campbell, and C. A. Watson, eds. 2001. *Sustainable Management of Soil Organic Matter*. CABI Publishing, 440 p.
Reicosky, D. C. and M. J. Lindstrom. 1995. Impact of fall tillage on short-term carbon dioxide flux, in Lal, R., Kimble, J., Levine, E., Stewart, B. A., eds. *Soils and Global Change*. CRC Press.
Robert, M. 2001. Soil carbon sequestration for improved land management. *World Soil Resources Report 96*. FAO, Rome, 57 p., traductions française et espagnole en 2002.
Robert, M. and B. Saugier. 2003. Contribution des écosystèmes à la séquestration du carbone. *C.R. Geosciences* 335: 577–595.
Rosenzweig, C. and D. Hillel. 2000. Soils and global climate change: challenges and opportunities. *Soil Science* 165 (1):45–56.
Schroeder, P. 1994. Carbon storage benefits of agroforestry systems. *Agroforestry Systems* 27:89–97.
Smith, K. A. 1999. After the Kyoto Protocol: can soil scientists make a useful contribution? *Soil Use and Management* 15:71–75.
Soberon, J., P. Rodriguez, and E. Vazquez-Dominguez. 2000. Implications of the hierarchical structure of biodiversity for the development of ecological indicators of sustainable use. *Ambio* 29,3:136.
Tisdall, J. M. and J. M. Oades. 1982. Organic matter and water stable aggregates. *Journal of Soil Science* 33:141–163.

CHAPTER 2

Soil Carbon Sequestration

Martial Bernoux, Christian Feller, Carlos C. Cerri, Vincent Eschenbrenner,
and Carlos E. P. Cerri

CONTENTS

2.1 Introduction 13
2.2 Available Definitions and Concepts 14
2.3 Discussion 15
    2.3.1 What Is the Form and Mean Residence Time of the Sequestered C at the
            Plot Level? 17
    2.3.2 Time Scales 17
    2.3.3 Taking into Account the C (or Equivalent C) Transferred Off-Site 19
    2.3.4 Assessing C Sequestration of the Soil–Plant System for the Emissions
            Balances at the National Level 20
2.4 Conclusion 20
Acknowledgments 21
References 21

2.1 INTRODUCTION

Concerns regarding global warming and increasing atmospheric greenhouse gas (GHG) concentrations (CO_2, CH_4, and N_2O) have led to questions about the role of soils as a carbon (C) source or sink (Houghton, 2003). Excluding the carbonated rocks, soils constitute the largest surface C pool, approximately 1500 Gt C, which is almost three times the quantity stored in the terrestrial biomass, and twice that in the atmosphere (Lal, 2003). Therefore, any modification of land use and management practices, even for the agricultural systems at the steady state, can change soil C stocks (Schuman et al., 2002). Locally, these stock variations concern mainly the topsoil horizon (between 0 and 30 cm depth) and occur because of different processes at the plot scale, such as modification of the organic matter rates and quality inputs (Jenkinson et al., 1992; Paustian et al., 1992; Trumbore et al., 1995), transfer (deposition, erosion, leaching, and run-off) in solid or soluble form (Chan, 2001; Lal, 2002), and losses by mineralization (CO_2, CH_4) of soil organic matter (Schimel, 1995; Shang and Tiessen, 1997). It is, therefore, apparent that soils play a significant role in the control of the C stocks and fluxes (King et al., 1997; Schlesinger, 2000). For tropical soils, these changes may represent up to 50% of the original C stock in the top 20-cm depth (Feller et al., 1991; Feller

Table 2.1 Number of References Indexed in the ISI-Web of Science (1945 to 2003) for the Word Queries "Soil," "Carbon," and "Sequestration" (Query 1) and "Soil" and "Carbon" (Query 2), Respectively, in the Topics and in the Title (in parentheses)

Years	Number of References Returned by the Queries		Query 1/Query 2 ‰
	Query 1	Query 2	
1945 to 1990	0	719	0
1991	1[a]	643	1.6
1992	5 (1)[b]	694	7.2
1993	14 (1)	816	17.2
1994	7	908	7.7
1995	21 (1)	985	21.3
1996	24	1220	19.7
1997	36 (2)	1398	25.7
1998	47 (3)	1520	30.9
1999	38 (3)	1565	24.3
2000	94 (9)	1616	58.2
2001	104 (18)	1725	60.3
2002	153 (15)	1850	82.7
2003	150 (13)	2133	70.3
Total (1945 to 2003)	694 (66)	17792	39.0

[a] Thornley et al., 1991.
[b] Dewar and Cannell, 1992.
Queries performed on January 6, 2004.

and Beare, 1997). Therefore, land-use management policies may significantly influence fluxes of C between continental ecosystems and the atmosphere (King et al., 1997; Schlesinger, 2000).

The world community has been preoccupied since the early 1990s with potential climatic change due to increasing atmospheric GHG concentrations. Two possible courses of action to alleviate climate change are: (1) limiting the GHG emissions, and (2) enhancing the removal (or uptake) of these gases from the atmosphere to stabilize the pools (for example, sediments, trees, soil organic matter). World soils are one such pool. Yet, some prefer to use other terms in relation to the capture and retention of GHGs from the atmosphere; thus the terms *sequester* and *sequestration* have gained importance not only because they represent innovative ideas but also because they have gained widespread publicity.

With regards to the potential of the soil to mitigate the greenhouse effect, and more generally with regards to land use, land-use change, and forestry (LULUCF), the correct term is *soil carbon sequestration*. Although the published literature dates back to 1945, scientific publications increasingly began to use "soil carbon sequestration" in the early 1990s (Table 2.1).

As shown in Table 2.1, these terms are increasingly being used, but a definition or at least the broad meaning of "soil carbon sequestration" is rarely given. This chapter reviews and discusses some current definitions, proposes an alternative one, and draws attention to some necessary cautions when referring to soil carbon sequestration.

2.2 AVAILABLE DEFINITIONS AND CONCEPTS

A list of available definitions in publication or on Web pages is given below:

- U.S. Department of Energy: "Carbon sequestration in terrestrial ecosystems is either the net removal of CO_2 from the atmosphere or the prevention of CO_2 net emissions from the terrestrial ecosystems into the atmosphere." (U.S. Department of Energy, 1999)

- U.S. Department of Agriculture (USDA): "What is soil carbon sequestration? Atmospheric concentrations of carbon dioxide can be lowered either by reducing emissions or by taking carbon dioxide out of the atmosphere and storing it in terrestrial, oceanic, or freshwater aquatic ecosystems." (USDA — FAQ, www.usda.gov/oce/gcpo/sequeste.htm)
- Oak Ridge National Laboratory (ORNL): "From the viewpoint of terrestrial ecosystems, carbon sequestration is the removal of carbon dioxide from the atmosphere by enhancing natural absorption processes and storing the carbon for a long time in vegetation and soils. Carbon sequestration may be accomplished by fixing more carbon in plants by photosynthesis, increasing plant biomass per unit land area, reducing decomposition of soil organic matter, and increasing the area of land covered by ecosystems that store [carbon]." (Jacobs, 1999)
- Soil Science Society of America (SSSA): "Carbon sequestration refers to the storage of carbon in a stable solid form … . The amount of carbon sequestered at a site reflects the long-term balance between carbon uptake and release mechanisms." (Position of the SSSA, dated October 25, 2001: www.soils.org/carbseq.html)

Intensive speculations are being made about a future C market. Moreover, LULUCF has been accepted as a credit-earning climate change mitigation option for the first five-year commitment period. International negotiations also recognize afforestation and reforestation as viable LULUCF sink activities (Bernoux et al., 2002) for the clean development mechanism (CDM), which is based on specific projects undertaken by an Annex I country in a non-Annex I country. Therefore, reported below are definitions given by the potential entities that formulate theses projects and are C traders:

- Ecoenergy International Corporation (EIC): "Carbon sequestration is a strategy to slow the accumulation of atmospheric carbon dioxide by absorbing carbon into soil and perennial vegetation. This can be achieved through reforestation, agroforestry, or forest management activities that preserve or increase an existing carbon 'sink.' Carbon sinks include forests and other ecosystems, as well as sustainable agriculture crops that sequester carbon in the soil and in long-lived harvested products." (www.eic-co.com/sequestration.htm)
- CO2e.com (www.CO2e.com): "Carbon Sequestration is a Category on the CO2e Trading Floor. It refers to projects that capture and store carbon in a manner that prevents it from being released into the atmosphere for a specified period of time, the storage area is commonly referred to as a carbon sink (A carbon sink is a reservoir that can absorb or "sequester" carbon dioxide from the atmosphere. Forests are the most common form of sink, as well as soils, peat, permafrost, ocean water and carbonate deposits in the deep ocean.). Carbon Sequestration projects include: Forest Sequestration; Land Conservation; Soil Conservation and Land Use; Waste CO_2 Recovery/Deep Injection." (www.co2e.com/common/glossary.asp)

In the Kyoto Protocol (UNFCCC 1998) the word "sequestration" appears only once in its Article 2.1: "Each Party included in Annex I, in achieving its quantified emission limitation and reduction commitments under Article 3, in order to promote sustainable development, shall: (a) Implement and/or further elaborate policies and measures in accordance with its national circumstances, such as: … (iv) Research on, and promotion, development and increased use of, new and renewable forms of energy, of carbon dioxide sequestration technologies and of advanced and innovative environmentally sound … " Moreover, it is cited within a section that has no direct relation with the LULUCF sector that is treated by the points a-ii and a-iii of the same article.

Another important document that deals with soils and their management is the International Panel on Climate Change (IPCC) special report on LULUCF. This report defines sequestration as "the process of increasing the carbon content of a carbon pool other than the atmosphere."

2.3 DISCUSSION

Most of these definitions (soil specific or not) are based on CO_2 removal from the atmosphere and storage in an organic form in the soil or plant pools. Only the SSSA and IPCC (in its Special

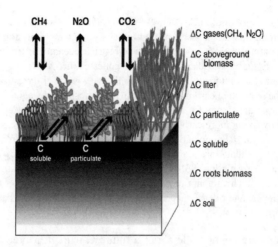

Figure 2.1 Fluxes (arrows) and changes (Δ) that need to be accounted, at the plot level, for a complete comparison of the C and GWP balances for a given agrosystem.

Report on LULUCF) give a definition based only on C storage, and that this should be a stable pool. Other definitions consider different pools, such as fresh water, oceans, and carbonated sediments.

A major flaw in these definitions is considering only the CO_2 fluxes. In addition to CO_2, soils are also characterized with methane (CH_4) and nitrous oxide (N_2O) fluxes. Furthermore, the Kyoto protocol covers all sources and sinks of those gases. The net emission calculations of the signatories of the United Nations Framework Convention on the Climate Change (UNFCCC) are expressed in equivalents of CO_2, by taking into account the global warming potential (GWP) of each gas: the Third Assessment Report of the IPCC (IPCC, 2001) expressed GWP at the secular horizon values (100 yr-GWP) of 23 for CH_4 and 296 for N_2O. GWPs are measurements of the relative radiative effect of a given substance (CO_2 here) compared to another and integrated over a specific time period. This means that 1 kg of CH_4 is as effective, in terms of radiative forcing, as 23 kg of CO_2. On a C or N mass base, 1 kg of C-CH_4 is equivalent to 8.36 kg of C-CO_2, and 1 kg of N-N_2O to 126.86 kg C-CO_2.

A recent review by Six et al. (2002) illustrates the importance of those considerations. They reported that in both tropical and temperate soils, a general increase in C pool (≈325 ± 113 kg C ha^{-1} yr^{-1}) was observed under no-till (NT) systems compared with conventional till (CT). But that, on average, in temperate soils under no-till, compared with conventional till, CH_4 uptake (0.42 ± 0.10 kg C-CH_4 ha^{-1} yr^{-1}) and N_2O emissions increased (≈2.91 ± 0.78 kg N-N_2O ha^{-1} yr^{-1}). The increased N_2O emissions led to a negative GWP when expressed on a C-CO_2 equivalent basis. Other changes in soil induced by NT showed that "from an agronomic standpoint NT is beneficial, but from a global change standpoint more research is needed to investigate the interactive effects of tillage, fertilizer application methodology, and crop rotation as they affect C-sequestration, CH_4-uptake, and N_2O-fluxes, especially in tropical soils, where data on this matter is still lacking." This is particularly true for the N_2O fluxes when leguminous crops are used as cover crops or green manure, because some studies show that N_2O emissions may be enhanced (Giller et al., 2002; Flessa et al., 2002; Millar et al., 2004). Figure 2.1 is a schematic of different C pools and fluxes among them at the plot level.

A holistic comparison of NT vs. CT must involve the computation of all contributors to the net GWP of these systems (Robertson et al., 2000; Flessa et al., 2002). Some potential contributors depend on the crop production cycle: GHG emissions from agricultural machines, direct and indirect GHG emission following liming (Bernoux et al., 2003), and direct and indirect GHG emission associated with pesticides and herbicides use.

Based on all these considerations, it appears that a concept of "soil carbon sequestration" must not be limited to C storage consideration or CO_2 balance. All GHG fluxes must be computed at the plot level in C-CO$_2$ or CO$_2$ equivalent, incorporating as many emission sources and sinks as possible for the entire soil-plant system. Moreover, the term soil appears to be too restrictive; and, the whole agronomic system must be considered. Finally, there is no absolute "soil carbon sequestration" potential for a given agronomic system. This raises different problems that need to be taken into account at the plot level, or at the national level, when drawing a national inventory. For the national inventory assessments, calculations are made by sectors. For instance, the emissions associated with fertilizers or lime manufacture is computed in CO$_2$ equivalent (Bernoux et al., 2002). These emissions, to avoid double accounting, cannot be taken into account in the balance of direct and indirect fluxes at the plot scale that would be extrapolated to the national scale. If, however, for a same fertilizer quantity, different emission fluxes are observed according to the plot management (e.g., no-till vs. conventional till), those differences must be computed during the national scale extrapolation. But, at the plot scale, it is only all the direct and indirect fluxes that need to be computed to enable an absolute comparison even for different agroecosystems (pasture against agroforestry, for instance) among them.

Therefore, a new definition is proposed that could be applied only to the soil pool, but that is more appropriate for the entire soil-plant pools of agroecosystems. This definition takes into account all the fluxes, in gaseous form, of GHG at the soil-plant-atmosphere interfaces expressed in equivalent CO$_2$ or equivalent C-CO$_2$ exchanges. These fluxes may originate from different ecosystem pools: solid or dissolved, organic or mineral.

"Soil carbon sequestration" or "Soil-plant carbon sequestration" for a specific agroecosystem, in comparison with a reference, should be considered as the result for a given period of time and portion of space of the net balance of all GHG expressed in C-CO$_2$ equivalent or CO$_2$ equivalent computing all emissions sources at the soil–plant–atmosphere interface.

In addition to that general definition, it is important to emphasize some other different aspects:

- What is the form and mean residence time of the sequestered C at the plot level?
- Which time scales have to be considered?
- How is the C (or equivalent C) transferred off-site?
- How is C sequestration determined at the plot scale for the emissions balances at the national level?

2.3.1 What Is the Form and Mean Residence Time of the Sequestered C at the Plot Level?

C sequestration is more effective when the mean residence time (MRT) of new C stored is long. It is, thus, absolutely necessary to evaluate the different C pools and to have an estimate of their respective turnover time. As it is not possible to undertake such determinations systematically, two approximations are possible based on the results of the literature:

- Measuring, by simple methods, variations in the soil organic C pools and their MRT. Different approaches of fractionation, chemical or physical, are possible; in particular granulometric separations that allow an adequate segregation of organic compartment with contrasted biostability: organic matter of the sand, silt, and clay fractions (Feller and Beare, 1997; Balesdent et al., 1998).
- Using simulation models of organic matter dynamics that incorporate conceptual pools defined by their biostability (e.g., active, slowly decomposable, refractory) like the Century (Parton et al. 1987), RothC (Jenkinson and Rayner, 1977) or Morgane Models (Arrouays et al., 1999).

2.3.2 Time Scales

Several issues with regards to time and space scales must be addressed. The desirable time scale has to be sufficient to cover the entire vegetation successions and husbandries for a specific agroecosystem (for example fallow-culture successions), and must be considered on multi-decadal scales. In the context of the Kyoto protocol, the first evaluation will be made in 2010

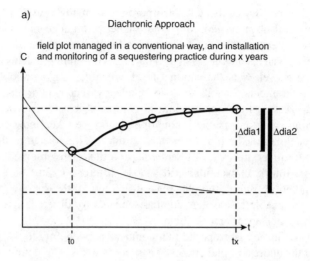

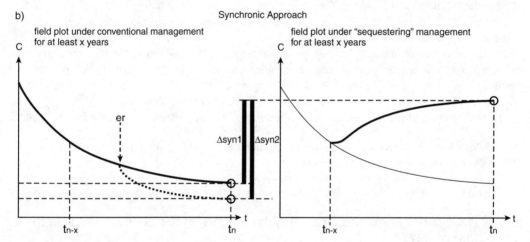

Figure 2.2 Comparison of the diachronic (a) and synchronic (b) approaches. Black circles correspond to C stocks determination; "er" stands for erosion.

using the year 1990 as reference or baseline. In other words, the corresponding time considered here is 21 years.

In addition, as the net balance for a given agroecosystem is always given compared to a reference system, this raises the problem of the choice for the year 0, or its equivalent, referring back to when the sequestering agroecosystem was established. Two approaches are possible: diachronic and synchronic (Figure 2.2).

The diachronic approach consists of measuring, Δt on the same field plot, soil C sequestration between time 0 (installation of the new system) and time x. C sequestered is then represented (Figure 2.2a) by $\Delta dia1$. This value is accurate only if soil C under the previous agrosystem of reference was at steady-state. If it was not the case, and the dynamics of C went toward an additional loss by mineralization during Δt ($t_x - t_0$), it would then be necessary to consider not $\Delta dia1$ but rather $\Delta dia2$. The measurement of $\Delta dia1$ is thus an approximation of C stored. It is then necessary to be able to evaluate the additional loss by mineralization ($\Delta dia2 - \Delta dia1$) that would have occurred during Δt for the original agrosystem without change of the practices.

The major disadvantage of the diachronic approach is that one must wait and measure over long periods of time before being able to evaluate the quantity of C sequestered. Therefore, research is generally based on a synchronic approach.

SOIL CARBON SEQUESTRATION

The synchronic approach consists, at a given time tn, of comparing the C stock of a field plot corresponding to the sequestering practice tested during x years to that of a field (control or conventional practices) under traditional management to represent t0 state or the reference point (Figure 2.2b). C sequestered is then represented by Δ_{syn1} (equivalent to $\Delta dia2$ in Figure 2.2a). However, it may be that the referenced plot was subjected to drastic alterations in its C stocks by accelerated erosion ("er" in Figure 2.2b). Losses of C occurred (in the reference field plot) in the form of solid transfer of C out of the field, and the difference of soil C between the two fields, is an apparent sequestration Δ_{syn2}. This difference is allotted to the only process of sequestration, where it is the sum of the net sequestration process Δ_{syn1} (capture of C-CO_2) and of a transfer process (deposition of C eroded: $\Delta_{syn2} - \Delta_{syn1}$), which, a priori, does not have to be considered as a sequestration or desequestration (see next section). In this case, Δ_{syn2} overestimates the C sequestration. Thus, it is necessary to be very careful in this type of approach with the existing risks of erosion for the reference plot. In addition, it is known that the quantities of C likely to be lost by erosion at field scale (from 0 to 1tC ha^{-1}yr^{-1}) are of the same order of magnitude as those susceptible to be gained by sequestration (conclusions of the conference Erosion and Sequestration of Carbon, Montpellier, September 2002).

2.3.3 Taking into Account the C (or Equivalent C) Transferred Off-Site

A significant problem rarely taken into account is the transfer of C in solid or soluble form among two adjacent ecosystems as is the case erosion/deposition cycle represented in Figure 2.3 along a toposequence. During the timeframe corresponding to the variations of C stocks on various situation of the toposequence, it may be that part of the observed variations (reduction or increase) that are due to a loss of solid or soluble C by erosion and run-off upstream and by an accumulation downstream. Determination of the sole variations of C stocks at plot level for a specific period does not reflect only variations of C-CO_2 fluxes. The variations due to the transfers of solid or soluble C should not then be considered in the assessment of GHG fluxes. This problem is particularly important for Mediterranean and tropical regions where erosion processes are very frequent, even on very gentle slopes. The conference Erosion and Sequestration of Carbon (Montpellier, September 2002) concluded that the quantities of solid C transferred by water erosion could be of the same order of magnitude, between 0 and 1tC ha^{-1}yr^{-1}, and that the amount is likely to be sequestered under the effect of an improving management. Therefore, the real level of C sequestration can be under- or overestimated in absence of measurements or at least estimates of the transfers of solid or soluble C.

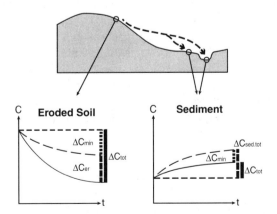

Figure 2.3 Soil C stock variations resulting from lateral transfers (solid or soluble) during Δt. Eroded soil: the decrease $\Delta Ctot$ corresponds to the sum of the erosion (ΔCer) and mineralization ($\Delta Cmin$) processes. Deposit: the increase ($\Delta Ctot$) is issued from the difference between the overall sedimentation ($\Delta Csed.tot$) and the mineralization during or after the transportation ($\Delta Cmin$).

Another aspect rarely documented is the change, or lack thereof, of the mean residence time (MRT) of the C transferred and deposited. For instance, in the case of eroded and redeposited C (ΔCsed.tot), the changes in term of C MRT throughout the transfer process are: soil aggregate breakdown, detachment of the soil particles from the initial site, transfer by water erosion, and deposition at the bottom of slope or sedimentation in fresh or marine waters. It is known that the breakdown of the soil aggregates tends to increase the potential of mineralization of soil organic C initially protected within the aggregates. Lal reports, in this volume, values of about 30% of additional C mineralized. Those values are to be taken into account in GHG inventories. Furthermore, is the C deposited in the solid form in the alluvium or sediments as stable as that in the original material? There are few data available on this subject.

Considering the scarcity of information relative to the MRT modification of C induced by and during detachment, transport, and deposition, it is prudent not to use C transferred in solid or soluble forms in computing C sequestration balance.

2.3.4 Assessing C Sequestration of the Soil–Plant System for the Emissions Balances at the National Level

The establishment of the GHG inventories, for both industry and agriculture sectors on a national scale, is an important step for the GHG fluxes management at the global scale. Several countries are establishing these inventories according to the guidelines provided by the IPCC (IPCC/UNEP/OECD/IEA, 1997).

The definition of the soil C sequestration, given above, implies all the GHG sources and sinks at plot level. This is, for instance, the case of GHG fluxes from N fertilizers use. It is estimated (IPCC/UNEP/OECD/IEA, 1997) that the application of 100 kg ha^{-1} of N-manure led to the emission of N_2O (a GHG with very high GWP), at an average rate of 1.25 kg ha^{-1} N-N_2O; that is an equivalent of 158 kg ha^{-1} C-CO_2. This 1.25 (±1)% emission rate applied to mineral and organic fertilizers, as suggested by IPCC/UNEP/OECD/IEA (1997), was established from the observation of a relation of proportionality established on a reduced dataset between the intensity of the N_2O emissions and the quantities of N applied to the soil (Bouwman, 1996).

At the plot scale, the GHG balance must consider those fluxes. However, if the results obtained at plot scale are used in national assessments, it is necessary to ascertain that these N_2O emissions are not accounted twice by (1) using only the national fertilizer consumption, and (2) using the extrapolation and generalization of the plot data at national scale. The examples may be extended to other types of contributions (lime, pesticides, etc.). However, any change in the contribution of N fertilizers induced by specific farming practices can result in variations of N_2O emissions by simple modification of the soil properties or modification of quality and quantities of the organic input. That is the case of potential sequestering practices in term of C-CO_2 like no-till practices associated with cover crop plants or agroforestry with leguminous plant integration. In these two cases, there is a significant risk of increased emissions of N-N_2O for this system compared to conventional tillage (Six et al., 2002; Choudhary et al., 2002; Millar et al., 2004), which can completely cancel the beneficial effect of the C-CO_2 capture by the soil. These emissions induced by specific management are not entered elsewhere than at the plot level, and therefore must be generalized at the national scale for a complete balance.

2.4 CONCLUSION

Soil carbon storage is only half of the story: land-use management is not a long-term solution for the global warming in terms of C storage but rather in terms of N_2O and CH_4 mitigation options. Therefore, a definition of soil carbon sequestration or of a sequestrating agroecosystem must address these issues. Due to its large 100-yr GWP, N_2O is perhaps the key point of the C sequestration concept! Most success of agricultural mitigation strategies would be linked to a careful management of the N cycle on the top of the crop cycle. In some cases, N fertilizer may have a positive effect on soil C storage by increasing plant productivity and organic matter restoration. In other situations, N fertilizer may lead to drastic N_2O emissions. If food security has to be insured, the maintenance of yield levels have to be achieved through improved N fertilizer use, and probably a careful management of cover crops. Moreover, the recommended land-use management must be beneficial from a global change standpoint, but also for the agronomic standpoint (erosion control, biodiversity, environmental, etc.), which is commonly achieved with increasing C stocks.

ACKNOWLEDGMENTS

Research that led to this definition and reflection was supported partly by the Fond Français pour l'Environnement Mondial (FFEM-Agro-écologie) and the Global Environment Facility (project number GFL/2740-02-4381).

REFERENCES

Arrouays, D., J. Balesdent, W. Deslais, J. Daroussin, and J. Gaillard. 1999. Construction d'un modèle d'évolution des stocks de carbone organique des sols et évaluation des données disponibles pour sa validation. Premières estimations des stocks français. Etude pour la Mission Interministérielle de l'Effet de Serre. *Lettre de Commande* 17/98. INRA Orléans, Final Report. Abril 1999. 40 p.

Balesdent, J., E. Besnard, D. Arrouays, and C. Chenu C. 1998. The dynamics of carbon in particle-size fractions of soil in a forest-cultivation sequence. *Plant and Soil* 201:49–57.

Bernoux, M., V. Eschenbrenner, C. C. Cerri, J. M. Melillo, and C. Feller. 2002. LULUCF-based CDM: too much ado for … a small carbon market. *Climate Policy* 2:379–385.

Bernoux, M., B. Volkoff, M. C. S. Carvalho, and C. C. Cerri. 2003. CO_2 emissions from liming of agricultural soils in Brazil. *Global Biogeochemical Cycles* 17(2):art. no.-1049.

Bouwman, A. F. 1996. Direct emissions of nitrous oxide from agricultural soils. *Nutrient Cycling in Agroecosystems* 46(1):53–70.

Chan, K. Y. 2001. Soil particulate organic carbon under different land use and management. *Soil Use and Management* 17:217–221.

Choudhary, M. A., A. Akramkhanov, and S. Saggar. 2002. Nitrous oxide emissions from a New Zealand cropped soil: Tillage effects, spatial and seasonal variability. *Agriculture, Ecosystems and Environment* 93:33–43.

Dewar, R. C. and M. G. R. Cannell. 1992. Carbon sequestration in the trees, products and soils of forest plantations — an analysis using UK examples. *Tree Physiology* 11(1):49–71.

Feller, C. and M. H. Beare. 1997. Physical control of soil organic matter dynamics in the tropics. *Geoderma* 79:69–116.

Feller, C., C. François, G. Villemin, J. M. Portal, F. Toutain, and J. L. Morel. 1991. Nature des matières organiques associées aux fractions argileuses d'un sol ferrallitique. *C.R. Acad. Sci. Paris* Sér. 2, 312:1491–1497.

Flessa, H., R. Ruser, P. Dörsch, T. Kamp, M. A. Jimenez, J. C. Munch, and F. Beese. 2002. Integrated evaluation of greenhouse gas emissions (CO_2, CH_4, N_2O) from two farming systems in southern Germany. *Agriculture, Ecosystems and Environment* 91:175–189.

Giller, K. E., G. Cadisch, and C. Palm. 2002. The North-South divide! Organic wastes, or resources for nutrient management? *Agronomie* 22 (7–8):703–709.

Houghton, R. A. 2003. Why are estimates of the terrestrial carbon balance so different? *Global Change Biology* 9:500–509.

IPCC. 2001. Climate Change 2001: The Scientific Basis. Contribution of Working Group 1 to the Third Assessment Report of the Intergovernmental Panel on Climate Change. Houghton, J. T., Y. Ding, D. J. Griggs, M. Noguer, P. J. van der Linden, X. Dai, K. Maskell, and C. A. Johnson, eds., Cambridge University Press, Cambridge, U.K., 881 pp.

IPCC. 2000. *Land-Use, Land-Use Change and Forestry.* Cambridge University Press, Cambridge.

IPCC/UNEP/OECD/IEA. 1997. *Revised 1996 IPCC Guidelines for National Greenhouse Gas Inventories: Reporting Instructions*, Vol. 1; *Workbook*, Vol. 2; *Reference Manual*, Vol. 3. Paris: Intergovernmental Panel on Climate Change, United Nations Environment Programme, Organization for Economic Co-Operation and Development, International Energy Agency.

Jacobs G. 1999. Earth's vegetation and soil: Natural scrubber for carbon emissions? an interview with ORNL's Gary Jacobs. *Oak Ridge National Laboratory Review Volume* 32 (3):21–23. (Electronic version at www.ornl.gov/ORNLReview/rev32_3/captur.htm.)

Jenkinson, D. S. and J. H. Rayner. 1977. The turnover of soil organic matter in some of the Rothamsted classical experiments. *Soil Science* 123:298–305.

Jenkinson, D. S., D. D. Harkness, E. D. Vance, D. E. Adams, and A. F. Harrison. 1992. Calculating net primary production and annual input of organic matter to soil from the amount and radiocarbon content of soil organic matter. *Soil Biology and Biochemistry* 24:295–308.

King, A. W., W. M. Post, and A. Wullschleger. 1997. The potential response of terrestrial carbon storage to changes in climate and atmospheric CO_2. *Climatic Change* 35:199–227.

Lal, R. 2003. Global potential of soil carbon sequestration to mitigate the greenhouse effect. *Critical Reviews in Plant Sciences* 22(2):151–184.

Lal, R. 2002. Soil carbon dynamic in cropland and rangeland. *Environmental Pollution* 116:353–362.

Millar, N., J. K. Ndufa, G. Cadish, and E. M. Baggs. 2004. Nitrous oxide emissions following incorporation of improved-fallow residues in the humid tropics. *Global Biogeochemical Cycles.* 18(1):GB 1032.

Parton, W. J., D. S. Schimel, C. V. Cole, and D. S. Ojima. 1987. Analysis of factors controlling soil organic matter levels in Great Plains grasslands. *Soil Science Society of America Journal* 51:1173–1179.

Paustian, K., W. J. Parton, and J. Persson. 1992. Modeling soil organic matter in organic-amended and nitrogen-fertilized long-term plots. *Soil Science Society of America Journal* 56:476–488.

Robertson, G. P., E. A. Paul, and R. R. Harwood. 2000. Greenhouse gases in intensive agriculture: Contributions of individual gases to the radiative forcing of the atmosphere. *Science* 289(5486):1922–1925.

Schimel, D. S. 1995. Terrestrial ecosystem and the carbon cycle. *Global Change Biology* 1:77–91.

Schlesinger, W. H. 2000. Carbon sequestration in soils: some cautions amidst optimism. *Agriculture Ecosystems and Environment* 82 (1–3):121–127.

Schuman, G. E., H. H. Janzen, and J. E Herrick. 2002. Soil carbon dynamics and potential carbon sequestration by rangelands. *Environmental Pollution* 116:391–396.

Shang, C. and H. Tiessen. 1997. Organic matter liability in a tropical Oxisol: Evidence from shifting cultivation, chemical oxidation, particle size, density, and magnetic fractionations. *Soil Science* 162:795–807.

Six, J., C. Feller, K. Denef, S. M. Ogle, J. C. M. Sá, and A. Albrecht. 2002. Soil organic matter, biota and aggregation in temperate and tropical soils — Effects of no-tillage. *Agronomie* 22 (7–8):755–775.

Thornley, J. H. M., D. Fowler, and M. G. R. Cannell. 1991. Terrestrial Carbon Storage Resulting from CO_2 and Nitrogen-Fertilization in Temperate Grasslands. *Plant Cell and Environment* 14(9):1007–1011.

Trumbore, S. E., E. A. Davidson, P. B. Camargo, D. C. Nepstad, and L. A. Martinelli. 1995. Below-ground cycling of carbon in forest and pastures of eastern Amazonia. *Global Biogeochemical Cycles* 9:515–528.

UNFCCC (United Nations Framework convention on Climate Change). 1998. Kyoto protocol to the United Nations Framework convention on Climate Change, in Report of the Conference of the Parties on its Third Session, Held at Kyoto from 1 to 11 December 1997 — Addendum: Part 2: Action Taken by the Conference of the Parties at its Third Session, pp. 7–27. UNFCC document FCCC/CP/1997/7/Add. 1. 60 pp.

U.S. Department of Energy. 1999. In Carbon Sequestration — Research and Development. U.S. Department of Energy. December 1999.

CHAPTER 3

Influence of Soil Erosion on Carbon Dynamics in the World

Rattan Lal

CONTENTS

3.1 Introduction 23
3.2 Soil Organic Carbon Pool 25
3.3 Soil Degradation and Climate Change 26
3.4 Soil Erosion and Carbon Dynamics 26
3.5 Fate of Eroded Soil Carbon 28
    3.5.1 Sedimentologist's View of the Deposition of Eroded Soil Carbon 29
    3.5.2 Soil Scientist's View of the Fate of Carbon Displaced by Erosion 30
3.6 Erosion-Induced Global Emission of Carbon 31
3.7 Summary and Conclusions 32
References 32

3.1 INTRODUCTION

Increase in atmospheric concentration of carbon dioxide (CO_2) and other greenhouse gases (GHGs) during the twentieth century (IPCC, 2001), linked to the observed and projected climate change, has raised concerns regarding sources and sinks of these gases. Land-use change and fossil fuel combustion are linked to the climate change, defined as a "change of climate attributed to human activity that alters the composition of the global atmosphere." The global average surface temperature has increased by 0.6 ± 0.2°C since 1850 (IPCC, 1995; 2001), which is attributed to an anthropogenic increase in atmospheric concentration of several GHGs. The concentration of CO_2 has increased by 31% from 280 ppmv in 1750 to 367 ppmv in 1999 at an average rate of 1.5 ppmv or 0.4%/yr (Etheridge et al., 1996; IPCC, 2001), that of CH_4 from 700 ppbv to 1760 ppbv (increase of 151%) (Etheridge et al., 1998; IPCC, 2001) and that of N_2O from 270 ppbv to 316 ppbv (increase of 17%) over the same period (IPCC, 2001). The concentration of CO_2 in 2003 is reported to be 379 ppm and it increased by 3 ppm during 2003. Land-use change, soil cultivation, and erosional processes have a strong impact on the carbon (C) cycle at pedon, soilscape, landscape, and watershed scales. The effects of land-use change, conversion of natural to managed ecosystems, and soil cultivation on the emissions of GHGs depend on complex interacting processes leading to: (1) decomposition of biomass, (2) mineralization of soil organic matter or humus exacerbated by increase in soil temperature and decrease in soil moisture, (3) increase in susceptibility to soil erosion, (4) displacement and redistribution of soil organic carbon (SOC) over the landscape including burial in depressional sites, and (5) possible increase in emission of CO_2 and other GHGs into the atmosphere due to increase in oxidation and mineralization.

The impacts of projected climate change on soil properties, SOC pool and dynamics, and susceptibility to erosion are not known. A principal unknown is the effect of increase in global temperature on SOC pool as affected by change in net primary productivity (NPP), possible increase in the rate of mineralization of SOC, and in susceptibility to soil erosion. Arnell and Liu (2001) reported that rise in concentrations of GHGs may lead to reduced soil moisture storage, increase in surface and shallow runoff, and increase in sediment loads in rivers. The scientific understanding of the processes set in motion by climate change and the attendant alterations in soil erodibility, rainfall/climate erosivity, redistribution of sediments, and the associated SOC over the landscape and susceptibility of displaced SOC to mineralization under aerobic or anaerobic conditions is rather sketchy and incomplete. Mineralization of SOC depends on temperature and moisture regimes (Leiros et al., 1999), which are in turn influenced by soil erosion.

Gaseous emissions from terrestrial ecosystems are exacerbated by soil degradation. Soil erosion is by far the most widespread form of soil degradation. Total land area affected by soil erosion is 1094 million hectares (Mha) by water erosion of which 751 Mha is at moderate plus level of severity, and 548 Mha by wind erosion of which 280 Mha is at moderate plus level of severity (Oldeman, 1994). Because of low density (1.2 to 1.5 Mg/m^3) and being concentrated in vicinity of the soil surface, the SOC fraction is strongly influenced by erosional processes. It is preferentially removed along with sediments by both runoff water and wind.

While the onsite effects of soil erosion lead to adverse impacts on soil quality and productivity, offsite impacts are even more drastic yet difficult to precisely quantify and comprehend. Over and above the problem of eutrophication and contamination of surface waters by nonpoint source pollution, emission of erosion-induced GHGs is a major concern that needs to be addressed.

Natural or geologic soil erosion is a slow but constructive process. It formed some of the most fertile soils of alluvial and loess origins. Many ancient civilizations (e.g., Nile, Indus, Euphrates, Yangtze), the so-called hydraulic societies thrived on alluvial soils. In contrast, the accelerated erosion by anthropogenic activities is a destructive process. It has caused the demise of once thriving civilizations. In addition to its decline in productivity, it also causes nonpoint source pollution, decline in air quality by dust, and emission of GHGs. Erosion by water and wind preferentially removes soil organic matter (SOM), a light fraction concentrated in the vicinity of the soil surface. As a key determinant of soil quality, depletion of SOM has numerous ecological, economic, and environmental consequences. Yet, the fate of eroded SOC is not known and is a debatable issue. Sedimentologists argue that eroded SOC transported into the aquatic ecosystems is buried and sequestered, and soil scientists believe that 20 to 30% of the SOC transported over the landscape is mineralized and released into the atmosphere. The debate about the fate of eroded SOC is accentuated by numerous uncertainties about the lack of understanding of the complex and interacting processes involved: (1) breakdown of aggregates by erosion leads to exposure of SOC to microbial processes and enhances mineralization, (2) change in soil moisture and temperate regimes of eroded soil enhances oxidation of SOM, (3) redistribution of SOC-enriched sediments over the landscape may also increase oxidation, (4) some of the SOC buried in depressional and protected sites may be reaggregated and sequestered, and (5) SOC transported into rivers may be mineralized depending on the climatic conditions. Resolving this issue is necessary to developing strategies for sustainable management of natural resources for erosion control and decreasing the rate of enrichment of CO_2.

The objective of this manuscript is to discuss the impact of erosional processes on SOC dynamics, describe the fate of eroded SOC, and identify soil/site conditions and soil processes which lead to emission of GHGs or sequestration of C displaced by erosion.

3.2 SOIL ORGANIC CARBON POOL

The SOM comprises the sum of all organic substances in soil. It consists of a mixture of plant and animal residues at various stages of decomposition, of substances synthesized microbiologically and chemically from the breakdown products, and of the bodies of live microorganisms and small animals and their decomposition by-products (Schnitzer, 1991). The SOM comprises about 58% of SOC and is a key determinant of soil quality — biomass productivity and environment moderating capacity. The beneficial impacts of SOC on soil quality are attributed to: (1) stabilization of soil structure through formation of organo-mineral complexes, and development of stable aggregates; (2) improvement in available water-holding capacity of the soil through increase in soil moisture retention at field capacity (0.3 bar suction); (3) improvement in soil biodiversity especially activity of soil fauna (e.g., earthworms); (4) biodegradation of contaminants; (5) buffering soil against sudden changes in pH and elemental concentrations; (6) minimizing leaching losses of fertilizer through chelation and absorption; (7) filtering and purification of water by sorption and degradation of pollutants; (8) strengthening mechanisms of elemental cycling; (9) improving soil quality and productivity; and (10) sequestering C and mitigating climate change.

These and other multifarious benefits led Albrecht (1938) to state that "soil organic matter is one of our most important national resources; its unwise exploitation has been devastating; and it must be given proper rank in any conservation policy as one of the major factors affecting the level of crop production in the future." The importance of Albrecht's statement cannot be overemphasized in the context of the issue of the projected climate change in the twenty-first century. Identifying options for mitigating climate change necessitates delineation of sources and sinks of atmospheric CO_2. Soil humus has historically been a source of CO_2 and CH_4, especially in case of low-input and subsistence agriculture that has been practiced throughout the world until the advent and widespread use of chemical fertilizers after World War II. Jenny (1980) wrote that "among the causes held responsible for the CO_2 enrichment, highest ranks are accorded to the continuous burning of fossil fuels and the cutting of forests. The contribution of soil organic matter appear underestimated." Indeed, IPCC (2001) lists "land use change" as a source of atmospheric CO_2 estimated at 1.7 ± 0.8 Pg (1 Pg petagram = 1 billion metric tons = 10^{15} g) during the 1980s and 1.6 ± 0.8 Pg during the 1990s. In this context, the term *land-use change* refers to tropical deforestation and the attendant release of CO_2 and CH_4 by decomposition/burning of biomass. The emission of C from soil due to anthropogenic perturbations is not specifically accounted for, partly because of the difficulty in evaluating small changes in SOC in a large global reservoir of 1550 Pg (Batjes, 1996), and partly because of lack of knowledge about the complex processes involved. Nonetheless, contribution of SOC pool as a source or sink for atmospheric CO_2 at the global scale cannot be ignored (Lal, 2000).

Soil humus and its dynamics are important components of the global C pools and fluxes. Because most agricultural, degraded, and drastically disturbed soils now contain a lower SOC pool than that under undisturbed conditions, there exists a potential sink to absorb more C from the biomass through the process of C sequestration. Soil C sequestration implies capturing atmospheric C and securely storing it in biota, soil, and other terrestrial and aquatic systems. Two principal processes of C sequestration are biotic and abiotic. Biotic processes include C storage including those in soil and biota. The latter comprises plants in terrestrial, aquatic, and oceanic ecosystems, and biofuel offset is a principal component of biotic C sequestration mechanisms. Formation of secondary carbonates, dissolution of atmospheric CO_2 to form carbonic acid, and its reaction with cations to form pedogenic carbonates, is also a chemical reaction facilitated by biotic processes. In contrast, abiotic processes of C sequestration involve compression of CO_2 from industrial sources and its injection in the geologic strata, saline aquifer, and deep ocean.

3.3 SOIL DEGRADATION AND CLIMATE CHANGE

Soil degradation is defined as "decline in soil quality by several degradative processes including erosion, salinization, soil structure, depletion of SOC pool and essential nutrients." Soil erodibility, susceptibility of soil to erosivity of rainfall and wind, depends on SOC concentration, soil structure, texture, and water infiltration characteristics. However, the effects of climate change on soil erodibility and erosion hazard are not known. Soil erodibility increases with decrease in SOC concentration, reduction in structural stability, and decline in water infiltration capacity. The SOC concentration, being in a dynamic equilibrium with input and output of biosolids and other climatic variables, may be sensitive to the projected increase in global temperature. Increase in global temperatures may lead to higher SOC turnover and increase in mineralization rate (Newton et al., 1996; Hungate et al., 1997; Leiros et al., 1999). The attendant depletion of SOC pool may adversely affect soil structure, decrease infiltration rate, and reduce available water capacity, leading to a possible increase in erodibility, increase in runoff, and high risks of soil erosion.

Soil degradation affects SOC pool directly and indirectly. Directly, it reduces the impact of biomass C into the system because of reduction in NPP and decrease in water and nutrient availability with attendant reduction in biodiversity. Indirectly, it leads to disruption in biogeochemical cycles and decline in soil resilience. Soil degradation also accentuates losses of SOC pool by exacerbating the rate of mineralization, leaching, and soil erosion. Soil erosion hazard may increase due to increase in erodibility and erosivity factors. Rains may become more intense and of high erosivity. Consequently, susceptibility to both water and rill erosion may increase, with attendant effects on depletion of SOC pool (Smith et al., 1997), which may be exacerbated by aridization of the climate, such as in the Mediterranean region (Lavee et al., 1998). The fragile ecosystems of arid and semiarid climates may be highly sensitive to desertification even with minor changes in rainfall distribution and temperature regime (Puigdefabregas, 1998; Villers-Rúiz and Trejo-Vázquez, 1998). Desertification and CO_2-induced climate change are intricately linked. Desertification leads to reduction in biological productivity of the ecosystem and long-term loss of natural vegetation, which reduces the biomass input into the soil. The land area prone to desertification is 7 million km^2 in Africa (assuming 25% reduction in productivity) and is increasing.

Increase in soil degradation and desertification may alter the efflux of GHGs from soil to the atmosphere. Erosion, both by water and wind, may enhance gaseous emissions onsite and offsite. Onsite erosion alters soil moisture regime and may accentuate mineralization. Offsite, erosion and deposition may alter soil moisture regime and increase susceptibility of SOM to mineralization. The eroded landscape continues to redistribute SOC and enhance its susceptibility to emission (Page et al., 2000). The effects of erosion on soil are closely linked to soil physical degradation, especially to aggregates leading to anaerobiosis either by compaction, water imbalance, or both. Soil chemical degradation and nutrient imbalance can also cause depletion of SOC pool. The integrated effect of soil physical and chemical degradation on soil biological quality is to decrease the SOC pool and increase in efflux of GHGs from soil to the atmosphere.

3.4 SOIL EROSION AND CARBON DYNAMICS

Soil erosion is a complex and global problem, and its environmental impact is now being debated as to whether the process is a source or sink for atmospheric C. Resolving this issue objectively requires a thorough understanding of the mechanisms involved. As a physical process,

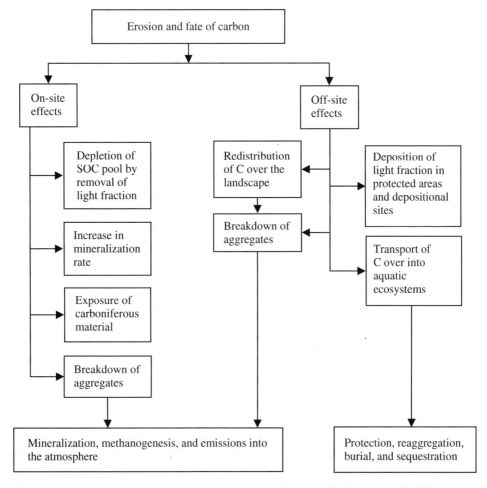

Figure 3.1 Soil carbon dynamics in relation to the accelerated soil erosion by water and wind.

soil erosion involves "work" in detachment and transport of soil particles. The energy for the work is supplied by agents of erosion including raindrop, runoff, wind, gravity, etc. The work involves: (1) detaching soil particles from clods; (2) breakdown of macro- into micro-aggregates and dispersion into soil separates or primary particles; (3) transporting particles called "sediments" over the landscape by water runoff, ice, wind, or gravity; and (4) depositing sediments into depressional sites or against barriers that retard the velocity/carrying capacity of runoff water or wind. All these stages of erosion have a strong impact on SOC and soil inorganic carbon (SIC) components. Severe soil erosion, by water or wind, can lead to truncation of the soil profile by removal of the surface layer and exposure the carbonate-rich subsoil. Carbonates thus exposed may react with acidiferous material (e.g., acid rain, fertilizers, etc.) and release C to the atmosphere. In contrast, burial of a carbonaceous soil layer by sediments may reduce the magnitude of CO_2 emission (Figure 3.1).

Soil particles detached are transported over the landscape and a proportion of them eventually redeposited in depressional sites and aquatic/loessial ecosystems. The distance particles are transported depends on the density/weight of the particles and velocity or carrying capacity of the fluid (wind or water). The lighter particles (e.g., SOC and clay fractions) are carried longer distances than heavier fractions (e.g., silt, sand, and gravel).

The fact that SOC is concentrated in the surface layer (0 to 20 cm depth) has numerous implications with regards to the erosion-induced emissions. First, particulate organic matter (POM) is easily transported by surface runoff or blowing wind, especially because it is also a light fraction.

Consequently, the enrichment ratio of C in sediments is more than one and often as high as five. Second, there is also an inverse relationship between the quantity of sediments generated by different erosional processes and the concentration of C displaced (Trustrum et al., 2002). Third, the SOC pool is depleted onsite and enhanced at the depositional site, leading to different processes at these locations. Liu and Bliss (2003) modeled the impact of erosion on SOC dynamics. They reported that soils were consistently C sources to the atmosphere at all landscape positions on a soil in the Mississippi basin during 1870 and 1950 with a range of emission of 13 to 49 g C/m^2/yr. In Colombia, Ruppenthal et al. (1997) observed that sediment-bound losses of SOC ranged from 26 to 1726 kg C/ha/yr in cassava-based systems. In comparison, losses of SOC from bare fallow (plowed but uncropped) treatments ranged from 4760 to 6530 kg C/ha/yr.

Similar complex processes are involved in erosion-induced SOC dynamics whereas aggregation involves formation of stable micro-aggregates or the organo-mineral complexes, slaking and breakdown of aggregates lead to dispersion of soil and emission of C (as CO_2 or CH_4) through microbial action on organic matter that had been encapsulated within aggregates. While the process of aggregation sequesters SOC (Edwards and Bremner, 1967; Tisdall and Oades, 1982; Chaney and Swift, 1986; Oades and Waters, 1991; Tisdall, 1996), that of dispersion releases C as shown in Equation 3.1.

$$(\text{Clay} - \text{P} - \text{SOC}) \xleftrightarrow{\text{aggregation}} (\text{Clay} - \text{P} - \text{SOC})_x \xleftrightarrow{\text{aggregation}} [(\text{Clay} - \text{P} - \text{SOC})_x]_y \quad (3.1)$$

$$\text{Clay} \longleftrightarrow \text{Domains} \xleftrightarrow[\text{dispersion}]{} \text{Micro-aggregates} \xleftrightarrow[\text{dispersion}]{} \text{Macroaggregates}$$

Note the P in Equation 3.1 refers to polyvalent cations such as Ca^{+2}, Al^{+3}, Fe^{+3}, Mn^{+3}, etc. Therefore, breakdown of aggregates releases C and its transport and redistribution over the landscape and deposition in depressional sites further exposes it to numerous interacting processes. Some of the SOC displaced is merely redistributed over the landscape and may never reach streams, reservoirs, or other aquatic systems (Gregorich and Anderson, 1985). A fraction of the SOC displaced, however, may reach streams and reservoirs. The time involved in the transport of SOC from upslope to aquatic ecosystems may range from days to centuries, and SOC is subjected to numerous interacting processes en route. It is the fate of C thus being transported over the landscape that has become a topic of debate among soil scientists and sedimentologists.

3.5 FATE OF ERODED SOIL CARBON

The fate of C translocated (e.g., detached, redistributed, and redeposited) from the original site is difficult to predict because of the confounding effect of numerous interacting factors such as moisture and temperature regimes, quality of the sediments, oxidation/reducing conditions, dispersion, or reaggregation. Consequently, a mass-balance approach is needed to assess the fate of displaced C at the watershed scale (Jacinthe and Lal, 2001). A schematic of the processes involved in SOC dynamics in a landscape influenced by accelerated soil erosion is shown in Figure 3.2. The erosion-induced perturbation alters SOC dynamics (Equation 3.2).

$$\Delta \text{SOC} = (\text{SOC}_a + A) - (E + L + M) \quad (3.2)$$

Where SOC_a is the antecedent pool, A is accretion or input, E is erosion, L is leaching, and M is mineralization. A part of the SOC redistributed over the landscape may be emitted into the atmosphere either a CO_2 or CH_4 depending on the degree of aeration. Some of the dissolved organic carbon (DOC) may be leached and precipitated in the subsoil or transported into the aquatic ecosystems and reprecipitated. Erosion may lead to C sink if $(SOC_a + A) > (E + L + M)$ or net

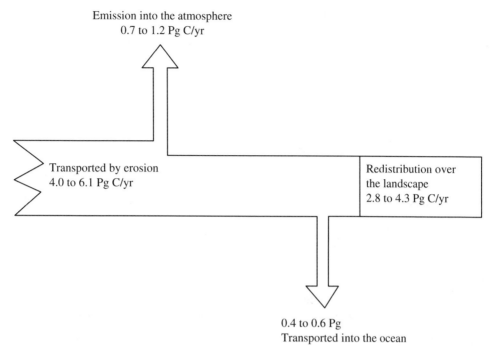

Figure 3.2 Global carbon budget as influenced by soil erosion.

source if $(E + L + M) > (SOC_a + A)$. In most cases, however, $(E + L + M) > (SOC_a + A)$ making erosion a net source. Consequently, there are numerous ramifications of the erosion-induced translocation of SOC:

- Onsite, the SOC pool is depleted, often severely, because of the preferential removal of SOM. Consequently, the enrichment ratio of SOM in sediments is often as much as 2:5 because SOC is concentrated in the surface layer where it is easily carried by surface or shallow flow and wind currents.
- The depletion of SOC pool leads to a decline in soil quality because of a reduction in available water capacity, decrease in effective rooting depth, and depletion of the reserves of essential plant nutrients. Decline in soil quality has strong adverse impacts upon biomass production and the quantity of crop residue (both above and below ground) returned to the soil.
- Restoration of eroded soil, through conversion to an ameliorative land use, and replacement of nutrients, can lead to SOC sequestration.
- The magnitude of SOC sink capacity thus created depends on the extent of SOC depletion, soil profile characteristics, landscape position, soil moisture and temperature regimes, prevalent climate, and the intended land use and management.

Two schools of thought exist, one proposed by sedimentologists and the other by soil scientists.

3.5.1 Sedimentologist's View of the Deposition of Eroded Soil Carbon

Sedimentologists argue that erosional processes transport C to the burial/depositional sites (Van Noordwijk et al., 1997; Stallard, 1998; Smith et al., 2001) and account for most of the so-called missing sink for CO_2 estimated at 0.5 to 2.0 Pg C (Tans et al., 1990). It is argued that erosion leads to C sequestration in at least two ways: (1) onsite erosion depletes SOC pool and creates a C sink that is filled by vegetation regrowth and residue return, and (2) offsite SOC is transported to depressional sites and is buried, sequestered, and otherwise taken out of circulation. Liu and Bliss

Table 3.1 Estimates of Particulate Organic Carbon Transported to the Ocean

Source	Annual Flux to the Ocean (Pg C/yr)
Berner (1992)	0.09–0.19
Chen et al. (2001)	0.24
Ittekot and Lane (1991)	0.231
Lal (1995)	0.57
Ludwig et al. (1996)	0.17
Meybeck (1993)	0.17
Meybeck and Vörösmarty (1999)	0.195
Smith et al. (2001)	0.4

Table 3.2 Global and U.S. Budget of Sediments and Soil Organic Carbon Displaced

	U.S.A.		Global	
Process	Sediment (Pg/yr)	SOC (Pg/yr)	Sediment (Pg/yr)	SOC (Pg/yr)
Erosion	7.4	0.05	200	1.4
River + wind flux to ocean	0.7	0.01	20	0.4
Land deposition	6.7	0.04	180	1.0

Note: Assuming no losses of SOC due to mineralization en route.

Adapted from Smith, S. V., W. H. Renwick, R. W. Buddemeier, and C. J. Crossland. 2001. Budgets of soil erosion and deposition for sediments and sedimentary organic carbon across the conterminous U.S. *Global Biogeochem. Cycles* 15:697–707.

(2003) observed that soils of the upper Mississippi basin became C sinks between 1950 and 1997, and the sink strength was the highest at a severely eroding site. Harden et al. (1999) also observed that soil erosion amplifies C loss and its recovery. McCarty and Ritchie (2002) reported that soils of the riparian zone are a major sink for C transported from agricultural watersheds. The annual transport of particulate organic carbon to the ocean is estimated at 0.09 to 0.57 Pg C/yr (Table 3.1). With this assumption, Smith et al. (2001) computed the sediment and SOC budget (Table 3.2). Assuming no gaseous loss due to mineralization of SOC en route, SOC transported to rivers and oceans is estimated at 0.01 Pg C for the U.S. and 0.4 Pg C for the world. Sedimentologists extrapolate these projections further and surmise that the so-called missing sink or fugitive carbon is buried annually into aquatic ecosystems and depressional sites and that accelerated soil erosion is a good thing to have.

3.5.2 Soil Scientist's View of the Fate of Carbon Displaced by Erosion

Soil scientists argue that accelerated soil erosion is a principal cause of the emission of GHGs from soil to the atmosphere. Onsite, accelerated soil erosion depletes SOC pool and degrades soil quality. Significant reductions in crop yield on eroded soils can occur (Monreal et al., 1997) especially in low-input and resource-based agriculture (Lal, 1998). The reduction in productivity may be due to decrease in effective rooting depth, reduction in available water capacity, and exposure of edaphologically inferior subsoil of poor structure and unfavorable elemental balance. With progressive decline in productivity of the above- and below-ground biomass, both the quality and quantity of residue returned to the soil are less, which further depletes the SOC pool. Indeed, SOC pool of eroded sites is drastically lower than those of uneroded sites (Rhoton and Tyler, 1990). Off-site, erosional processes transport SOC over the landscape and redistribute eroded sediment selectively depending on the density. During the process, aggregates are subjected to disruptive forces of runoff or wind. Some of the sediments are deposited down slope in depressional sites or protected areas. The SOC budget of an eroded landscape can be computed by Equation 3.3.

Table 3.3 Estimates of Soil Organic Loss by Oxidation during the Erosional Process

Source	Fraction Lost by Oxidation (%)
Lal (1995)	20
Jacinthe and Lal (2001)	25–30
Beyer et al. (1993)	70
Schlesinger (1995)	100
Smith et al. (2001)	0

$$(SOC)_L = (SOC)_A - (SOC)_D - (SOC)_R + (SOC)_M \qquad (3.3)$$

Where L represents mean SOC pool over the landscape after erosional event, A is the antecedent pool prior to the event, D is SOC deposited in depressional sites, R is SOC transported in rivers and aquatic ecosystems, and M is the fraction mineralized emitted into the atmosphere. Mineralization of SOM may occur over the landscape, in depressional sites and in aquatic ecosystems.

The principal discrepancy between sedimentologists and soil scientists lies in the assumption with regards to the magnitude of oxidation of SOM. Some sedimentologists assume that the oxidation flux is insignificant (Smith et al., 2001) and others believe that SOM lost during erosion is largely oxidized (Schlesinger, 1995). The magnitude of oxidation of eroded material may depend on the composition of particulate organic material. While humins are preserved (Hatcher et al., 1985; Hatcher and Spiker, 1988), some of the material may be reaggregated and protected against mineralization (Gregorich et al., 1998), but easily decomposable labile fraction is mineralized (Beyer et al., 1993). The proportion of eroded SOM that is mineralized en route to and in the depositional sites may be 20 to 30% (Table 3.3). Therefore, about 10% of the eroded SOC is transported to the ocean, 20 to 30% is emitted into the atmosphere, and 60 to 70% is redistributed over the landscape. The magnitude of the emission may also depend on soil moisture and temperature regime (Bajracharya et al., 2000) and soil reflectance properties as influenced by erosion (Wagner-Ridle et al., 1996).

3.6 EROSION-INDUCED GLOBAL EMISSION OF CARBON

Several estimates have been made of transport of C into the ocean (Lyons et al., 2002; Milliman and Meade, 1983). There are two approaches to estimating erosion-induced C emission. First, sediment by world's rivers is estimated at 15 to 20 billion Mg (Walling and Web, 1996). Assuming delivery ratio of 13 to 20% (Walling and Web, 1996) and SOC content of 2%, total SOC displaced by erosion is 4.0 to 6.1 Pg. Assuming 20% is emitted into the atmosphere (Lal, 1995; 2003), erosion-induced C emission is 0.7 to 1.2 Pg C/yr (Figure 3.2). Based on these assumptions, the global sediment and C budget is shown in Table 3.4.

Numerous uncertainties regard the fate of the eroded SOC, and little quantitative data assesses erosion-induced emissions. The data on continental basis is shown in Table 3.5. Estimates of erosion-induced emissions are shown in Table 3.5. Total erosion-induced emissions are estimated

Table 3.4 Global Budget of Sediment and Soil Organic Carbon

Process	Erosion (Pg/yr)	SOC Dynamic (Pg/yr)
Erosion over the landscape	88–135	4.0–6.1
Transport to oceans	15–20	0.4–0.6
Redistribution over the landscape	73–115	2.8–4.3
Emission to the atmosphere	0	0.8–1.2

Table 3.5 Continental Distribution of Eroded Soil Organic Carbon and Emissions by Decomposition

Continent	Estimates of SOC Displaced by Erosion (Tg C/yr)	C Emission by Erosion (20%) (Tg C/yr)
North America	456–700	91–140
South America	563–866	113–173
Africa	235–362	47–72
Asia	2220–3415	444–683
Europe	330–509	66–102
Oceania	153–236	31–47
Total	3957–6083	792–1217

Note: Assuming sediment load estimates by Walling and Web (1996), delivery ratio of 13 to 20% and SOC concentration of 2% and emission of 20%.

from 0.8 to 1.2 Pg C/yr (with a mean of 1.0 Pg C/yr) are comparable to those reported earlier at 1.14 Pg C/yr (Lal, 1995). Such large emissions must be accounted for in the global C cycle.

3.7 SUMMARY AND CONCLUSIONS

Geologic/natural soil erosion is a constructive process. It is an important soil-forming factor that created the world's most fertile soils in the flood plains and deltas of major rivers. Annual renewal of soil fertility by sediments deposited in flood plains sustained agricultural production and supported dense populations on alluvial soils. Accelerated soil erosion, however, due to anthropogenic activities involving land misuse and soil mismanagement, is a destructive process. Offsite, it causes nonpoint source pollution and emission of CO_2 and other GHGs into the atmosphere. For transport of C to the oceans estimated at 0.4 to 0.6 Pg/yr, emissions as CO_2 by oxidation of eroded SOC is estimated at 0.8 to 1.2 Pg C/yr.

Sustainable management of soil and water resources is needed not only for food/biomass production but also for maintaining environment quality including mitigation of climate change. In fact, SOC sequestration is also needed for desertification control (Squire et al., 1995) and restoration of degraded ecosystems.

Several hot spots of accelerated soil erosion exist around the world. Principal among these are South Asia, especially the Himalayan-Tibetan ecosystem, Central Asia, the Loess Plateau of China, sub-Saharan Africa and the Maghreb region of northwest Africa, the Andean region of South America, the Dominican Republic and the Caribbean, and the highlands of Central America. Resource-poor farmers who practice low-input and subsistence farming accentuate the problems of soil erosion in these regions. A coordinated effort is needed to facilitate widespread adoption of science-based and conservation-effective agriculture.

There are numerous options for mitigating projected climate change. Sustainable management of soil and water resources is an important option. It is a win-win strategy. Adopting effective measures of soil and water conservation can enhance productivity, improve water quality, reduce erosion-induced emissions of GHGs, and sequester C in soil and biomass to reduce the rate of enrichment atmospheric CO_2 and other GHGs.

REFERENCES

Albrecht, W. A. 1938. Loss of soil organic matter and its restoration, in *Soils and Men*, Yearbook of Agriculture, USDA, U.S. Govt. Printing Office, Washington, D.C., pp. 347–360.

Arnell, N. and C. Liu. 2001. Hydrology and water resources. Ch. 4 in *Climate Change 2001: Impacts, Adaptation and Vulnerability*, Climate Change 2001, IPCC, Cambridge Univ. Press, U.K., pp. 191–233.

Bajracharya, R. M., R. Lal, and J. M. Kimble. 2000. Erosion effects on CO_2 concentration and C flux from an Ohio Alfisol. *Soil Sci. Soc. Am. J.* 64:694–700.

Batjes, N. H. 1996. The total carbon and nitrogen in soils of the world. *Eur. J. Soil Sci.* 47:151–163.

Berner, R. A. 1992. Comments on the role of marine sediment burial as a repository of anthropogenic CO_2. *Global Biogeochem. Cycles* 6:1–2.

Beyer, L., C. Köbbemann, J. Finnern, D. Elsner, and W. Schluß. 1993. Colluvisols under cultivation in Schleswig-Holstein: 1. Genesis, definition, and geo-ecological significance. *J. Plant Nutr. Soil Sci.* 156:197–202.

Chaney, K. and R. S. Swift. 1986. Studies on aggregate stability: Reformation of soil aggregates. *J. Soil Sci.* 35:223–230.

Chen, C. -T. A., K. K. Liu, and R. McDonald. 2001. Continental margins and seas as carbon sinks. *Stockholm, Global Change Newsletter* 46 (June 2001):11–13.

Edwards, A. P. and J. M. Bremner. 1967. Microaggregates in soils. *J. Soil Sci.* 18:64–73.

Etheridge, D. M., L. P. Steele, R. L. Langenfelds, R. J. Francey, J. M. Barnola, and V. I. Morgan. 1996. Natural and anthropogenic changes in atmospheric CO_2 over the last 1000 years from air in Antarctic ice and fire. *J. Geophys. Res. Atmos.* 101:41415–41428.

Etheridge, D. M., L. P. Steele, R. J. Francey, and R. L. Langenfelds. 1998. Atmospheric methane between 1000 AD and present: Evidence of anthropogenic emissions and climatic variability. *J. Geophys. Res. Atmos.* 103:15979–15993.

Gregorich, E. G. and D. W. Anderson. 1985. The effects of cultivation and erosion on soils of four toposequences in the Canadian prairies. *Geoderma* 36:343–354.

Gregorich, E. G., K. J. Greer, D. W. Anderson, and B. C. Liang. 1998. Carbon distribution and losses: Erosion and deposition effects. *Soil Tillage Res.* 47:291–302.

Harden, J. W., J. M. Sharpe, W. J. Parton, D. S. Ojima, T. L. Fries, T. G. Huntington, and S. M. Dabney. 1999. Dynamic replacement and loss of soil carbon on eroding cropland. *Global Biogeochem. Cycles* 13:885–901.

Hatcher, P. G. and E. C. Spiker. 1988. Selective degradation of plant biomolecules in F. H. Frimmel and R. F. Christman, eds., *Humic Substances and Their Role in the Environment*. John Wiley & Sons, New York, pp. 59–74.

Hatcher, P. G., I. A. Berger, E. G. Maciel, and N. M. Szeverenyi. 1985. Geochemistry of humin, in G. R. Aiken et al., eds., *Humic Substances in Soil Sediment and Water*. John Wiley & Sons, Chichester, U.K., pp. 275–302.

Hungate, B.A., E. A. Holland, R. B. Jackson, F. S. Chapin, H. A. Mooney, and C. B. Field. 1997. The fate of carbon in grasslands under CO_2 enrichment. *Nature* 388: 576–579.

IPCC. 1995. *Revised IPCC Guidelines for National Greenhouse Gas Inventories, Reference Manual*. Intergovernmental Panel on Climate Change, vol. 3. Cambridge Univ. Press, U.K.

IPCC. 2001. *Climate Change 2001: The Scientific Basis. Intergovernmental Panel on Climate Change*. Cambridge Univ. Press, U.K.

Ittekot, V. and R. W. P. M. Lane. 1991. Fate of riverine particulate organic matter, in E. T. Degens et al., eds., *Biogeochemistry of Major Rivers*. John Wiley & Sons, Chichester, U.K., pp. 233–243.

Jacinthe, P. and R. Lal. 2001. A mass balance approach to assess carbon dioxide evolution during erosional events. *Land Degrad. Dev.* 12:329–339.

Jenny, H. 1980. *The Soil Resource: Origin and Behavior*. Springer-Verlag, New York, pp. 377.

Lal, R. 1995. Global soil erosion by water and carbon dynamics, in R. Lal, J. M. Kimble, E. Levine, and B. A. Stewart, eds., *Soils and Global Change*. CRC/Lewis Publishers, Boca Raton, FL, pp. 131–141.

Lal, R. 1998. Soil erosion impact on agronomic productivity and environment quality. *CRC Crit. Rev. Plant Sci.* 17:319–464.

Lal, R. 2000. World cropland soils as a source or sink for atmospheric. *C. Adv. Agron.* 71:145–191.

Lal, R. 2003. Soil erosion and the global carbon budget. *Env. Intl.* 29:437–450.

Lavee, H., A. C. Imeson, and P. Sarah. 1998. The impact of climate change on geomorphology and desertification along a Mediterranean-arid transect. *Land Degradation and Development* 9:407–422.

Leiros, M. C., C. Trasar-Cepeda, S. Seoane, and F. Gil-Sotres. 1999. Dependence of mineralization of soil organic matter on temperature and moisture. *Soil Biol. Biochem.*

Liu, S. and N. Bliss. 2003. Modeling carbon dynamics in vegetation and soil under the impact of soil erosion and deposition. *Global Biogeochem. Cycles* 17:1–24.

Ludwig, W., P. Amiotte-Suchet, and J. L. Probst. 1996. River discharges of carbon to the world's oceans: Determining local inputs of alkalinity and of dissolved and particulate organic carbon. *C.R. Acad. Sci., Series 2., Earth Plan Sci.* 323:1007–1014.

Lyons, W. B., C. A. Nezat, A. E. Carey, and D. M. Hicks. 2002. Organic carbon fluxes to the ocean from high standing islands. *Geology* 30:443–446.

McCarty, G. W. and J. C. Ritchie. 2002. Impact of soil movement on carbon sequestration in agricultural ecosystems. *Env. Pollution* 116:423–434.

Meybeck, M. 1993. C, N, P, and S in rivers: from sources to global inputs, in: R. Wollast et al., eds. *Interactions of C, N, P, and S Biogeochemical Cycles and Global Change*. Springer-Verlag, Berlin, pp. 163–193.

Meybeck, M. and C. Vörösmarty. 1999. Global transport of carbon by rivers. *Global Change Newsletter* 37:181–19.

Milliman, J. D. and R. H. Meade. 1983. Worldwide delivery of river sediments to the oceans. *J. Geogr.* 91:1–21.

Monreal, C. M., R. P. Zentner, and J. A. Robinson. 1997. An analyses of soil organic matter dynamics in relation to management, erosion and yield of wheat in long-term crop rotation plots. *Can. J. Soil Sci.* 77:553–563.

Newton, P. C. D., H. Clark, C. C. Bell, and E. M. Glasgow. 1996. Interaction of soil moisture and elevated CO_2 on above-ground growth rate, root length density, and gas exchange of turves from temperate pastures. *J. Expl. Botany* 47:771–779.

Oades, J. M. and A. G. Waters. 1991. Aggregate hierarchy in soils. *Aust. J. Soil Res.* 29:815–828.

Oldeman, L. R. 1994. The global extent of soil degradation, in D. J. Greenland and I. Szabolcs, eds., *Soil Resilience and Sustainable Land Use*. CAB International, Wallingford, pp. 99–118.

Page, M. J., N. A. Trustrum, and B. Gomez. 2000. Implications of a century of anthropogenic erosion for future land use in the Gisborne-East Coast region of New Zealand. *New Zealand Geographer* 56:13–24.

Puigdefabregas, J. 1998. Ecological impacts of global change on drylands and their implications for desertification. *Land Degradation and Development* 9:393–406.

Rhoton, F. E. and D. D. Tyler, 1990. Erosion-induced changes in the properties of a Fragipan soil. *Soil Sci. Soc. Am. J.* 54:223–228.

Ruppenthal, M., D. E. Leihner, N. Steinmüller, and M. A. El-Sharkawy. 1997. Losses of organic matter and nutrients by water erosion in cassava-based cropping systems. *Expl. Agric.* 33: 487–498.

Schlesinger, W. H. 1995. Soil respiration and changes in soil carbon stocks, in G. M. Woodwell and F. T. Mackenzie, eds. *Biotic Feedback in the Global Climatic System: Will the Warming Feed the Warming?* Oxford Univ. Press, New York, pp. 159–168.

Schnitzer, M. 1991. Soil organic matter: The next 75 years. *Soil Sci.* 151:41–58.

Smith, P., J. U. Smith, D. S. Powlson, J. R. M. Arah, O. G. Chertov, K. Coleman, U. Franko, S. Frolking, D. S. Jenkinson, L. S. Jensen, R. H. Kelvy, H. Parton, J. H. M. Thornley, and A. P. Whitmore. 1997. A comparison of the performance of nine soil organic matter models using data sets from seven long-term experiments. *Geoderma* 81:153–225.

Smith, S. V., W. H. Renwick, R. W. Buddemeier, and C. J. Crossland. 2001. Budgets of soil erosion and deposition for sediments and sedimentary organic carbon across the conterminous U.S. *Global Biogeochem. Cycles* 15:697–707.

Squires, V., E. P. Glenn, and A. T. Ayoub, eds. 1995. Combating global climate change by combating land degradation. Proc. of a Workshop held in Nairobi, Kenya, 4–8 Sept. 1995, UNEP, Nairobi, Kenya.

Stallard, R. F. 1998. Terrestrial sedimentation and the carbon cycle: coupling weathering and erosion to carbon burial. *Global Biogeochem. Cycles* 12:231–237.

Tans, P. P., I. Y. Fung, and T. Takahashi. 1990. Observational constraints on the global atmospheric CO_2 budget. *Science* 247:1431–1438.

Tisdall, J. M. 1996. Formation of soil aggregates and accumulation of soil organic matter, in M. R. Carter and B. A. Stewart, eds., *Structure and Organic Matter Storage in Agricultural Soils*. CRC/Lewis Publishers, Boca Raton, FL, pp. 57–96.

Tisdall, J. M. and J. M. Oades. 1982. Organic matter and water stable aggregates in soil. *J. Soil Sci.* 33:141–163.

Trustrum, N. A., K. R. Tate, M. J. Page, A. Sidorchuk, and W. T. Baisden. 2002. Toward a national assessment of erosion-related soil carbon losses in New Zealand. 12th ISCO Conference, 26–31 May 2002, Beijing, China, vol. 3:182–186.

Van Noordwijk, M., C. Cerri, P. L. Woomer, K. Nugroho K., and M. Bernoux. 1997. Soil carbon dynamics in the humid tropical forest zone. *Geoderma* 79:187–225.

Villers-Rúiz, L. and I. Trejo-Vazquez. 1998. Climate change on Mexican forests and natural protected areas. *Global Environmental Change: Human and Policy Dimensions* 8:141–157.

Wagner-Ridle, C., T. J. Gillepsi, and C. J. Swanton. 1996. Rye mulch characterization for the purpose of micro-climatic modelling. *Agric. Forest Meteor.* 78:67–81.

Walling, D. E. and B. W. Webb. 1996. Erosion and sediment yield: A global overview. Erosion and sediment yield: global and regional perspectives. *Proc. Exeter Symp.* July 1996. IAHS Publ. vol. 236: 3–19.

CHAPTER 4

Modeling Soil Erosion Induced Carbon Fluxes between Soil and Atmosphere on Agricultural Land Using SPEROS-C

Kristof Van Oost, Tim A. Quine, Gerard Govers, and Goswin Heckrath

CONTENTS

4.1 Introduction 37
4.2 SPEROS-C Model Description 39
    4.2.1 Multisoil-Layer Landscape Model Structure 40
    4.2.2 Soil Redistribution Model (SPEROS) 40
        4.2.2.1 Water 40
        4.2.2.2 Tillage 42
4.3 Soil Carbon Decomposition 44
    4.3.1 Carbon Redistribution and Profile Evolution 45
    4.3.2 Model Implementation 46
4.4 Model Application 47
4.5 Discussion and Conclusion 49
References 50

4.1 INTRODUCTION

Soil organic carbon (SOC) has an important influence on the chemical and physical properties of the soil and is one of the key components for assessing soil quality (e.g., Gregorich et al., 1994; Lal, 2004). Being located in the vicinity of the soil surface, soil organic carbon is drastically impacted by erosion processes. Many studies on soil erosion–carbon relationships have reported high losses of soil carbon on eroding sites (Dejong and Kachanoski, 1988; Lal, 2004; e.g., Quine and Zhang, 2002). Soil erosion is therefore usually associated with a decline in soil carbon content. However, soil erosion is a multi-stage process, which involves (1) the detachment of particles, (2) the transport and redistribution of eroded sediment over the landscape, and (3) the deposition or export into the fluvial system. It is therefore clear that the overall impact of soil erosion on the SOC budget should be assessed at a spatial scale that incorporates erosion, transport, and deposition. Lal (2003) gives an overview of the influences of soil erosion on carbon dynamics during these stages: onsite depletion of the SOC pool may increase local mineralization rates due to changes in temperature and moisture regimes and decrease soil productivity which in turn reduces the amount of biomass reduced to the soil. The detachment and transport of sediment by water accelerates the mineralization of SOC, mainly due to the breakdown of aggregates. On the other hand, deposition and burial of soil and SOC in the soil profile and reaggregation of the dispersed clay and silt with SOC may decrease mineralization rates.

Some assessments have indicated that soil erosion has an overall negative influence on C storage in the terrestrial ecosystem (Lal, 2001; 2003; Lal et al., 2004). This expectation for increased CO_2 emissions due to erosion is largely based on the assumption that a large part of the SOC mobilized with eroded soils is likely to be mineralized. Although it was initially asserted that between 20% (Lal, 2003) and 100% (Schlesinger, 1995) of the eroded carbon could be mineralized, water erosion–induced mineralization has been observed and rates have been derived in recent studies: Jacinthe and Lal (2001) calculated CO_2 emissions rates from a series of long-term plot experiments. They estimated a global annual flux of 0.37 Pg CO_2-C to the atmosphere due to water erosion. Based on runoff experiments and incubation, Jacinthe et al. (2002) showed that between 29 and 46% of the C exported in runoff was potentially mineralizable. Based on the incubation of runoff samples from small agricultural watersheds, Jacinthe et al. (2004) illustrated that rainfall intensity characteristics largely influence the impact of erosional events on terrestrial C cycling. Thirty to 40% of the eroded SOC was potentially mineralizable during low-intensity storms while the figure decreased to only 13% for high-intensity summer storms. These studies clearly showed that erosion and transport of SOC has the potential to represent a significant source of atmospheric CO_2. It is estimated that erosion-induced emission may be between 0.37 and 1.2 Pg C yr^{-1} (Lal, 2003; 2004).

On the other hand, arguing that eroded SOC is more likely to remain immobilized rather than mineralized, Stallard (1998) proposed that depositional areas could act as a C sink: the burial of eroded SOC in depositional areas may substantially constrain the SOC decomposition process. Quine and Zhang (2002) presented a detailed study of SOC profile evolution under intense soil redistribution at a very fine scale. These authors concluded that that carbon storage within sloping agricultural land is spatially heterogeneous and highlighted the importance of redistribution processes, especially by tillage erosion, at the field scale. They argued that the plow layer becomes depleted in carbon at eroding sites as carbon-poor subsoil is incorporated into the plow layer when the same plow depth is maintained. At aggrading sites, substantial amounts of carbon-rich soil are buried below the plow layer due to continuous soil deposition. The significance of these processes is illustrated in Figure 4.1. The extent to which erosion and deposition lead to an atmospheric C sink depends, however, on how much of the depositional accumulation is replaced by newly produced SOC at eroding sites. Harden et al. (1999, 2002) showed the potential for dynamic replacement of SOC at eroding sites by combining C, N, and ^{14}C measurements with the Century model. This study concluded that about 100% of the SOC had been lost from eroding sites and that about 30% of the C had been replaced with increased fertilization since the 1950s. Harden et al. (2002) also presented field evidence of dynamic replacement of eroded carbon. Stallard (1998) analyzed a range of scenarios for global erosion and sedimentation and estimated that soil erosion can induce formation of a C sink in the order of 0.5 to 1.5 Pg C yr^{-1}.

There is now growing recognition that the role of erosion within the global carbon budget may be significant but is highly uncertain, as it is sensitive to the balance between (1) increased mineralization during and after transport and (2) the combination of reduced mineralization in depositional areas and dynamic replacement at eroding sites. Owing to the difficulty of deploying field experiments for direct assessment and the large spatial and temporal variability of the processes involved, the dynamic responses of carbon fluxes to due the redistribution of eroded carbon are still poorly understood (Liu et al., 2003; McCarty and Ritchie, 2002; Stallard, 1998). In order to overcome this problem, biogeochemical models have been developed to quantify the linkages between the carbon cycle and soil redistribution processes on land. While most models focus on

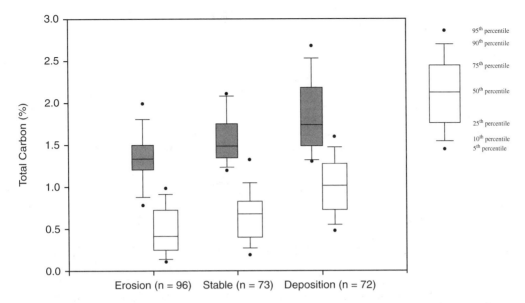

Figure 4.1 Observed variation in carbon content for the plow (0 to 0.25 m, grey) and sub-plow layer (0.25 to 0.5 m, white) at an agricultural field in Devon, U.K. for eroding, stable, and aggrading sites. Erosion classes are derived from ^{137}Cs measurements: values significantly lower or higher than the reference value (at the 0.05 probability level) were classified as erosion and deposition respectively. When the residual was not significantly different from the reference value, it was classified as stable. (Data from Quine, T. A. and Y. Zhang. 2002. An investigation of spatial variation in soil erosion, soil properties, and crop production within an agricultural field in Devon, United Kingdom. *Journal of Soil and Water Conservation* 57(1):55–65.)

the erosion part of the process (Harden et al., 1999; Manies et al., 2001; Polyakov and Lal, 2004), models that account for both erosion and deposition have only been developed recently (e.g., EDCM [Liu et al., 2003]). Although these modeling studies allow to gain more insight in soil–atmosphere interactions under soil erosion and deposition, an overall budget of C between the soil and atmosphere could not be made. This is because the impacts of erosion and deposition were analyzed separately at a few points only in the landscape. An overall C budget at the watershed scale requires an integrated spatial analysis of the fate of eroded SOC during transport, the impact of soil erosion on the SOC remaining at eroding sites, and burial of SOC at depositional sites.

This chapter seeks to address the need for such integration by combining current generation geomorphological models with established carbon dynamics models to examine the interaction between soil redistribution and carbon fluxes between soil and atmosphere on agricultural land. Therefore, a spatially distributed soil erosion model (SPEROS) is linked with a model simulating carbon dynamics (ICBM). This approach extends the classical point-based or plot-scale approach to a grid-based, landscape-scale approach so that the transfers of C within and between landscape elements are explicitly taken into account. We illustrate the potential of the model and discuss its limitations.

4.2 SPEROS-C MODEL DESCRIPTION

SPEROS-C integrates a spatially distributed water and tillage erosion model, as described by Van Oost et al. (2003a, 2003b) with a carbon model, based on the Introductory Carbon Balance Model (ICBM; Andren and Katterer, 1997). SPEROS-C has been briefly described in Van Oost et al. (2005). Here, we present a detailed description of the main characteristics, process formulation, main assumptions, and implementation of the model.

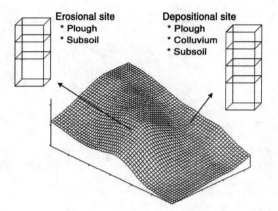

Figure 4.2 Three-dimensional representation of the soil-landscape in SPEROS-C.

4.2.1 Multisoil-Layer Landscape Model Structure

The main characteristic of the SPEROS-C model is the three-dimensional representation of the soil landscape. The model represents space by a uniform grid of cells (Figure 4.2) and explicitly simulates the pathways of eroded sediment and SOC toward zones of deposition. The soil profile is characterized by using different layers and concentrations of SOC are assumed to be uniform within one layer (Figure 4.3). The current version allows to specify four layers up to a depth of 1 m. The soil erosion model is used to redistribute soil and SOC within the landscape and to predict changes in the SOC profile due to soil truncation or aggradation. The carbon module is then used to estimate changes SOC contents and carbon fluxes for all grid cells and soil layers. The model uses a time-step of 1 year and the output is used as input for the next simulation year. SPEROS-C can be used to assess soil erosion induced carbon fluxes between soil and atmosphere at the scale of an individual field, i.e., 1 to 20 ha.

4.2.2 Soil Redistribution Model (SPEROS)

4.2.2.1 Water

The water erosion model used within SPEROS-C aims to represent the effects of topography and catchment heterogeneity on the dynamics of sediment redistribution by overland flow. The most important feature of the water erosion model is the mapping of predictions into space. Low numbers of parameters are used to provide a flexible and transparent model structure.

The local erosion rate is considered equal to the sum of the rill and interrill detachment unless the local transport capacity is exceeded. The model describes the potential for rill erosion as a power function of contributing area and slope gradient and the potential for interrill erosion as a power function of slope gradient:

$$E_{rill} = k_1 \rho_b S^a A_s^b \tag{4.1}$$

$$E_{irill} = k_2 \rho_b S^c \tag{4.2}$$

Where E_{rill} is the rill potential (kg m^{-2}), E_{irill} is the interrill potential, ρ_b is the dry bulk density of the soil (kg m^{-3}), S is the slope (m m^{-1}), A_s is the unit contributing area, i.e., contributing area per unit contour width (m² m^{-1}), k_1, k_2, a, b, and c are coefficients.

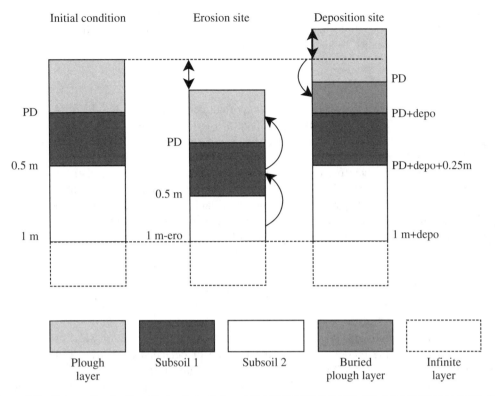

Figure 4.3 Schematic of soil profile implementation of the SPEROS-C model. PD is the local plow depth. Soil profile characteristics change with continued erosion/deposition.

The k_1 and k_2 coefficients regulate the rate of water erosion while the spatial pattern is controlled by the exponents in Equation 4.1 and Equation 4.2. The use of the contributing area implies that the effects of flow convergence and divergence on topographically complex landscapes are explicitly simulated. Field observations indicate that this two-dimensional approach not only accounts for interrill and rill erosion but also to some extent for (ephemeral) gully erosion (Desmet and Govers, 1997). The transport capacity on a given slope segment is considered to be directly proportional to the potential for rill erosion.

$$T_c = k_3 E_{rill} \tag{4.3}$$

where T_c is the transport capacity (kg m^{-1}) and k_3 is a coefficient.

If the sediment inflow exceeds the transport capacity, deposition occurs, so that the amount of material equals the transport capacity. The transport capacity, coefficient k_3, regulates the amount and spatial pattern of deposition. This approach is conceptually similar to the process description used by Foster and Meyer (1975). However, flow detachment is assumed to be proportional to the transporting capacity and not the transport capacity deficit. This modeling concept appears to be appropriate for cohesive soils (e.g., Desmet and Govers, 1995) but may not work so well in different conditions.

Equation 4.1 through Equation 4.3 are implemented in a grid-based structure to estimate the local erosion potential. Slope gradient and contributing area are derived from a grid-based digital elevation model. After calculation of the local soil erosion rate, the eroded soil is transferred down slope over the grid using a flux decomposition algorithm (Desmet and Govers, 1996). It is assumed that the area is an isolated hydrological unit, i.e., no overland flow (represented by the contributing area) or sediment enters the study area at the boundaries. In many cases, not all the sediment that is produced is redeposited in the same field. The model therefore keeps track of the amount of sediment that is exported. If, during the routing of sediment, no lower grid cells are found within the study area, it is assumed to leave the study area.

4.2.2.2 Tillage

Tillage erosion is recently identified as an important contributor to soil redistribution in rolling topographies (e.g., Govers et al., 1994). Tillage erosion experiments have shown that soil is translocated and dispersed during soil tillage. Most soil particles are typically displaced in the range 0.1 to 0.5 m while some are displaced as far as 2 m. Most models of soil and soil constituent redistribution by tillage therefore explicitly account for a soil displacement distribution (see Van Oost et al., 2000 for a discussion). In SPEROS-C, the tillage model consists of (1) the tillage displacement distribution (G), which represents the tillage process, and (2) a map describing the spatial distribution of SOC in the plow layer (S). In a grid-based model, the resulting spatial distribution of SOC in the plow layer (P) after a tillage pass can be described as:

$$P(a) = \sum_{x=-\infty}^{+\infty} S(x)G(a-x) \tag{4.4}$$

In this equation, G is the displacement density distribution, evaluated at a distance $a - x$. Using Equation 4.4, the model calculates the fraction of detached SOC from different source cells that is transferred to a destination cell. As tillage is an anisotropic process, which redistributes soil both in and perpendicular to the direction of tillage, Equation 4.4 is extended to include a second dimension:

$$P(a,b) = \sum_{x=-\infty}^{+\infty}\sum_{y=-\infty}^{+\infty} S(x,y)G(a-x,b-y) \tag{4.5}$$

The model thus requires the prediction of G at each grid cell. It is important to account for variation of soil translocation with slope gradient, which is a basic characteristic of tillage, leading to loss or accumulation of soil constituents on specific landscape positions. When simulating redistribution of soil on sloping land, G must therefore be linked with the local slope gradient. Van Oost et al. (2000, 2003b) showed, from tillage experiments, how the moments of the displacement probability distributions are affected by topography. The longitudinal and lateral components of the tillage displacement distribution (G) are estimated from the local slope in the direction of tillage (S_{long}) and lateral direction (S_{lat}) using an empirical equation of the form (Van Muysen et al., 2002):

$$d_{long} = \left(g + k_{4,long} S_{long}\right)\left(\frac{TD}{D_{ref}}\right)^{\alpha}\left(\frac{V}{V_{ref}}\right)^{\beta} \tag{4.6}$$

$$d_{lat} = \left(g + k_{4,lat} S_{lat}\right)\left(\frac{TD}{D_{ref}}\right)^{\alpha}\left(\frac{V}{V_{ref}}\right)^{\beta} \tag{4.7}$$

where d_{long} and d_{lat} (m) are the average displacement distance in the tillage and lateral direction, respectively; TD is the tillage depth (m); V is the tillage speed (m s^{-1}) and g, $k_{4,long}$, $k_{4,lat}$, α and β

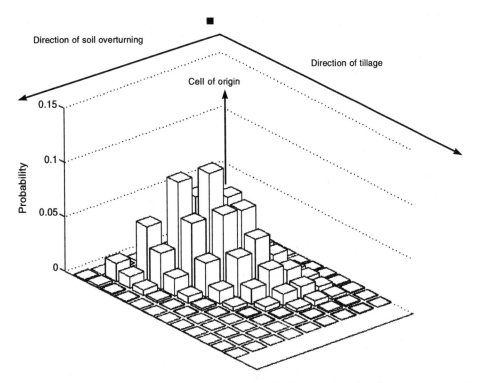

Figure 4.4 Illustration of the tillage soil displacement distributions used in the tillage erosion model. The distribution for a single tillage operation are calculated based on tillage speed, depth, and slope and are discretized in order to fit the grid. Grid size is 0.25 m.

are regression coefficients obtained from tillage experiments, D_{ref} and V_{ref} are reference tillage depth and speed respectively. The procedures proposed by Van Oost et al. (2003b) are used to predict the second and third moments of G. The calculation of the displacement distribution is illustrated in Figure 4.4. Van Oost et al. (2003b) showed that this approach is able to describe soil translocation and dispersion by tillage for a variety of tillage implements and agroenvironmental conditions.

Under most agricultural systems, multiple tillage operations with different tillage tools, performed in different directions, are carried out during a year. Rather than simulating all individual tillage operations, SPEROS-C calculates an average annual displacement distribution for each implement, whereby the contribution of an individual operation, characterized by a specific tillage direction, is weighted according to the frequency of its occurrence.

Tillage not only leads to the redistribution of soil and soil constituents, it also results in a net loss or gain of soil at specific landscape positions. This is due to the spatial variation of soil translocation with slope gradient. Tillage erosivity can therefore be characterized by the slope of the regression equation of the relationship between average soil displacement and slope gradient (i.e., $k_{4,long}$ and $k_{4,lat}$ coefficients). The k_4 values regulate the tillage erosion intensity while the spatial pattern is controlled by topography and tillage direction. The tillage soil redistribution rate (kg m^{-2}) at location (a,b) is computed as:

$$E_{til}(a,b) = \rho_b D \left[\left(\sum_{x=-\infty}^{+\infty} \sum_{y=-\infty}^{+\infty} G(a-x, b-y) \right) - 1 \right] \quad (4.8)$$

The summation term represents the probability of finding initial soil mass at location (a,b) after a tillage pass.

In real life, field boundaries act as lines of zero-flux, as no soil material is translocated over a field boundary. The model therefore applies zero-flux boundary conditions: no soil material enters the field and if the destination cell is located outside the field, the soil material remains in the destination cell.

4.3 SOIL CARBON DECOMPOSITION

SPEROS-C uses a relatively simple carbon dynamics model structure. The carbon model is based on ICBM (Andren and Katterer, 1997). It consists of two state variables (a young [Y] and an old pool [O] [g C m^{-2}]) and outflow from the pools follows first-order kinetics. There are four fluxes: (1) input, (2) mineralization from the young pool, (3) mineralization from the old pool, and (4) transformation of Y into O. The fluxes are governed by four rate-determining parameters. The differential equations describing the dynamics are:

$$\frac{dY}{dt} = i - k_y rY \tag{4.9}$$

$$\frac{dO}{dt} = hk_y rY - k_o rO \tag{4.10}$$

where i is the input of carbon (g C m^{-2} year^{-1}), h is the humification coefficient and is in the range [0.1], r is a climate coefficient, and k_y and k_o are turnover rates of young and old carbon pools, respectively (year^{-1}). Note that the external factors are condensed into one parameter, r, which affects the decomposition rates of Y and O equally. The climate effect (r) is assumed to depend primarily on temperature using a correction factor r_T:

$$r_T = 2.07^{\frac{T-5.4}{10}} \tag{4.11}$$

where T is annual mean air temperature (°C). The Q_{10} dependency on temperature is based on the mean annual temperature of 5.4 °C for the site in Sweden where the k_y and k_o values were estimated (Andren and Katterer, 1997). The Q_{10} value is based on Katterer et al. (1998).

The humification coefficient depends on carbon source and soil clay content (Katterer and Andren, 1999):

$$h = \frac{i_c h_c + i_m h_m}{i} \exp(0.0112(cl - 36.5)) \tag{4.12}$$

where cl is soil clay content (%), i_c and i_m are inputs of carbon in crop residues and manure (g m^{-2} yr^{-2}), respectively, and h_c and h_m are humification coefficients for crop residues and manure, respectively.

The carbon input into the soil profile is modeled by an exponential root density profile:

$$\varphi(z) = \begin{cases} 1 & z \leq z_r \\ \exp(-c(z-z_r)) & z > z_r \end{cases} \tag{4.13}$$

where z is soil depth (m) and z_r is a reference depth. The proportion of total root dry matter to 1 m depth that can be found in the soil is calculated as:

$$p_t = \begin{cases} \dfrac{z_t}{z_r + (1 - \exp(-c(1-z_r)))/c} & z_t \leq z_r \\ \dfrac{z_r + (1 - \exp(-c(z_t-z_r)))/c}{z_r + (1 - \exp(-c(1-z_r)))/c} & z_t > z_r \end{cases} \qquad (4.14)$$

Additional carbon input into the plow layer by residue incorporation and manure can be specified.

For cereal crops the proportion of carbon allocated to roots may be assumed to constitute 30% of total carbon assimilation (Andren and Katterer, 1997). The grain dry matter yield is assumed to constitute 45% of above ground dry matter, and stubble and other losses of above ground crop residues can be set to 15% of above ground biomass. The root dry matter thus constitutes 95% of grain dry matter, and stubble dry matter constitutes 33% of grain dry matter. The carbon content in dry matter was set to 45%. The final model of carbon input as a function of depth and grain yield is then:

$$CI_{PD} = 0.45 \left[0.33 Y . D_{residue} + p_{PD} 0.95 Y \right] + man \qquad (4.15)$$

$$CI_{x-y} = 0.45 \left[(p_y - p_x) 0.95 Y \right] \qquad (4.16)$$

where CI_{PD} and CI_{x-y} are the carbon input into the plow layer and x–y soil layer; x and y represent the depth boundaries of the soil layer; Y is the dry grain yield; $D_{residue}$ is a dummy variable and equals 1 (0) when residues are (not) incorporated into the plow layer; man is the carbon input by manuring; p_{PD} and p_x are the proportion of root dry matter to plow and x m depth, respectively.

The transport of eroded SOC by water erosion may result in a source of atmospheric CO_2 due to the additional mineralization of the displaced SOC (e.g., Lal, 2003). A fixed fraction of the C transported in runoff (C_{ero}) is therefore assumed to be mineralized (f_{trans}) so that the C loss due to mineralization of SOC in soil eroded by water (C_{trans}) can be calculated as

$$C_{trans} = C_{ero} * f_{trans} \qquad (4.17)$$

f_{trans} is in the range [0.1]. A separate value for f_{trans} can be applied to the different carbon pools, i.e., young and old. In contrast to water erosion, there is no clear reason to expect the soil transport associated with tillage erosion to lead to additional C mineralization (although the mineralization caused by tillage independently of transport must be accounted for).

4.3.1 Carbon Redistribution and Profile Evolution

The rate of soil erosion is used to calculate the amount of young ($C_{ero,Y}$) or old ($C_{ero,O}$) C eroded from the top layer:

$$C_{ero,Y} = Y_{PD} \frac{M_{ero}}{M_{PD}} \qquad (4.18)$$

where Y_{PD} is the amount of young C in the plow layer (g), M_{ero} is the mass of the eroded soil (g), and M_{PD} is the total mass of the plow layer (g). A similar expression is used to calculate the amount of old C in the eroded soil. In the case of water erosion, an enrichment factor can be applied to both pools to simulate selective erosion/deposition of carbon. Soil and carbon redistribution by tillage is simulated as a nonselective process.

As plow layer depth is maintained, a fraction of C from the first subsoil layer is incorporated into the plow layer while some carbon from the second subsoil layer is assigned to the first subsoil layer, in proportion to the erosion height (Figure 4.3). Carbon eroded by water is transferred over the landscape using a flux decomposition algorithm until it reaches an area of deposition or is exported from the study area. Carbon translocated by tillage is redistributed over its neighboring grid cells using the tillage displacement distributions.

The rate and location of deposition is derived from the soil redistribution model and is used to simulate a change in soil depth. As plow layer depth is maintained, a fraction of the C from the plow layer is transferred toward a buried plow layer. The amount of C transferred is proportional to the deposition height. The depth of the buried plow layer is dynamic and equals the total deposition height (Figure 4.3). The subsoil layers are also buried in the soil profile.

It is also possible that a fraction of the eroded soil is exported from the field by water erosion from the study area under consideration, and this must be taken into account when calculating carbon budgets. This exported C may be deposited on other agricultural fields or may reach the fluvial network. This aspect is not spatially implemented in the model and therefore a k_{exp} term is introduced, which allows the user to manipulate the decomposition of the water erosion exported carbon. Essentially, this term is a simple adjustment factor. When this term equals one, the exported carbon is deposited in an environment where it is fully sequestered (for example, when it is not allowed to decompose in wetlands or is deposited in an area with a higher net primary production). Alternatively, the exported carbon by water erosion may be assumed to decompose at rates of depositional areas. The k_{exp} term is then calculated as

$$1 - \frac{CD}{C_{Tdepo}} \qquad (4.19)$$

where CD is the carbon mineralized due to deposition and C_{Tdepo} is the total amount of carbon deposited in depositional areas.

4.3.2 Model Implementation

The SPEROS-C model is written in Delphi™, uses the Idrisi32 data format for the raster maps, and has a Windows interface where the user can change the main input parameters of the model. Parameter values for the carbon module described above must be specified for the plow layer. As carbon dynamics are substantially different for the subsoil layers (e.g., differences in aeration, temperature, moisture content) the model allows the application of a correction factor for the turnover rates in the subsoil. This is implemented by reducing the climatic variable (r) for the subsoil layers. The model requires that soil texture, bulk density, and the SOC content for all grid cells, depth layers, and pools are known at the beginning of the simulation period. In addition, a digital elevation model of the study area must be provided in order to calculate soil redistribution rates.

The output of the model consists of (1) maps of SOC and soil redistribution rates and (2) total sediment and SOC budgets at the landscape scale. The output maps include water and tillage soil redistribution rates, rates of soil erosion/deposition induced carbon fluxes between the soil and atmosphere and SOC content for all layers after the simulation period. The landscape scale budgets include (1) the C efflux due to mineralization of buried carbon rich material below the plow layer

(*CE*), (2) the atmospheric C that is bound at eroding sites due to formation of new soil organic matter (*CB*), (3) the fraction of the eroded carbon will be exported from the field by water erosion (*Cexp*), and (4) the additional carbon that is mineralized during and after transport by overland flow (*Ctrans*). These sources and sinks are combined to calculate the net landscape scale C flux induced by soil redistribution (*Cnet*).

4.4 MODEL APPLICATION

SPEROS-C is applied to a study site of 5.8 ha with rolling topography (Figure 4.5). We apply the model assuming only water erosion and use various assumptions for the carbon model parameters in order to demonstrate the potential of the model and to investigate to what extent the landscape carbon budget is sensitive to initial model assumptions. Model parameters controlling erosion and deposition are taken from Van Oost et al. (2003a) and result in a field average annual erosion rate of 8.5 Mg ha^{-1}. Subsoil reduction r and C mineralization rate during and after erosion were varied (Table 4.1). We also assume that sediment is primarily transported in aggregated form, i.e., the enrichment ratio is unity, and that f_{trans} is identical for the young and old carbon pools.

The model is run using a spatially uniform carbon content at steady state at the start of the simulation. The steady state SOC contents for the soil profile are calculated with the carbon module

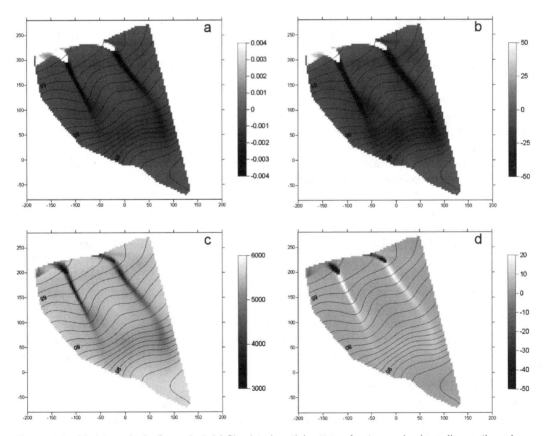

Figure 4.5 Model results for Scenario 1. (a) Simulated spatial pattern of water erosion (m yr^{-1}), negative values indicate erosion, positive values deposition, (b) carbon redistribution rates after 50 years of erosion (g C m^{-2} yr^{-1}), (c) carbon inventory of the plow layer after 50 years of erosion (g C m^{-2} yr^{-1}), and (d) net carbon fluxes between soil and atmosphere (*Cnet*, g C m^{-2} yr^{-1}), negative values indicate a net flux to the atmosphere, positive values a flux to soils.

Table 4.1 Input Parameters for the SPEROS-C Model Scenarios

Scenario	Erosion Rate[a] (Mg/ha yr)	Subsoil r Reduction[b] (%)	f_{trans} (%)
1	8.5	20	20
2	8.5	20	0
3	8.5	20	40
4	8.5	40	20

[a] Field average, gross soil erosion rate.
[b] The r-value for the subsoil are assumed to be 20 and 40% of those of the plow layer.

using input parameters that are representative for present-day agriculture in the United Kingdom. Modeled steady state carbon contents are 1.6% for the plow layer and 0.7% for the sub-plow layer. If no soil erosion occurs, these conditions lead to constant SOC levels over time and space and C flux equilibrium between the soil and the atmosphere. SPEROS-C is then run for a simulation period of 50 years to assess soil erosion induced carbon fluxes.

In Figure 4.5a, the model results for Scenario 1 are shown. High rates of water erosion occur on the steep slopes and where water concentrates, i.e., in the two hollows. Deposition occurs in the lowermost part of the field where the slopes are below 3%. The average erosion rate on eroding sites equals ca. 850 g m^{-2} yr^{-1} and approximately 56% of the eroded sediment is deposited on the lower slopes while 44% is exported from the study area. Note that soil erosion occurs over a large portion of the field (ca. 95%) while deposition is spatially limited (ca. 5%). After 50 years of soil erosion and deposition by overland flow, the model estimates that, on average, 13 g m^{-2} yr^{-1} of carbon is lost on eroding sites and that 159 g m^{-2} yr^{-1} of carbon accumulate in depositional areas (Figure 4.5b).

When erosion occurs, soil carbon in the plow layer strongly declines as soil with a lower carbon content from the lower soil horizons is incorporated and mixed into the plow layer. The eroded carbon is transported toward the depositional areas and 20% of the displaced carbon is lost due to increased mineralization. Thus, the soil that is deposited at colluvial sites also has a lower SOC content. Erosion therefore leads to low SOC levels in the plow layer at both eroding and aggrading sites (Figure 4.5c). This does not necessarily imply that there is a net loss of C to the atmosphere. SPEROS-C predicts a net carbon sink at eroding sites (Figure 4.5d). As the SOC content below the plow layer is below the equilibrium level the formation of new soil organic matter is induced. Erosion enhances carbon uptake at the eroding sites by continuously taking away a fraction of SOC that is replenished with C input. Conversely, the model predicts a net carbon source at sites of deposition as carbon-rich sediment from the plow layer is buried in an environment with substantial lower turnover rates and consequently SOC levels at equilibrium. This leads to additional mineralization of buried C. These observations are in qualitative agreement with previous modeling results (Harden et al., 1999; Liu et al., 2003).

The spatial implementation of SPEROS-C allows us to calculate the net effect of soil erosion and deposition on the landscape scale C budget (Table 4.2). In Scenario 1, the carbon replacement at eroding sites (CB) is larger than the loss in depositional areas (CE). This must be attributed to the differences in carbon dynamics between the plow layer and sub-plow layers. Turnover rates for the plow layer are typically higher than for the sub-plow layers, leading to storage of C in the subsoil at depositional sites. This leads to a positive C balance, even if it is assumed that a relative large fraction of the eroded SOC is lost to the atmosphere due to increased mineralization during and after transport ($Ctrans$). The overall budget for Scenario 1 and Scenario 2 is toward a small net carbon flux from the atmosphere to the soil, irrespective of the fate of the exported carbon.

In Scenario 2, we assumed that f_{trans} equals 0% or that the transport of soil by water erosion does not accelerate mineralization. In this case, the net C flux to soils ($Cnet$) is substantially higher

Table 4.2 Field Averaged Fluxes of Carbon Due to Soil Erosion for Different Scenarios (g C m^{-2} yr^{-1})

Scenario	Cexp	Ctrans	CE	CB	Cnet[a]
1	4.72	−2.62	−0.81	3.87	0.22/0.44
2	5.90	0.0	−0.91	3.67	1.86/2.77
3	3.54	−5.24	−0.72	4.07	−1.89/1.98
4	4.72	−2.62	−1.39	4.30	−0.08/0.29

[a] Maximum values indicate net flux when the exported SOC by water erosion is fully sequestered; minimum value indicates net flux when the exported carbon by water erosion is assumed to decompose at rates of depositional areas.

(Table 4.2) as C_{trans} becomes zero. On the other hand, if f_{trans} increases to 40%, soil erosion may result in a net carbon source or sink term, depending on the fate of the exported sediment (Scenario 3). The model is very sensitive for the turnover rates in the subsoil. Increasing the decomposition rate for the sub-plow layer from 20 to 40% of the of the plow layer leads to a sharp increase of the *CE* term (Scenario 4).

4.5 DISCUSSION AND CONCLUSION

In this paper, we described the SPEROS-C model in detail. The model can be used as an experimental tool for hypothesis testing and to gain more insight in soil erosion–carbon dynamics relationships. The main advantage of our approach over existing models is the explicit consideration of spatial transfers between landscape elements due to water and tillage erosion. This approach accounts for both soil erosion and deposition effects on carbon dynamics and to calculate the overall C budget at the landscape scale. The model was applied on a small agricultural field with rolling topography. The results indicated that carbon dynamics are spatially heterogeneous and that erosion induces both carbon sink and source terms. A sensitivity analysis clearly showed that soil erosion has the potential to affect the carbon balance of agricultural fields substantially but that the model is very sensitive to several key model parameters.

Several assumptions were made all of which were built into the model but which must be tested independently and refined in subsequent work. At present, various aspects of the dynamic responses of carbon fluxes to redistribution of soil are still poorly understood, which hampers an accurate quantification of carbon fluxes between soil and atmosphere induced by soil erosion. We discuss the most important issues: (1) *quantification of CO_2 release during and after transport by water erosion*. Based on the incubation of runoff-samples, Jacinthe et al. (2002, 2004) presented the first quantitative evidence that a significant fraction of the SOC released through water erosion is mineralizable. However, the extrapolation of these findings to the landscape scale, where a variety of processes are active, remains problematic. In contrast to transport of soil by water erosion, which is associated with selective transport and the breakdown of aggregate structure, no transport related increase in mineralization is associated with tillage erosion. (2) *SOC dynamics in the deeper layers of the soil profile*. Traditionally, soil scientists have focused on the upper soil profiles when examining the controls on organic matter dynamics, although significant amounts of carbon are stored below this layer (e.g., Stallard, 1998; Quine and Zhang, 2002). While the controls over soil carbon dynamics are well established for the upper layers, controls over decomposition rates in the lower soil profile remain poorly understood. The amount of carbon that can be sequestered in depositional areas largely depends on the decomposition rates in the subsoil and, therefore, until this is better known, more confident simulation is problematic. (3) *The variety of SOC forms*. Another important issue is that different pools of SOC might behave differentially when soil is eroded, transported, and deposited while the reactivity of specific pools may also change during the transport process. At present there exists very little experimental studies that address these issues. However, it is reasonable to assume that this only applies to soil erosion by water. Tillage erosion is a nonselective process as it displaces the whole plow layer. (4) *Crop production.* Finally, the linkages between soil erosion and crop productivity need further attention. Dynamic replacement of SOC at eroding sites may be seriously hampered if high fertilization levels can no longer compensate for the productivity losses.

REFERENCES

Andren, O. and T. Katterer. 1997. ICBM: The introductory carbon balance model for exploration of soil carbon balances. *Ecological Applications* 7 (4):1226–1236.

Dejong, E. and R. G. Kachanoski. 1988. The importance of erosion in the carbon balance of prairie soils. *Canadian Journal of Soil Science* 68 (1):111–119.

Desmet, P. J. J. and G. Govers. 1995. GIS-based simulation of erosion and deposition patterns in an agricultural landscape — a comparison of model results with soil map information. *Catena* 25 (1–4):389–401.

Desmet, P. J. J. and G. Govers. 1996. Comparison of routing algorithms for digital elevation models and their implications for predicting ephemeral gullies. *International Journal of Geographical Information Systems* 10 (3):311–331.

Desmet, P. J. J. and G. Govers. 1997. Two-dimensional modelling of the within-field variation in rill and gully geometry and location related to topography. *Catena* 29 (3–4):283–306.

Foster, G. R. and L. D. Meyer. 1975. Mathematical simulation of upland erosion by fundamental erosion mechanics, in *Present and Perspective Technology for Predicting Sediment Yields and Sources —* Proceedings of Sediment-Yield Workshop, USDA Agricultural Research Service Report ARS-S-40, pp. 190–206, United States Department of Agriculture Sedimentation Laboratory, Oxford, Mississippi.

Govers, G., K. Vandaele, P. Desmet, J. Poesen, and K. Bunte. 1994. The role of tillage in soil redistribution on hillslopes. *European Journal of Soil Science* 45:469–478.

Gregorich, E. G., M. R. Carter, D. A. Angers, C. M. Monreal, and B. H. Ellert. 1994. Toward a minimum data set to assess soil organic-matter quality in agricultural soils. *Canadian Journal of Soil Science* 74 (4):367–385.

Harden, J. W., T. L. Fries, and M. J. Pavich. 2002. Cycling of beryllium and carbon through hillslope soils in Iowa. *Biogeochemistry* 60 (3):317–335.

Harden, J. W., J. M. Sharpe, W. J. Parton, D. S. Ojima, T. L. Fries, T. G. Huntington, and S. M. Dabney. 1999. Dynamic replacement and loss of soil carbon on eroding cropland. *Global Biogeochemical Cycles* 13 (4):885–901.

Jacinthe, P. A. and R. Lal. 2001. A mass balance approach to assess carbon dioxide evolution during erosional events. *Land Degradation and Development* 12 (4):329–339.

Jacinthe, P. A., R. Lal, and J. M. Kimble. 2002. Carbon dioxide evolution in runoff from simulated rainfall on long-term no-till and plowed soils in southwestern Ohio. *Soil and Tillage Research* 66 (1):23–33.

Jacinthe, P. A., R. Lal, L. B. Owens, and D. L. Hothem. 2004. Transport of labile carbon in runoff as affected by land use and rainfall characteristics. *Soil and Tillage Research* 77(2):111–123.

Katterer, T. and O. Andren. 1999. Long-term agricultural field experiments in Northern Europe: Analysis of the influence of management on soil carbon stocks using the ICBM model. *Agriculture Ecosystems and Environment* 72(2):165–179.

Katterer, T., M. Reichstein, O. Andren, and A. Lomander. 1998. Temperature dependence of organic matter decomposition: a critical review using literature data analyzed with different models. *Biology and Fertility of Soils* 27 (3):258–262.

Lal, R. 2001. Soil degradation by erosion. *Land Degradation and Development* 12 (6): 519–539.

Lal, R. 2003. Soil erosion and the global carbon budget. *Environment International* 29 (4):437–450.

Lal, R. 2004. Soil carbon sequestration impacts on global climate change and food security. *Science* 304 (5677):1623–1627.

Lal, R., M. Griffin, J. Apt, L. Lave, and M. G. Morgan. 2004. Ecology — Managing soil carbon. *Science* 304(5669):393.

Liu, S. G., N. Bliss, E. Sundquist, and T. G. Huntington. 2003. Modeling carbon dynamics in vegetation and soil under the impact of soil erosion and deposition. *Global Biogeochemical Cycles* 17(2):1074.

Manies, K. L., J. W. Harden, L. Kramer, and W. J. Parton. 2001. Carbon dynamics within agricultural and native sites in the loess region of western Iowa. *Global Change Biology* 7(5):545–555.

McCarty, G. W. and J. C. Ritchie. 2002. Impact of soil movement on carbon sequestration in agricultural ecosystems. *Environmental Pollution* 116(3):423–430.

Polyakov, V. and R. Lal. 2004. Modeling soil organic matter dynamics as affected by soil water erosion. *Environment International* 30:547–556.

Quine, T. A. and Y. Zhang. 2002. An investigation of spatial variation in soil erosion, soil properties, and crop production within an agricultural field in Devon, United Kingdom. *Journal of Soil and Water Conservation* 57(1):55–65.

Schlesinger, W. H. 1995. Soil respiration and changes in soil carbon stocks, in G. M.Woodwell and G. M. Mackenzie, eds., *Biotic Feedbacks in the Global Climatic System: Will the Warming Feed the Warming?* Oxford University Press, New York, pp. 159–168.

Stallard, R. F. 1998. Terrestrial sedimentation and the carbon cycle: Coupling weathering and erosion to carbon burial. *Global Biogeochemical Cycles* 12(2):231–257.

Van Muysen, W., G. Govers, and K. Van Oost. 2002. Identification of important factors in the process of tillage erosion: The case of mouldboard tillage. *Soil and Tillage Research* 65(1):77–93.

Van Oost, K., G. Govers, and W. Van Muysen. 2003a. A process-based conversion model for caesium-137 derived erosion rates on agricultural land: An integrated spatial approach. *Earth Surface Processes and Landforms* 28(2):187–207.

Van Oost, K., G. Govers, W. Van Muysen, and T. A. Quine. 2000. Modeling translocation and dispersion of soil constituents by tillage on sloping land. *Soil Science Society of America Journal* 64(5):1733–1739.

Van Oost, K., W. Van Muysen, G. Govers, G. Heckrath, T. A. Quine, and J. Poesen. 2003b. Simulation of the redistribution of soil by tillage on complex topographies. *European Journal of Soil Science* 54(1): 63–76.

Van Oost, K., G. Govers, G. Heckrath, T. A. Quine, J. Olesen, and R. Merckx, 2005. Soil erosion and carbon dynamics: The role of tillage erosion. *Global Biogeochemical Cycles*, in press.

Section 2

Erosion at the Plot Scale

CHAPTER 5

Soil Carbon Erosion and Its Selectivity at the Plot Scale in Tropical and Mediterranean Regions

Eric Roose and Bernard Barthès

CONTENTS

5.1 Introduction..55
5.2 Sites, Materials, and Methods...57
    5.2.1 Sites..57
    5.2.2 Measurement of Soil Erosion...57
    5.2.3 Sediment and Soil Analysis 60
5.3 Results..60
    5.3.1 Erosion..60
    5.3.2 Eroded Carbon..61
    5.3.3 Relationship between Eroded SOC, Erosion, and Topsoil SOC Content.......61
    5.3.4 Carbon Enrichment Ratio of Sediments (CER) as Affected by Erosion
          and Eroded SOC...62
    5.3.5 CER as Affected by Land Use...62
    5.3.6 The CER as Expressed by the Relation between Eroded SOC and the
          Product Erosion × Topsoil SOC...64
5.4 Discussion..64
    5.4.1 Erosion..64
    5.4.2 Eroded SOC and Carbon Enrichment Ratio of Sediments (CER).................66
    5.4.3 Erosion Selectivity and Preferential Removal 68
    5.4.4 Relationship between Erosion, Eroded SOC, and CER 68
5.5 Conclusion 69
Acknowledgments 69
References 70

5.1 INTRODUCTION

Global warming is one of the greatest challenges of the twenty-first century (Robert, 2001). At the human time scale, climate change is closely linked to the increasing atmospheric concentration of greenhouse gases (GHGs), which are mainly carbon dioxide (CO_2), methane (CH_4), and nitrous oxide (N_2O) (IPCC, 2001). It is estimated that land use and land-use change and forestry (LULUCF) represent 34% of GHG emissions globally, and 50% of GHG emissions in the tropics and subtropics (IPCC, 2001). Due to the importance of carbon (C) fluxes in the processes relating to LULUCF, there is a renewed interest in studies dealing with the effects of land use and management on C balances. However, such studies generally focus on changes in soil and biomass C or on GHG emissions (mainly CO_2), but rarely take into account C fluxes resulting from erosion. In experiments conducted to determine the effects of cropping systems on C budgets, erosion is generally considered as negligible. Several authors have suggested that this assumption is not valid even when experiments are conducted on relatively flat land with slope gradient of less than 1% (Boli, 1996; Roose, 1996). Thus, erosion and eroded C cannot be ignored (Voroney et al., 1981; Gregorich et al., 1998; Mitchell et al., 1998). Indeed, erosion is considered the most widespread form of soil degradation (Gregorich et al., 1998), with water being its most common agent. Land areas affected by water and wind erosion are estimated at 1100 and 550 Mha respectively (Lal, 2003); those affected by tillage erosion are not precisely known, but are equally important (Govers et al., 1999). Additionally, as soil organic carbon (SOC) has a low density and is concentrated near the soil surface, it is one of the first soil constituents removed by erosion (Roose, 1977; Lowrance and Williams, 1988). Furthermore, erosion is one of the only soil processes that can remove stable soil SOC in large quantities (Starr et al., 2000).

Soil erosion consists of detachment, transport, and deposition of soil particles (Roose, 1981; Lal, 2001). The main mechanisms of water erosion detachment are disintegration of soil aggregates by slaking, cracking, dispersion, and shearing by raindrop impact or runoff. Particles are transported by runoff and splash resulting from the raindrop impact (Lal, 2001). The shearing and transport capacities of runoff increase with the increase in slope length and steepness. Water erosion then transforms from sheet (interrill) erosion, in which detachment and transport are caused by raindrops and shallow surface flow, to rill erosion, dominated by runoff concentrated into discernible channels (Jayawardena and Bhuiyan, 1999). As surface soil is enriched in SOC, its erosion also results in SOC erosion. The carbon enrichment ratio (CER) is defined as the ratio of SOC content in sediments to that in the topsoil (0 to 10 cm depth in general) (Roose, 1977).

The quantification of SOC erosion requires the quantification of soil losses and the determination of sediment SOC content. Soil losses may be assessed at different scales: catchment ($> 10^4$ m^2), plot (10 to 10^4 m^2) and microplot (< 10 m^2) (Mutchler et al., 1988; Hudson, 1993). Catchments are generally heterogeneous in terms of soil and land management, and the contributions of the spatial subunits are often difficult to distinguish (Roose, 1981; Le Bissonnais et al., 1998). Measurements at the microplot scale underestimate soil losses because runoff flow cannot gain velocity and concentrate on a short slope (Le Bissonnais et al., 1998). At the intermediate plot scale, slope length is sufficient for runoff to concentrate, and most of the sedimentation is avoided as long as slope gradient and soil surface roughness are uniform. Additionally, such plots can be easily established in homogeneous edaphic and vegetal conditions. Consequently runoff plots have been widely used for erosion studies (Mutchler et al., 1988; Roose and Sarrailh, 1989; Hudson, 1993). Other methods based on ^{137}Cs analysis and measurement of magnetic susceptibility have also been used to estimate soil redistribution over the landscape. In addition to being innovative and sophisticated, these methods provide pluri-decennial balances of soil movements. Nonetheless, these methods are not the most suitable to assessing water erosion and the effects of some of its determinants (e.g., land management). Moreover, variability in ^{137}Cs assessment from noneroded references remains an important concern as regards the first method, as well as variability in parent materials (with different magnetic properties) regarding the second one (Sutherland, 1996; De Jong et al., 1998).

The objective of this paper is to collate and synthesize data of numerous experiments regarding the effect of land use and management on SOC losses by water erosion in runoff plots representing a wide range of tropical and Mediterranean environments, with a range of climate, slope, soil, and management conditions. Factors affecting CER of sediments are also discussed.

5.2 SITES, MATERIALS, AND METHODS

5.2.1 Sites

Table 5.1 and Table 5.2 present the location, altitude, slope gradient, annual rainfall, soil type, topsoil texture (stoniness and clay content), topsoil SOC content, and land use of the 54 runoff plots under study in tropical and Mediterranean regions, respectively. Tropical regions may be divided into three groups: humid West Indies, humid West Africa, and subhumid West and Central Africa.

In the humid West Indies (2200-mm yr^{-1} rainfall), seven runoff plots were established near St. Joseph in Martinique. Soils are very acidic and clayey Inceptisols developed from volcanic ashes on steep slopes (10 to 40%). Uncultivated areas are forested (Khamsouk, 2001).

In the humid West Africa (1400- to 1900-mm yr^{-1} rainfall), six runoff plots were established in Adiopodoumé (Roose, 1979b), Azaguié (Roose and Godefroy, 1977), and Divo (Roose, 1981), in southern Ivory Coast. Soils are acidic and well-drained sandy loam Ultisols developed from sands, schist, or granite on half orange-shaped hills. Uncultivated areas are forested.

In the subhumid West and Central Africa (600- to 1300-mm yr^{-1} rainfall, concentrated within few months), 18 runoff plots were established in Korhogo (Roose, 1979a) and Bouaké (Roose, 1981) in northern Ivory Coast, Gonsé (Roose, 1978) and Saria (Roose, 1981) in western Burkina Faso, Djitiko in southwestern Mali (Diallo, 2000), Séfa in southwestern Senegal (Roose, 1967), and Mbissiri in northern Cameroon (Boli, 1996). Soils are either sandy Ultisols, Inceptisols, and Alfisols developed from granite and sandstone on long slopes (declivity < 3%) below residual hills with ironstone, or Alfisols with Vertic properties developed from schist in plains. Uncultivated areas are under bush or woody savannas that are traditionally burnt during the dry season.

In the Mediterranean regions, 23 runoff plots were established near Médéa, Mascara, and Tlemcen in northwestern Algerian highlands (Arabi, 1991; Roose et al., 1996; Morsli et al., 2004). Total annual rainfall is low (350- to 550-mm yr^{-1}) but intense rain storms often occur at the end of the summer. Soils are Mollisols, Inceptisols, and Vertisols developed on steep slopes (10 to 40%), with high clay and calcium contents, and often containing a high proportion of gravels and stones. Uncultivated areas are generally covered by matorral due to grazing by sheep and goats.

In total, 54 runoff plots established represent a wide range of land uses and vegetation covers: uncultivated plots with little disturbance (i.e., forests or savannas) or some disturbances (i.e., burnt savanna or matorral used as rangeland); plots previously cultivated but now under fallow, or orchard and vineyard; cultivated plots based on cropping systems involving either intensive tillage and some organic inputs (cereals e.g., maize *Zea mays*, cotton *Gossypium sp.*), direct drilling with residue mulch (idem.), or infrequent tillage with large biomass (banana *Musa sp.*, sugar cane *Saccharum officinarum*); and bare tilled soil as baseline for soil erodibility assessment.

5.2.2 Measurement of Soil Erosion

Most runoff plots were 100-m^2 in area (20 × 5 m), except in Martinique where cultivated plots were 200 m^2 (20 × 10 m), and in Médéa (Algeria) where plot area ranged from 80 to 220 m^2. Each runoff plot was surrounded by half-buried metal sheets and fitted out with a collector draining runoff and sediments into tanks arranged in series. The first tank trapped coarse sediments (aggregates, gravels, coarse sands, litter). When full, the overflow moved through divisors into two tanks in series, which were used to measure the runoff amount and suspended sediments (the third tank was necessary for rainfall events resulting in large runoff volume).

Wet coarse sediments were collected in the first tank after each rainfall event or sequence of events, and weighed. The weight of dry coarse sediments was either determined by oven drying of aliquots, or by using calibration curves drawn up by weighing increasing amounts of dry topsoil in a bucket filled up with water. These different determinations were supposed to give equivalent

Table 5.1 Description of the Runoff Plots Established in Tropical Regions (Soil Properties Refer to 0 to 10 cm)

Location		Altitude m	Slope %	Rainfall mm yr^{-1}	Soil Type	Gravels %	Clay %	Land Management or Vegetation Cover	Soil OC g kg^{-1}	Erosion Mg ha^{-1} yr^{-1}	Eroded OC kg ha^{-1} yr^{-1}	CER	Years
St Joseph Mart.[1]	14°40′N, 61°00′W	70	10	2220	oxic Dystrudept	0	68	bare tilled soil	15.1	85.8	1249	1.0	2
St Joseph Mart.[1]	14°40′N, 61°00′W	70	10	2220	oxic Dystrudept	0	68	pineapple, ridged	16.2	17.2	294	1.1	2
St Joseph Mart.[1]	14°40′N, 61°00′W	70	10	2220	oxic Dystrudept	0	68	banana, mulched	17.9	0.5	12	1.5	2
St Joseph Mart.[1]	14°40′N, 61°00′W	70	10	2220	oxic Dystrudept	0	68	sugar cane, mulched	16.5	0.1	2	1.7	2
St Joseph Mart.[1]	14°40′N, 61°00′W	70	10	2220	oxic Dystrudept	0	68	flat pineapple, mulched	16.7	0.0	1	2.1	2
St Joseph Mart.[1]	14°40′N, 61°00′W	70	25	2220	oxic Dystrudept	0	68	bare tilled soil	17.0	127.5	2274	1.0	2
St Joseph Mart.[1]	14°40′N, 61°00′W	70	40	2220	oxic Dystrudept	0	68	bare tilled soil	18.8	147.4	2999	1.1	2
Adiopodoumé IC[2]	05°20′N, 04°08′W	30	7	1900	typic Hapludult	0	11	maize	10.8	99.1	1982	1.9	2
Adiopodoumé IC[2]	05°20′N, 04°08′W	30	11	1800	typic Hapludult	0	13	forest	18.6	0.1	13	14.0	10
Adiopodoumé IC[2]	05°20′N, 04°08′W	30	65	1800	typic Kandiudult	0	15	forest	21.8	0.5	42	4.3	10
Azaguié IC[3]	05°33′N, 04°03′W	80	14	1640	typic Kandiudult	0	13	forest	11.3	0.2	12	7.0	7
Azaguié IC[3]	05°33′N, 04°03′W	80	14	1640	typic Kandiudult	0	15	banana	19.1	1.8	105	3.0	7
Divo IC[4]	05°48′N, 05°18′W	<100	10	1400	oxyaquic Kandiudult	0	26	forest	12.4	0.1	8	4.9	7
Korhogo IC[5]	09°25′N, 05°39′W	390	3	1280	typic Kandiustult	69	16	bush savanna (burnt)	15.8	0.1	5	3.4	8
Korhogo IC[5]	09°25′N, 05°39′W	390	3	1280	typic Kandiustult	69	16	maize	8.1	5.4	63	1.5	8
Séfa Sénégal[6]	13°10′N, 15°30′W	<50	2	1200	typic Kandiustalf	0	15	peanut, rice, sorghum	4.6	9.3	73	1.7	9
Bouaké IC[7]	07°46′N, 05°06′W	370	4	1200	typic Kandiustult	15	6	woody savanna	15.8	0.1	2	2.6	4
Djitiko Mali[8]	12°05′N, 08°25′W	350	2	1080	typic Haplustalf	0	25	old (bush) fallow	11.0	4.8	125	2.4	1
Djitiko Mali[8]	12°05′N, 08°25′W	350	2	1080	typic Haplustalf	0	25	maize/cotton plowed	6.9	18.4	330	2.6	1
Djitiko Mali[8]	12°05′N, 08°25′W	350	2	1080	typic Haplustalf	0	25	maize/cotton no-tilled	8.9	7.4	154	2.3	1
Djitiko Mali[8]	12°05′N, 08°25′W	350	2	1080	vertic Haplustept	0	26	old (bush) fallow	38.4	1.7	190	2.9	1
Djitiko Mali[8]	12°05′N, 08°25′W	350	2	1080	vertic Haplustept	0	26	maize/cotton plowed	38.1	14.1	280	0.5	1
Djitiko Mali[8]	12°05′N, 08°25′W	350	2	1080	vertic Haplustept	0	26	maize/cotton no-tilled	40.5	6.0	358	1.5	1
Mbissiri Cam.[9]	08°23′N, 14°33′E	370	2	860	typic Haplustalf	0	7	cotton plowed (recent)	3.5	8.7	90	3.0	1
Mbissiri Cam.[9]	08°23′N, 14°33′E	370	2	860	typic Haplustalf	0	7	idem with manure	3.0	12.2	111	3.0	1
Mbissiri Cam.[9]	08°23′N, 14°33′E	370	2	860	typic Haplustalf	0	7	cotton minitilled (recent)	5.0	6.0	57	1.9	1
Mbissiri Cam.[9]	08°23′N, 14°33′E	370	2.5	860	typic Haplustalf	0	8	cotton plowed (old)	3.0	40.4	160	1.3	1
Mbissiri Cam.[9]	08°23′N, 14°33′E	370	2.5	860	typic Haplustalf	0	8	idem with manure	3.0	15.2	85	1.9	1
Mbissiri Cam.[9]	08°23′N, 14°33′E	370	2.5	860	typic Haplustalf	0	8	cotton minitilled (old)	3.0	2.4	12	1.8	1
Gonsé BF[10]	12°22′N, 01°19′W	300	0.5	700	kanhaplic Haplustalf	0	8	woody savanna (burnt)	5.5	0.2	9	10.7	6
Saria BF[11]	12°16′N, 02°09′W	300	0.7	640	typic Plinthustalf	0	13	young grass fallow	4.7	0.5	9	3.6	3

[1] Khamsouk (2001), Blanchart et al. (this issue); [2] Roose (1979b); [3] Roose and Godefroy (1977); [4] Roose (1981); [5] Roose (1979a); [6] Roose (1967); [7] Roose (1981); [8] Diallo et al. (2004); [9] Boli (1996), Bep A Ziem et al. (2004); [10] Roose (1978); [11] Roose (1981).

Abbreviations for locations: Mart. Martinique; IC Ivory Coast; Cam. Cameroon; BF Burkina Faso.

Table 5.2 Description of the Runoff Plots Established in the Mediterranean Regions (Algeria; Soil Properties Refer to 0 to 10 cm)

Location	Altitude m	Slope %	Rainfall mm yr⁻¹	Soil Type	Gravels %	Clay %	Land Management or Vegetation Cover	Soil OC g kg⁻¹	Erosion Mg ha⁻¹ yr⁻¹	Eroded SOC kg ha⁻¹ yr⁻¹	CER	Years
Médéa[1] 36°14'N, 02°51'E	900	14	550	typic Haploxerert	4	64	improved cereals	5.2	0.0	0	1.8	5
Médéa[1] 36°14'N, 02°51'E	900	14	550	typic Haploxerert	4	65	cereal with legume	6.8	0.0	0	1.3	5
Médéa[1] 36°14'N, 02°51'E	900	40	550	typic Haploxeroll	16	43	matorral with shrubs	9.6	0.1	1	2.1	5
Médéa[1] 36°14'N, 02°51'E	900	40	550	typic Haploxeroll	16	50	idem but regrassed	7.1	0.0	0	3.1	5
Médéa[1] 36°14'N, 02°51'E	900	35	550	typic Haploxerept	0	51	orchard on bare soil	7.1	1.4	16	1.6	5
Médéa[1] 36°14'N, 02°51'E	900	35	550	typic Haploxeroll	20	39	vineyard, cereal, legume	8.2	0.1	1	1.2	5
Mascara[2] 35°20'N, 00°17'E	640	20	470	typic Haploxeroll	0	17	rangeland (matorral)	10.2	1.6	42	2.5	1
Mascara[2] 35°20'N, 00°17'E	640	20	470	typic Haploxeroll	0	17	cereals	10.3	0.8	22	2.7	1
Mascara[2] 35°20'N, 00°17'E	640	20	470	typic Haploxeroll	0	17	bare tilled soil	10.0	8.5	136	1.6	1
Mascara[2] 35°20'N, 00°17'E	640	20	470	typic Haploxeroll	0	17	protected fallow	12.3	0.5	24	3.9	1
Mascara[2] 35°20'N, 00°17'E	670	40	470	vertic Haploxeroll	0	57	rangeland (matorral)	11.3	1.2	31	2.3	1
Mascara[2] 35°20'N, 00°17'E	670	40	470	vertic Haploxeroll	0	57	cereals	11.6	1.1	25	2.0	1
Mascara[2] 35°20'N, 00°17'E	670	40	470	vertic Haploxeroll	0	57	bare tilled soil	10.2	6.8	95	1.4	1
Mascara[2] 35°20'N, 00°17'E	670	40	470	vertic Haploxeroll	0	57	protected fallow	12.3	0.6	21	2.8	1
Tlemcen[2] 34°50'N, 01°10'W	450	21	450	typic Haploxeroll	46	20	bare tilled soil	18.6	3.9	78	1.1	1
Tlemcen[2] 34°50'N, 01°10'W	450	21	450	typic Haploxeroll	46	20	rangeland (matorral)	23.0	0.7	28	1.7	1
Tlemcen[2] 34°50'N, 01°10'W	450	21	450	typic Haploxeroll	46	20	protected fallow	33.3	0.7	34	1.4	1
Tlemcen[2] 34°50'N, 01°10'W	450	15	420	vertic Haploxeroll	7	50	bare tilled soil	8.0	1.8	19	1.3	1
Tlemcen[2] 34°50'N, 01°10'W	450	15	420	vertic Haploxeroll	7	50	cereals/rangeland	9.3	1.6	27	1.8	1
Tlemcen[2] 34°50'N, 01°10'W	450	15	420	vertic Haploxeroll	7	50	fertilized cereal (contour)	10.3	1.6	33	2.0	1
Tlemcen[2] 34°50'N, 01°10'W	450	10	360	typic Haploxerept	42	37	bare tilled soil	6.3	3.2	31	1.6	1
Tlemcen[2] 34°50'N, 01°10'W	450	10	360	typic Haploxerept	42	37	rangeland (matorral)	6.8	1.8	26	2.1	1
Tlemcen[2] 34°50'N, 01°10'W	450	10	360	typic Haploxerept	42	37	protected fallow	9.4	1.0	18	1.9	1

[1] Arabi (2004); [2] Morsli et al. (2004).

results. Suspended sediment concentration in runoff was assessed by flocculation and oven drying of aliquots collected in the second tank or in every tank, and was used in conjunction with the runoff amount to determine the weight of dry suspended sediment (runoff amount was assessed by measuring the volume of water in each tank and multiplying it by the coefficients depending on the number of divisors). Erosion (Mg ha^{-1} yr^{-1}) was calculated as the sum of dry-coarse and dry-suspended sediment amounts for all rainfall events over the year.

5.2.3 Sediment and Soil Analysis

The SOC contents of coarse and suspended sediments were determined separately, on individual samples resulting from "representative" rainfall events, or more frequently, on composite samples resulting from all events that occurred over a given period (e.g., month or season). Topsoil samples (0 to 10 cm) were also collected for SOC analysis. The SOC content was determined using either the Walkley and Black method (Ivory Coast, Senegal, Burkina Faso) or dry combustion in an Elemental Analyser (Martinique, Mali, Cameroon), after possible destruction of carbonates by hydrochloric acid (HCl), or using the Anne method (Algeria) (for these methods, see Nelson and Sommers, 1996). It was assumed that the three methods would give equivalent results. Eroded SOC (kg C ha^{-1} yr^{-1}) was defined as the sum of coarse sediment SOC and suspended sediment SOC over one year.

The gravel content (> 2 mm) of topsoil was determined by dry sieving of air-dry samples. Particle-size analysis of air-dry topsoil samples (< 2 mm) was determined by a combination of dry sieving and sedimentation (pipette method), after destruction of the organic matter and total dispersion (Gee and Bauder, 1986). The clay fraction was defined as < 2 µm.

5.3 RESULTS

5.3.1 Erosion

In general, soil erosion ranged from 0 to 150 Mg ha^{-1} yr^{-1} (Table 5.1 and Table 5.2). The maximum erosion rate was ca. 10 Mg ha^{-1} yr^{-1} in Mediterranean areas, 40 Mg ha^{-1} yr^{-1} in subhumid areas, 100 Mg ha^{-1} yr^{-1} in humid African areas, and 150 Mg ha^{-1} yr^{-1} in humid West Indies areas. It increased with increase in annual rainfall. Similarly, maximum erosion was 40 Mg ha^{-1} yr^{-1} for slope < 5%, 100 Mg ha^{-1} yr^{-1} for 5 to 15% slope, 130 Mg ha^{-1} yr^{-1} for 15 to 30% slope, and 150 Mg ha^{-1} yr^{-1} for 30 to 65% slope. Therefore, maximal soil erosion also increased with increase in slope gradient. In addition to climate and slope gradient, land use also had an important influence on erosion. Soil erosion was divided into the following groups:

- Under 2000-mm yr^{-1} rainfall, bare tilled soils eroded at a rate of 80 to 150 Mg ha^{-1} yr^{-1}, which was also the case under maize for a sandy soil on 7% slope
- Under rainfall > 860 mm yr^{-1}, conventionally tilled cereals and cotton on slope < 3% eroded at a rate of 9 to 40 Mg ha^{-1} yr^{-1} (9 to 20 Mg ha^{-1} yr^{-1} in general), as did pineapple (*Ananas comosus*) on 10% slope (clayey soil)
- Maize and cotton grown with reduced or no till under rainfall of 860 to 1300 mm yr^{-1} and bare tilled soils in Mediterranean regions eroded at a rate of 2 to 9 Mg ha^{-1} yr^{-1}
- Under Mediterranean climate, cultivated plots and rangelands eroded at a rate of 0.7 to 2 Mg ha^{-1} yr^{-1}, as did a banana plantation on a sandy soil with 14% slope in Ivory Coast
- Under all eco-regions, forest, savanna, fallow, and crops with thick mulch (sugar cane, pineapple, banana) eroded at a rate of ≤ 0.7 Mg ha^{-1} yr^{-1}, as did Mediterranean rangelands and cultivated plots on clayey or stony soils (≥ 40% clay or gravels)

Of the 54 plots under study, three fallows (two in Mali and one near Tlemcen, Algeria) did not fit this pattern, and eroded at a rate of 1 to 5 Mg ha^{-1} yr^{-1} (instead of ≤ 0.7 Mg ha^{-1} yr^{-1}). This trend was probably related to previous cropping history, burning, and grazing systems. In general, cultivated plots lost less than 40 Mg ha^{-1} yr^{-1} (and generally less than 20 Mg ha^{-1} yr^{-1}) and plots under natural vegetation less than 5 Mg ha^{-1} yr^{-1} (and generally less than 2 Mg ha^{-1} yr^{-1}).

5.3.2 Eroded Carbon

In general, eroded SOC ranged from 0 to 3 Mg C ha^{-1} yr^{-1} (Table 5.1 and Table 5.2), and its variations with regard to annual rainfall, topography, and vegetation were similar to those for erosion. Indeed, maximum eroded SOC increased with increase in annual rainfall: it was 140, 360, 2000, and 3000 kg C ha^{-1} yr^{-1} in Mediterranean, subhumid African, humid African, and humid West Indies eco-regions, respectively. Maximum eroded SOC also increased with increase in slope gradient, and was 360, 2000, 2300, and 3000 kg C ha^{-1} yr^{-1} for slope < 5%, 5 to 15%, 15 to 30%, and 30 to 65%, respectively. The magnitude of eroded SOC was influenced by vegetation and land management, and can be divided into the following groups:

- Under 2000-mm yr^{-1} rainfall, bare tilled soils (and maize on a sandy soil with 7% slope) eroded SOC between 1000 and 3000 kg C ha^{-1} yr^{-1} (median 2130 kg C ha^{-1} yr^{-1}), and the rate corresponded to the maximum erosion
- Under rainfall > 860-mm yr^{-1}, cereals and cotton on slope < 3%, pineapple on 10% slope (clayey soil), banana plantation on 14% slope (sandy soil), and several Mediterranean bare plots eroded SOC between 50 and 400 kg C ha^{-1} yr^{-1} (median 105 kg C ha^{-1} yr^{-1}); among row crops (cereals, cotton) it was not possible to separate conventional and conservation tillage into two different classes (as had been done for erosion classes), though at a given site conservation tillage generally resulted in less eroded SOC than conventional tillage
- Under Mediterranean climate, cultivated plots, rangelands, and some bare plots eroded SOC between 15 and 50 kg C ha^{-1} yr^{-1} (median 26 kg C ha^{-1} yr^{-1}), in accord with a similar erosion class
- Under all eco-regions, forest, savanna, fallow, and crops with thick mulch (sugar cane, pineapple, banana), as well as Mediterranean rangelands and cultivated plots on clayey or stony soils ($\geq 40\%$), eroded OC < 15 kg C ha^{-1} yr^{-1} (median 10 kg C ha^{-1} yr^{-1}), in accord with a class of minimum erosion rate

Of the 54 plots under study, five fallows did not fit this pattern: two burnt fallows in Mali lost SOC between 100 and 200 kg C ha^{-1} yr^{-1}, and three grazed fallows in Algeria lost SOC between 18 and 25 kg C ha^{-1} yr^{-1} (instead of < 15 kg C ha^{-1} yr^{-1}). Additionally, one forest plot from Ivory Coast lost OC at 42 kg C ha^{-1} yr^{-1}. Moreover, Mediterranean bare tilled soils were separated into two classes (this separation was not clearly defined), whereas conventional and conservation tillage could not be separated. Thus the relationship between land use and eroded SOC was less defined than that reported between erosion and land use.

In general, cultivated plots lost SOC at a rate of less than 400 kg C ha^{-1} yr^{-1} (median 90 kg C ha^{-1} yr^{-1}; 110 and 20 kg C ha^{-1} yr^{-1} for tropical and Mediterranean areas, respectively), and those under natural vegetation less than 50 kg C ha^{-1} yr^{-1} (median 10 kg C ha^{-1} yr^{-1}).

5.3.3 Relationship between Eroded SOC, Erosion, and Topsoil SOC Content

Eroded SOC was strongly correlated with erosion, but this relation was markedly influenced by the few highly erodible plots, and became weaker when they were not taken into account (Table 5.3). A correlation existed between eroded SOC and the product of erosion multiplied by topsoil SOC content (0 to 10 cm), which became much closer than the former when highly erodible plots were not taken into account. Except for the few highly erodible plots, eroded SOC was better

Table 5.3 Correlations between Eroded OC, Erosion, and Topsoil OC Content (0 to 10 cm)

Plots under Consideration	Correlation between Eroded OC and Topsoil OC	Correlation between Eroded OC and Erosion	Correlation between Eroded OC and Erosion × Topsoil OC
All 54 plots	r = 0.157, p > 0.1	r = 0.979, p < 0.001	r = 0.976, p < 0.001
The 50 plots having eroded OC < 400 kg C ha^{-1} yr^{-1} (erosion < 80 Mg ha^{-1} yr^{-1})	r = 0.409, p < 0.01	r = 0.631, p < 0.001	r = 0.739, p < 0.001
The 46 plots having eroded OC < 200 kg C ha^{-1} yr^{-1}	r = 0.116, p > 0.1	r = 0.633, p < 0.001	r = 0.904, p < 0.001

correlated with the product erosion × topsoil SOC than with erosion only. In contrast, eroded SOC was weakly correlated with topsoil SOC.

5.3.4 Carbon Enrichment Ratio of Sediments (CER) as Affected by Erosion and Eroded SOC

In general, CER ranged from 0.5 to 14 (one conventionally tilled maize/cotton plot in Mali yielded 0.5). The highest CER (> 3.0) values were always measured on plots with low erosion (≤ 0.5 Mg ha^{-1} yr^{-1}) and the smallest CER (1.0) on plots with high erosion (> 20 Mg ha^{-1} yr^{-1}). However, the relationship between erosion and CER was not unique, as low CER values were also observed on plots with low erosion (Figure 5.1a):

- Plots with erosion < 1 Mg ha^{-1} yr^{-1} had CER between 1.2 and 14.0
- Plots with erosion between 1 and 20 Mg ha^{-1} yr^{-1} had CER between 1.1 and 3.0
- Plots with erosion > 20 Mg ha^{-1} yr^{-1} had CER between 1.0 and 2.0

More precisely, the five erosion groups (i.e., > 80, 9 to 40, 2 to 9, 0.7 to 2, and < 0.7 Mg ha^{-1} yr^{-1}) had average CER of 1.2 (±0.4), 1.9 (±0.9), 1.7 (±0.4), 2.2 (±0.5), and 3.9 (±3.4), respectively. A similar relationship was observed between eroded SOC and CER (Figure 5.1b). The four eroded SOC groups (i.e., 1000 to 3000, 50 to 400, 15 to 50, and < 15 kg C ha^{-1} yr^{-1}) had average CER of 1.2 (±0.4), 1.9 (±0.8), 2.3 (±0.8), and 3.9 (±3.7), respectively. Linear correlations computed between CER and erosion or eroded SOC were not significant, but they were significant between Ln (CER) and Ln (erosion) or Ln (eroded SOC) (r = 0.482, p < 0.001, and r = 0.295, p < 0.05, respectively; Figure 5.1c and d). In general, erosion selectivity for SOC (expressed by CER) increased with a decrease in erosion and eroded SOC.

5.3.5 CER as Affected by Land Use

The preferential removal of SOC was also related to land use. The CER values > 3.0 were observed under forest, savanna, or fallow only, and 12 out of the 15 plots (i.e., 80%) under natural vegetation had CER > 2.4. For these 15 plots, mean CER was 4.6 (±3.5) and median CER 3.4. Mean and median CER were 7.5 (±4.5) and 6.0 under forest, 4.2 (±3.2) and 3.0 under woody savanna or bush fallow, and 2.7 (±1.0) and 2.8 under grass fallow or herbaceous savanna, respectively. In contrast, most of the plots (80%) with CER ≤ 1.1 corresponded to bare tilled soils. Of the eight bare plots, four had CER of 1.0 to 1.1, and four of 1.3 to 1.6. For these bare plots, mean CER was 1.3 (±0.2) and median CER 1.2. Mean CER was 1.0 for bare tilled soils in Martinique, which lost large amount of soil and SOC (> 80 Mg ha^{-1} yr^{-1} and > 1000 kg C ha^{-1} yr^{-1}, respectively), but it was 1.4 for Algerian soils, where soil and SOC losses were much less (2 to 9 Mg ha^{-1} yr^{-1} and 20 to 140 kg C ha^{-1} yr^{-1}, respectively). Plots under cultivation or used as rangeland had CER of 1.1 to 3.0, mean CER was 2.0 (±0.5) and median CER was 1.9. It was difficult to distinguish subgroups:

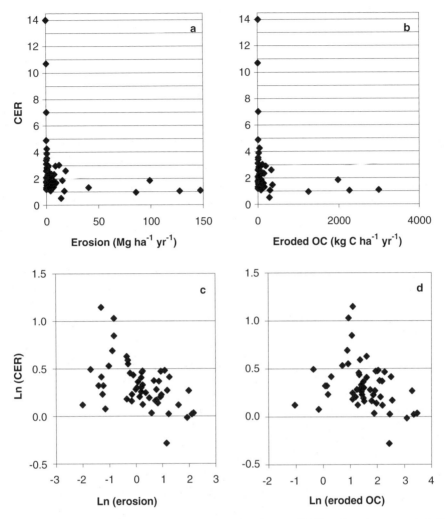

Figure 5.1 Relations between carbon enrichment ratio of sediments (CER) and erosion or eroded OC on the 54 runoff plots under study.

- Fifteen plots under conventionally tilled row crops (cereals, cotton) had CER of 1.1 to 3.0; mean CER was 1.9 (± 0.6) and median was 1.8 (Mediterranean orchard on bare soil was included in this group due to its poor surface cover during the rainy season)
- Four plots under no-till row crops (a fifth plot with contour tillage was included in the same subgroup) had CER of 1.5 to 2.3, and mean and median CER were 1.9 (±0.3)
- Four plots where cultivation practices involved thick mulching had CER ranging from 1.5 to 3.0; mean CER was 2.1 (± 0.7) and median was 1.9
- Six Mediterranean rangelands had CER of 1.7 to 2.5, and mean and median CER were 2.1 (±0.3)

Considering 54 plots, CER values were finally divided into three groups according to land use: bare tilled soils (mean CER 1.3); cultivated plots and rangelands (mean CER 2.0); forest, savanna, and fallows (mean CER 4.6; 2.7 for grass fallow, 4.2 for woody savanna and bush fallow, 7.5 for forest). Mean and median CER did not differ within bare plots and within cultivated plots and rangelands, but the difference was more marked under forest, savanna, and fallow, due to the presence of extreme (i.e., more variable) CER values.

5.3.6 The CER as Expressed by the Relation between Eroded SOC and the Product Erosion × Topsoil SOC

The CER of sediments was also assessed through the relationship between eroded SOC and the product erosion × topsoil SOC content. The eroded SOC is defined as the product of erosion multiplied by SOC content of sediments. Thus plotting eroded SOC against erosion × topsoil SOC is similar to plotting erosion × sediment SOC against erosion × topsoil SOC, and the slope of the regression lines corresponds to the ratio sediment SOC/topsoil SOC (i.e., CER). Plotting eroded SOC against erosion × topsoil SOC produced the following regression equations (Figure 5.2, which includes three scale levels):

- For bare tilled soils, slope of the regression line was 1.1 ($r = 0.999$, $p < 0.001$)
- For cultivated plots and rangelands, the slope was 1.8 ($r = 0.990$, $p < 0.001$; $r = 0.996$ when excluding ridged pineapple from Martinique)
- For forests, savannas, and fallows, the slope was 2.6 ($r = 0.975$, $p < 0.001$), but 2.7 ($r = 0.986$) when excluding one outlier (protected fallow near Tlemcen, Algeria); this group could be further divided into three subgroups:
  - For grass fallow, slope of the regression line was 1.7 ($r = 0.240$, $p > 0.1$); excluding one outlier (protected fallow near Tlemcen, Algeria, where CER was 1.4), the slope was 2.6 ($r = 0.395$, $p > 0.1$)
  - For woody savanna or bush fallow, the slope was 2.7 ($r = 0.992$, $p < 0.001$)
  - For forest, the slope was 4.4 ($r = 0.927$, $p < 0.1$)

Within each group, slope of the regression line was often close to mean or median CER, especially for bare and cultivated plots and rangelands. The difference between mean (or median) CER and slope may be explained by the smaller CER of plots having greater SOC losses, which had more influence in the determination of regression equations (whereas all plots had the same weight when calculating means or medians). Thus slopes of the regression lines tended to be smaller than mean and median CER.

5.4 DISCUSSION

5.4.1 Erosion

The importance of rainfall and slope in water erosion is widely recognized (Wischmeier and Smith, 1978; Roose, 1977; 1996; Lal, 2001), despite some contradictions with regard to the effects of slope steepness (Roose et al., 1996; El-Swaify, 1997; Lal, 1997; Fox and Bryan, 1999). The effect of land use on erosion has also been extensively reported (Wischmeier and Smith, 1978; Roose, 1996; Lal, 2001). Considering the 54 runoff plots in tropical and Mediterranean regions, five erosion groups were distinguished: 80 to 150 Mg ha^{-1} yr^{-1} for bare tilled soils in very humid regions; 9 to 40 Mg ha^{-1} yr^{-1} for conventionally tilled cereals and cotton in humid and subhumid regions; 2 to 9 Mg ha^{-1} yr^{-1} for no-tilled cereals and cotton in subhumid regions and bare tilled soils in Mediterranean regions; 0.7 to 2 Mg ha^{-1} yr^{-1} for crops and rangelands in Mediterranean regions; ≤ 0.7 Mg ha^{-1} yr^{-1} for forests, savannas, fallows, and crops with thick mulch.

The data presented were in agreement with those published in the literature. In Malaysia, Hashim et al. (1995) reported more than 100 Mg ha^{-1} yr^{-1} of erosion on bare tilled soils (3000-mm yr^{-1} rainfall, 18% slope). In Cabo Verde, Smolikowski et al. (2001) estimated erosion of 84 Mg ha^{-1} yr^{-1} on bare tilled soil, and 34 and 0.1 Mg ha^{-1} yr^{-1} under maize-bean association without and with mulch cover, respectively. Though annual rainfall was less (300-mm yr^{-1}), high rainfall erosivity and steep slope gradient (45%) resulted in high erosion rates similar to those presented herein (> 80, 9 to 40 and ≤ 0.7 Mg ha^{-1} yr^{-1}, respectively). In northern Cameroon, Thébé (1987) reported

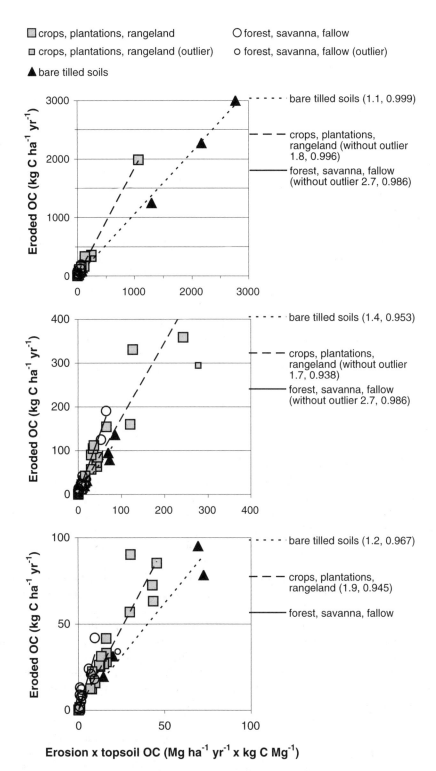

Figure 5.2 Relation between eroded OC and the product erosion-topsoil OC (0 to 10 cm) at three scales (data into brackets are slopes of the regression lines and correlation coefficients).

erosion rates of 11 and 21 Mg ha^{-1} yr^{-1} under conventional cultivation, within the range of present data (9 to 40 Mg ha^{-1} yr^{-1}). In the southern U.S. (1400-mm yr^{-1} rainfall), mean erosion measured in runoff plots (5% slope) by McGregor et al. (1999) was 48.5 and 5.2 Mg ha^{-1} yr^{-1} under conventionally and no-tilled cotton, respectively, and was consistent with present data for cotton in Cameroon (for slope < 3% and rainfall < 1300 mm yr^{-1}, 9 to 40 and 2 to 8 Mg ha^{-1} yr^{-1} under conventionally and no-tilled cotton, respectively). In Nigeria (1400-mm yr^{-1} rainfall, 1 to 15% slope), Lal (1997) measured mean erosion under no-till maize-legume rotations (two cropping seasons per year) between 0.5 and 2.2 Mg ha^{-1} yr^{-1}. Perhaps due to better soil cover over the year, this rate was less than the present data for no-till plots under row crops in subhumid areas with one rainy season (2 to 8 Mg ha^{-1} yr^{-1}). Additionally, Lal (1997) underlined the interest of creating rough soil surface (e.g., by residue mulching and no-till) in order to control erosion. In the Australian contour bay catchments (550-mm yr^{-1} rainfall, 2% slope) studied by Carroll et al. (1997), erosion under row crops was threefold greater for conventional than for zero tillage (4.0 vs. 1.4 Mg ha^{-1} yr^{-1}). Though these values were somewhat more than those in the Mediterranean cropped plots (< 2 Mg ha^{-1} yr^{-1}), perhaps due to a scale effect, they confirmed that reducing tillage also reduced erosion. In Syria, erosion measured by Shinjo et al. (2000) on runoff plots under matorral (280-mm yr^{-1} rainfall, 4 to 19% slope, 40 to 45% clay) ranged from 0.1 to 0.3 Mg ha^{-1} yr^{-1} when the plots were grazed, and from 0.0 to 0.1 Mg ha^{-1} yr^{-1} when they were protected from grazing. This was less than in the Mediterranean rangelands (0.7 to 1.8 Mg ha^{-1} yr^{-1}) and protected fallows (0.1 to 1.0 Mg ha^{-1} yr^{-1}), perhaps due to higher annual rainfall (360-550 mm) and either smaller clay content or greater slope gradient in these plots. In semiarid Spain (360-mm yr^{-1} rainfall, 23% slope), Castillo et al. (1997) measured 0.1 Mg ha^{-1} yr^{-1} erosion under natural shrubland vegetation and 0.3 Mg ha^{-1} yr^{-1} on a counterpart plot where vegetation had been removed but litter left intact. These values were comparable to the present data under protected fallows in Mediterranean areas (0.1 to 1.0 Mg ha^{-1} yr^{-1}).

5.4.2 Eroded SOC and Carbon Enrichment Ratio of Sediments (CER)

The data from 54 runoff plots were divided into four eroded OC groups: 1 to 3 Mg C ha^{-1} yr^{-1} for bare tilled soils in very humid regions; 50 to 400 kg C ha^{-1} yr^{-1} for cereals and cotton in humid and subhumid regions, and some bare plots in Mediterranean regions; 15 to 50 kg C ha^{-1} yr^{-1} for crops, rangelands, and other bare plots in Mediterranean regions; < 15 kg C ha^{-1} yr^{-1} for forests, savannas, fallows, and crops with thick mulch. This grouping indicates that in plots vulnerable to erosion, eroded SOC may be the same order of magnitude as the changes in SOC, especially under row crops in tropical areas. In their review on SOC dynamics under no-till, Six et al. (2002) reported that SOC increase for 0 to 30 cm depth was 325 ± 113 kg C ha^{-1} yr^{-1} under no-till systems in tropical regions (i.e., within the range of eroded SOC in the present cereal and cotton plots, which was 50 to 400 kg C ha^{-1} yr^{-1}). The relative importance of eroded SOC in C budgets is generally more on tilled plots because (1) SOC is generally lesser than on no-till plots (Balesdent et al., 2000) and (2) changes in SOC are either positive but generally less than on no-till plots, or negative. This observation confirms that eroded SOC cannot be neglected when conditions accentuate soil erosion risk (Voroney et al., 1981; Gregorich et al., 1998; Mitchell et al., 1998). In contrast, under forest, savanna, and crops with thick mulch, eroded SOC is small and negligible compared with changes in SOC resulting from large residue biomass.

The data from 54 plots is grouped into three CER classes: bare soils with CER ranging from 1 to 1.6 and averaging 1.3 (1.0 in the tropics, 1.4 in Algeria); cultivated plots and rangelands with CER ranging from 1.1 and 3.0 and averaging 2.0; forests, savannas, and fallows with CER of > 2.4 and averaging 4.6 (2.7 in grass fallows, 4.2 in woody savannas and bush fallows, and 7.5 in forests).

The distinction between conventional and no-till row crops was not possible for both eroded SOC and CER though it was relevant to erosion: no-till resulted in less erosion than conventional tillage, but the effect of tillage on eroded OC and CER was not clear. This trend indicates that the processes involved in SOC enrichment of sediments did not depend on tillage, and to a larger extent, on cropping system. However, it is likely that under no-till a noticeable proportion of eroded SOC consisted of coarse plant debris or litter, which may have been trapped by obstacles on the hillside and not transported to long distances. Thus SOC erosion under no-till may be considered less severe than under conventional tillage though the present data at the plot scale were not explicit. Determining the size distribution of organic matter in sediments may help verify this assumption. Additionally, measurements at the hillside scale may reveal differences in eroded SOC between conventional and no-till cultivation that were not apparent on 100-m^2 plots.

Comparison with published data was not easy due to the scarcity of references regarding SOC erosion at the plot scale in tropical and Mediterranean regions. On clayey Inceptisols in Colombia (2000-mm yr^{-1} rainfall, 7 to 20% slope), eroded SOC measured on runoff plots by Ruppenthal et al. (1997) averaged 5700 kg C ha^{-1} yr^{-1} on bare tilled soil and 200 kg C ha^{-1} yr^{-1} on cassava plots with forage legume intercropping whose effect on eroded SOC was not clear. These losses were more than those reported herein especially for bare soil, probably due to the very high SOC content (60 g C kg^{-1} in 0 to 20 cm). However, CER did not differ significantly among treatments and averaged 1.0, indicating that erosion was not selective for SOC probably due to strong soil aggregation and steep slope. On sandy Alfisols in Zimbabwe (500-mm yr^{-1} rainfall, 5% slope), measurements carried out on runoff plots by Moyo (1998) on bare tilled fallow, conventional and no-till maize showed erosion rates of 81.8, 34.3, and 0.2 Mg ha^{-1} yr^{-1}, eroded SOC of 210, 180, and 5 kg C ha^{-1} yr^{-1}, and CER of 1.1, 1.5, and 6.6, respectively. On bare soil, erosion was consistent with that of the present tropical plots (> 80 Mg ha^{-1} yr^{-1}), but smaller SOC content in Zimbabwe (< 3 g C kg^{-1}) resulted in smaller eroded SOC. Nevertheless CER was similar to the present data (i.e., close to 1). Data were also consistent with the present report for conventional maize, but for no-till maize there were similar to plots under forest or savanna (CER > 3), indicating the effect of thick mulch cover in no-till in Zimbabwe. Thus in particular conditions (e.g., thick mulch cover), cropping system could have an important effect on eroded SOC and CER. On sandy clay loam Alfisols in India (660-mm yr^{-1} rainfall, 1.5 to 2% slope) cropped with cereals, Cogle et al. (2002) measured eroded SOC ranging from 46 to 178 kg C ha^{-1} yr^{-1} and CER of 1.5 to 4, depending on tillage depth and mulch. Straw mulching decreased eroded SOC but the effect of tillage was unclear, as in the tropical plots under row crops in the present study. Overall, the values presented in this paper were consistent with those measured in Cameroon plots under cotton (57 to 160 kg C ha^{-1} yr^{-1} and $1.9 \leq CER \leq 3.0$ for 2% slope and 860-mm yr^{-1} rainfall). On clayey Alfisols in Kenya (1000-mm yr^{-1} rainfall, 30% slope), eroded SOC measured on runoff plots by Gachene et al. (1997) was 650 and 2370 kg C ha^{-1} yr^{-1} for conventional-till maize with and without fertilizers, respectively. This rate was much higher than in most cereal plots in tropical areas (50 to 400 kg C ha^{-1} yr^{-1}), though erosion data were consistent (7 and 29 Mg ha^{-1} yr^{-1}, respectively, vs. 9 to 40 Mg ha^{-1} yr^{-1}). The difference may be explained by rather high SOC content (ca. 30 g C kg^{-1} at 0 to 10 cm) and erosion rates (due to the 30% slope) of Kenyan plots, whereas the cultivated plots with high SOC content generally had low soil erosion. Indeed, CER was 1.3 in both Kenyan plots vs. 1.1 to 3.0 in the present cultivated plots, indicating that sediments were not particularly enriched in SOC on Kenyan steep slopes. In Malaysia (3000-mm yr^{-1} rainfall, 18% slope), Hashim et al. (1995) studied 1000-m^2 plots in cocoa plantations with possible legume cover crop on soils derived from sandstone and shale (20% clay). Erosion was ten times smaller with than without cover crop; but CER was 1.6 and 1.4, respectively, i.e., in the range for the cultivated plots reported herein, and confirmed the limited influence of cultivation practices on CER.

From these references and from the present data, it is concluded that land use has more influence on erosion than on eroded SOC and on CER. In similar climate, soil and slope conditions, land uses that increased soil OC generally reduced erosion, due to better soil aggregate stability and better infiltration (Barthès et al., 2000; Roose and Barthès, 2001). Because an increase in topsoil SOC content also increases sediment SOC content, it results in low soil losses and high SOC content, and the impact on eroded SOC and CER is not evident because of increase in both soil and sediment SOC. Thus differences in land use that affect SOC only are perhaps not sufficient to clearly affect eroded SOC and CER, though they affect erosion. Land use affects eroded OC and CER through effects on preferential erosional processes (e.g., due to change in soil surface cover). Indeed, the three classes consisting of (1) bare soil, (2) cultivated and rangelands, and (3) virgin soils and fallows, differed in soil surface cover and in eroded SOC and CER. To a larger extent, important differences in climate, slope, and soil conditions (texture especially) may also affect erosion selectivity along with eroded SOC and CER, as indicated e.g., by lower CER on bare soils in Martinique (high erosivity) than in Algeria (low erosivity).

5.4.3 Erosion Selectivity and Preferential Removal

Erosion is a selective process that removes the smallest or lightest soil particles faster than sand and gravels (Roose, 1977; Gregorich et al., 1998; Starr et al., 2000). Indeed, sheet erosion is generally selective because shallow runoff can only transport small or light particles usually produced by macroaggregate disintegration, such as organic particles, clay and silt, or microaggregates (Roose, 1996; Wan and El-Swaify, 1997; Cogle et al., 2002). Sheet erosion also results in selective deposition, which occurs when the flow velocity decreases due to vegetation, litter, surface roughness, or decrease in slope angle. Raindrop splash can transport larger or heavier particles such as sands, in all directions (including upslope), but is not considered an important interrill transport process under normal field conditions (Sutherland et al., 1996; Wan and El-Swaify, 1998). Sheet erosion is however less selective in some cases, especially on steep slopes with well-aggregated soils, e.g., with high clay or organic matter content, such as volcanic soils (De Noni et al., 2001; Khamsouk, 2001) or Vertisols with high calcium content (Roose, 2004). Indeed, on steep slopes, stable macroaggregates can be displaced by shallow runoff flows, resulting in comparable sediment and topsoil composition.

Rill erosion is less selective due to the greater shearing and transport capacities of concentrated runoff flow, which can incise and scour the whole topsoil; thus sediments and topsoil do not differ much in composition (Roose, 1977; Wan and El-Swaify, 1997). However, it results in selective deposition: stones and gravels first, followed by sands, and finally fine particles along the flood plains (Roose, 1996).

Thus the selectivity of erosion for SOC decreases with increase in soil losses (Avnimelech and McHenry, 1984; Sharpley, 1985; Cogle et al., 2002). Low soil loss results from sheet erosion only, which affects SOC-rich top layers mainly and transports small or light particles preferentially, organic matter. Consequently sediments are enriched in SOC. In contrast, high soil loss also involves rill erosion, which affects subsoil layers with less SOC and can transport heavier (and less organic) particles, thereby decreasing the sediment enrichment ratio. CER may be of < 1 when erosion affects soil layers that contain less SOC than the reference layer (i.e., 0 to 10 cm). In general, factors that decrease runoff velocity increase the erosion selectivity (e.g., thick litter and mulch). In contrast, factors that increase runoff velocity decrease erosion selectivity (e.g., bare soil surface or steep slope).

5.4.4 Relationship between Erosion, Eroded SOC, and CER

The relationship between erosion and eroded SOC is generally recognized (Gregorich et al., 1998) and is evident from the fact that eroded SOC is included in sediments. The influence of topsoil SOC content (in addition to that of erosion) is also evident because topsoil provides the materials that are eroded, although this relationship has never been explicitly described. The present data indicate that taking erosion and topsoil SOC into account allows a better prediction of eroded SOC than considering erosion only, especially with regards to the principal land uses.

At the watershed scale and considering individual events, Starr et al. (2000) suggested a power law relationship between erosion and eroded SOC (eroded SOC = a × erosion^b). Such a relationship was not observed with the present data. Starr et al. (2000) also reported a logarithmically linear but inverse relationship between erosion and CER, which was also the case with the present data (Figure 5.1c). Sharpley (1985) reported a similar relationship from rainfall simulations on 2-mm sieved topsoil samples. The fact that investigations at very different scales (individual events on watersheds, annual rainfall on runoff plots, simulated rainfall on sieved samples) result in a comparable relationship between erosion and CER validate the general trend.

5.5 CONCLUSION

On the basis of the data from 54 plots, five erosion classes were defined as follows: bare soil in very humid regions (maximum); conventional-tillage row crops in humid and subhumid regions; no-till row crops in subhumid regions and bare soil in Mediterranean regions; crops and rangelands in Mediterranean regions; and forest, savanna, fallows, and crops with thick mulch (minimum). In contrast, only four eroded SOC classes (conventional-tillage and no-till plots behaving similarly) and three CER classes (bare soils, crops and rangelands, forest, savannas, and mulched crops) were identified. Factors that affected soil erosion did not necessarily affect carbon erosion or the enrichment ratio. In particular, changes in topsoil SOC did not clearly influence changes in eroded SOC or CER, which required changes in erosion selectivity (e.g., resulting from changes in soil surface cover or roughness). Indeed, erosion selectivity for SOC was low on bare soils (CER $\leq$ 1.6), especially in humid conditions (CER $\leq$ 1.1), indicating that subsoil layers contributed to sediment production by rill erosion, or intact aggregates were eroded. In contrast, erosion selectivity for SOC was high on plots covered by thick litter ($2.4 \leq$ CER ≤ 14.0), suggesting the preferential removal of organic particles by sheet erosion. The preferential removal of SOC was intermediate under cultivation and rangeland ($1.1 \leq$ CER ≤ 3.1).

The annual SOC losses by water erosion generally ranged between 50 and 400 kg C ha^{-1} yr^{-1} in tropical regions and between 15 and 50 kg C ha^{-1} yr^{-1} in Mediterranean regions under cultivation and rangeland. This is the same order of magnitude as annual changes in SOC, indicating that on plots vulnerable to erosion, SOC erosion cannot be neglected when assessing carbon balances at the plot scale. Eroded SOC was less ($<$ 15 kg C ha^{-1} yr^{-1}) under mulch farming and under forest, savanna, and fallows, where the soil surface is covered by litter. In such conditions, eroded SOC is generally negligible as compared to SOC input as residues or litter. Besides losses of SOC in sediments, soluble SOC is also lost in water runoff. Preliminary and partial data indicated that runoff SOC was generally 4 to 20 times less than eroded SOC, but that it could sometimes be of the same order of magnitude such as under mulched crops in very humid conditions (Roose, 2004). This observation underlines the need for complementary research taking into account all SOC losses resulting from water erosion. The same preliminary data also indicate the need for studying losses of soluble SOC through deep/vertical drainage, which can be 30 to 75 kg C ha^{-1} yr^{-1} in plots where eroded SOC is less than 15 kg C ha^{-1} yr^{-1}.

Nested studies on plots and watersheds are also necessary to determine the fate of SOC removed from plots, which does not necessarily reach the river. Such studies may address specific questions with regard to the differences in eroded SOC or CER among land use and management (e.g., between conventional and no-till farming).

ACKNOWLEDGMENTS

The authors thank M. Arabi, J. Arrivets, B. Bep A Ziem, R. Bertrand, E. Blanchart, Z. Boli, D. Diallo, J. Godefroy, P. Jadin, B. Khamsouk, M. Mazour, and B. Morsli for their contribution to the collection of data, E. Kouakoua for his help with collecting literature references, and G. Bourgeon for his help regarding soil classification.

REFERENCES

Arabi, M. 1991. Influence de quatre systèmes de production sur le ruissellement et l'érosion en milieu montagnard Méditerranéen (Médéa, Algérie). Ph.D. dissertation, Université de Grenoble, France.

Arabi, M. and E. Roose. 2004. Influence du système de production et du sol sur l'érosion en nappe, le ruissellement, le stock du sol et les pertes de carbone par érosion en zone de montagne méditerranéenne (Médéa, Algérie). Proceedings of the International Colloquium Land Use, Erosion and Carbon Sequestration, Montpellier, France, 23–28 September 2002. Bulletin du Réseau Erosion 22:166–175.

Avnimelech, Y. and J. R. McHenry. 1984. Enrichment of transported sediments with organic carbon, nutrients, and clay. *Soil Science Society of America Journal* 48:259–266.

Balesdent, J., C. Chenu, and M. Balabane. 2000. Relationship of soil organic matter dynamics to physical protection and tillage. *Soil and Tillage Research* 53:215–230.

Barthès, B., A. Azontonde, B. Z. Boli, C. Prat, and E. Roose. 2000. Field-scale run-off and erosion in relation to topsoil aggregate stability in three tropical regions (Benin, Cameroon, Mexico). *European Journal of Soil Science* 51:485–495.

Bep, A. Ziem, B., Z. B. Boli, and E. Roose. 2004. Influence du labour, du fumier et de l'âge de la défriche sur le stock de carbone du sol et les pertes de carbone par érosion et drainage dans une rotation intensive coton/maïs sur un sol ferrugineux tropical sableux du Nord Cameroun (Mbissiri, 1995). Proceedings of the International Colloquium Land Use, Erosion and Carbon Sequestration, Montpellier, France, 23–28 September 2002. Bulletin du Réseau Erosion 22:176–192.

Blanchart, E., E. Roose, and B. Khamsouk. 2006. Soil carbon dynamics and losses by erosion and leaching in banana cropping systems with different practices (Nitisol, Martinique, West Indies), in E. Roose, R. Lal, C. Feller, B. Barthès, and B. A. Stewart, eds., *Soil Erosion and Carbon Dynamics*. Taylor & Francis, Boca Raton, FL.

Boli, Z. 1996. Fonctionnement de Sols Sableux et Optimisation des Pratiques Culturales en Zone Soudanienne Humide du Nord Cameroun. Ph.D. dissertation, Université de Dijon, France.

Carroll, C., M. Halpin, P. Burger, K. Bell, M. M. Sallaway, and D. F. Yule. 1997. The effect of crop type, crop rotation, and tillage practice on runoff and soil loss on a Vertisol in central Queensland. *Australian Journal of Soil Research* 35:925–939.

Castillo, V. M., M. Martinez-Mena, and J. Albaladejo. 1997. Runoff and soil loss response to vegetation removal in a semiarid environment. *Soil Science Society of America Journal* 61:1116–1121.

Cogle, A. L., K. P. C. Rao, D. F. Yule, G. D. Smith, P. J. George, S. T. Srinivasan, and L. Jangawad. 2002. Soil management for Alfisols in the semiarid tropics: Erosion, enrichment ratios, and runoff. *Soil Use and Management* 18:10–17.

De Jong, E., P. A. Nestor, and D. J. Pennock. 1998. The use of magnetic susceptibility to measure long-term soil redistribution. *Catena* 32:23–35.

De Noni, G., M. Viennot, J. Asseline, and G. Trujillo. 2001. *Terres d'altitude, terres de risques. La lutte contre l'érosion dans les Andes équatoriennes*. IRD Editions, Paris.

Diallo, D. 2000. Erosion des sols en zone soudanienne du Mali. Transfert des matériaux érodés dans le bassin de Djitiko (Haut, Niger). Ph.D. dissertation, Université de Grenoble, France.

Diallo, D., E. Roose, and D. Orange. 2004. Influence du couvert végétal et des sols sur les risques de ruissellement et d'érosion et sur les stocks et les pertes de carbone en zone soudanienne du Mali. Proceedings of the International Colloquium Land Use, Erosion and Carbon Sequestration, Montpellier, France, 23–28 September 2002. Bulletin du Réseau Erosion 22:193–207.

El-Swaify, S. A. 1997. Factors affecting soil erosion hazards and conservation needs for tropical steeplands. *Soil Technology* 11:3–16.

Fox, D. M., and R. B. Bryan. 1999. The relationship of soil loss by interrill erosion to slope gradient. *Catena* 38:211–222.

Gachene, C. K. K., N. J. Jarvis, H. Linner, and J. P. Mbuvi. 1997. Soil erosion effects on soil properties in a highland area of Central Kenya. *Soil Science Society of America Journal* 61:559–564.

Gee, G. W. and J. W. Bauder. 1986. Particle-size analysis, in A. Klute, ed., *Methods of Soil Analysis: Part 1 — Physical and Mineralogical Methods*, 2nd ed. American Society of Agronomy, Madison, WI, pp. 383–411.

Govers, G., D. A. Lobb, and T. A. Quine. 1999. Preface — Tillage erosion and translocation: Emergence of a new paradigm in soil erosion research. *Soil and Tillage Research* 51:167–174.

Gregorich, E. G., K. J. Greer, D. W. Anderson, and B. C. Liang. 1998. Carbon distribution and losses: Erosion and deposition effects. *Soil and Tillage Research* 47:291–302.

Hashim, G. M., C. A. A. Ciesiolka, W. A. Yusoff, A. W. Nafis, M. R. Mispan, C. W. Rose, and K. J. Coughlan. 1995. Soil erosion processes in sloping land in the east coast of Peninsular Malaysia. *Soil Technology* 8:215–233.

Hudson, N. W. 1993. Field Measurement of Soil Erosion and Runoff. *FAO Soils Bulletin* 68, FAO, Rome.

IPCC (Intergovernmental Panel on Climate Change). 2001. Climate change 2001: The scientific basis. Contribution of working group I to the third assessment report of the IPCC. Cambridge University Press, UK.

Jayawardena, A. W. and R. R. Bhuiyan. 1999. Evaluation of an interrill soil erosion model using laboratory catchment data. *Hydrological Processes* 13:89–100.

Khamsouk, B. 2001. Impact de la culture bananière sur l'environnement. Influence des systèmes de cultures bananières sur l'érosion, le bilan hydrique et les pertes en nutriments sur un sol volcanique en Martinique (Cas du Sol Brun Rouille à Halloysite). Ph.D. dissertation, Université de Montpellier, France.

Lal, R. 1997. Soil degradative effects of slope length on Alfisols in western Nigeria. IV. Plots of equal land area. *Land Degradation and Development* 8:343–354.

Lal, R. 2001. Soil degradation by erosion. *Land Degradation and Development* 12:519–539.

Lal, R. 2003. Soil erosion and the global carbon budget. *Environment International* 29:437–450.

Le Bissonnais, Y., H. Benkhadra, V. Chaplot, D. Fox, D. King, and J. Daroussin. 1998. Crusting, runoff, and sheet erosion on silty loamy soils at various scales and upscaling from m^2 to small catchments. *Soil and Tillage Research* 46:69–80.

Lowrance, R. and R. G. Williams. 1988. Carbon movement in runoff and erosion under simulated rainfall. *Soil Science Society of America Journal* 52:1445–1448.

McGregor, K. C., S. M. Dabney, and J. R. Johnson. 1999. Runoff and soil loss from cotton plots with and without stiff-grass hedges. *Transactions of the ASAE* 42:361–368.

Mitchell, P. D., P. G. Lakshminarayan, T. Otake, and B. A. Babcock. 1998. The impact of soil conservation policies on carbon sequestration in agricultural soils of the central U.S., in R. Lal, J. M. Kimble, R. F. Follett, and B. A. Stewart, eds., *Management of Carbon Sequestration in Soil*. CRC Press, Boca Raton, FL, pp. 125–142.

Morsli, B., M. Mazour, N. Mededjel, A. Halitim, and E. Roose. 2004. Effet des systèmes de gestion des terres sur l'érosion et le stock du carbone dans les monts du Tell Occidental — Algérie. Proceedings of the International Colloquium Land Use, Erosion and Carbon Sequestration, Montpellier, France, 23–28 September 2002. Bulletin du Réseau Erosion 22:144–165

Moyo, A. 1998. The effects of soil erosion on soil productivity as influenced by tillage: With special reference to clay and organic matter losses. *Advances in GeoEcology* 31:363–368.

Mutchler, C. K., C. E. Murphree, and K. C. McGregor. 1988. Laboratory and field plots for erosion studies, in R. Lal, ed., *Soil Erosion Research Method*. Soil and Water Conservation Society, Ankeny, IA, pp. 9–36.

Nelson, D. W. and L. E. Sommers. 1996. Total carbon, organic carbon, and organic matter, in D. L. Sparks, A. L. Page, P. A. Helmke, R. H. Loeppert, P. N. Soltanpour, M. A. Tabatabai, C. T. Johnson, and M. E. Sumner, eds., *Methods of Soil Analysis: Part 3 — Chemical Methods*. Soil Science Society of America and American Society of Agronomy, Madison, WI, pp. 9–36.

Robert, M. 2001. Soil carbon sequestration for improved land management. *World Soil Resources Reports* 96, FAO, Rome.

Roose, E. 1967. Dix années de mesure de l'érosion et du ruissellement au Sénégal. *Agronomie Tropicale* 22:123–152.

Roose, E. 1977. Erosion et ruissellement en Afrique de l'Ouest: Vingt années de mesures en petites parcelles expérimentales. *Travaux et Documents ORSTOM* 78, ORSTOM, Paris.

Roose, E. 1978. Pédogenèse actuelle d'un sol ferrugineux complexe issu de granite sous une savane arborescente du Plateau Mossi (Haute-Volta): Gonsé, 1968–74. *Cahiers ORSTOM, série Pédologie* 16:1993–223.

Roose, E. 1979a. Dynamique actuelle d'un sol ferrallitique gravillonnaire issu de granite sous culture et sous savane arbustive soudanienne du nord de la Côte d'Ivoire. *Cahiers ORSTOM, série Pédologie* 17:81–118.

Roose, E. 1979b. Dynamique actuelle d'un sol ferrallitique sablo-argileux très désaturé sous maïs et sous forêt dense humide: Adiopodoumé, 1964–75. *Cahiers ORSTOM, série Pédologie* 17:259–282.

Roose, E. 1981. Dynamique Actuelle de Sols Ferrallitiques et Ferrugineux Tropicaux d'Afrique Occidentale. *Travaux et Documents ORSTOM 130*, ORSTOM, Paris.

Roose, E. 1996. Land Husbandry Components and Strategy. *FAO Soils Bulletin 70*, FAO, Rome.

Roose, E. 2004. Erosion du carbone et indice de sélectivité à l'échelle de la parcelle dans les régions tropicales et méditerranéennes. Proceedings of the International Colloquium Land Use, Erosion and Carbon Sequestration, Montpellier, France, 23–28 September 2002. Bulletin du Réseau Erosion 22:74–94.

Roose, E., M. Arabi, K. Brahamia, R. Chebbani, M. Mazour, and B. Morsli. 1996. Erosion en nappe et ruissellement en montagne méditerranéenne d'Algérie: Synthèse des campagnes 1984–95 sur un réseau de 50 parcelles. *Cahiers ORSTOM, série Pédologie* 28:289–308.

Roose, E. and B. Barthès. 2001. Organic matter management for soil conservation and productivity restoration in Africa: a contribution from Francophone research. *Nutrient Cycling in Agroecosystems* 61:159–170.

Roose, E. and J. Godefroy. 1977. Pédogenèse actuelle comparée d'un sol ferrallitique remanié sur schiste sous forêt et sous bananeraie fertilisée de Basse Côte d'Ivoire (Azaguié, 1967–75). *Cahiers ORSTOM, série Pédologie* 15:409–436.

Roose, E. and J. M. Sarrailh. 1989. Erodibilité de quelques sols tropicaux. Vingt années de mesure en parcelles d'érosion sous pluies naturelles. *Cahiers ORSTOM, série Pédologie* 25:7–30.

Ruppenthal, M., D. E. Leihner, N. Steinmuller, and M. A. El-Sharkawy. 1997. Losses of organic matter and nutrients by water erosion in cassava-based cropping systems. *Experimental Agriculture* 33:487–498.

Sharpley, A. N. 1985. The selective erosion of plant nutrients in runoff. *Soil Science Society of America Journal* 49:1527–1534.

Shinjo, H., H. Fujita, G. Gintzbuger, and T. Kosaki. 2000. Impact of grazing and tillage on water erosion in northeastern Syria. *Soil Science and Plant Nutrition* 46:151–162.

Six, J., C. Feller, K. Denef, S. M. Ogle, J. C. de Moraes Sá, and A. Albrecht. 2002. Soil organic matter, biota and aggregation in temperate and tropical soils — Effect of no-tillage. *Agronomie* 22:755–775.

Smolikowski, B., H. Puig, and E. Roose. 2001. Influence of soil protection techniques on runoff, erosion and plant production on semiarid hillslopes of Cabo Verde. *Agriculture, Ecosystems and Environment* 87:67–80.

Starr, G. C., R. Lal, R. Malone, D. Hothem, L. Owens, and J. Kimble. 2000. Modeling soil carbon transported by water erosion processes. *Land Degradation and Development* 11:83–91.

Sutherland, R. A. 1996. Caesium-137 soil sampling and inventory variability in reference locations: a literature survey. *Hydrological Processes* 10:43–53.

Sutherland, R. A., Y. Wan, A. D. Ziegler, C. T. Lee, and S. A. El-Swaify. 1996. Splash and wash dynamics: an experimental investigation using an Oxisol. *Geoderma* 69:85–103.

Thébé, B. 1987. Hydrodynamique de Quelques Sols du Nord-Cameroun. Bassin Versants de Mouda. Contribution à l'Etude des Transferts d'Echelles. Ph.D. dissertation, Université de Montpellier, France.

Voroney, R. P., J. A. Van Veen, and E. A. Paul. 1981. Organic C dynamics in grassland soils. 2. Model validation and simulation of the long-term effects of cultivation and rainfall erosion. *Canadian Journal of Soil Science* 61:211–224.

Wan, Y. and S. A. El Swaify. 1997. Flow-induced transport and enrichment of erosional sediment from a well-aggregated and uniformly-textured Oxisol. *Geoderma* 75:251–265.

Wan, Y. and S. A. El-Swaify. 1998. Characterizing interrill sediment size by partitioning splash and wash processes. *Soil Science Society of America Journal* 62:430–437.

Wischmeier, W. H. and D. D. Smith. 1978. *Predicting rainfall erosion losses: A guide to erosion planning*. Agriculture Handbook 537, U.S. Department of Agriculture, Washington DC.

CHAPTER 6

Organic Carbon in Forest Andosols of the Canary Islands and Effects of Deforestation on Carbon Losses by Water Erosion

Antonio Rodríguez Rodríguez, C. D. Arbelo, J. A. Guerra, J. L. Mora, and C. M. Armas

CONTENTS

6.1   Introduction 73
6.2   Material and Methods 75
    6.2.1   Study Area and Environmental Conditions 75
    6.2.2   Soil Genesis and Soil Properties 75
    6.2.3   Field Sampling and Field Plots 75
        6.2.3.1   Field Sampling 75
        6.2.3.2   Field Plots 76
    6.2.4   Laboratory Analyses 77
        6.2.4.1   Soil Samples 78
        6.2.4.2   Sediment Samples 79
        6.2.4.3   Water Runoff Samples 79
    6.2.3   Statistical Analysis 79
6.3   Results 79
    6.3.1   Organic Carbon Content and Forms 79
    6.3.2   Organic Carbon Losses by Water Erosion 80
6.4   Discussion 82
6.5   Conclusions 83
Acknowledgments 83
References 84

6.1 INTRODUCTION

Soil organic carbon (SOC) is one of the main components and a basic parameter for soil quality, since SOC content correlates strongly with many soil properties and functions: soil porosity and water holding capacity, nutrient availability, soil biodiversity, soil structural stability, etc. (Karlen and Andrews, 2000; Singer and Ewing, 2000). Therefore, increasing SOC levels is a desirable strategy in all terrestrial ecosystems, not only because of its favorable effects on soil dynamics, but also because of its contribution to mitigating the greenhouse effect.

The SOC contents in most soils is the result of several processes, some of which accumulate carbon in the soil surface and others reduce it. One of the most important of the latter processes is interrill erosion. Interrill erosion is the process of detachment and transport of soil by raindrops and very shallow flows (Laflen and Roose, 1997).

Many studies have been conducted to evaluate the impact of soil erosion on SOC dynamics and vice versa, although there are still numerous knowledge gaps about the importance of SOC, and the behavior of eroded SOC as a sink or a source of CO_2 and, in general, the net effect of erosion on atmospheric CO_2 evolution (Lal et al., 1998; Lal, 2001a).

Since interrill erosion is a selective process which preferentially removes the finest components of soil (Walling, 1994; Morgan, 1995; Starr et al., 2000; Kimble et al., 2001) the sediments are enriched in fine silt and clay-size particles that contain the most stable SOC pools in soils, because of the physical protection afforded in soil aggregates (Lal, 1995). Moreover, a significant part of C mobilized in soils by erosive processes is dissolved in runoff water as dissolved organic carbon (DOC).

Harden et al. (1999) and other authors proposed that 70 to 80% of SOC in eroded soils may be decomposed during transport and deposition. In contrast, Lal (1995, 2001b), Jacinthe et al. (2001), and Jacinthe and Lal (2001) observed that only 20% of displaced SOC is mineralized. Van Noordwijk et al. (1997) indicated that, at least potentially, erosion may contribute to carbon sequestration since eroded SOC is more protected from decomposition in acidic swamp environments or in freshwater and marine sediments.

A synthesis of the available literature shows that an important factor in the subsequent dynamics of SOC mobilized by erosion is its stability. Eswaran et al. (1995) considered four pools of SOC: dissolved, labile, slowly oxidizable, and passive or recalcitrant. The stability of SOC pools is closely related to the degree of complexity of organic molecules (biochemical recalcitrance), with their position in the soil aggregates (physical protection), and with the mode of their association with metals and secondary soil minerals (chemical stabilization) (Jastrow and Miller, 1998).

Volcanic ash soils, and more specifically Andosols (Soil Survey Staff, 1999), are characterized by a high SOC contents (80 to 300 g kg^{-1}) that imparts very dark color to the humic surface horizons (Nanzyo et al., 1993; Kimble et al., 2000). This high SOC content is attributed to the fact that in these soils the organic matter is stabilized by short-range ordered minerals (allophanes, imogolite, ferrihydrite), which have a large specific surface capable of adsorbing organic molecules, or by the formation of Al-humus complexes (Parfitt et al., 1997; Dahlgren et al., 1993). In both cases, SOC becomes highly resistant to microbial attack, such that the mean residence time of SOC in Andosols is very high and the turnover rates of SOC are very low (Saggar et al., 1994).

Hence, Andosols are characterized by a structure with very stable crumbly and granular microaggregates (0.1 nm to 1 μm to 1 to 3 mm) where SOC, in the form of organomineral and organometallic complexes, is physically protected from microbial mineralization (Warkentin and Maeda, 1980). Thus Andosols are soils with a large potential to sequester C, although little is yet known regarding the dynamics of SOC in these soils and the mechanisms governing stabilization of organic compounds (Lal et al., 1998).

In general, Andosols have a strong resistance to water erosion, and the low erodibility is strongly related to the specific physical properties of Andosols (Roose et al., 1999; Khamsouk et al., 2002; Rodríguez Rodríguez et al., 2002a). However, this behavior of volcanic ash soils is not in complete agreement with field observations, particularly when land use changes leading to high soil losses (Poulenard et al., 2001; Rodríguez Rodríguez et al., 2002a).

The organic matter content and forms and the unusual mineralogical composition of andic soils result in specific physical and mechanical properties, such as fluffy consistency, high porosity, low bulk density, high capacity for water retention, microaggregation, changes on properties after drying, critical limit of stability under pressure, etc. (Warkentin and Maeda, 1980; Meurisse, 1985). These properties, in turn, give rise to specific mechanisms in the process of erosion. The predominance of short-range ordination minerals in the fine soil fraction is responsible for erosion of Andosols by transport of stable microaggregates. In contrast, in soils with a mineralogy dominated by crystalline clays, runoff generally transports dispersed particles resulting from aggregate breakdown (Egashira et al., 1983; Kubota et al., 1990; Pla, 1992).

Thus, the objectives of this study were to determine: (1) the SOC content of Andosols under evergreen forest vegetation (laurel and heather forest) and (2) SOC losses by water erosion in Andosols after deforestation, and the forms of SOC preferentially affected by these losses.

6.2 MATERIAL AND METHODS

6.2.1 Study Area and Environmental Conditions

Andosols in the mountainous Canary Islands occur at the altitude between 700 and 1500 m above sea level in the area of influence influenced by the trade winds. These are characterized by a thermomediterranean mesophytic subhumid bioclimate (Rivas Martínez et al., 1993), with annual precipitation ranging between 650 and 900 mm, a mean annual temperature of 14 to 16°C and a potential evapotranspiration of 750 to 800 mm yr^{-1}. These conditions lead to a udic soil moisture and thermic soil temperature regimes (Soil Survey Staff, 1999).

The region is characterized by land smoothly sloping down to the sea, with slopes between 15 and 35%, longitudinally dissected by cliffs of variable depth carved in the basaltic pyroclasts and outflows produced by volcanic activity during the Upper Pleistocene. The natural potential vegetation corresponds to laurel and heather forest (*Lauro-Perseeto indicae* and *Myrico-Ericetum arboreae*).

6.2.2 Soil Genesis and Soil Properties

Predominant Andosols in these areas (Melanudands, Fulvudands, and Hapludands) have developed over a complex geologic material constituted by Pliocene-Pleistocene basaltic lava flows and successive deposits of volcanic ash of varied granulometries, generally from the eruptive stage of the same eruptions.

Weathering of these materials in a climate with high soil moisture content occurs rapidly, leading to the formation of allophane-like noncrystalline minerals. The rapid development of vegetation provides the plant residues leading to the formation of thick, dark humus-rich epipedons and Al/Fe humus complexes.

Illuviation of clay-size materials does not occur in these soils, because noncrystalline materials do not disperse under field conditions (Shoji et al., 1993). The presence of a clayey horizon at depth is, therefore, interpreted as an in situ alteration of the crystalline basaltic lava flows at the base of the soils, leading to the formation of halloysite and smectite-type clay minerals.

6.2.3 Field Sampling and Field Plots

6.2.3.1 Field Sampling

In assessment of the SOC contents and forms under forested Andosols, a study area of some 40 km^2 was chosen. Ecological characteristics of the region are discussed below.

Some soil properties of representative profile of allophanic Andosols (Melanudands) of this area are presented in Table 6.1. In general, these soils have low color values in the surface organic horizon and high organic matter contents and base saturation.

Additionally, soil samples were obtained using a 500 × 1000 m grid over the entire surface. A total of 163 sampling points were selected and soil samples were collected from 0 to 30 cm at three randomly distributed sampling points. At each point, soil samples were obtained manually,

Table 6.1 Selected Morphological, Physical, and General Chemical Properties of the Soils under Study

Horizon (depth)	Allophanic Andosols Ultic Melanudands		Methods
	A_a (0 to 65 cm)	B_w (65 to 85 cm)	
Color (moist)	5YR 2/2	7.5 YR 2/2	Munsell Soil Color Chart
Soil structure	Crumb	Massive	Morphology
Water content (g kg^{-1})			
at 33 kPa	631	686	Ceramic plate
at 1500 kPa	307	346	(Klute, 1986)
Bulk density at 33 kPa (Mg m^{-3})	0.60	0.70	Cylinder method
Coarse fragments > 2 mm (g kg^{-1})	61	62	Sieving
Clay (g kg^{-1})	190	472	Na-resin
Silt (g kg^{-1})	714	443	method
Sand (g kg^{-1})	63	39	(Bartoli et al., 1991)
pH H$_2$O	6.0	5.8	Soil:water ratio 1:2.5
pH KCl	4.9	4.9	Soil:KCl ratio 1:2.5
SOC (g kg^{-1})	145	69	Walkley-Black (Nelson and Sommers, 1982)
C/N	21	14	Calculated
Cation exchange capacity (cmol$_c$ kg^{-1})	57	48	Ammonium acetate 1 N pH7 (Bower et al., 1952)
Base saturation (%)	42	17	Ammonium acetate 1 N pH7 (Bower et al., 1952)

mixed, and air-dried at room temperature. One part of the sample was passed through a 2-mm sieve and analyzed, and the other part was used for aggregate stability analysis. Three separate samples were collected for bulk density at each point by inserting tin cylinders into the soil.

6.2.3.2 Field Plots

Runoff plots were established to assess the rates of SOC loss by water erosion in forest allophanic Andosols after deforestation. The study was conducted on three experimental plots established in 1993 on the northern slopes of the island of Tenerife, Canary Islands. Experimental plots were set up 3 years after clear-cutting, because these practices generate important erosive processes in the Adosols. The environmental conditions of the zone are similar to those described earlier and the soils are allophanic Andosols (Ultic Melanudands) represented by the profile described in Table 6.1 and Table 6.2, with a slope of 24%. The soil of the plots was kept bare by removing weeds manually.

The plots were enclosed to facilitate soil and runoff losses, and to assess the movement measurement of nutrients, (dissolved, adsorbed) and particulate organic carbon and other chemicals. Each plot had an area of 200 m^2 (25 × 8 m), and all plots were equipped with collection tank and H-flumes to collect sediments and runoff.

The contents of the tanks were carefully decanted in the field after each significant rainfall event, and the sediments manually collected, weighed, and transported to laboratory. The runoff amount was determined using an ultrasound device located inside the tank, which was previously calibrated, and which continuously recorded the height reached by the liquid in the tank (every 2 mm). After each significant rainfall event, a sample of water runoff was collected using an automatic sampling device. Rainfall was recorded by using two pluviographs with measurements taken recordings every 5 minutes. All instruments that generate digital data (ultrasound device and pluviographs) were connected to a data logger.

Table 6.2 Chemical Characteristics Relating to Andic Properties

	Allophanic Andosols Ultic Melanudands		
Horizon (depth)	A_a (0 to 65 cm)	B_w (65 to 85 cm)	Methods
P-retention (%)	95	96	Blakemore et al. (1981)
Al_o (g kg^{-1})	38	49	Blakemore et al. (1981)
Fe_o (g kg^{-1})	15	18	Blakemore et al. (1981)
Si_o (g kg^{-1})	31	41	Blakemore et al. (1981)
Al_p (g kg^{-1})	17	14	Blakemore et al. (1981)
Fe_p (g kg^{-1})	4	4	Blakemore et al. (1981)
C_p (g kg^{-1})	88	nd	Blakemore et al. (1981)
$Al_o + 1/2 Fe_o$ (g kg^{-1})	45	58	Calculated
Al_p/Al_o	0.45	0.29	Calculated
$(Al_p + Fe_p)/C_p$ (M)	0.10	nd	Calculated
Melanic index	1.34	nd	Honna et al. (1987)
Allophane content (g kg^{-1})	140	190	Parfitt and Wilson (1985); Mizota and van Reeuwijk (1989)

Note: nd: not determined; P-retention: phosphate retention; Al_o, Fe_o, Si_o: acid-oxalate extractable Al, Fe, Si; Al_p, Fe_p, C_p: pyrophosphate extractable Al, Fe, C.

During the nine years under study, rainfall was 660 ± 187 mm yr^{-1} (mean ± SD) and displayed high monthly and interannual variability (minimum: 396 mm yr^{-1}, maximal: 1035 mm yr^{-1}). The highest maximum rainfall intensity (I_{max}) was 240 mm h^{-1} in 15 min, but I_{max} generally ranged between 30 and 85 mm h^{-1} in 3 to 10 minutes. Mean interannual value of the R-USLE factor (Wischmeier and Smith, 1978) was 640 MJ mm ha^{-1} h^{-1}. Thus, rains always had a moderate degree of erosivity, with frequent, high-intensity events and periods of concentration of precipitations.

Soil erodibility was calculated according to the method of Wischmeier et al. (1971) using the following regression equation:

$$\text{K-USLE} = 2.8 \times 10^{-7} M^{1.14} (12 - a) + 4.3 \times 10^{-3} (b - 2) + 3.3 \times 10^{-3} (c - 3) \quad (6.1)$$

(in t ha h ha^{-1} MJ^{-1} mm^{-1}), where M = particle size parameter [(% silt + % very fine sand) × (100 − % clay)], a = % organic matter, b = soil structure code, c = permeability class. K-USLE for these soils is very low, with a mean value of 0.10 ± 0.11 t ha h ha^{-1} MJ^{-1} mm^{-1}. This factor represents the erodibility of the soil expressed as soil loss from a unit plot per unite erosivity index. Given their specific aggregation properties, however, the erodibility of these Andosols is not accurately determined by means of the K-USLE factor (Rodríguez Rodríguez et al., 2002a, b).

Soil surface samples were collected for at 0 to 5 cm depth, at ten randomly distributed sampling points. At each point, soil samples were collected manually, mixed, air-dried at room temperature and analyzed. Three separate samples for bulk density were obtained at each sampling point by the core method.

All the samples were collected and determinations were carried out during 2002, with the exception of SOC at the soil surface which was also studied in 1999, 2000, and 2001, from which the enrichment ratio (ER) presented in Table 6.4 was calculated.

6.2.4 Laboratory Analyses

The morphological description and the physical and chemical description of the horizons of the soil profiles were performed according to the methods described in Table 6.1. Soil samples were taken for each morphological horizon and described in the field, and the analyses were made in triplicate.

6.2.4.1 Soil Samples

Soil samples collected in the study area (0 to 30 cm) and in the plots (0 to 5 cm) were sieved using 2 mm-meshes, and the following properties were determined.

Walkley-Black SOC

The SOC was determined by the wet oxidation of the carbon by the 1 N dichromate Walkley-Black procedure (Nelson and Sommers, 1982). Using this method on 23 representative forest Andosols, SOC values were 98 to 103% of those determined by the dry combustion method using a LECO CSN 1000 Analyzer (unpublished data). The Walkley and Black procedure (with compensation) was thus considered adequate for determining SOC content of the soils under study.

Pyrophosphate Extractable SOC (C_p)

An extraction with 0.1 M sodium pyrophosphate solution selectively dissolves organic matter and associated Al and Fe (Blakemore et al., 1981; van Reeuwijk, 1993). Therefore, the SOC dissolved in the pyrophosphate extract corresponds to the SOC associated (complexed) with the active forms of Al and Fe. The SOC dissolved in the pyrophosphate extract was determined by the wet oxidation method, using diluted dichromate (0.05 N) instead of 1 N dichromate (Horwath and Paul, 1994).

Potassium Sulfate Extractable SOC (C_{ps})

The SOC extracted with 0.5 M potassium sulfate solution (1:5 soil:solution ratio) corresponds to the more labile forms of SOC. The SOC extracted was determined as above (Horwath and Paul, 1994).

Dissolved SOC (C_d)

200 g of 2-mm sieved soil was weighed and the exact amount of water required to saturate the sample was added. The resultant pastes were left overnight and then transferred to Buchner funnels. The filtrate was subsequently analyzed for organic carbon by oxidation by 0.05 N dichromate, as above. The carbon thus extracted was assumed to be DOC.

Acid Oxalate Extractable Iron, Aluminum, and Silicon (Fe_o, Al_o, Si_o)

These fractions were extracted with acid (pH 3) 0.2 M ammonium oxalate solution (Blakemore et al., 1981; van Reeuwijk, 1993). This extraction dissolves all active Al and Fe components as well as associated Si (Mizota and van Reeuwijk, 1989). This includes noncrystalline minerals (allophanes, imogolite, ferrihydrite) and Al/Fe-humus complexes (Parfitt and Henmi, 1982; Nanzyo et al., 1993). Andic soil properties in Andosols showing a high degree of weathering are determined according to $Al_o + 1/2 Fe_o$ (Shoji et al., 1993).

Pyrophosphate Extractable Iron and Aluminum (Fe_p, Al_p)

Pyrophosphate extraction was used to quantify active Al and Fe complexed to organic components. Al_p and Fe_p were measured by extracting soil with 0.1 M sodium pyrophosphate solution (Blakemore et al., 1981; van Reeuwijk, 1993). Pyrophosphate only dissolves humus-Al/Fe. The Al_p/Al_o ratio is used as an index of allophanic or nonallophanic soil properties.

Allophane Content

The allophane content was estimated by a method proposed by Parfitt and Wilson (1985), modified by Mizota and van Reeuwijk (1989), and based on the Al/Si compositional ratio, $(Al_o - Al_p)/Si_o$, and the Si content (Si_o).

Soil Moisture

Soil moisture content was monitored in the plots over the study period. Ten samples each of 20 g were collected twice a month and their moisture content was determined gravimetrically. All samples were analyzed in triplicate and the mean soil moisture contents reported on an oven-dried basis.

6.2.4.2 Sediment Samples

Fourteen rainfall events resulted in significant amounts of sediments between 1999 and 2002. A sample of the sediments collected in each event was air-dried at room temperature and passed through a 2-mm sieve for the following determinations.

Particle-size Analysis

Particle-size distribution was determined by sieving and sedimentation (pipette method) after dispersion and 16-h shaking in Na Amberlite IR-120 resin (Bartoli et al., 1991).

Walkley-Black

Pyrophosphate extractable and potassium sulfate extractable soil organic carbon. These determinations involved the methods that have been described above for soil samples.

All the samples were analyzed in triplicate and average results are reported on an oven-dry basis.

6.2.4.3 Water Runoff Samples

The samples of runoff water were filtered through a 0.45-µm pore size filter and analyzed for DOC by oxidation using a mixture of 0.05 M potassium dichromate and concentrated sulfuric and phosphoric acids, according to the Walkley-Black procedure. All samples were analyzed in triplicate.

6.2.3 Statistical Analysis

The statistical analysis of the results was done using SPSS software (Anon, 1990). Comparisons between pairs of sample groups were done using the U-Mann-Whitney test since the variables studied did not meet previously described criteria of normality and homocedasticity.

6.3 RESULTS

6.3.1 Organic Carbon Content and Forms

Table 6.2 shows the andic characteristics of the soil profile of which morphological, physical and chemical properties are presented in Table 6.1.

Table 6.3 Mean ± Standard Deviation Values of Different Forms of Soil Organic Carbon in Allophanic Andosols (0 to 30 cm)

	SOC		C_p		C_{ps}		C_d		C/N Ratio
	g kg^{-1}	kg m^{-2}	g kg^{-1}	kg m^{-2}	g kg^{-1}	kg m^{-2}	g kg^{-1}	kg m^{-2}	
Allophanic Andosols	149 ± 5	18.8 ± 2.4	67 ± 2	8.4 ± 1.3	4.6 ± 1.7	0.58 ± 0.04	82 ± 2	10.3 ± 2.0	15.1 ± 0.9

Note: SOC: total soil organic carbon; C_p: pyrophosphate extractable soil organic carbon; C_{ps}: potassium sulfate extractable soil organic carbon; C_d: dissolved soil organic carbon.

The existence of $Al_o + 1/2Fe_o$ values higher than 20 g kg^{-1}, a bulk density measured at 33 kPa water retention < 0.90 Mg m^{-3}, and a phosphate retention higher than 85% indicate andic soil properties in the A_a and B_w horizons. Moreover, the A_a horizon has a color value, moist, and chroma of 2, a melanic index less than 1.70 and more than 60 g kg^{-1} of SOC, which characterizes a melanic epipedon (Soil Survey Staff, 1999). The Al_p/Al_o ratio (lower than 0.5) and the high content in allophane in the Bw horizon (190 g kg^{-1}) classifies these soils as allophanic Andosols (Mizota and van Reeuwijk, 1989) (Table 6.2).

Allophanic Andosols under forest have a high capacity to accumulate SOC (149 ± 5 g C kg^{-1} and 18.8 ± 2.4 kg C m^{-2} at 0 to 30 cm). A high proportion of SOC (43 to 45%) was found in pyrophosphate extractable forms (8.4 ± 1.3 kg C m^{-2}), i.e., complexed with the active forms of Al and Fe (Table 6.3).

6.3.2 Organic Carbon Losses by Water Erosion

The runoff generated in these soils (allophanic Andosols) was relatively small, with mean values of 13 ± 9.8% (85.8 ± 54.5 mm yr^{-1}). Thus soil loss due to interrill erosion was relatively small (for bare soils), with a mean value of 9.6 ± 9.2 t ha^{-1} yr^{-1}; significant soil loss (up to 317 kg ha^{-1}) may occur during certain erosive events (Rodríguez Rodríguez et al., 2002b).

Moreover, there was a high interannual variability in rainfall, soil loss and runoff, with no clear relationship between them. However, soil loss was closely related to the antecedent moisture content, i.e., the highest runoff and soil losses were observed from initially dry soil (Rodríguez Rodríguez et al., 2002b). Figure 6.1 shows the significant correlation between the sediment yield for the 14 events considered and topsoil moisture in the days before the corresponding rainfall event. This suggests that aggregate slaking is an important mechanism in particle detachment and soil erosion in these environments (Figure 6.1).

The total SOC content of the sediments ranged between 86 ± 4 and 145 ± 2 g C kg^{-1}, and averaged 116 ± 18.1 g C kg^{-1} (Table 6.4). The erosion-induced losses of SOC bonded to the solid phase ranged between 0.10 and 3.11 g C m^{-2} and averaged 0.86 ± 0.94 g C m^{-2} per individual event, depending on the amount of sediments generated during each event.

The enrichment ratio of SOC in the sediments was always close to one, implying a low selectivity of the erosion process in relation to SOC (Table 6.4). No relationship was observed between the enrichment ratio and the amount of sediments produced, with the concentration of SOC in the sediments, or with the rainfall intensity ($R^2 < 0.01$).

The concentration of DOC in runoff also varied among events, and ranged between 2 ± 0.3 and 44.9 ± 4.3 mg C L^{-1} (Table 6.4). Losses of DOC ranged from 0.8 to 18.0 mg C m^{-2}, and averaged 4.92 ± 5.30 mg C m^{-2} per event (Table 6.4).

The total SOC content in the topsoil (0 to 5 cm) was 120.0 ± 1.5 g C kg^{-1} (2.4 kg C m^{-2}), 73.3 ± 0.7% of which (88.2 ± 0.4 g C kg^{-1}) occurred in pyrophosphate extractable form (i.e., associated with the active forms of Al and Fe), 2.6 ± 0.1% (3.10 ± 0.15 g C kg^{-1}) in potassium sulfate extractable form, 0.1% (188.0 ± 1.5 mg C L^{-1}) dissolved in the soil solution, and the remainder in other nonextractable forms (Table 6.5).

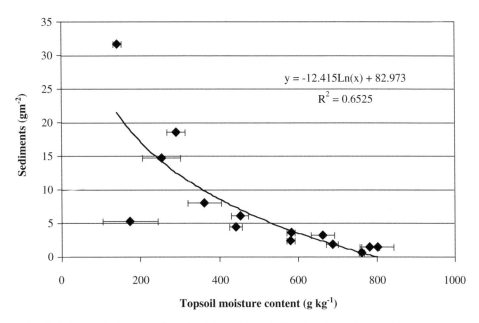

Figure 6.1 Relationship between sediments yield and topsoil (0 to 5 cm) previous moisture.

Table 6.4 Organic Carbon in Sediments and Runoff

Events	P (mm)	Sediments (g m^{-2})	SOC in Sediments, Mean ± SD (g kg^{-1})	SOC in Sediments (g m^{-2})	SOC Enrichment Ratio*	C_d in Runoff, Mean ± SD (mg L^{-1})	C_d in Runoff (mg m^{-2})
1-99	2.2	1.5	86 ± 4	0.13	0.67	4.6 ± 0.7	1.8
2-99	0.1	2.5	115 ± 4	0.29	0.89	18.3 ± 1.0	7.3
3-99	17.0	31.7	98 ± 2	3.11	0.76	38.0 ± 2.3	15.2
1-00	2.5	1.5	114 ± 3	0.17	0.93	11.9 ± 0.9	4.8
2-00	10.0	3.3	101 ± 3	0.33	0.82	2 ± 0.4	0.8
3-00	83.8	4.5	112 ± 3	0.50	0.91	6.2 ± 0.3	2.5
4-00	43.6	3.7	99 ± 4	0.37	0.80	2 ± 0.3	0.8
1-01	73.4	0.7	145 ± 2	0.10	1.18	8.7 ± 1.7	3.5
2-01	3.8	14.8	139 ± 3	2.06	1.13	3.5 ± 0.4	1.4
3-01	13.2	5.4	127 ± 2	0.69	1.03	44.9 ± 4.3	18.0
1-02	9.0	6.1	139 ± 1	0.85	1.16	8.4 ± 0.3	3.4
2-02	12.1	8.0	131 ± 3	1.05	1.09	6.4 ± 0.5	2.6
3-02	13.0	1.9	101 ± 3	0.19	0.84	3.8 ± 0.5	1.5
4-02	8.3	18.6	120 ± 2	2.23	1.00	13.3 ± 0.4	5.3

Note: P: rainfall amount; SOC: total soil organic carbon; C_d: dissolved organic carbon; SD: standard deviation.

* Enrichment ratio was determined by dividing total SOC in sediments and in the topsoil (0 to 5 cm).

The data indicate that in these soils, the largest proportion of total SOC occurs is in the form of relatively stable allophane-humus and metal-humus complexes (Dahlgren et al., 1993; Powers and Schlesinger, 2002). However, the comparison between topsoils and sediments showed that the proportion of SOC in pyrophosphate extractable form was significantly smaller in sediments than in topsoils (46.6 ± 11.7% vs. 73.3 ± 0.7%, Mann-Whitney test, $p < 0.01$). Therefore, SOC in sediments basically occurs in nonpyrophosphate extractable forms. Particle size distribution in sediments (clay = 383 ± 26, silt = 572 ± 29, sand = 47 ± 7) indicates that clay particles are the most easily eroded ones (ER = 2.02). However, the ER values close to 1 for total SOC in sediments suggests that SOC is not selectively associated with any particle fraction in the surface soil.

Table 6.5 Forms of Organic Carbon in Topsoils (0 to 5 cm), Sediments, and Runoff

	In Topsoils (0 to 5 cm), Mean ± SD	In Sediments (n = 14), Mean ± SD	In Runoff (n = 14), Mean ± SD	Enrichment Ratio
SOC (g kg^{-1})	120.0 ± 1.5	116.0 ± 18	nd	0.97
C_p (g kg^{-1})	88.0 ± 0.4	53.0 ± 11	nd	0.60
C_p/SOC (%)	73.3 ± 0.7	46.6 ± 11.7	nd	nd
C_{ps} (g kg^{-1})	3.10 ± 0.15	1.01 ± 0.52	nd	0.33
C_d (mg L^{-1})	188.0 ± 1.5	nd	12.0 ± 13.6	0.06
Rainfall (mm yr^{-1}) (mean ± SD, n = 9 yr)		660 ± 187		
Runoff (mm yr^{-1}) (mean ± SD, n = 9 yr)		85.8 ± 54.5		
Soil losses (kg m^{-2} yr^{-1}) (mean ± SD, n = 9 yr)		0.96 ± 0.91		

Note: nd: not determined; SOC: total soil or sediment organic carbon; C_p: pyrophosphate extractable soil or sediment organic carbon; C_{ps}: potassium sulfate extractable soil or sediment organic carbon; C_d: dissolved organic carbon.

6.4 DISCUSSION

The high potential of Andosols to function as a sink for organic carbon is related in some way to the formation of complexes with the active forms of Fe and Al or with highly reactive minerals such as allophane, imogolite, and ferrihydride (Dahlgren et al., 1993; Parfitt et al., 1997; Lal et al., 1998). Indeed, the Andosols under study had high SOC contents (60 to 280 g C kg^{-1}), only slightly lower than those reported by Mizota and van Reeuwijk (1989) in different Andosols worldwide (80 to 300 g C kg^{-1}). The SOC stocks for 0 to 30 cm depth ranged from 8 to 32 kg C m^{-2}, and were similar to those reported by other authors such as Morisada et al. (2003) in forest Andosols in Japan (10.6 to 13.8 kg C m^{-2}) and Robert (2001) for a wide range of Andosols (11.4 kg C m^{-2}).

The proportion of SOC adsorbed onto the mineral fraction (pyrophosphate extractable SOC, C_p) represented 43 to 45% of total SOC (67 ± 2 g kg^{-1} in allophanic Andosols). This fraction is higher than those reported by other authors, between 17 and 34% (Pinheiro, 1990; Cecchini et al., 2002; Donisa et al., 2003). The large proportion of C_p in these soils suggests a high degree of humic complexation and evolution of the epipedon. Wada (1985) reported that in most Andosols in Japan, C_p content < 50 g kg^{-1}, and C_p content between 50 and 110 g kg^{-1} occur primarily in older and more evolved horizons.

The interrelationship between SOC and erosion is well established and SOC content is a key factor in the resistance of the soil to interrill erosion. Interrill erosion results from particle detachment and transport by raindrop and sheet runoff (Morgan, 1995). Several mechanisms are involved in soil's susceptibility to particle detachment, and are mainly related to the topsoil aggregate stability, and more specifically, to aggregate's resistance to slaking and to mechanical breakdown by raindrop impact (Barthès and Roose, 2002). Although aggregate slaking is an important mechanism in particle detachment and soil erosion in these environments, an erosion mechanism in the form of floating stable, hydrophobic, small-size, and low bulk density andic aggregates, with no apparent relationship to the intensity of rainfall or the amount of runoff, has been proposed for these soils by Rodríguez Rodríguez et al. (2002a, b). The same mechanism occurs in some Andosols of Ecuador (Poulenard et al., 2001) and Costa Rica (Collinet et al., 1998).

In erosion plots, mean annual soil loss was 0.96 kg m^{-2} yr^{-1}. Considering the SOC content of sediments determined for 14 rainfall events, which averaged 116 g C kg^{-1}, the mean annual losses of SOC bound to the solid phase was 111 g C m^{-2} yr^{-1}, i.e., ca. 1 t C ha^{-1} yr^{-1}. These values are consistent with those reported by Jacinthe and Lal (2001) for cultivated soils (30 to 260 g C m^{-2}

yr^{-1}), but much higher than those cited by Lal (1995) for tropical arable and forest lands (30 to 37 g C m^{-2} yr^{-1}).

Mean annual runoff our plots was 85.8 mm yr^{-1}, i.e., 13% of annual rainfall. Considering values of DOC content for 14 rainfall events, losses of DOC were 1.03 g C m^{-2} yr^{-1}. These results are within the limits reported by other authors at the catchment scale, for a wide range of geographical areas (1 to 10 g C m^{-2} yr^{-1}) (Hope et al. 1994). In the present study, the DOC represented 0.9%, of total eroded SOC (dissolved SOC plus SOC in sediments).

Similarly, pyrophosphate extractable SOC represented a significantly smaller proportion of total SOC in sediments (46.6 ± 11.7%) than in topsoils (73.3 ± 0.7%). Thus, a preferential erosion of SOC in the form with a small degree of complexation with the mineral fraction occurs in these soils.

Experimental plots were set up 3 years after clear-cutting. At the beginning of field experiences, soil profile in the plot had a SOC content of 145 g kg^{-1} (Table 6.1), which is equivalent to a 4.3 kg C m^{-2} stock in surface (0 to 5 cm). In adjacent soils under forest, the respective values were 179 g kg^{-1} (5.4 kg C m^{-2} at 0 to 5 cm). As the last measurements in 2002 showed SOC contents equal to 120 g kg^{-1} (3.6 kg C m^{-2} at 0 to 5 cm), that SOC losses in deforested soils during 9 years were equal to 1.8 kg C m^{-2} in the upper 5-cm depth. As 112.03 g C m^{-2} yr^{-1} (i.e., 1.0 kg C m^{-2}) are known to be lost by erosion during the period of study, the rest (0.8 kg C m^{-2}) may have been lost via mineralization after deforestation. Therefore, the removal of forest vegetation implies SOC losses in these soils up to 2 t C ha^{-1} yr^{-1}, of which 55% is due to erosion, and 45% due to mineralization.

6.5 CONCLUSIONS

The Andosols under the study have a high capacity to accumulate carbon (149 ± 5 g C kg^{-1} and 18.8 ± 2.4 kg C m^{-2} at 0 to 30 cm in allophanic Andosols). A large proportion of this organic carbon (43 to 45%) occurs in pyrophosphate extractable forms (8.4 ± 1.3 kg C m^{-2}), i.e., complexed with the active forms of Fe and Al. The amount of dissolved organic carbon is about 10 kg C m^{-2} in these soils (0 to 30 cm).

Water erosion resulted in high losses of organic carbon on bare soil (112 g C m^{-2} yr^{-1}), 99% of lost SOC being bound to the mineral particles, whereas losses of DOC in runoff water were negligible (1 g C m^{-2} yr^{-1}).

Complexed forms of SOC such as allophane-Al-humus and Al-humus complexes predominated in the topsoil (73% of total SOC at 0 to 5 cm), but nonpyrophosphate extractable and noncomplexed forms of SOC were dominant in the sediments (53%). These trends suggest a preferential erosion of SOC forms having a small degree of complexation with the mineral fraction. However, additional research is needed on the susceptibility of the different forms of SOC to erosion and on the dynamics of these forms during transport and after deposition.

ACKNOWLEDGMENTS

This work has been supported by the agreement, "Spanish Contribution to UN Convention to Combat Desertification. I. Network of Experimental Stations for the Monitoring and Evaluation of Erosion and Desertification (RESEL)." (Spanish Ministerio de Medio Ambiente and Universidad de La Laguna) and by the Research Project REN2000-1178GLO: "Methodological design for soil degradation assessment at detailed scale" (Spanish Ministerio de Ciencia y Tecnología). We are indebted to Drs. B. Barthès, C. Feller, and E. Roose for their invaluable comments and suggestions on an earlier version of the manuscript.

REFERENCES

Anon. 1990. SPSS/PC+ V.6.0. Base manual. SPSS Inc., Chicago, Illinois, U.S.

Barthès, B. and E. Roose. 2002. Aggregate stability as an indicator of soil susceptibility to runoff and erosion; validation at several levels. *Catena* 47 (2):133–149.

Bartoli, F., G. Burtin, and A. J. Herbillon. 1991. Disaggregation and clay dispersion of Oxisols: Na-resin, a recommended methodology. *Geoderma* 49:301–317.

Blakemore, L. C., P. L. Searle, and B. K. Daly. 1981. Soil Bureau Laboratory Methods. A: Methods for chemical analysis of soils. Newland Soil Bureau Scientific Report 10 A, CSIRO, Lower Hutt, New Zealand.

Bower, C. A., F. F. Reitemeier, and M. Fireman. 1952. Exchangeable cation analysis of alkaline and saline soils. *J. Soil Sci.* 73:251–261.

Cecchini, G., S. Carnicelli, A. Mirabella, F. Mantelli, and G. Sanesi. 2002. Soil conditions under *Fagus sylvatica* CONECOFOR stand in Central Italy: An integrated assessment through combined solid phase and solution studies, in R. Mosello, B. Petriccione, and A. Marchet, eds., *Long-Term Ecological Research in Italian Forest Ecosystems, J. Limnol.* 61 (Suppl. 1):36–45.

Collinet, J., J. Asseline, F. Jimenez, A. T. Bermudez, and S. Dromard. 1998. Comportements hydrodynamiques et érosifs de sols volcaniques au Costa Rica. Rapport CATIE, Turrialba.

Dahlgren, R., S. Shoji, and M. Nanzyo. 1993. Mineralogical characteristics of volcanic ash soils, in S. Shoji, M. Nanzyo, and R. A. Dahlgren, eds., *Volcanic Ash Soils: Genesis, Properties and Utilization*. Developments in Soil Science 21, Elsevier Science Publishers, Amsterdam, pp. 101–143.

Donisa, C., R. Mocanu, and E. Steinnes. 2003. Distribution of some major and minor elements between fulvic and humic fractions in natural soils. *Geoderma* 111:75–84.

Egashira, K., Y. Kaetsu, and K. Takuma. 1983. Aggregate stability as an index of erodibility of Andosols. *Soil Sci. Plant Nutr.* 29:473–481.

Eswaran, H., E. Van der Berg, P. Reich, and J. Kimble. 1995. Global soil carbon resources, in R. Lal, J. Kimble, E. Levine, and B. A. Stewart, eds., *Soils and Global Change*. CRC Press/Lewis Publishers, Boca Raton, FL, pp. 27–43.

Harden, J. W., J. M. Sharpe, W. J. Parton, D. S. Ojima, T. L. Fries, T. G. Huntington, and S. M. Dabney. 1999. Dynamic replacement and loss of soil carbon on eroding cropland. *Global Biogeochemical Cycles* 13:885–901.

Honna, T., S. Yamamoto, and K. Matsui. 1987. A simple procedure to determine melanic index useful to separation of melanic and fulvic andisols. *Pedologist* 32:69–78.

Hope, D., M. F. Billet, and M. S. Cresser. 1994. A review of the export of carbon in river water: Fluxes and processes. *Environmental Pollution* 84:301–324.

Horwath, W. R. and E. A. Paul. 1994. Microbial biomass, in R. W. Weaver et al., eds., *Methods of Soil Analysis. Part. 2: Microbiological and Biochemical Properties*. SSSA Book Series no. 5, Madison, pp. 753–773.

Jacinthe, P. A. and R. Lal. 2001. A mass balance approach to assess carbon dioxide evolution during erosional events. *Land Degradation and Development* 12:329–339.

Jacinthe, P. A., R. Lal, and J. M. Kimble. 2001. Assessing water erosion impacts on soil carbon pools and fluxes, in R. Lal, J. M. Kimble, R. F. Follett, and B. A. Stewart, eds., *Assessment Methods for Soil Carbon*. CRC Press/Lewis Publishers, Boca Raton, FL, pp. 427–449.

Jastrow, J. D. and R. M. Miller. 1998. Soil aggregate stabilization and carbon sequestration: Feedbacks through organomineral associations, in R. Lal, J. M. Kimble, E. Follett, and R. F. Follett, eds., *Soils Processes and the Carbon Cycle*. CRC Press/Lewis Publishers, Boca Raton, FL, pp. 207–223.

Karlen, D. L. and S. S. Andrews. 2000. The soil quality concept: A tool for evaluating sustainability, in S. Elmholt, B. Stenberg, A. Gronlund, and V. Nuutinen, eds., *Soil Stresses, Quality and Care*. DIAS Report no. 38, Danish Institute of Agricultural Sciences, Tjele, Denmark.

Khamsouk B., G. De Noni, and E. Roose. 2002. New data concerning erosion processes and soil management on Andosols from Ecuador and Martinique. Proceedings of the 12th ISCO Conference, Tsinghua University Press, Beijing, China, II:73–79.

Kimble, J. M., C. L. Ping, M. L. Sumner, and L. P. Wilding. 2000. Andisols, in M. E. Sumner, ed., *Handbook of Soil Science*. CRC Press, Boca Raton, FL, pp. E209–E224.

Kimble, J. M., R. Lal, and M. Mausbach. 2001. Erosion effects on soil organic carbon pool in soils of Iowa, in D. E. Stot, R. H. Mohtar, and G. C. Steinhardt, eds., *Sustaining the Global Farm.* Selected Papers of 10 ISCO, Purdue University. Purdue, pp. 472–475.

Klute, A. 1986. Water retention: Laboratory methods, in A. Klute, ed., *Methods of Soil Analysis. Part 1: Physical and Mineralogical Methods*, Agronomy Monograph 9, ASA, Madison, WI, pp. 635–662.

Kubota. T., A. Ishihara, I. Taniyama, H. Katou, and S. Osozawa. 1990. Erodibility of Andosols in Japan. Nat. Inst. of Agro-Environ. Sci. Ibaraki (Japan) (mimeo).

Laflen, J. M. and E. J. Roose. 1997. Methodologies for assessment of soil degradation due to water erosion, in R. Lal, W. H. Blum, C. Valentin, and B. A. Stewart, eds., *Methods for Assessment of Soil Degradation*, CRC Press/Lewis Publishers, Boca Raton, FL, pp. 31–55.

Lal, R. 1995. Global soil erosion by water and carbon dynamics, in R. Lal, J. Kimble, E. Levine, and B. A. Stewart, eds., *Soils and Global Change*, CRC Press/Lewis Publishers, Boca Raton, FL, pp. 131–142.

Lal, R. 2001a. Myths and facts about soils and the greenhouse effect, in R. Lal, ed., *Soil Carbon Sequestration and the Greenhouse Effect*, SSSA Special Publication no. 57, Soil Science Society of America, Madison, WI, pp. 9–26.

Lal, R. 2001b. Fate of eroded soil organic carbon: Emission or sequestration, in R. Lal, ed., *Soil Carbon Sequestration and the Greenhouse Effect*, SSSA Special Publication no. 57, Soil Science Society of America, Madison, WI, pp. 173–182.

Lal, R., J. M. Kimble, and R. Follett. 1998. Knowledge gaps and researchable priorities, in R. Lal, J. M. Kimble, R. F. Follett, and B. A. Steward, eds., *Soils Processes and the Carbon Cycle*, CRC Press/Lewis Publishers, Boca Raton, FL, pp. 595–604.

Meurisse, R. T. 1985. Properties of Andisols important to forestry. Taxonomy and management of Andisols. Proc. 6th Int. Soil Classif. Workshop. Soc. Chilena de la Ciencia del Suelo, pp. 53–67.

Mizota, C. and L. P. van Reeuwijk. 1989. Clay mineralogy and chemistry of soils formed in volcanic material in diverse climatic regions. Soil Monograph 2, ISRIC, Wageningen.

Morgan, R. P. C. 1995. *Soil Erosion and Conservation.* 2nd Ed., Longman Group Limited, Essex, UK, 198 p.

Morisada, K., O. Kenji, and H. Kanomata. 2003. Organic carbon stock in forest soils in Japan. *Geoderma* 119:21–32.

Nanzyo, M., R. Dahlgren, and S. Shoji. 1993. Chemical characteristics of volcanic ash soils, in S. Shoji, M. Nanzyo, and R. A. Dahlgren, eds., *Volcanic Ash Soils: Genesis, Properties and Utilization*, Developments in Soil Science 21, Elsevier Science Publishers, Amsterdam, pp. 145–187.

Nelson, D. W. and L. E. Sommers. 1982. Total carbon, organic carbon and organic matter, in A. L. Page et al., eds., *Methods of Soil Analysis. Part. 2: Chemical and Microbiological Properties*, Agronomy Monograph no. 9, ASA, Madison, WI, pp. 539–579.

Parfitt, R. L. and T. Henmi. 1982. Comparison of an oxalate extraction method and an infrared spectroscopic method for determining allophane in soils clays. *Soil Sci. Plant Nutr.* 28:183–190.

Parfitt, R. L. and A. D. Wilson. 1985. Estimation of allophane and halloysite in three sequences of volcanic soils, New Zealand, in E. Fernández-Caldas and D. H. Yaalon, eds., *Volcanic Soils. Weathering and Landscape Relationships of Soils on Tephra and Basalt. Catena* Supplement 7, Catena Verlag, Giessen, West Germany, pp. 1–8.

Parfitt, R. L., B. K. G. Theng, J. S. Whitton, and T. G. Shepherd. 1997. Effects of clay minerals and land use on organic matter pools. *Geoderma* 75:1–12.

Pinheiro, J. A. 1990. Etudo dos principais tipos de solos da ilha Terceira (Açores). Tese de Doutoramento (unpublished), Universidade dos Açores. 206 p.

Pla, I. 1992. La erodabilidad de los Andisoles en Latinoamerica. *Suelos Ecuatoriales* 22(1):33–43.

Poulenard, J., P. Podwojewski, J. L. Janeau, and J. Collinet. 2001. Runoff and soil erosion under rainfall simulation of Andisols from the Ecuadorian Paramo: effect of tillage and burning. *Catena* 45:185–207.

Powers, J. S. and W. H. Schlesinger. 2002. Relationships among soil carbon distribution and biophysical factors at nested spatial scales in rain forest of northeastern Costa Rica. *Geoderma* 109:165–190.

Rivas-Martínez, S., W. Wildpret, T. E. Díaz, P. L. Pérez de Paz, M. Del Arco, and O. Rodríguez. 1993. Outline vegetation of Tenerife Island (Canary Islands). *Itinera Geobotánica* 7:5–168.

Robert, M. 2001. Soil carbon sequestration for improved land management. FAO, World Soil Resources Report 96, 57 p.

Rodríguez Rodríguez, A., S. P. Gorrín, J. A. Guerra, C. D. Arbelo, and J. L. Mora. 2002a. Mechanisms of soil erosion in andic soils of the Canary Islands, in J. Juren, ed., *Sustainable Utilization of Global Soils and Water Resources*, Tsinghua University Press, Beijing, vol. I, pp. 342–348.

Rodríguez Rodríguez, A., J. A. Guerra, S. P. Gorrín, C. D. Arbelo, and J. L. Mora. 2002b. Aggregates stability and water erosion in Andosols of the Canary Islands. *Land Degradation and Development* 13:515–523.

Roose, E., B. Khamsouk, A. Lassoudiere, and M. Dorel. 1999. Origine du ruissellement et de l'érosion sur sols bruns à halloysite de Martinique. Premières observations sous bananiers. *Bull. Réseau Erosion* 19:139–147.

Saggar, S., K. R. Tate, C. W. Feltham, C. W. Childs, and A. Parshotam. 1994. Carbon turnover in a range of allophanic soils amended with ^{14}C-labelled glucose. *Soil Biol. Biochem.* 26:1263–1271.

Shoji, S., R. Dahlgren, and M. Nanzyo. 1993. Classification of volcanic ash soils, in S. Shoji, M. Nanzyo, and R. A. Dahlgren, eds., *Volcanic Ash Soils: Genesis, Properties and Utilization*, Developments in Soil Science 21, Elsevier Science Publishers, Amsterdam, pp. 73–100.

Singer, M. J. and S. Ewing, 2000. Soil quality, in M. E. Sumner, ed., *Handbook of Soil Science*, CRC Press, Boca Raton, FL, pp. G271–G298..

Soil Survey Staff. 1999. Soil taxonomy. A basic system of soil classification for making and interpreting soil surveys, 2nd ed., USDA-NRCS-US Gov. Printing Office, Washington, DC.

Starr, G. C., Lal, R., Malone, R., Hothem, D., Owens, L., and Kimble, J. 2000. Modeling soil carbon transported by water erosion processes. *Land Degradation and Development* 11:83–91.

Van Noordwijk, M., C. Cerri, P. L. Woomer, K. Nugroho, and M. Bernoux. 1997. Soil carbon dynamics in the humid tropical forest zone. *Geoderma* 79:187–225.

Van Reeuwijk, L. P. 1993. Procedures for soil analysis, 4th ed., ISRIC Technical Paper no. 9, Wageningen.

Wada, K. 1985. The distinctive properties of Andosols, in B. A. Stewart, ed., *Advances in Soil Science*, vol. 2, Springer-Verlag New York Inc. pp. 173–229.

Walling, D. E. 1994. Measuring sediment yield from river basins, in R. Lal, ed., *Soil Erosion. Research Methods*, 2nd ed. Soil and Water Conservation Soc. and St. Lucie Press, Ankeny IA, pp. 39–80.

Warkentin, B. P. and T. Maeda. 1980. Physical and mechanical characteristics of Andisols, in: B. K. G. Theng, ed., *Soils with Variable Charge*. New Zealand Society of Soil Science, Lower Hutt, New Zealand, pp. 281–299.

Wischmeier, W. H., C. B. Johnson, and B. V. Cross. 1971. A soil erodibility nomograph for farmland and construction sites. *Journal of Soil and Water Conservation* 26 (5):189–193.

Wischmeier, W.H. and D.D. Smith. 1978. Predicting Rainfall Erosion Losses: A Guide to Conservation Planning. Agriculture Handbook N° 537, U.S. Department of Agriculture, Washington, DC.

CHAPTER 7

Soil Carbon Dynamics and Losses by Erosion and Leaching in Banana Cropping Systems with Different Practices (Nitisol, Martinique, West Indies)

Eric Blanchart, Eric Roose, and Bounmanh Khamsouk

CONTENTS

7.1 Introduction 88
7.2 Materials and Methods 88
    7.2.1 Study Site 88
    7.2.2 Experimental Plots..89
    7.2.3 Soil C Stock Measurements 89
    7.2.4 C Losses by Runoff and Erosion 90
    7.2.5 C Losses by Leaching 91
7.3 Results...91
    7.3.1 Soil C Stocks Changes (Equivalent Mass Basis) 91
    7.3.2 Soil C Losses by Erosion and Leaching...91
        7.3.2.1 Soil Losses by Erosion 91
        7.3.2.2 C Losses by Erosion (Coarse and Suspended Sediments).........91
        7.3.2.3 C Losses by Leaching and Runoff 96
        7.3.2.4 Total C Losses 96
7.4 Discussion...96
    7.4.1 Effects of Agricultural Systems on C Losses 96
    7.4.2 Soil C Budgets at the Plot Scale...97
        7.4.2.1 Soil C Outputs...98
        7.4.2.2 Soil C Stock Changes 99
        7.4.2.3 Soil C Inputs...99
        7.4.2.4 C Budgets...99
7.5 Conclusion 99
References 100

7.1 INTRODUCTION

A strong link between increase in concentration of greenhouse gases (GHGs) in the atmosphere and the global temperature warrant: (1) assessment of soil as a potential sink for atmospheric CO_2 (Schlesinger, 1984; Detwiler, 1986; Bouwmann, 1989; Lugo and Brown, 1993), (2) determination of the fate of C translocated by erosion or leaching in C budgets, and (3) calculation of C budgets at different spatial scales (global, regional, local) (Turner et al., 1997; Lal, 2003). Furthermore, the maintenance or increase of soil C stocks deeply affects soil fertility (Feller et al., 1996). A change in soil C stocks at medium- or long-term has often been described as an indicator of agrosystem sustainability and environmental quality. This explains why agricultural practices, which increase soil C stocks, generally result in high crop productivity, decrease erosion, increase biodiversity, and also act as a sink for atmospheric C (Lal et al., 1998).

In Martinique, banana (*Musa spp.*) is the main cropping system. Banana crops occupied an area of 11,800 ha and produced more than 305,000 Mg of fruits in 1999 (Agreste DOM, 1999). Agricultural practices are very intensive and bananas are often cropped on steep slopes (from 10 to 40%), which can cause severe erosion due to intense tropical rains. This crop is also very sensitive to a range of pests or diseases (nematodes, insects, fungi). The traditional practices used to combat diseases and pests are based on the use of massive amounts of pesticides and frequent replanting. Every year an average crop is characterized by two or three applications of nematicides, one or two applications of insecticides, three to five applications of herbicides, four to twelve aerial applications of fungicides, and high doses of fertilizers (Chabrier and Dorel, 1998). The impact of such practices on the environment and human health are dramatic, and widespread incidence of soil and water pollution have been reported (Balland et al., 1998). Practices that improve soil and water quality, and favor agriculture sustainability, have been developed since 1990s. In 1999, the CIRAD (Centre de Coopération Internationale en Recherche Agronomique pour le Développement) set up an experimental site to test the effect of different cropping systems (rotations) on soil erosion, nutrient and pesticide erosion, and leaching (Khamsouk, 2001). Three rotations were tested: (1) a rotation with sugarcane (residues left on the soil surface), (2) a rotation with pineapple cropped on the flat and covered with organic residues, and (3) a rotation with conventional pineapple (intensive-till, residues buried, and ridged). Before the study, all plots were cropped with perennial banana. Runoff plots were set up for each rotation and those were compared to soil losses measured in perennial banana crops and "bare soil" treatments (Khamsouk, 2001; Khamsouk et al., 2002b). These experimental plots were used to measure soil C losses by erosion (in a solid or dissolved form) and leaching (in a dissolved form) and to follow C stock changes over two consecutive years. Input of C (rain, litter, roots) was used to prepare C budgets and to calculate the contribution of erosion and leaching to total soil C loss.

7.2 MATERIALS AND METHODS

7.2.1 Study Site

Martinique is a volcanic island of the West Indies in the Atlantic Ocean (14 to 16° N, 60 to 62° W, 1080 km^2). It has a hilly relief with high young volcanic mountains in the north (maximum 1393 m) and old volcanic mountains in the south (maximum 505 m). The climate is humid tropical with mean annual rainfall ranging from 1200 to 8000 mm, depending on the altitude, and mean annual temperature around 26°C. The climate is characterized by two contrasting seasons: a dry season from January to June and a rainy season marked by tropical storms and hurricanes. This study was established at a CIRAD experimental Station (Rivière-Lézarde) in the central part of the island (rainfall 2000 mm yr^{-1}, 70 m above sea level). Soil developed from volcanic pumices and ashes is described as a Nitisol (FAO classification) (Colmet-Daage and Lagache, 1965). It is acidic,

Table 7.1 Soil Characteristics of the Surface Layer (0 to 10 cm) of the Nitisol

Parameter	Value
Bulk density (Mg m^{-3})	0.77–0.92
pH (water)	4.9–5.7
Sand content (g 100 g^{-1} soil)	16
Silt content (g 100 g^{-1} soil)	16
Clay content (g 100 g^{-1} soil)	68
Organic matter content (g 100 g^{-1} soil)	2.7–3.3
Erodibility index K[a]	0.08–0.1

[a] According to the K index nomograph (Wischmeier et al., 1971).

clayey (mainly halloysite), with high organic matter content, low bulk density, strong resistance to sheet erosion and a low erodibility index K (Table 7.1) (Khamsouk et al., 2002b).

7.2.2 Experimental Plots

The present study comprised of different experimental plots representing: (1) conventional, perennial banana (*Musa paradisiacal*) crops with intensive practices and without rotations, and (2) new banana cropping practices involving rotations with other crops such as sugarcane (*Saccharum officinarum*) or pineapple (*Ananas comosus*) (Khamsouk, 2001). Bare fallow plots were also established to assess soil erodibility (Wischmeier and Smith, 1978). Three slope levels studied were: 10%, 25%, and 40%. A total of ten plots established at the beginning of 1999 comprised the following treatments:

- Three bare soil plots located on three slope levels 10, 25, and 40%, soil was tilled to 20-cm depth and levelled each year.
- Sugarcane, with residues left on the soil surface, was a new rotation proposed in banana crops in order to reduce soil erosion. Three plots were established on three slope levels; sugarcane was planted in 13 horizontal rows spaced 1.5 m apart, with reduced soil tillage and residues left as mulch in the interrows.
- Two perennial banana crops were studied (banana 1 and banana 2); crop residues were left in strips perpendicular to the slope. These two plots had a plant density of 1800 trees ha^{-1} and were located on 10% slope. These crops were tilled for 2 years prior to and during the experiment.
- Intensive-till pineapple (tilled twice with romeplow, residues buried, and ridged) was characterized by intensive practices conventionally used in Martinique. This plot was established on a 10% slope. Previous banana residues were buried. The plant density was 42,500 plants ha^{-1} in seven rows parallel to the slope.
- No-till pineapple was cropped on the flat and soil covered with organic residues, representing a new cropping system designed to reduce soil erosion. This plot was established on 10% slope. Previous banana residues were left at the soil surface in the interrows parallel to the slope.

A runoff plot was established on each of these 10 plots in order to measure soil and C losses by erosion. These plots were 200 m^2 (20 × 10 m), except bare soil runoff plots, which were 20 × 5 m (100 m^2). Plot replication was not implemented in this demonstration trial, as it is usually difficult in long-duration trials (Shang and Tiessen, 2000), especially when these include runoff plots, as was the case in this experiment.

Details concerning erosion, runoff, aggregate stability, water balance and nutrient losses are given by Roose et al. (1999), Khamsouk (2001), and Khamsouk et al. (2002a, b).

7.2.3 Soil C Stock Measurements

Soil C stocks were measured just before the establishment of these experimental plots (at the beginning of 1999) when all plots were cropped with perennial banana crops, and two years after the installation of experimental plots (at the beginning of 2001).

In 1999, eight pits were dug (just outside the location of next runoff plots): four pits in the 10% slope level, two pits in the 25% slope level, and two pits in the 40% slope level. These pits were 1-m deep except one pit in the 10% slope level, which was 2-m deep to measure soil C stock under the root zone. Soil C stocks were measured at different depths: 0 to 10, 10 to 20, 20 to 30, 30 to 40, and 50 to 60 cm depths. For the deepest pit, C stocks were also measured at 70 to 80, 110 to 120, and 150 to 160 cm. For each of these layers, soil bulk density was measured using five replications (1000 cm³ each). Soil was then oven-dried at 105°C and weighed. Three composite samples were taken from each layer for C content analysis. Soil was oven-dried at 60°C and ground to pass through a 200 µm sieve. Soil C content was analyzed using a CNS analyzer (Carlo Erba 1500 NS). Soil C stocks were calculated as follows:

$$\text{Soil C stock (Mg C ha}^{-1}) = \text{C content (g kg}^{-1}) \times \text{bulk density (Mg m}^{-3}) \times d \text{ (layer thickness, m)} \times 10 \qquad (7.1)$$

Soil C stocks were calculated both on a volume basis and on an equivalent mass basis, the latter being more adapted to compare situations with changes in bulk density, which is the case in the present study (Ellert and Bettany, 1995). To compare soil C stocks on an equivalent mass basis, the chosen reference was the mass of soil in the first 30 cm under banana grown on 10% slope gradient in 1999 (i.e., 2205 Mg soil ha⁻¹). In 2001, a pit was dug toward the upstream side of each plot. Soil C stocks were measured the same way as in 1999.

Because C stock changes were detectable in the surface layers, results presented herein are from the 0 to 30 cm depth only.

Soil C stock was calculated from soil C content (Co) and bulk density (BD). Standard deviation of C stocks (SD_S) for each horizon was calculated by the following formula:

$$SD_{SH} = S \times [(SD_{Co}/Co)^2 + (SD_{BD}/BD)^2]^{1/2} \qquad (7.2)$$

where Co is the mean C content, BD is the mean bulk density, and SD_{SH}, SD_{Co}, and SD_{BD} are the standard deviations of C stock of the horizon, C content and bulk density, respectively (Pansu et al., 2001). The standard deviation of C stock for the upper 30 cm was calculated by the following equation:

$$SD_S = [(SD_{SH1})^2 + (SD_{SH2})^2 + (SD_{SH3})^2]^{1/2} \qquad (7.3)$$

where SD_{SH1}, SD_{SH2}, SD_{SH3} are the standard deviations of C stock of the three horizons (0 to 10, 10 to 20, and 20 to 30 cm).

7.2.4 C Losses by Runoff and Erosion

Runoff plots were rectangular, and the surface runoff (with sediments) discharged into calibrated storage tanks (Roose, 1980). Rainfall, runoff, and erosion were measured after each erosive event. Rainfall was characterized by its amount (mm), 30-min maximum rainfall intensity (Ipmax30, mm h⁻¹) and its erosivity (MJ mm (ha h)⁻¹) using an automatic meteorological station (Khamsouk, 2002b). Runoff was defined by the mean annual runoff coefficient Cram (%) which corresponds to the annual runoff water divided by the annual rainfall, and by the maximum runoff coefficient Crmax (%) which is the maximum runoff depth divided by its generating rainfall event. Annual erosion E (Mg ha⁻¹ yr⁻¹) was determined by the total dry weight of whole soil loss, i.e., coarse (particles deposited at the bottom of storage tanks) and suspended sediments, for every erosive event. A part of the coarse and suspended sediments was separately sampled in order to analyze sediment C content for each erosive event. Dissolved C content in runoff water was analyzed using a Shimadzu TOC-5000. Because these measurements were made over two consecutive years with different rainfall amounts, C losses are reported as Mg C ha^{-1} 2 yr^{-1}.

7.2.5 C Losses by Leaching

Lysimeters were installed in perennial banana crops in order to measure C losses by leaching (Khamsouk, 2001). Five cylindrical lysimeters (90 cm diameter) were placed at 80 cm depth under banana trees. Seepage water was collected from storage tanks on a weekly basis. The amounts of seepage water and its dissolved C content were measured. Lysimeters were not installed in other plots. Thus, the dissolved C content for other plots was assumed similar to that of the banana.

7.3 RESULTS

7.3.1 Soil C Stocks Changes (Equivalent Mass Basis)

Soil C stocks in banana crops (slope 10%) were highly variable during 1999, just before the installation of experimental plots. Stocks in the 0 to 30 cm depth ranged from 27.5 to 33.5 Mg C ha^{-1} (mean 30.0 ± 2.9 Mg C ha^{-1}) in the upper 2205 Mg soil ha^{-1} (Table 7.2). Mean C stock was equal to 30.6 Mg C ha^{-1} for 25% slope, and it increased to 34.6 Mg C ha^{-1} for 40% slope.

There were no differences ($P < 0.05$) in C stock among treatments for the same plot gradient after 2 years of experimentation (Table 7.3). Irrespective of slope gradient, soil C stocks under sugarcane were higher than in bare soil treatment (with significant differences in the C content, $P < 0.05$). For 10% slope, there were only slight differences in C stock among plots growing sugarcane, bananas, no-till pineapple and intensive-till pineapple (29.5 in intensive-till pineapple to 33.7 Mg C ha^{-1} in banana plot). These values were higher than mean soil C stock in bare soil treatment (23.8 Mg C ha^{-1}).

Comparing data of 1999 and 2001, soil C stocks decreased in bare soil treatment (by 1.8 to 3.1 Mg C ha^{-1} yr^{-1} depending on slope gradient) and did not change significantly in intensive-till pineapple, no-till pineapple, sugarcane, and banana treatments (Table 7.4). The C stock in sugarcane treatment (40% slope) increased between 1999 and 2001 (by 3.1 Mg C ha^{-1} yr^{-1}).

7.3.2 Soil C Losses by Erosion and Leaching

7.3.2.1 Soil Losses by Erosion

The amounts of runoff and seepage water and of soil eroded during 2 years of the experiment are given by Khamsouk et al. (2002b). Only a small fraction of the rainfall was lost as runoff. Thus, the mean annual runoff coefficient Cram was < 15% (this value was measured in 2000 in the intensive-till pineapple treatment), and ranged between 0 and 4% in mulched plots (sugarcane, no-till pineapple, and banana). Total soil losses (coarse and suspended sediments) for the 2 years of measurements are given in Table 7.5. Soil losses were important in bare soil treatment (between 170 Mg ha^{-1} for 10% slope and 300 Mg ha^{-1} for 40 % slope during the 2 years). Soil losses were also relatively high in the intensive-till pineapple plot but low (< 1 Mg ha^{-1}) in other treatments.

7.3.2.2 C Losses by Erosion (Coarse and Suspended Sediments)

Soil C losses by erosion followed the same trend as soil losses, i.e., losses were more due to coarse than suspended sediments (Table 7.6). For bare soil treatments in which soil losses were mainly due to coarse sediments, C losses in coarse sediments increased with slope gradient from 2.5 to 6.0 Mg C ha^{-1} 2 yr^{-1}. Soil C losses were 580 kg C ha^{-1} 2 yr^{-1} in the intensive-till pineapple

Table 7.2 Bulk Density (Mg m^{-3}), Soil C Content (g C 100 g^{-1} Soil), and Soil C Stock (Mg C ha^{-1}) Measured at Three Depths at the Beginning of the Experiment (1999) for Each Slope Level (Mean and Standard Deviation SD)[a]

Soil Depth (cm)	10% Slope						25% Slope						40% Slope					
	Bulk Density		C Content		C Stock		Bulk Density		C Content		C Stock		Bulk Density		C Content		C Stock	
	Mean	SD	Mean	SD	Mean	SD	Mean	SD	Mean	SD	Mean	SD	Mean	SD	Mean	SD	Mean	SD
0 to 10	0.772	0.036	1.677	0.253	12.95	3.31	0.83	0.054	1.545	0.191	12.82	2.55	0.802	0.047	2.045	0.064	16.40	1.30
10 to 20	0.728	0.053	1.245	0.037	9.06	0.59	0.749	0.032	1.360	0.071	10.19	0.79	0.876	0.039	1.345	0.007	11.78	0.47
20 to 30	0.705	0.061	1.137	0.067	8.02	0.73	0.748	0.096	1.215	0.078	9.09	1.12	0.838	0.058	1.220	0.028	10.22	0.66
Stock (volume basis)					30.03						32.10						38.41	
Stock (mass basis)[b]					30.03	3.44					30.62	2.90					34.61	1.53

[a] For bulk density, n = 20 at the 10% slope level and n = 10 at the other slope levels; for C content, n = 12 at the 10% slope level and n = 6 at the other slope levels.
[b] Reference being the soil mass of 0 to 30 cm layer under banana on 10% slope in 1999.

Table 7.3 Bulk Density (Mg m^{-3}), Soil C Content (g C 100 g^{-1} Soil), and Soil C Stock (Mg C ha^{-1}) Measured at Three Depths at the End of the Experiment (2001) for Each Cropping System (Mean and Standard Deviation SD) (n = 5 for Bulk Density, n = 3 for C Content)

	Sugarcane						Banana 1						Banana 2					
	Bulk Density		C Content		C Stock		Bulk Density		C Content		C Stock		Bulk Density		C Content		C Stock	
Soil Depth (cm)	Mean	SD	Mean	SD	Mean	SD	Mean	SD	Mean	SD	Mean	SD	Mean	SD	Mean	SD	Mean	SD
							10% Slope											
0 to 10	0.800	0.039	1.62	0.18	12.96	2.39	0.750	0.031	1.99	0.13	14.93	1.99	0.803	0.078	1.83	0.36	14.69	5.41
10 to 20	0.822	0.079	1.29	0.10	10.60	1.35	0.738	0.056	1.34	0.12	9.89	1.31	0.809	0.047	1.25	0.03	10.11	0.56
20 to 30	0.861	0.025	1.24	0.09	10.68	1.00	0.716	0.067	1.24	0.07	8.88	0.86	0.783	0.036	1.18	0.05	9.24	0.57
Stock (volume basis)					34.24						33.69						34.05	
Stock (mass basis)[a]					30.79	2.92					33.70	2.53					31.81	5.47

	No-Till Pineapple						Intensive-Till Pineapple						Bare Soil					
	Bulk Density		C Content		C Stock		Bulk Density		C Content		C Stock		Bulk Density		C Content		C Stock	
Soil Depth (cm)	Mean	SD	Mean	SD	Mean	SD	Mean	SD	Mean	SD	Mean	SD	Mean	SD	Mean	SD	Mean	SD
							10% Slope											
0 to 10	0.864	0.094	1.66	0.13	14.34	2.30	0.732	0.051	1.57	0.31	11.49	3.61	0.758	0.084	1.34	0.30	10.16	3.16
10 to 20	0.808	0.047	1.26	0.02	10.18	0.52	0.738	0.093	1.25	0.03	9.23	0.90	0.729	0.046	1.06	0.03	7.73	0.42
20 to 30	0.782	0.066	1.12	0.15	8.76	1.44	0.786	0.038	1.19	0.20	9.35	1.90	0.592	0.050	0.83	0.10	4.91	0.55
Stock (volume basis)					33.28						30.07						22.80	
Stock (mass basis)[a]					30.49	2.76					29.47	4.18					23.84	3.23

Table 7.3 Bulk Density (Mg m^{-3}), Soil C Content (g C 100 g^{-1} Soil), and Soil C Stock (Mg C ha^{-1}) Measured at Three Depths at the End of the Experiment (2001) for Each Cropping System (Mean and Standard Deviation SD) (n = 5 for Bulk Density, n = 3 for C Content) (Continued)

25% Slope

Soil Depth (cm)	Sugarcane						Bare Soil					
	Bulk Density		C Content		C Stock		Bulk Density		C Content		C Stock	
	Mean	SD	Mean	SD	Mean	SD	Mean	SD	Mean	SD	Mean	SD
0 to 10	0.733	0.023	1.92	0.3	14.07	4.23	0.733	0.023	1.85	0.15	13.56	2.06
10 to 20	0.743	0.035	1.30	0.16	9.66	1.58	0.772	0.121	1.11	0.12	8.57	1.46
20 to 30	0.748	0.063	0.97	0.16	7.26	1.25	0.817	0.017	0.69	0.29	5.64	1.64
Stock (volume basis)					30.99						27.77	
Stock (mass basis)[a]					30.80	4.69					26.96	3.01

40% Slope

Soil Depth (cm)	Sugarcane						Bare Soil					
	Bulk Density		C Content		C Stock		Bulk Density		C Content		C Stock	
	Mean	SD	Mean	SD	Mean	SD	Mean	SD	Mean	SD	Mean	SD
0 to 10	0.833	0.029	2.44	0.29	20.33	5.92	0.887	0.032	1.71	0.15	15.17	2.33
10 to 20	0.846	0.047	1.56	0.09	13.20	1.34	0.711	0.039	1.24	0.35	8.82	3.10
20 to 30	0.782	0.026	1.4	0.140	10.95	1.56	0.788	0.026	0.72	0.02	5.67	0.19
Stock (volume basis)					44.47						29.66	
Stock (mass basis)[a]					40.88	6.27					28.35	3.88

[a] Reference being the soil mass of 0 to 30 cm layer under banana on 10% slope in 1999.

Table 7.4 C Stock Changes (on an Equivalent Mass Basis) (in Mg C ha^{-1} yr^{-1}) after 2 Years of Experiment (1999 and 2000)[a]

Slope Gradient (%)	Bare Soil	Sugarcane	Banana 1	Banana 2	No-Till Pineapple	Intensive-Till Pineapple
10	−3.09	+0.38	+1.83	+0.89	+0.23	−0.28
25	−1.83	+0.09				
40	−3.13	+3.13				

[a] None of the changes is significantly ($P < 0.005$) different among treatments.

Table 7.5 Total Soil Losses (Coarse and Suspended Sediments) (in Mg ha^{-1}) during 2 Years of Experiment (1999, 2000) in Different Experimental Plots

Slope Gradient (%)	Bare Soil	Sugarcane	Banana 1	Banana 2	No-Till Pineapple	Intensive-Till Pineapple
10	171.6	0.11	0.8	1.0	0.08	34.3
25	254.9	0.10				
40	294.8	0.22				

Data from Khamsouk, B. 2001. Impact de la culture bananière sur l'environnement. Influence des systèmes de cultures bananières sur l'érosion, le bilan hydrique et les pertes en nutriments sur un sol volcanique en Martinique (cas du sol brun-rouille à halloysite). PhD ENSA-Montpellier, France, pp. 174.

Table 7.6 C Losses (in kg C ha^{-1} 2 yr^{-1}) by Erosion and Leaching during 2 Years of Experiment (1999, 2000) in Different Experimental Plots

Slope Gradients (%)	Bare Soil	Sugarcane	Banana 1	Banana 2	No-Till Pineapple	Intensive-Till Pineapple
Coarse Sediments						
10	2488	2	23	19	2	580
25	4537	2				
40	5984	4				
Suspended Sediments						
10	9.2	1.1	4	2.9	0.8	8.5
25	11.2	1.2				
40	13.1	2.1				
Runoff (Dissolved)						
10	25.4	3.2	18.1	13.2	9	52.4
25	24.2	2.8				
40	19.7	3.3				
Leaching (Dissolved)						
10	53	80	62.5	62.5	79.1	51.9
25	57.3	85.8				
40	59.3	84.1				
Total						
10	2575.6	86.3	107.6	97.6	90.9	692.8
25	4629.7	91.8				
40	6076.1	93.5				

plot and only few kg C ha^{-1} 2 yr^{-1} in other plots. Soil C losses in suspended sediments were low with a high value of 13 kg C ha^{-1} 2 yr^{-1} in bare soil treatment for 40% slope. Finally, dissolved C losses in runoff water were generally low, but were high in the intensive-till pineapple plot (52.4 kg C ha^{-1} 2 yr^{-1}), low under sugarcane (3 kg C ha^{-1} 2 yr^{-1}) and intermediate in other treatments.

7.3.2.3 C Losses by Leaching and Runoff

There was low runoff under sugarcane and no-till pineapple, and most rainfall infiltrated. Thus, soil C losses by leaching were relatively important for these treatments and were about 80 kg C ha^{-1} 2 yr^{-1}. Dissolved C losses by leaching were slightly lower in other treatments (between 50 and 60 kg C ha^{-1} 2 yr^{-1}; Table 7.6). Losses of dissolved C in runoff water were much lower than those by leaching (from 3 in sugarcane to 25 kg C ha^{-1} 2 yr^{-1} for bare soil) except for intensive-till pineapple in which case losses by leaching and runoff were similar.

7.3.2.4 Total C Losses

Total C losses in solid (coarse and suspended sediments) and dissolved (in runoff and leaching) forms were high in bare soil treatments, and they increased with slope gradient from 2.5 to more than 6 Mg C ha^{-1} 2 yr^{-1}. Losses were relatively high in intensive-till pineapple treatment (693 kg C ha^{-1} 2 yr^{-1}) and low in mulched crops with similar losses in sugarcane, no-till pineapple, and banana plots (ca. 90 to 100 kg C ha^{-1} 2 yr^{-1}; Table 7.6).

7.4 DISCUSSION

7.4.1 Effects of Agricultural Systems on C Losses

Aside from bare soils treatments, three soil responses can be discerned concerning C losses by erosion and leaching:

1. Treatments that lost high amounts of C, especially in the form of coarse sediments (84% of total C losses) and in the dissolved form (15%). Intensive-till pineapple in which soil tillage and ridging parallel to the slope increased runoff and erosion.
2. Treatments in which C losses were not important but where coarse sediments and runoff water represented a relatively important part of C losses (21 and 15% of total losses, respectively). Established banana crops in which surface litter limited soil erosion. These results are in agreement with those obtained for banana crops in Burundi, since the authors reported that C and soil losses are proportional to the amount of litter on the soil surface (Rhishirumuhirwa and Roose, 1998). Moreover C losses by erosion and leaching reported herein are similar to those reported by Roose and Godefroy (1977) from Ivory Coast.
3. Treatments in which C losses were small, and coarse sediments and runoff water played a minor role in total C losses (3 and 3 to 10% of total losses, respectively). Most of C losses occurred in dissolved form in seepage water (87 to 92% of total C losses): sugarcane and no-till pineapple for which organic residues located in the inter-rows protected soil surface against runoff and erosion.

These results are in accord with those synthesized by Roose (2002), and show the importance of mulch in controlling soil and C erosion and especially particulate C erosion. In mulched crops (sugarcane, banana, no-till pineapple), C losses by erosion were mainly in dissolved form in seepage water, while in crops without mulch or in bare soil, most of eroded C was made up of particulate (associated with sediments) C (Rodriguez et al., 2002). The amount of dissolved carbon (in seepage and runoff) in mulched crops was 40 kg ha^{-1} yr^{-1} (i.e., 4 g m^{-2} yr^{-1}), which is in the range of the common value of continental export (3 to 10 g m^{-2} yr^{-1}; Moore, 1998).

7.4.2 Soil C Budgets at the Plot Scale

In order to compute the C budgets at the plot scale, several data such as C inputs, C outputs, and C stock changes in the system for a given period are needed (Figure 7.1 and Figure 7.2).

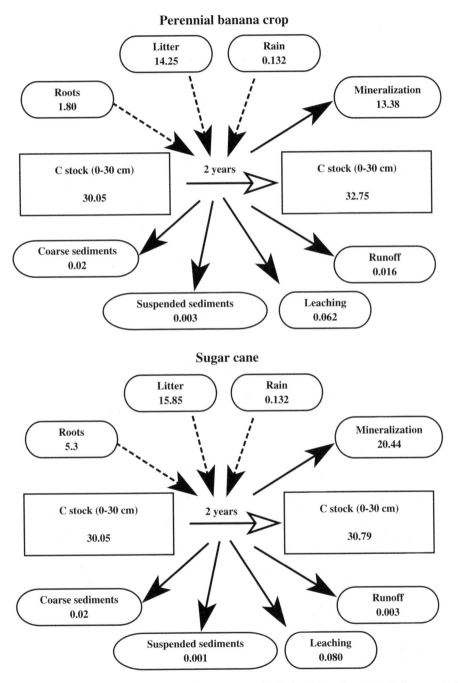

Figure 7.1 C budget in banana and sugarcane treatments (10% slope) showing C stock changes in 2 years (equivalent mass basis) (in Mg C ha^{-1}), C inputs (litter, root, and rain) and C outputs (erosion, leaching, and mineralization) (in Mg C ha^{-1} 2 yr^{-1}).

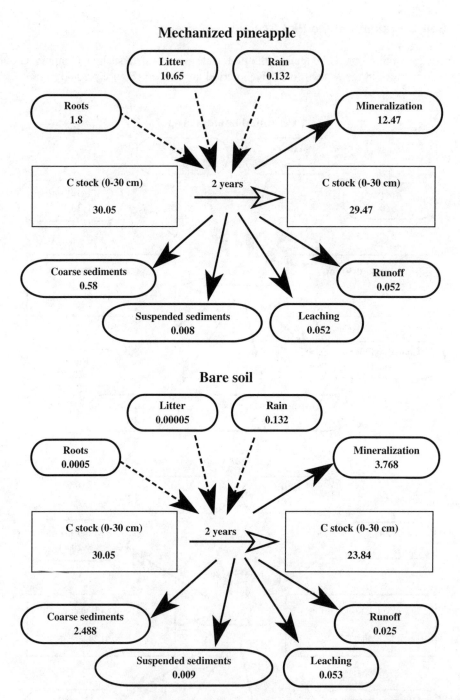

Figure 7.2 C budget in intensive-till pineapple and bare soil treatments (10% slope) showing C stock changes in 2 years (equivalent mass basis) (in Mg C ha^{-1}), C inputs (litter, root, and rain) and C outputs (erosion, leaching, and mineralization) (in Mg C ha^{-1} 2 yr^{-1}).

7.4.2.1 Soil C Outputs

The amounts of solid and dissolved C losses were studied in our experiment; nevertheless, gaseous C losses (i.e., emission of CO_2 and CH_4) were not. These data can be calculated from the other budget components.

7.4.2.2 Soil C Stock Changes

C stock changes during 2 years of experiment were measured, and can be used to calculate C budgets.

7.4.2.3 Soil C Inputs

Aerial C inputs were calculated from the measurement of litter amount at the soil surface (initial inputs from banana crops and inputs during the 2 years of experiment) (Kamsouk, 2001), from literature data, and from litter C content (Marchal and Mallessard, 1979; Lassoudière, 1980). These calculations were based on the assumption that all residues are decomposed within a year, C inputs by litter were estimated to be as high as 14.3, 15.9, 10.7, and 10.7 Mg C ha^{-1} 2 yr^{-1} in banana, sugarcane, no-till pineapple, and intensive-till pineapple crops, respectively. In bare soil, C inputs by weeds were estimated at 50 g C ha^{-1} 2 yr^{-1}.

Nonetheless, root C inputs cannot be precisely estimated since they can be lower or higher than aerial C inputs. In banana, because of a weak root development, they are relatively low and estimated at 1.8 Mg C ha^{-1} 2 yr^{-1} (Godefroy, 1974; Lassoudière, 1980). In pineapple, root C inputs were calculated at 1/6th of aerial C input, i.e., 1.8 Mg C ha^{-1} 2 yr^{-1} (Godefroy, 1974). In sugarcane, root production within a year is equal to about one third that of aerial production (Cerri, 1986; Brouwers, pers. comm.), i.e., 5.3 Mg C ha^{-1} 2 yr^{-1} in our experiment. In bare soil treatments, root C input was estimated to be as high as aerial C input (i.e., 50 g C ha^{-1} 2 yr^{-1}).

Rainfall is another source of C input in ecosystems (Roose, 1980). Based on local measurements and literature data, we used a mean value of 3 ppm, which in our two-year experiment, and considering the total amount of rainfall, gave a C input as high as 0.13 Mg C ha^{-1} 2 yr^{-1}.

7.4.2.4 C Budgets

Knowing C stock changes (increase, decrease, or no change) in 2 years due to C inputs (by litter, roots, and rain) and C outputs (by erosion and leaching), it is possible to build C budgets and to estimate gaseous C losses:

$$\Delta SOC = (SOCa + A) - (E + L + M) \qquad (7.3)$$

where ΔSOC is the change in stock (for a given period), SOCa is the antecedent stock, A is C inputs, E is erosion, L is leaching and M is mineralization (Lal, 2003).

Gaseous C losses reached 12.5 Mg C ha^{-1} 2 yr^{-1} in banana crops, 20.4 Mg C ha^{-1} 2 yr^{-1} in sugarcane, and 12.5 Mg C ha^{-1} 2 yr^{-1} in intensive-till pineapple (Figure 7.1 and Figure 7.2).

Thus, for the cropping systems studied in this pedoclimatic area, C losses were mainly due to mineralization of organic matter (more than 10 Mg C ha^{-1} 2 yr^{-1}) while solid and dissolved C losses were small: ca. 0.1 Mg C ha^{-1} 2 yr^{-1} for sugarcane, banana, and no-till pineapple, and 0.7 Mg C ha^{-1} 2 yr^{-1} for intensive-till pineapple. In bare soil treatments (for which C inputs, C outputs and C stock changes were precisely known), gaseous C losses were as high as 4.8 Mg C ha^{-1} 2 yr^{-1}, and were similar to C losses by erosion and leaching, which ranged from 2.5 to 6 Mg C ha^{-1} 2 yr^{-1} depending on slope gradient (Figure 7.2).

7.5 CONCLUSION

In Martinique, banana cropped on Nitisol are characterized by relatively low soil and C losses by erosion as long as soils are mulched with crop residues. This is the case in banana, sugarcane, and no-till pineapple. Apart from significant gaseous C losses, C losses in mulched crops are mainly due to leaching and runoff in a dissolved form. Rotations with sugarcane or pineapple (either on the flat with residues or intensive-tilled and ridged) induce changes in C losses. For sugarcane and pineapple grown on the flat with crop residue mulch, C losses (in dissolved and solid form) decreased compared with that from banana crops. Conversely, conventional intensive-till pineapple increased C losses (seven times more), because of soil tillage and the lack of crop residue mulch on the surface.

In bare soil, C losses by erosion and leaching were similar to those by mineralization. Losses by gaseous emission are estimated at 2.4 Mg C ha^{-1} yr^{-1} (4.8 Mg C ha^{-1} 2 yr^{-1}), and similar losses, due to the mineralization of soil organic matter, may also occur in cropped treatments.

REFERENCES

Agreste, DOM. 1999. Regard sur l'agriculture dans les Départements d'Outre-Mer. Ministère de l'Agriculture et de la Pêche, pp. 55.

Balland, P., R. Mestres, and M. Fagot. 1998. Rapport sur l'évaluation des risques liés à l'utilisation des produits phytosanitaires en Guadeloupe et en Martinique. Ministère de l'Aménagement du Territoire et de l'Environnement, Ministère de l'Agriculture et de le Pêche, Rapport no. 1998-0054-01, pp. 96.

Bouwmann, A. F. 1989. *Soils and the Greenhouse Effect*. John Wiley & Sons, New York.

Cerri, C. C. 1986. Dinâmica da matéria orgânica do solo no agrossistema cana-de-açucar. Ph.D. thesis, Piracicaba, Brazil, pp. 197.

Chabrier, C. and M. Dorel. 1998. Projet d'étude – Impact des pesticides sur l'environnement: étude de la contamination des eaux de ruissellement. Document CIRAD-FHLOR, Martinique, pp. 7.

Colmet-Daage, F. and P. Lagache. 1965. Caractéristiques de quelques groupes de sols dérivés de roches volcaniques aux Antilles Françaises. *Cahiers ORSTOM, Série Pédologie* 3:91–121.

Detwiler, R. P. 1986. Land use change and the global carbon cycle: the role of tropical soils. *Biogeochemistry* 2:67–93.

Ellert, B. J. and J. R. Bettany. 1995. Calculation of organic matter stored in soils under contrasting management regimes. *Can. J. Soil Sci.* 75:529–538.

Feller, C., A. Albrecht, and D. Tessier. 1996. Aggregation and organic carbon storage in kaolinitic and smectitic soils, in M. R. Carter, ed., *Structure and Organic Matter Storage in Agricultural Soils*. CRC Press, Boca Raton, FL, pp. 309–359.

Godefroy, J. 1974. Evolution de la matière organique du sol sous culture du bananier et de l'ananas. Relations avec la structure et la capacité d'échange cationique. Thèse de docteur-ingénieur, Université de Nancy I, France, pp. 166.

Khamsouk, B. 2001. Impact de la culture bananière sur l'environnement. Influence des systèmes de cultures bananières sur l'érosion, le bilan hydrique et les pertes en nutriments sur un sol volcanique en Martinique (cas du sol brun-rouille à halloysite). Ph.D. ENSA-Montpellier, France, pp. 174.

Khamsouk, B., E. Roose, M. Dorel, and E. Blanchart. 2002a. Effets des systèmes de culture bananière sur la stabilité structurale et l'érosion d'un sol brun-rouille à halloysite en Martinique. *Bulletin du Réseau Erosion* 19: 206–215.

Khamsouk B., G. De Noni, and E. Roose. 2002b. New data concerning erosion processes and soil management on Andosols from Ecuador and Martinique. Proceedings, 12th International Soil Conservation Organization Conference, May 26–31, 2002, Beijing, China Tsinghua University Press.

Lal, R., J. M. Kimble, R. F. Follett, and B. A. Stewart, eds. 1998. *Soil Processes and the Carbon Cycle*. CRC Press, Boca Raton, FL, pp. 609.

Lal, R. 2003. Soil erosion and the global carbon budget. *Environment International* 29:437–450.

Lassoudière, A. 1980. Matière végétale élaborée par le bananier Poyo depuis la plantation jusqu'à la récolte du deuxième cycle. *Fruits* 35:405–446.

Lugo, A. E. and S. Brown. 1993. Management of tropical soils as sinks or sources of atmospheric carbon. *Plant and Soil* 149:27–41.

Marchal, J. and R. Mallessard. 1979. Comparaison des immobilisations minérales de quatre cultivars de bananiers à fruits pour cuisson et de deux "Cavendish." *Fruits* 34:373–392.

Moore, T. R. 1998. Dissolved organic carbon: Sources, sinks, and fluxes and role in the soil carbon cycle, in R. Lal, J. M. Kimble, R. F. Follett, and B. A. Stewart, eds. *Soil Processes and the Carbon Cycle.* CRC Press, Boca Raton, FL, pp. 281–292.

Pansu, M., J. Gautheyrou, and J. Y. Loyer, eds. 2001. *Soil Analysis. Sampling, Instrumentation, Quality Control.* A. A. Balkema Publishers, Lisse, The Netherlands, pp. 489.

Rishirumuhirwa, T. and E. Roose. 1998. The contribution of banana farming systems to sustainable land use in Burundi. *Adv. GeoEcol.* 31: 1197–1204.

Rodriguez, R. A., J. A. Guerra, C. D. Arbelo, and S. P. Gorrin. 2002. Eroded soil organic carbon in Andosols of the Canary Islands. International colloquium Gestion de la biomasse, érosion et séquestration du carbone, Montpellier, 23–28 September 2002. Bulletin du Réseau Erosion 22:114–133.

Roose, E. 1980. Dynamique actuelle des sols ferrallitiques et ferrugineux tropicaux d'Afrique occidentale. Travaux et Documents ORSTOM, 130, pp. 587.

Roose, E. 2002. Perte de carbone par érosion hydrique et indice de sélectivité: Influences du couvert végétal, du sol et du type d'érosion en régions tropicales. International colloquium, Gestion de la biomasse, érosion et séquestration du carbone, Montpellier, 23–28 September 2002. Bulletin du Réseau Erosion 22:74–94.

Roose, E. and J. Godefroy. 1977. Pédogenèse actuelle comparée d'un sol ferrallitique remanié sur schiste sous forêt et sous une bananeraie fertilisée de basse Côte d'Ivoire, Azaguié: 1968 à 1973. *Cahiers ORSTOM, Série Pédologie* 15: 409–436.

Roose, E., B. Khamsouk, A. Lassoudière, and M. Dorel. 1999. Origine du ruissellement et de l'érosion sur sols bruns à halloysite de Martinique. Premières observations sous bananiers. *Bulletin du Réseau Erosion* 19: 139–147.

Schlesinger, W. H. 1984. Soil organic matter: a source of atmospheric CO_2, in G. M. Woodwell, ed., *The Role of Terrestrial Vegetation in the Global Carbon Cycle: Measurement by Remote Sensing.* John Wiley & Sons, London, pp. 11–27.

Shang, C. and H. Tiessen. 2000. Carbon turnover and carbon-13 natural abundance in organo-mineral fractions of a tropical dry forest soil under cultivation. *Soil Sci. Soc. Am. J.* 64: 2149–2155.

Turner, D. P., J. K. Winjum, T. P. Kolchugina, and M. A. Cairns. 1997. Accounting for biological and anthropogenic factors in national land-base carbon budgets. *Ambio* 26: 220–226.

Wischmeier, W. H., C. B. Johnson, and B. V. Cross. 1971. A soil erodibility monograph for farmland and construction sites. *Journal of Soil and Water Conservation* 26:189–192.

Wischmeier, W. H. and D. D Smith. 1978. *Predicting rainfall erosion losses — A guide to conservation planning.* U.S. Department of Agriculture, Agriculture Handbook no. 282.

CHAPTER 8

Influence of Land Use, Soils, and Cultural Practices on Erosion, Eroded Carbon, and Soil Carbon Stocks at the Plot Scale in the Mediterranean Mountains of Northern Algeria

Boutkhil Morsli, Mohamed Mazour, Mourad Arabi, Nadjia Mededjel, and Eric Roose

CONTENTS

8.1 Introduction	104
8.2 Materials and Methods	105
8.2.1 Description of the Sites	105
8.2.2 Description of the Treatments	106
8.2.3 Measurement of Runoff and Erosion	106
8.2.4 Measurement of Eroded and Soil Organic Carbon	107
8.3 Results	107
8.3.1 Rainfalls	107
8.3.2 Runoff	108
8.3.3 Sheet Erosion	108
8.3.4 Organic Carbon Content and Stock of Surface Soil	110
8.3.5 Eroded Organic Carbon	111
8.3.6 Seasonal Variations in Runoff, Erosion, and Eroded Organic Carbon	113
8.4 Discussion	116
8.4.1 Rainfalls	116
8.4.2 Runoff	116
8.4.3 Sheet Erosion	117
8.4.4 Soil Organic Carbon	118
8.4.5 Eroded Organic Carbon and Erosion Selectivity for Carbon (CER)	118
8.5 Conclusions	119
References	120

8.1 INTRODUCTION

Following successive colonizations and the rapid increase of the population in northern Algeria, large forested areas of the mountains were assigned to grazing or cropping. Therefore, various erosion processes have been observed for centuries on the mountains around the Mediterranean basin, from the Roman author Tite-Live to the recent geographers, soil scientists, hydrologists, and agronomists (in Tunisia, Cormary and Masson, 1964; Dumas, 1965; in Morocco, Heusch, 1970; Laouina, 1992; in Algeria, Roose et al., 1993; Kouri et al., 1997). In northern Algeria, mountains are a socioeconomic stake under demographic pressure (Benchetrit, 1972). With the failure of industry, the forests have been overgrazed, cleared, and cropped even on steep hillslopes, resulting in soil degradation and leading to the continuous decrease in cereal production, particularly during the dry years of the 1990s.

Presently, 6 million hectares are affected by an excessive rate of erosion and 120 million tons of sediments are exported yearly by the rivers (Heddadj, 1997). About 20 million m^3 of water are replaced each year by sediments in the reservoirs (Remini, 2000). Finally, the population is affected by soil fertility degradation and the scouring of organic topsoil, by gullies on hillslopes and by floods in the plains, and mudflows and landslides occurring both in the countryside and cities. The rainstorm of the November 10, 2001 amounted to 200 mm in 26 hours with an intensity of 75 mm h^{-1} during 2 hours: more than 850 people died and 5500 houses were destroyed, and mud flows covered many km of drains, streets, and roads around Algiers (newspaper *Liberté*, November 11, 2002). In the semiarid mountains of northern Algeria, even if erosion indicators do not always reach spectacular values, they are linked with the decrease in cereal production, the reduction in soil organic carbon (SOC) in cropped fields, and with the selective depletion of nutrients. The damages are very severe when tilled fields are located on steep hillslopes, as the crop residues are grazed by the herds of goats and sheep, and the fields remain bare after harvest (between May and July) until the next cropping period (between September and January). Cropped vegetation limits erosion only from the spring. In summer the main crops are already harvested, leaving the soil bare and sealed.

SOC plays an important role in soil fertility. It has a major impact on the water budget under perennial crops like vineyards (Pla Sentis, 2002). Moreover, SOC stock can be strongly altered by change in land use (Batjes, 1996). Indeed, after forest clearing and a few years of cropping, the SOC stock decreases by more than 50% (Roose and Barthès, 2001). Lal et al. (2004) observed that SOC loss is caused by plowing, which turns the soil over, making it susceptible to accelerated erosion. Leaving crop residues on the soil surface after harvest increases SOC and controls erosion, but the benefits are lost if the biomass is buried because microorganisms quickly degrade residue carbon into CO_2 (Lal et al., 2004). Additionally, soils constitute a great reservoir of OC (1500 to 2000 Gt C) and are an important sink to control CO_2 fluxes at the global level (Doran et al., 1996; Lal et al., 1998; Lal, 2004; Bernoux et al., this volume; Robert, this volume). Therefore, soil restoration and conservation techniques must aim at increasing SOC stocks to improve soil fertility and mitigate the greenhouse effect (Hien, 2002; Lal, 2002).

The soils of Maghreb were subjected to a rapid land use change during 1990s, especially those in Algeria (Coelho et al., 2002). Considering the effects of these changes in the Tell mountains of northern Algeria, a research program was developed by the Algerian INRF (Institut National de la Recherche Forestière) and the French IRD (Institut de Recherche pour le Développement), to study the influence of land uses and cultural practices on runoff, erosion, soil fertility, and SOC dynamics at the scale of runoff plots (100 to 220 m^2). The study included comparisons between traditional and improved land management systems for the principal soils of northern Algeria. The data on runoff and erosion have already been published (Roose et al., 1993; Morsli et al., 2004). This report summarizes these results and focuses on eroded carbon, carbon enrichment ratio of sediments (CER), and SOC storage in the topsoil, which is the most affected by land-use changes and cultural practices. While the research on carbon erosion is scanty in the tropics (Lal, 2002), it is rare in the

Mediterranean region (Arabi and Roose, 2002; Roose and Barthès, this volume). In general, sheet erosion is not as important as gully erosion on the steep hillslopes (Heusch, 1970; Laouina et al., 2000; Roose et al., 2000). Therefore, the carbon enrichment ratio (CER) is of limited value.

8.2 MATERIALS AND METHODS

8.2.1 Description of the Sites

Field experiments were conducted: (1) from 1993 to 1998 in the Beni-Chougran mountains near Mascara, in western Algeria, 200 km away from Algiers (35°20'N, 00°17'E); (2) from 1991 to 2001 in the Tlemcen mountains in western Algeria, 300 km away from Algiers (34°50'N, 01°10'W); and (3) from 1988 to 1992 around Medea in central Algeria, 90 km away from Algiers (36°14'N, 02°51'E). These hilly regions have been severely degraded by a long history of grazing, burning, cropping, and colonization (by Romans, Arabs, Turkish, French; Gsell, 1913). These regions are representative of the Tell mountains with regards to landscape, erosion manifestations (sheet erosion, gullies, floods, and mass movements), and the various programs of soil conservation since 1950s.

The climate is Mediterranean semiarid with cool winter and annual rainfall ranging from 280 to 620 mm, falling mainly during the cool season (October to May), with some short but intensive rainstorms (intensity up to 80 mm h^{-1} during 30 minutes) during the very hot and dry summer (June to September). Every 10 years an exceptional rainstorm (100 to 200 mm within 3 days) may saturate the landscape and cause severe damages as seen in Algiers on November 10, 2001.

The landscape is mountainous, strongly dissected, with a very dense drainage, convex hillslopes with grazed or cultivated fields covering the whole hills even on the steepest slopes (15 to 40%). Soft rocks (marl, argillite, soft calcareous sandstone) alternate with hard calcareous rocks developing steep slopes and diverse erosion processes. Natural vegetation is mainly constituted by overgrazed scrub (*Olea sp.*, *Quercus sp.*, *Pinus alepensis*, and various bushes), which cover the soil surface partially, by some *Pinus sp.* or *Eucalyptus globulus* plantations, and by overgrazed bush lands. The main crops are cereals (winter or spring wheat, *Triticum durum* and *Triticum aestivum*, respectively and oats, *Avena sativa*), legumes (peas, *Pisum sativum*, chickpeas, *Cicer arietum*), onions (*Allium cepa*), vegetables and orchards of almond trees (*Prunus dulcis*), olive trees (*Olea europaea*), and fig trees (*Ficus carica*), and some contour lines of prickly pears (*Opuntia ficus indica*). If natural conditions played a major role on the development of erosion processes, deforestation, overgrazing, and extensive cultural practices have also accelerated soil degradation.

The plots under study were set up on three soil types representative of the northern mountains of Algeria: (1) clayey brown Vertic soils on marl (vertic Haploxeroll in Mascara and Tlemcen, typic Haploxerert in Medea), hereafter called Vertic soils; (2) brown calcareous soils on sandstone or limestone (typic Haploxeroll in Mascara, Tlemcen and Medea), hereafter called brown calcareous soil; (3) red Fersiallitic soils on sandstone (typic Haploxerept in Tlemcen and Medea), hereafter Fersiallitic soils.

Ten runoff plots were established near Mascara: five on Vertic soils and five on brown calcareous soils. Nine runoff plots were observed near Tlemcen: three on Vertic soils, three on brown calcareous soils, and three on Fersiallitic soils. Twelve runoff plots were observed near Medea: three on Vertic soils, six on brown calcareous soils, and three on Fersiallitic soils.

Some properties of the soils are presented in Table 8.1. These soils are rich in free calcium and magnesium carbonates (10 to 30%) except the Fersiallitic soils. SOC content is generally low (10 to 12 g C kg^{-1}) and decreases with depth. The C/N ratio (10 to 12) indicates that SOC decomposition is fast. Even in the topsoil, nitrogen and available phosphorus are deficient for cereals. The soil pH is neutral to slightly basic. Exchangeable sodium is not negligible in some fragile soils, but the presence of calcium carbonate and of stones in the topsoil increases the soil resistance to rain and runoff erosivity (Roose et al., 1993).

Table 8.1 General Properties of the Soils under Study (Northern Algeria)

	Brown Calcareous (Mascara)		Vertic (Tlemcen)		Brown Calcareous (Tlemcen)		Fersiallitic (Tlemcen)
Depth (cm)	0–15	15–45	0–15	15–45	0–10	10–30	0–10
Carbonates (%)	25.2	32.5	19.6	24.5	10.8	15.4	3.2
Clay (%)	17.2	16.1	57.1	57.2	20.2	28.3	37.3
Silt (%)	56.2	60.4	32.6	33.1	56.2	49.2	20.5
Sand (%)	25.8	23.3	10.1	9.1	22.1	21.5	41.4
Bulk density	1.5	1.6	1.3	1.5	1.1	1.2	1.5
Organic carbon (g kg^{-1})	10.4	7.5	11.6	10.8	26.7	18.6	10.4
Total Nitrogen (g kg^{-1})	0.9	0.7	1.1	1.0	2.4	1.6	0.9
C/N	11.5	10.7	10.5	10.8	11.1	11.6	11.6
P_2O_5 Olsen (mg kg^{-1})	10	4	13	6	nd	nd	nd
pH in water	7.5	7.6	8.2	8.1	7.2	7.5	7.0
Exch. Ca (cmol(+) kg^{-1})	20.3	19.2	28.4	26.4	21.3	23.8	21.3
Exch. Mg (cmol(+) kg^{-1})	2.8	2.7	10.6	9.4	0.9	1.2	1.6
Exch. K (cmol(+) kg^{-1})	0.9	0.8	1.1	1.0	1.1	1.0	1.2
Exch. Na (cmol(+) kg^{-1})	0.3	0.3	0.2	0.2	0.3	0.6	0.3
CEC (cmol(+) kg^{-1})	24.9	23.8	40.2	38.2	24.9	27.3	27.2

Note: CEC: cation exchange capacity. nd: not determined.

8.2.2 Description of the Treatments

The experimental treatments are representative of the most frequent land uses observed in this region, including traditional practices and some possible improved practices: plowing on the contour and ridging to improve the water storage and soil roughness, fertilizer use to enhance vegetation growth, vegetated fallow protection with the introduction of bushes and trees in the grazing lands, and legume and cereal rotations with recommended fertilizer use under vineyard and orchard. The land uses under study were:

- International standard bare plowed fallow (Wischmeier and Smith, 1978)
- Cereals (winter wheat, oats) alone or in rotation with a legume (broad beans, *Vicia faba*) or with grazed fallow, with up and down or contour tillage, with low or appropriate fertilizer rate
- Peas or chickpeas on ridges
- Grazed fallow (not tilled)
- Protected fallow (neither grazed nor burned)
- Protected fallow with the introduction of legumes
- Grazed scrub land
- Grazed scrub land but protected against fire
- Vineyard (traditional, i.e., bare soil and up and down tillage, or improved, i.e., with cereal-legume association and contour tillage)
- Orchards of apricot (*Prunus armeniaca*; traditional, i.e., bare soil and up and down tillage, or improved, i.e., with cereal-legume association and contour tillage)

Plots were not replicated for specific land use and soil type, but were replicated through the measurement of runoff and erosion over several years.

8.2.3 Measurement of Runoff and Erosion

Runoff and erosion were measured on standard runoff plots 22.2-m long and 4.5- to 10-m wide, set up on representative hillslopes with 12 to 40% slope gradient. The plots were surrounded by half-buried sheets and fitted out with a collector channel trapping the coarse sediments and draining runoff and fine sediments in suspension toward two tanks connected in series. When the first thank (0.2 m^3) was full, additional runoff flowed through a divisor (including three to nine slots) into a second tank having a 0.2- to 1-m^3 capacity, depending on the expected runoff volume (Fournier, 1967).

Runoff and erosion were measured from 1993 to 1998 in Mascara, from 1991 to 2001 in Tlemcen, and from 1988 to 1992 in Medea. Rainfall amount was determined using recording gauge rain. Runoff was estimated after each rainfall event by taking into account the volume of each tank and the divisors. It was expressed as annual runoff rate (in % of annual rainfall) and maximum event runoff rate over the period under study (denoted maximum runoff rate, in % of the event rainfall). Erosion was calculated as the sum of coarse sediments trapped in the channel collector and fine particles in suspension sampled from the runoff stored in the first tank. Coarse sediments were collected from the channel, dried, and weighted. Fine suspensions were determined from aliquots collected after homogenization of the runoff in the first tank (without coarse sediments), flocculated using 1 cm^3 of an aluminum sulfate solution (5%), dried and weighted. Erosion was expressed on an annual basis (Mg ha^{-1} yr^{-1}). Accuracy of runoff and erosion measurements was estimated at 10%.

8.2.4 Measurement of Eroded and Soil Organic Carbon

Eroded organic carbon (OC) is the sum of OC in the coarse sediments trapped in the channel and in the fine sediments in suspension in the runoff. Eroded OC (kg C ha^{-1} yr^{-1}) was calculated as the product of sediment OC content (g C kg^{-1}, i.e., kg C Mg^{-1}) multiplied by sediment amount (i.e., erosion in Mg ha^{-1} yr^{-1}). Sediment OC content was determined on sediment samples collected after each rainstorm during the seasons 1995 and 1996 in Mascara and 2000 and 2001 in Tlemcen mountains. In Medea, sediment OC content was determined on samples collected from 1988 to 1992.

Soil samples (0 to 10 cm depth) were collected during the dry season (September). Samples for the determination of bulk density were obtained using 250-ml cylinders, with three replicates per plot. Samples for OC determination were composited from eight original core samples per plot. Soil and sediment OC contents (g C kg^{-1}) were determined following the Anne method (Nelson and Sommers, 1996). The SOC stock (Mg C ha^{-1}) at 0 to 10 cm depth was calculated as the product of SOC content multiplied by bulk density. Soluble carbon in runoff water was not determined. The carbon enrichment ratio (CER) was computed as the ratio of OC content in sediments to that in the topsoil (0 to 10 cm depth).

8.3 RESULTS

8.3.1 Rainfalls

During the measurement period, the maximum daily rainfall ranged from 43 to 45 mm in Mascara and Tlemcen to 85 mm in Medea (Table 8.2). The annual rainfall was 470 mm in Mascara in 1995 and 1996 and 422 mm in Tlemcen in 2000 and 2001 (when eroded OC was determined).

Table 8.2 Annual Rainfall (1913 to 1970, 1971 to 2001, and 1991 to 2001) and Maximum Daily Rainfall (1991 to 2001) in Mascara, Tlemcen, and Medea (Northern Algeria)

Location	Annual Rainfall for the Period 1913 to 1970 mm yr^{-1}	Annual Rainfall for the Period 1971 to 2001 mm yr^{-1}	Annual Rainfall for the Period 1991 to 2001 mm yr^{-1}	Maximum Daily Rainfall for the Period 1991 to 2001 mm d^{-1}
Mascara	511 (143)	380 (102)	289 (85)	43
Tlemcen	496 (100)	347 (110)	331 (100)	45
Medea	618 (180)	510 (170)	461 (167)	85

Note: Figures in parentheses are standard deviations.

In Medea, the annual rainfall ranged from 408 to 621 mm during the 1988 through 1991 period, with an average of 533 mm yr^{-1}. Annual rainfall ranged from 240 to 540 mm in the Tlemcen region during the 1991 through 2001 period. This is 33 to 40% less than the long-term average for 1913 through 1971 (Morsli et al., 2004). The minimum rainfall amount producing runoff ranged from 22 mm d^{-1} in dry conditions (5 days without rain) to 2 to 4 mm d^{-1} in humid conditions or on compacted and sealed soil.

8.3.2 Runoff

The mean annual runoff rates were moderate during the dry years under study, with very few abundant storms: mean annual runoff rate ranged from 1 to 7% in Mascara, 2 to 13% in Tlemcen, and 0.4 to 19% in Medea (Table 8.3). On bare soil, it tended to be higher in Medea (10 to 20%) than in Mascara and Tlemcen (4 to 13%), probably due to more annual rainfall in Medea (540 mm yr^{-1} on average vs. 300 to 400 mm yr^{-1}). There was no distinct effect of soil type: for bare plots, the highest annual runoff rate was observed on the brown calcareous soil in Mascara, on the Fersiallitic soil in Tlemcen, and on the Vertic soil in Medea. However, runoff rate was higher on Fersiallitic than on brown calcareous soils. The effect of slope gradient was also not well defined: on bare soil, annual runoff rate increased with an increase in slope gradient on brown calcareous and Fersiallitic soils, but with a decrease in slope gradient on Vertic soils. Mean annual runoff rate was generally higher on bare soil than on crops, fallows, and scrub. Excluding bare soils, differences in annual runoff rate between treatments were generally small for a given location and soil type in Mascara (nevertheless annual runoff was 25 to 45% more on grazed than protected fallows) and Tlemcen (e.g., mean annual runoff rate on overgrazed and protected scrub was 2.4 and 2.2% on the brown calcareous soil and 11 and 11.2% on the Fersiallitic soil, respectively). In contrast, there was a distinct effect of treatments in Medea, annual runoff being smaller for improved than for traditional practices.

Maximum runoff rate per event (which determines the linear erosion risk) ranged from 10 to 30% in Mascara, 20 to 40% in Tlemcen, and 2 to 80% in Medea. The effects of soil type and slope gradient were slight in general, as maximum runoff rate on bare soil ranged between 32 and 42% (except for one plot in Medea). For a given location and soil type, maximum runoff rate was always higher on bare than on cropped soils, fallows, and scrub. Excluding bare plots, differences among treatments were generally small for a given location and soil type in Mascara and Tlemcen (however maximum runoff rate was twice lower for crops than for fallows on the brown calcareous soil in Mascara). In contrast, maximum runoff rate was particularly low for improved practices in Medea ($\leq 5\%$), probably due to a better soil surface cover. Indeed, runoff risks are significantly reduced when the soil surface is covered by vegetation, mulch, or pebbles (Blavet et al., 2004).

8.3.3 Sheet Erosion

Mean annual erosion was moderate: it was less than 14 Mg ha^{-1} yr^{-1}, and less than 6 Mg ha^{-1} yr^{-1} for 30 out of the 31 plots under study (Table 8.3). Erosion ranged from 0.5 to 5.9 Mg ha^{-1} yr^{-1} in Mascara, 0.4 to 3.4 Mg ha^{-1} yr^{-1} in Tlemcen, and 0 to 14 Mg ha^{-1} yr^{-1} in Medea (0 to 3 Mg ha^{-1} yr^{-1} when excluding one plot). Thus it tended to be more in Mascara than in Tlemcen and Medea, and this was confirmed by soil losses measured on bare plots: indeed, mean annual erosion on bare soils in Mascara, Tlemcen, and Medea was 4.0, 2.0, and 2.2 Mg ha^{-1} yr^{-1} on Vertic soils, and 5.9, 3.6, and 2.8 to 3.0 Mg ha^{-1} yr^{-1} on brown calcareous soils, respectively. For a given location, erosion on bare plots was always smaller on Vertic soils than on Fersiallitic (–41 to –84%) and brown calcareous soils (–21 to –44%) whereas differences between brown calcareous and Fersiallitic soils were less clear. The effect of slope gradient was not well defined: on bare plots, when the slope gradient increased, erosion increased on Vertic and Fersiallitic soils but tended to decrease on brown calcareous soils. The highest soil loss was measured on bare soil with 0% stone cover, but the effect

Table 8.3 Slope Gradient, Stone Cover, Rainfall, Runoff Rates, and Erosion on Runoff Plots Located near Mascara (1993 to 1998), Tlemcen (1991 to 2001), and Medea (1988 to 1992)

Location, Soil Type, and Treatment	Slope %	Stone Cover %	Annual Rainfall mm yr^{-1}	Rainfall Erosivity U.S. units[a]	Altitude m	Annual Runoff Rate %	Maximum Runoff Rate %	Erosion Mg ha^{-1} yr^{-1}
Mascara, Vertic soil	40	5	306 (97)	46 (15)	670			
Bare soil						3.8 (1.8)	32.6	4.0 (2.3)
Cereals						1.4 (0.9)	22.2	0.9 (0.3)
Chickpeas on ridges						2.1 (0.9)	23.5	1.4 (0.5)
Grazed fallow						2.1 (0.6)	22.1	1.0 (0.3)
Protected fallow						1.7 (0.9)	22.3	0.6 (0.1)
Mascara, brown calc. soil	20	8	306 (87)	46 (15)	640			
Bare soil						6.5 (0.8)	32.3	5.9 (2.7)
Cereals						2.2 (0.7)	13.1	0.7 (0.2)
Peas on ridges						2.0 (0.5)	11.0	0.5 (0.3)
Grazed fallow						3.9 (0.6)	25.2	1.0 (0.5)
Protected fallow						2.7 (1.0)	25.7	0.6 (0.5)
Tlemcen, Vertic soil	15	7	330 (97)	58 (11)	520			
Bare soil						6.2 (2.2)	38.6	2.0 (1.8)
Cer./fall., downslope till.						5.6 (1.8)	30.0	1.4 (1.2)
Fertil. cer., contour till.						4.7 (2.3)	24.0	1.0 (0.9)
Tlemcen, brown calc. soil	21	42	387 (77)	58 (11)	730			
Bare soil						3.9 (1.3)	42.1	3.6 (1.4)
Overgrazed scrub						2.4 (1.6)	19.4	0.5 (03)
Protected scrub						2.2 (1.4)	20.0	0.4 (0.3)
Tlemcen, Fersiallitic soil	10	42	411 (76)	63 (11)	980			
Bare soil						12.7 (2.1)	38.0	3.4 (1.4)
Overgrazed scrub						11.0 (3.2)	38.0	1.9 (0.6)
Protected scrub						11.2 (3.4)	30.0	1.4 (0.5)
Medea, Vertic soil	12	4	540 (79)	46 (6)	900			
Bare soil						19.4 (6.8)	80.0	2.2 (2.1)
Cereals/grazed fallow						7.8 (8.0)	14.0	0.4 (0.3)
Fertil. cer.-leg. assoc.						1.2 (1.1)	5.0	0.0 (0.1)
Medea, brown calc. soil 1	40	16	540 (79)	46 (6)	900			
Bare soil						10.5 (5.4)	33.0	3.0 (1.3)
Overgrazed scrub						14.3 (4.8)	24.0	2.0 (0.5)
"Regrassed" scrub						0.6 (0.6)	2.0	0.0 (0.0)
Medea, brown calc. soil 2	35	20	540 (79)	46 (6)	900			
Bare soil						9.9 (3.5)	40.0	2.8 (1.3)
Traditional vineyard						2.1 (1.1)	13.0	0.2 (0.1)
Fertil. vine[d] + cer.-leg.						0.4 (0.4)	3.0	0.0 (0.0)
Medea, Fersiallitic soil	35	0	540 (79)	46 (6)	900			
Bare soil under orchard						18.7 (8.4)	32.0	14.0 (8.3)
Traditional orchard						2.7 (1.3)	8.0	1.2 (0.7)
Orch[d]+ cer.-leg., contour						0.7 (0.8)	2.0	0.3 (0.4)

Note: Figures in parentheses are standard deviations. calc.: calcareous; cer.: cereal; leg.: legume; vine[d]: vineyard; orch[d]: orchard; fall.: fallow; till.: tillage; contour: contour tillage; fertil.: fertilized; assoc.: association.

[a] Hundreds of foot-tons per acre times inches per hour (to allow comparison with literature data), which may be converted into MJ mm (ha h)$^{-1}$ when multiplied by 17.35.

of stone cover was not well defined otherwise. Erosion was always more on bare (2 to 14 Mg ha^{-1} yr^{-1}) than on cropped soil (0 to 1.4 Mg ha^{-1} yr^{-1}), fallows (0.6 to 1 Mg ha^{-1} yr^{-1}), and scrub (0 to 2 Mg ha^{-1} yr^{-1}). Excluding bare plots, the effect of treatment on erosion followed distinct trends. Indeed, for a given location and soil type, mean annual erosion was always greater on grazed than on protected fallow, or scrub, and for traditional than for improved practices. As compared with conventional systems, improved systems (better fertilization, ridging, intercropping, protected fallow, or scrub) reduced erosion risks by 20 to 30% in Tlemcen and by 75 to 99% in Medea.

Two parameters affected erosion in opposite ways: topsoil properties (texture, stoniness) and slope steepness. Even on steep slopes, soils with high silt content (brown calcareous) were less stable than clayey Vertic soils or stony soils. Similarly, the deep Fersiallitic soils of Medea were more fragile than the stony Fersiallitic soils of Tlemcen, and suffered more erosion when tilled (Morsli et al., 2004). The influence of slope steepness was not evident: in Mediterranean areas, slope position is sometimes more important than slope steepness in case of soil saturation at the bottom of the hillslopes (Heusch, 1970; Roose et al., 1993; Roose, 1996; Mazour and Roose, 2002).

8.3.4 Organic Carbon Content and Stock of Surface Soil

The SOC content (0 to 10 cm depth) was low, and ranged from 6 to 12 g C kg^{-1} except for brown calcareous soils in Tlemcen where it was 19 to 33 g C kg^{-1} (Table 8.4). The SOC content of surface soil was 10 to 12 g C kg^{-1} in Mascara, 6 to 10 g C kg^{-1} in Tlemcen (excluding the brown calcareous soil), and 7 to 10 g C kg^{-1} in Medea. SOC content ranged from 8 to 12 g C kg^{-1} in Vertic soils, 7 to 12 g C kg^{-1} in brown calcareous soils (excluding Tlemcen), and 6 to 9 g C kg^{-1} in Fersiallitic soils. Thus location and soil type did not clearly affect SOC content of the surface soil, except that it was two to three times more for Tlemcen's brown calcareous soil than for the other location × soil type combinations. The SOC content of surface soil was more clearly affected by land use. It was lower in bare soils than under corresponding crops, fallows, and scrub (–3 to –44%). It was also lower under grazed than under protected fallows or scrub (–8 to –31% for a given location and soil type), whereas improved cropping systems were not always associated with increasing SOC content (difference in SOC content between improved and traditional systems was 15 to –6%). However, the effect of treatments was often small: for a given location and soil type, the maximum difference in SOC content between treatments was 16 and 23% in Medea and Mascara, respectively, but was 30, 50, and 80% for Vertic, Fersiallitic and brown calcareous soils in Tlemcen, respectively.

Changes in SOC content of the surface soil were measured in Tlemcen from 1991 to 2001 (Figure 8.1). Over the decade, the SOC content decreased markedly in the bare soils (–15% for the Vertic and Fersiallitic soils, –28% in the brown calcareous soil) and under grazed scrub (–14%). In contrast, it increased under protected scrub (12 and 25% for the Fersiallitic and brown calcareous soils, respectively). Changes in SOC content were small under cereals (–1% under cereal/grazed fallow, 7% under fertilized cereal).

SOC stock (0 to 10 cm depth) ranged from 9 to 18 Mg C ha^{-1} except for the brown calcareous soil in Tlemcen where it was 21 to 37 Mg C ha^{-1}. It was generally higher in Mascara (14 to 18 Mg C ha^{-1}) than in Medea (9 to 12 Mg C ha^{-1}), and Tlemcen (9 to 14 Mg C ha^{-1} when excluding the brown calcareous soil). Furthermore, for a given location it tended to be higher in brown calcareous than in Vertic soils. SOC stock was lower in bare soils than under cropped, fallows, and scrub (–5 to –44%), and was also lower under grazed than protected fallows or scrub (–4 to –11% in Mascara and Medea, –27 to –28% in Tlemcen). The effect of improved cropping systems on SOC stock was either positive (15% for the wheat in Tlemcen, 12% for the orchard in Medea) or negative (–9% for the cereals and –10% for the vineyard in Medea). The effect of treatments was less in Mascara and Medea than in Tlemcen: for a given location and soil type, the maximum difference in SOC stock among treatments was 12 and 15% in Medea and Mascara, respectively, but was 22 to 44% in Tlemcen.

Changes in SOC stock were measured from 1993 to 1998 in Mascara and from 1991 to 2001 in Tlemcen, for Vertic and brown calcareous soils (Figure 8.2). Over the periods under study, the SOC stock in bare soils decreased by 160 (Tlemcen's Vertic soil), 500 to 700 (Mascara), and 820 kg C ha^{-1} yr^{-1} (Tlemcen's brown calcareous soil). Under fallows or scrub, it decreased by 330 to 460 kg C ha^{-1} yr^{-1} in grazed plots, but increased by 570 to 750 kg C ha^{-1} yr^{-1} in protected plots. In crops, SOC stock decreased under legumes (50 to 80 kg C ha^{-1} yr^{-1}) but generally increased under cereals (–30 to 180 kg C ha^{-1} yr^{-1}). As compared with initial stocks, annual SOC change in

Table 8.4 Soil Organic Carbon (SOC) Content, Bulk Density, and SOC Stock at 0 to 10 cm Depth, Sediment Organic Carbon (OC) Content, Carbon Enrichment Ratio, and Eroded OC on Runoff Plots Located near Mascara (1996), Tlemcen (2001), and Medea (1988 to 1992)

Location, Soil Type, and Treatment	SOC Content g C kg^{-1}	Bulk Density Mg m^{-3}	SOC Stock Mg C ha^{-1}	Sediment OC Content g C kg^{-1}	Carbon Enrichment Ratio	Eroded OC kg C ha^{-1} yr^{-1}
Mascara, Vertic Soil						
Bare soil	10.2 (0.3)	1.35	13.8	14.0	1.4	95.2
Cereals	12.0 (0.8)	1.35	16.2	23.0	1.9	25.3
Chickpeas on ridges	11.0 (1.0)	1.32	14.5	27.0	2.5	29.7
Grazed fallow	11.3 (2.0)	1.36	15.4	26.0	2.3	31.2
Protected fallow	12.3 (2.2)	1.31	16.1	35.0	2.8	21.0
Mascara, Brown Calcareous Soil						
Bare soil	10.0 (0.4)	1.51	15.1	16.0	1.6	136.0
Cereals	10.3 (1.0)	1.52	15.7	28.0	2.7	22.4
Peas on ridges	10.3 (1.3)	1.50	15.5	29.0	2.8	17.4
Grazed fallow	10.3 (1.3)	1.53	15.8	26.0	2.5	41.6
Protected fallow	12.3 (1.6)	1.44	17.7	48.0	3.9	24.0
Tlemcen, Vertic Soil						
Bare soil	8.0 (0.3)	1.31	10.5	10.6	1.3	19.1
Wheat/fallow, downslope till.	9.0 (0.3)	1.30	11.7	16.8	1.9	26.9
Fertilized wheat, contour till.	10.3 (0.3)	1.30	13.4	20.6	2.0	33.0
Tlemcen, Brown Calcareous Soil						
Bare soil	18.6 (0.8)	1.12	20.8	20.1	1.1	78.4
Overgrazed scrub	23.0 (1.7)	1.16	26.7	39.0	1.7	27.3
Protected scrub	33.3 (1.2)	1.12	37.3	48.0	1.4	33.6
Tlemcen, Fersiallitic Soil						
Bare soil	6.3 (0.4)	1.49	9.4	9.8	1.6	31.4
Overgrazed degraded scrub	6.8 (0.3)	1.51	10.3	14.5	2.1	26.1
Protected scrub	9.4 (0.2)	1.50	14.1	18.0	1.9	18.0
Medea, Vertic Soil						
Bare soil	nd	nd	nd	nd	nd	nd
Cereals/grazed fallow	7.2	1.30	9.4	10.1	1.4	3.7
Fertilized cereal-legume association	6.8	1.30	8.8	9.0	1.3	0.4
Medea, Brown Calcareous Soil 1						
Bare soil	nd	nd	nd	nd	nd	nd
Overgrazed scrub	6.1	1.70	10.2	18.0	2.9	36.0
"Regrassed" scrub	7.1	1.50	10.7	22.1	3.1	0.4
Medea, Brown Calcareous Soil 2						
Bare soil	nd	nd	nd	nd	nd	nd
Traditional vineyard	8.3	1.20	10.0	8.9	1.1	1.7
Fertil. vine[d] with cereal-legume	8.2	1.10	9.0	9.8	1.2	0.1
Medea, Fersiallitic Soil						
Bare soil under orchard	nd	nd	nd	nd	nd	nd
Traditional orchard	7.1	1.50	10.7	11.4	1.6	13.9
Orch + cereal-legume, contour	8.0	1.50	12.0	13.4	1.7	3.9

Note: Figures in parentheses are standard deviations. nd: not determined; till.: tillage; fertil.: fertilized; vine: vineyard; orch: orchard.

the surface soil was –1.2 to –4.2% in bare soils, –1.2 to –2.8% under grazed plots, 2.3 to 4.1% under protected fallows or scrub, –0.3 to –0.5% under legumes, and –0.2 to 1.1% under cereals.

8.3.5 Eroded Organic Carbon

Except on bare soils where eroded OC was 19 to 136 kg C ha^{-1} yr^{-1}, it ranged from 0.1 to 42 kg C ha^{-1} yr^{-1}, which is a moderate level especially considering the slope steepness of some plots (Table 8.4). On bare soils, it was greater in Mascara than in Tlemcen (95 to 136 vs. 19 to 78 kg

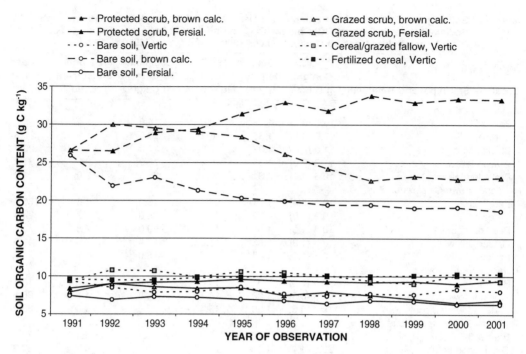

Figure 8.1 Changes in soil organic carbon content at 0 to 10 cm depth in the runoff plots located near Tlemcen (1991 to 2001).

C ha^{-1} yr^{-1}), but differences among locations were not distinct on vegetated plots (17 to 42 vs. 18 to 34 kg C ha^{-1} yr^{-1}). The amount of eroded OC was generally less in Medea (0.1 to 14 kg C ha^{-1} yr^{-1} except on one overgrazed plot where it was 36 kg C ha^{-1} yr^{-1}; however, eroded OC was not determined on bare soils). On average, eroded OC was more on brown calcareous than on Vertic and Fersiallitic soils for bare (107, 57, and 31 kg C ha^{-1} yr^{-1}, respectively) and grazed plots (35, 31, and 26 kg C ha^{-1} yr^{-1}, respectively). However, soil type had no effect on the amount of eroded OC for protected fallows or scrub (19, 21, and 18 kg C ha^{-1} yr^{-1}, respectively) and for cereal or legume cropping systems (20 kg C ha^{-1} yr^{-1} for both Vertic and brown calcareous soils). Considering vineyards and orchards in Medea, the amount of eroded OC was lesser on brown calcareous than on Fersiallitic soils (< 2 vs. 4 to 14 kg C ha^{-1} yr^{-1}). For a given location and soil type, eroded OC was generally 20 to 80% more on bare soil that on vegetated plots (except for Tlemcen's Vertic soils where it was 30 to 40% smaller). It was generally 20 to 90% greater on grazed than on protected fallows or scrub (except for Tlemcen's brown calcareous soils where it was 20% smaller). For a given soil and crop type, eroded OC was 70 to 90% lesser for improved than for traditional cropping systems in Medea (cereal, vineyard, and orchard systems), but was 20% more for the improved than for the traditional cereal system in Tlemcen (Vertic soil).

Comparing SOC contents of sediments and topsoil (0 to 10 cm depth), the carbon enrichment ratio of sediments (CER) was more on average for Mascara (2.4) than Tlemcen and Medea (1.7), whereas differences among soil types were small (2.1, 1.9, and 1.8 in average for brown calcareous, Vertic, and Fersiallitic soils, respectively; Table 8.4). With regard to the effect of land use, the mean CER was 1.4 on bare soils (ranging from 1.1 to 1.6) and on vineyards and orchards (1.1 to 1.7), 2.1 on cereal or legume systems (1.3 to 2.8), and on grazed plots (1.7 to 2.5), and 2.6 on protected fallows or scrub (1.4 to 3.9). Excluding vineyards and orchards (Medea), erosion selectivity for OC increased with increase in soil surface cover, from bare plots to protected fallows and scrub lands.

In comparison with the annual change in SOC stock (0 to 10 cm depth), which was computed for Vertic and brown calcareous soils in Mascara and Tlemcen (Figure 8.2), annual eroded OC was negligible on protected fallows and scrub land (< 5% of annual SOC change, which was positive

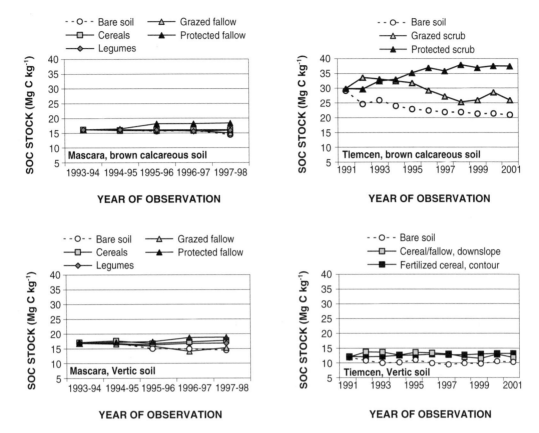

Figure 8.2 Changes in soil organic carbon (SOC) stocks at 0 to 10 cm depth in the runoff plots located on Vertic and brown calcareous soils near Mascara (1993 to 1998) and Tlemcen (1991 to 2001).

in that case). For two out of the three cereal plots on Vertic soils where SOC increased markedly (100 to 180 kg C ha^{-1} yr^{-1}), the consequences of SOC erosion (25 to 33 kg C ha^{-1} yr^{-1}) were also limited. In contrast, eroded OC contributed greatly to decrease in SOC in some cereal and legume plots (60% for chickpea on Vertic soil and 90% for cereals on brown calcareous soil in Mascara, 180% for traditional wheat/fallow on Vertic soil in Tlemcen). However, loss of eroded OC generally represented a small proportion of decrease in SOC in grazed (7 to 13%) and bare plots (10 to 14% in general, but 26% for the brown calcareous soil in Mascara). Therefore, the consequences of SOC erosion on change in SOC were limited in general and on bare soils, fallows, and scrub especially, but could be high on cropped plots.

For 27 of the 31 plots (eroded OC was not determined for the four bare plots in Medea), annual eroded OC was neither correlated with topsoil OC nor with sediment OC contents, but was correlated with the annual erosion rate ($r = 0.950$, $p < 0.01$; Figure 8.3). It was more closely correlated with the product of erosion multiplied by topsoil OC content ($r = 0.964$), especially when highly erodible plots were not taken into account ($r = 0.908$ for the 24 plots with erosion < 3.5 Mg ha^{-1} yr^{-1}, vs. $r = 0.711$ for the 27 plots; $p < 0.001$ in both cases). Thus considering topsoil SOC content in addition to annual erosion provided a good prediction of annual eroded OC.

8.3.6 Seasonal Variations in Runoff, Erosion, and Eroded Organic Carbon

Figure 8.4 depicts rainfall amount, runoff rate, erosion, sediment OC content, and eroded OC on a monthly basis from September 1995 to May 1996, for the brown calcareous soil in Mascara (monthly runoff rate is averaged over the period 1993 to 1998). Rainfall and runoff were maximum

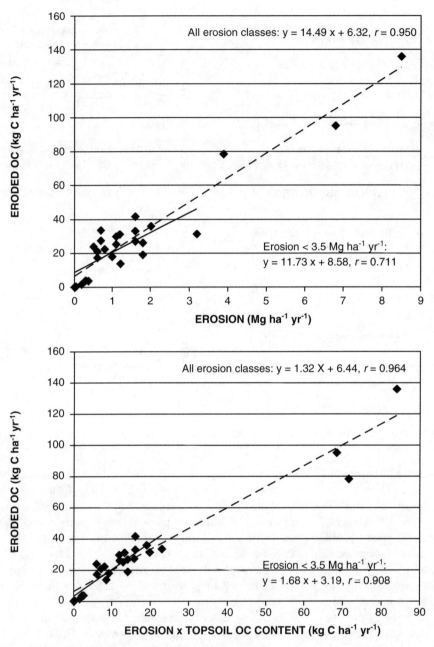

Figure 8.3 Relationships between eroded organic carbon (OC) and erosion, or between eroded OC and the product of erosion multiplied by topsoil (0 to 10 cm depth) OC content, for the runoff plots under study in northern Algeria.

during the winter (rainfall was more than 60 mm month^{-1} from November to February, and runoff more than 7% on bare soil). The peaks of erosion and eroded OC were observed for a short duration only during November and December. Sediment OC content was low for these both months, but was high in October and between January and March. Indeed, organic residues accumulated on vegetated plots during the summer were eroded by the first rainstorms in autumn, but due to small rainfall amounts, erosion was low and the sediment OC content was high. In November and December, the soil surface being partly covered by vegetation, high rainfall amounts resulted in

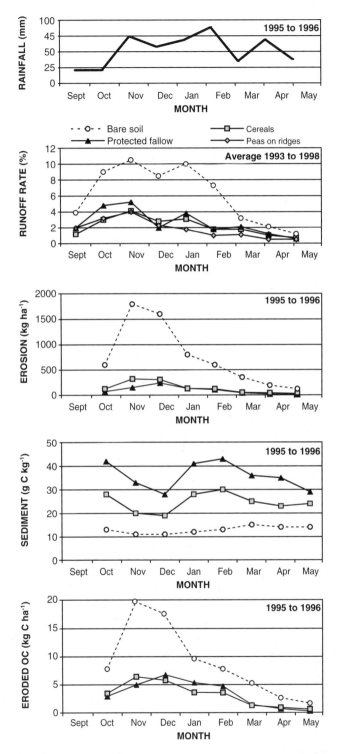

Figure 8.4 Seasonal variations in rainfall, runoff rate, erosion, sediment organic carbon (OC) content, and eroded OC in 1995 to 1996 for Mascara plots (average 1993 to 1998 for the runoff rate).

high soil losses. However, because organic residues had already been eroded, the sediment OC content decreased. In January and February, easily erodible soil particles had already been removed, and despite the high rainfall amounts, erosion decreased, but was more selective for OC. Due to increasing soil surface cover by vegetation, erosion remained low in the spring, but its selectivity decreased resulting in very small OC erosion. Seasonal changes in sediment OC content were less on bare plots due to the absence of vegetation and organic residues, with similar changes in erosion and eroded OC.

8.4 DISCUSSION

8.4.1 Rainfalls

For the study period, annual rainfall in the region was 30 to 40% lower that the long-term average for the 1913 to 1971 period. Storms with high intensities were also few: in 2000 there was only one intense storm with 84-mm hr^{-1} 30-min intensity. Therefore, the erosivity index of the USLE (Wischmeier and Smith, 1978) was moderate: for the study period, it varied from 30 to 80 U.S. units (hundreds of foot-tons per acre times inches per hour, which may be converted into MJ mm ha^{-1} h^{-1} when multiplied by 17.35; this index is expressed in U.S. units to allow comparison with literature data). The ratio of erosivity index to annual rainfall was about 0.10 in the Mediterranean mountainous zone: this is much smaller than in the tropics (0.50) or in tropical mountains (0.25) (Roose, 1996).

Contrary to the general opinion that Mediterranean climate is characterized by erosive rains, these data confirmed those obtained for Medea in Algeria (Arabi and Roose, 1992; Roose et al., 1993) and in Morocco (Heusch, 1970; Arnoldus, 1981). Thus, the principal reasons for the severe erosion of these landscapes are: (1) the weak vegetation cover in these semiarid areas, and (2) the occurrence once in 10 years of one exceptional rainstorm. Such extreme rainstorms completely saturate the thin soils compacted by overgrazing or rendered bare after crop harvest, leading to large amounts of runoff, which, on steep slopes, cause numerous gullies and landslides characteristic of the landscapes (Roose et al., 2000).

8.4.2 Runoff

Annual runoff rates were moderate during the dry period under study (< 20%). However, for Mascara and Tlemcen stations, the maximum (daily) runoff rate was 30 to 40% during exceptional rainstorms saturating the bare topsoil, sealed, compacted, and/or already wet. Similar results were observed on Medea's runoff plots, but the maximum runoff rate was up to 80% for bare plots on Vertic soils (Roose et al., 1993).

Annual runoff rate was correlated with the annual rainfall, as indicated by comparisons made in Tlemcen for bare plots and protected scrub (data not shown). The antecedent soil moisture content was an important determinant of runoff: the runoff began after 22 mm of intense rain when the topsoil was dry (5 days without rain) but after 2 to 4 mm when the soil was wet, compacted or crusted (Morsli et al., 2004; Sabir et al., 2004). These thresholds depended on rainfall (intensity) and soil surface characteristics (initial soil moisture, bulk density, roughness, stone and vegetation cover, macro-aggregation and clods; Sabir et al., 2002).

On bare plowed fallows, even low intensity and amount of rainfall (< 5 mm) produced runoff when rains were successive and the topsoil was wet or sealed. On row-cropped systems, maximum runoff rate occurred between October and January, when the soil surface conditions were susceptible (low vegetation cover, sealing crust, high soil moisture) and the rains were abundant and relatively intense (Figure 8.4). On overgrazed and degraded scrub, runoff events were frequent and large because of a weak tuft vegetation cover, surface crust, abundance of stones, and topsoil compaction by grazing (Sabir et al., 1994). Grazing lands are an important source of runoff degrading the fields downstream, and causing gullies, landslides, and floods (Roose et al., 2000). Around Tlemcen, runoff occurred for 80% of the rainfall events for bare and already scoured soils covered by grazed scrub or fallows (Mazour, 1992). The runoff generally decreased when these natural fallows were protected against fire and grazing, due to slowly increasing litter and vegetation cover.

8.4.3 Sheet Erosion

All measurements conducted in the Maghreb mountains at the scale of runoff plots have shown that sheet and rill erosion are moderate on the hillslopes (from 1 to 20 Mg ha^{-1} yr^{-1}): in Algeria, Arabi and Roose (1992), Gomer (1992), Mazour (1992), Brahamia (1993), Roose (1993), Roose et al. (1993), Chebbani (1996), Morsli (1996), Kouri et al. (1997), and Roose et al. (2000); in Morocco, Heusch (1970), Al Karkouri et al. (2000), Laouina et al. (2000), and Moufaddal (2002); and in Tunisia, Masson (1971), Bourges et al. (1977), Delhumeau (1981), and Delhoume (1987). This conclusion was confirmed by the data presented in this chapter.

This low erosion is probably related to low rainfall energy and to soil resistance provided by high contents of clay, exchangeable calcium and stones in the surface soil. Nevertheless the energy of large amounts of runoff on steep slopes causes gullies which are very active during exceptional storms falling on saturated soils during the winter and the spring (Roose, 1991; Roose et al., 1993; Kouri and Vogt, 1994). Indeed, according to Heusch (1970), Demmak (1982), and Moukhchane (2002) sheet erosion represents only a small proportion (0.2 to 10 Mg ha^{-1} yr^{-1}) of the sediments transported by the wadies (rivers) in the Mediterranean mountains. In these semiarid areas, severe erosion occurs only during exceptional storms cumulating a tremendous runoff energy on steep slopes. The concentrated runoff scarves gullies, increases the flow in wadies, causes floods, embankment degradation, landslides on hillslopes, and rapid siltation of reservoirs (Roose et al., 1993).

The effect of land use on erosion was well defined: improved cropping systems and protection against grazing caused less erosion than traditional cropping practices and grazing. Conservation effective cultural practices like ridging, rough tillage on the contour, appropriate fertilizer rate, and their interactions with the vegetation growth reduced the risks of runoff and erosion. This positive but limited influence of cultural practices in semiarid Mediterranean areas is confirmed by results obtained under natural (Arabi, 1991; Brahamia, 1993; Roose et al., 1993) and simulated rainfalls (Morsli et al., 2002). Higher runoff and soil losses on bare soil and grazed fallow or scrub than on protected scrub have already been reported (Mazour and Roose, 2002).

The influence of slope steepness on sheet erosion is not distinctively observed in the Mediterranean mountains. Heusch (1970) showed for Vertisols on marl in the Pré-Rif hillslopes of Morocco that the slope steepness (12 to 35%) was less important than the position on the toposequence. Indeed, rain falling on these Vertic clayey soils percolated into the cracks and bypassed to the concave or bottom of the hillslopes where it saturated the soil, producing gullies that progressed upslope from the valley to the hilltop. For runoff plots established on a range of soil types near Tlemcen and Medea, Mazour (1992) and Roose et al. (1993) reported that the runoff decreased and the sheet erosion did not increase with increase in slope steepness from 12 to 40%. Nevertheless, the gully erosion increased on steeper slopes, as runoff energy (for detachment and transport) increased with the square of the flow velocity (Roose et al., 2000).

The analysis of the erosion factors showed that the combination of maximum rainfall erosivity and soil erodibility in the cultivated fields occurred at the beginning of the rainy season and during the cool period of winter–spring (Mazour and Roose, 2002). Even if the erosion risks remain moderate according to the tolerable limit, the sheet erosion scours the organo-mineral topsoil, exports selectively the fine or light particles (clay, silts, SOC), and reduces the soil fertility potential, the SOC stock and the potential for SOC sequestration (Meddi and Morsli, 2001; Lal, 2004). Finally, by degrading the topsoil structure, sheet erosion paves the way for rills and gullies (Le Bissonnais, 1996).

8.4.4 Soil Organic Carbon

SOC content of the topsoil (0 to 10 cm depth) was low (< 12 g C kg^{-1} in general, and always < 35 g C kg^{-1}), due to low overall biomass production and residue return, and rapid mineralization, accelerated by erosion. Experiments conducted at the landscape level in the mountains of northern Algeria showed that SOC content varied according to topsoil characteristics such as soil texture (r = 0.55), land use, biomass produced, and the landscape position (Morsli, 1996). The SOC content of surface soil was about 10 g C kg^{-1} on the summit and 30 g C kg^{-1} on the fast slopes and valleys. The decrease in SOC content with depth depended on the soil type and was sharp in the brown calcareous soils, but gentle in the Vertic soils due to natural mixing in swelling horizons. These observations confirmed the results reported by Batjes (1996) that SOC content varies at the world level with soil texture and mineralogy.

Land use plays an important role in the variability of SOC and its dynamics (Lal, 2004). For the runoff plots under study, the SOC in the surface soil was lower in bare than vegetated plots (–3 to –44%), and under grazed than under protected fallows or scrub (–8 to –31%). The differences in SOC of surface soil between improved and traditional cropping systems were generally small (< 15%). For the study duration, the SOC of surface soil decreased in bare and grazed plots (–1 to –4% per year), increased under protected fallows or scrub (1 to 3% per year), and changed slightly under crops (–0.5 to 1% per year). Increasing SOC content of surface soil under protected natural fallow (28%) has been reported for the eastern Rif in Morocco (Tribak, 1988). This practice involves large biomass production and residue return, which contribute to the restoration of SOC, soil fertility, and biomass productivity (Sabir et al., 2002; Lal, 2004). However, the restoration is much slower when the fallow is grazed (at an appropriate stocking rate). Improved cropping systems (including appropriate fertilizer rate, intercropping, contour tillage, reduced tillage, etc.) generally result in increasing topsoil SOC, due to increasing biomass production and decreasing OC erosion (Campbell and Zentner, 1993). However, this trend was not clear in the paired plots under study (traditional vs. improved system). On slopes less than 20%, reduced tillage and ridging have a strong influence on soil degradation risks, as they delay the start of runoff, increase infiltration rate, and maintain SOC stock (Sabir et al., 2002). Nevertheless ridges on steep slopes may be broken leading to formation of gullies during intensive rainstorms. However, when it is well managed and on moderate slopes and permeable soils, and tied at 2 to 5 m distance, ridging may effectively reduce erosion and improve infiltration and biomass production (Azontonde, 1993; Roose, 1996; Morsli et al., 2004). Bare soils, overgrazed fallows, and degraded scrub had a high risk of erosion and depletion of SOC stock in the surface layer.

Additional data (not shown) indicated that hill-slope aspect (northern vs. southern slope) had an important influence on SOC content. SOC content was relatively higher on northern than on southern slopes, which received less rainfall, were less covered by vegetation, and were thus more easily eroded (Mazour and Benmansour, 2002).

8.4.5 Eroded Organic Carbon and Erosion Selectivity for Carbon (CER)

For the plots under study, annual erosion of OC ranged from 19 to 136 kg C ha^{-1} yr^{-1} on bare plots compared to 0.1 to 42 kg C ha^{-1} yr^{-1} on vegetated plots. Eroded OC was affected by land use and it was generally more on bare than on vegetated plots, on grazed than on protected fallows or scrub, and on conventional than on improved cropping systems.

Eroded OC contributed to decrease in SOC, but effects on topsoil SOC changes were limited in general, except on some cereal and legume plots. Thus, SOC changes depended primarily on the amount of biomass returned to the soil and on residue decomposition and SOC mineralization, and secondarily on erosion, as reported for tropical areas (Blanchart et al., 2002; Blanchart et al., this volume).

Sheet erosion not only scours the topsoil, it also preferentially removes the fine and light particles (clay, silts, SOC). Compared to the SOC content of 0 to 10 cm depth, sediment enrichment in OC (CER) ranged from 1.1 to 3.9. It was lower for bare soils, vineyards, and orchards (1.4 on average) than for protected fallows and scrub (2.6 on average), and was intermediate for cropped and grazed plots (2.1 on average). The presence of a thick litter layer reduced the velocity hence the detachment and transport capacities of the runoff flow, but provided easily transportable organic particles. Therefore, sediments from protected fallow or scrub were particularly enriched in OC, as they included relatively less heavy mineral particles and more light organic particles than sediments from bare, grazed, and cropped plots. This effect of litter explains why the increase in topsoil SOC content is generally associated with an increase in CER (Roose and Barthès, this volume). However, ratios of removed OC to topsoil OC were somewhat underestimated, because the dissolved OC in runoff and drainage water were not determined, though it may represent noticeable losses of OC (Blanchart et al., this volume).

The amount eroded OC correlated with erosion, and even more strongly with the product of erosion and topsoil SOC content, especially for plots with erosion rate of < 3.5 Mg ha^{-1} yr^{-1} (i.e., excluding most of the bare plots). In contrast, the amount of eroded OC did not depend on sediment OC content. Comparing the seasonal variations, Bep A Ziem et al. (2002) reported that eroded OC was more closely correlated with erosion than with sediment OC content. Indeed, most of the OC losses occur during large rainstorms that produce large amounts of sediments with low OC content. Such events produce large runoff amounts that carve rills or gullies in the subsoil where SOC content is low, so that sediment OC content is reduced. In contrast, high sediment OC contents are associated with small soil losses resulting from less erosive rainfall events or from rainfalls that occur when the soil surface is covered by vegetation. Such rains have low runoff energy and erosion affects only the most superficial soil layers, which are the richest in OC, and this results in high sediment OC content (Roose, 1981). In Algeria, the greatest OC losses were measured between November and January, when erosion was maximum and sediment OC content was minimum.

Eroded sediments and OC are partially redistributed in the fields or deposited as colluvium at the foot slopes and in the talwegs (Morsli, 1996). The amount of OC retained on the hillslopes depends on soil surface roughness, topography, and land use. Large sediment deposits have been observed in some parts of the landscapes, and they exert a strong influence on the distribution of soil fertility. Lal (2002) observed that a part of the eroded OC is deposited in the concavity of the hillslopes, where it is somewhat protected, and where it causes strong variations in crop yields. The study of OC dynamics over the landscape is necessary to improve the management of soil potential and to restore SOC stocks, in order to optimize sustainable production and to increase SOC sequestration.

8.5 CONCLUSIONS

The data presented in this chapter show that, at the plot scale (100 to 200 m^2), runoff and sheet erosion risks were generally moderate in the semiarid mountains of northern Algeria, even when the fields were cropped on steep slopes (mean annual runoff rate and erosion $< 20\%$ and < 15 Mg ha^{-1} yr^{-1}, respectively). However, a few exceptional rainstorms caused spectacular erosion features: rills, gullies, landslides, and floods. These observations confirm the conclusions of numerous researchers in northern Africa. Nevertheless this moderate erosion degraded the surface soil, depleted soil fertility, and reduced productivity. The losses of eroded OC were also moderate and ranged from 0.1 to 42 kg C ha^{-1} yr^{-1} on vegetated plots and 19 to 136 kg C ha^{-1} yr^{-1} on bare plots. Losses were influenced by landscape position, soil type, land use, and cultural practices. The eroded OC was generally more on bare than on vegetated plots, grazed than protected fallows or scrub, and improved than traditional cropping systems. Additionally, annual losses of eroded OC correlated with annual erosion, and more closely, with the product of erosion multiplied by topsoil SOC. The data also indicate that sediments were richer in OC than the topsoil (0 to 10 cm depth), and that this enrichment increased with soil surface cover (i.e., bare plots $<$ grazed and cropped plots $<$

protected fallow and scrub). Over the study periods, SOC in the surface layer decreased in bare and grazed plots, increased under protected fallows or scrub land, and changed only slightly under crops. However, the contribution of erosion to annual changes in SOC was limited in general, except on some cereal and legume plots. These results confirm that for most of the land uses, changes in SOC depend primarily on the amount of biomass returned to the soil and on residue decomposition and SOC mineralization, and secondarily on erosion. Such studies, carried out on plots, should be completed by measurements involving complete hillsides, in order to address mechanisms that occur at this scale, such as gully erosion, and sediment and OC redistribution along the slopes.

Considering land management, the results indicate that continuous cultivation did not increase the erosion risks, even under traditional systems. However, bare and compacted soils, abandoned fields, overgrazed fallows, or degraded scrub lands can produce high runoff and increase risk of gully formation and landslides. Ridge cropping and protecting fallows and scrub land can reduce erosion risks and increase the SOC and the biomass production. Use of innovative and improved cultural practices (appropriate fertilizer rate, seed selection, ridging, and adapted cultural practices like crop rotations and intercropping) incorporated in the best traditional farming systems, as land husbandry, increased production (yield of cereals multiplied by 2 to 4 on improved plots), decreased erosion risk, and increased SOC (up to 28% increase) (Roose, 1993). Other antierosive techniques (gully management, stone walls, hedges, planting on the contour, etc.) are effective in trapping sediments and organic matters. The land husbandry approach has demonstrated that it is possible to improve crop production in the hilly regions (intensification and diversification) while enhancing the environment (Arabi and Roose, 1992; Roose, 1993; Mazour and Roose, 2002; Hamoudi and Morsli, 2003). In Algeria, where land use is rapidly changing, the national projects on rural development are based on the land husbandry strategy with the participatory approach (Morsli et al., 2004). Fruit tree plantations, improved cultural practices and water management, agroforestry, and grazing regulation are the most widely used strategies for the rural development. These improved practices have a positive effect on productivity, soil erosion control, and carbon sequestration. Nevertheless, Algeria needs additional research regarding the effects of conservation cultural practices (like no-till and mulch farming) on carbon sequestration and soil productivity.

REFERENCES

Al Karkouri, J., A. Laouina, E. Roose, and M. Sabir. 2000. Capacité d'infiltration et risques d'érosion des sols dans la vallée des Béni Boufrah — Rif central (Maroc). *Bulletin du Réseau Erosion* 20:342–356.

Arabi, M. 1991. *Influence de quatre systèmes de production sur le ruissellement et l'érosion en milieu montagnard Méditerranéen (Médéa, Algérie)*. Ph.D. Dissertation, Université de Grenoble, France.

Arabi, M. and E. Roose. 1992. Water and soil fertility management (GCES). A new strategy to fight erosion in Algerian mountains, in P. G. Haskins and B. M. Murphy, eds., *Proceedings of the 7th ISCO (International Soil Conservation Organization) Conference*, Sydney, 27–30 September 1992, pp. 341–347.

Arabi, M. and E. Roose. 2002. Influence des systèmes de production et du sol sur l'érosion en nappe et rigole, le stock de carbone du sol et les pertes par érosion en moyenne montagne méditerranéenne du NO de l'Algérie. Communication at the International Colloquium Land Use, Erosion and Carbon Sequestration, Montpellier, France, 23–28 September 2002.

Arnoldus, H. 1981. An approximation of the rainfall factor in the USLE, in M. De Boodt and D. Gabriels, eds., *Assessment of Erosion*, John Wiley & Sons, New York, pp. 127–132.

Azontonde, A. 1993. Dégradation et restauration des terres de barre au Bénin. *Cahiers ORSTOM, série Pédologie* 28:217–226.

Batjes, N. H. 1996. Total carbon and nitrogen in the soils of the world. *European Journal of Soil Science* 47:151–163.

Benchetrit, M. 1972. *l'érosion actuelle et ses conséquences sur l'aménagement en Algérie*. Presses Universitaires de France, Paris.

Bep A Ziem, B., Z. B. Boli, and E. Roose. 2002. Pertes de carbone par érosion hydrique et évolution des stocks de carbone sous rotation intensive coton/maïs sur des sols ferrugineux sableux du Nord Cameroun. Communication at the International Colloquium Land Use, Erosion and Carbon Sequestration, Montpellier, France, 23–28 September 2002.

Bernoux, M., C. Feller, C. C. Cerri, V. Eschenbrenner, and C. E. P. Cerri. 2006. Soil carbon sequestration, in E. Roose, R. Lal, C. Feller, B. Barthès, and B. A. Stewart, eds., *Soil Erosion and Carbon Dynamics*, Taylor & Francis, Boca Raton, FL

Blanchart, E., E. Roose, B. Khamsouk, M. Dorel, J. Y. Laurent, C. Larré-Larrouy, L. Rangon, J. P. Tisot, and J. Louri. 2002. Comparaison des pertes de carbone par érosion et drainage aux variations du stock de c du sol en deux années: Cas d'une rotation bananiers-cannes-ananas-sol nu sur un nitisol argileux sur cendres volcaniques de Martinique. Communication at the International Colloquium Land Use, Erosion and Carbon Sequestration, Montpellier, France, 23–28 September 2002.

Blanchart, E., E. Roose, and B. Khamsouk. 2006. Soil carbon dynamics and losses by erosion and leaching in banana cropping systems with different practices (Nitisol, Martinique, West Indies), in E. Roose, R. Lal, C. Feller, B. Barthès, and B. A. Stewart, eds., *Soil Erosion and Carbon Dynamics*, Taylor & Francis, Boca Raton, FL.

Blavet, D., G. De Noni, E. Roose, L. Maillo, J. Y. Laurent, and J. Asseline. 2004. Effets des techniques culturales sur les risques de ruissellement et d'érosion en nappe sous vigne en Ardèche (France). *Sécheresse* 15:111–120.

Bourges, J., C. Floret, and R. Pontanier. 1977. *Etude d'un milieu représentatif du sud Tunisien: Citerne Telman, Type Segui, Campagnes 1972–74*. ORSTOM, Tunis.

Brahamia, K. 1993. *Essai sur la dynamique actuelle dans la moyenne montagne Méditerranéenne: Oued Mina, Algérie*. Ph.D. Dissertation, Université de Grenoble, France.

Campbell, C. A., and R. P. Zentner. 1993. Soil organic matter as influenced by crop rotations and fertilization. *Soil Science Society of America Journal* 57:1034–1040.

Chebbani, R. 1996. *Etude à différentes échelles des risques d'érosion dans le bassin Versant de l'Isser (Tlemcen)*. Magister Thesis, INA, Algiers.

Coelho, C. O. A., A. Laouina, A. J. D. Ferreira, K. Regaya, M. Chaker, R. Nafaa, R. Naciri, T. M. M. Carvalho, and A. K. Boulet. 2002. The dynamics of land use changes in Moroccan and Tunisian sub-humid and semiarid regions and the impact on erosion rates and overland flow generation. Communication at the International Colloquium Land Use, Erosion and Carbon Sequestration, Montpellier, France, 23–28 September 2002.

Cormary, Y. and J. M. Masson. 1964. Etude de la conservation des eaux et du sol au centre de recherche du génie rural de Tunisie: Application à un projet type de la formule de perte de sol de Wischmeier. *Cahiers ORSTOM, série Pédologie* 2:3–26.

Delhoume, J. P. 1987. Ruissellement et érosion en bioclimat méditerranéen semi-aride de Tunisie Centrale, in A. Godard and A. Rapp, eds., *Processus et Mesure de l'Erosion*, Editions CNRS, Paris, pp. 487–507.

Delhumeau, M. 1981. *Recherches en milieu Méditerranéen humide: Etude de la dynamique de l'eau sur parcelles du Bassin Versant de l'Oued Sidi Ben Naceur, Nord Tunisie*. Ministère de l'Agriculture (DRES) and ORSTOM, Tunis.

Demmak, A. 1982. *Contribution à l'étude de l'érosion et des transports solides en Algérie Septentrionale*. Ph.D. Dissertation, Université de Paris VI.

Doran, J. W., M. Sarrantonio, and M. A. Liebig. 1996. Soil health and sustainability. *Advances in Agronomy* 56:1–53.

Dumas, J. 1965. Relation entre l'érodibilité des sols et leurs caractéristiques analytiques. *Cahiers ORSTOM, série Pédologie* 3:307–333.

Fournier, F. 1967. La recherche en érosion et conservation des sols sur le continent africain. *Sols Africains* 12:5–53.

Gomer, D. 1992. *Ecoulement et érosion dans des petits Bassins Versants à sols marneux sous climat semi-aride Méditerranéen*. GTZ-ANRH, Algiers.

Gsell, S. 1913. *Histoire ancienne de l'Afrique du Nord. Volume 1: Les conditions du développement historique. Les temps primitifs. La colonisation Phénicienne et l'Empire de Carthage*. Hachette, Paris.

Hamoudi, A. and B. Morsli. 2003. Erosion et spécificité de l'agriculture de montagne: réflexion sur la conservation et la gestion de l'eau et du sol en milieu montagneux Algérien. *La Forêt Algérienne* 12:18–25.

Heddadj, D. 1997. La lutte contre l'érosion en Algérie. *Bulletin du Réseau Erosion* 17:168–175.

Heusch, B. 1970. L'érosion du Pré-Rif. Une étude quantitative de l'érosion hydraulique dans les collines marneuses du Pré-Rif Occidental (Maroc). *Annales de la Recherche Forestière du Maroc* 12:9–176.

Hien, E. 2002. Effet de la déforestation et de l'érosion sur le statut organique du sol: cas d'un sol ferrugineux tropical sableux du SO du Burkina Faso. Communication at the International Colloquium Land Use, Erosion and Carbon Sequestration, Montpellier, France, 23–28 September 2002.

Kouri, L., and H. Vogt. 1994. Gully processes and types in oued Mina basin, Algeria. Communication at the ESSC (European Society for Soil Conservation) Workshop Soil Erosion in Semiarid Mediterranean Areas, 28–30 October 1993, Taormina, ESSC and CSEI, Catania, Italy.

Kouri, L., H. Vogt, and D. Gomer. 1997. Analyse des processus d'érosion hydrique linéaire en terrain marneux, bassin versant de l'oued Mina, Tell oranais, Algérie. *Bulletin du Réseau Erosion* 17:64–73.

Lal, R. 2002. Soil conservation and restoration to sequester carbon and mitigate the greenhouse effect, in J. L. Rubio, R. P. Morgan, S. Asins, and V. Andreu, eds., Proceedings of the 3rd International Congress of the ESSC (European Society for Soil Conservation) *Man and Soil at the Third Millenium*, Geoforma Ediciones, Logrono, Spain.

Lal, R. 2004. Soil carbon sequestration impacts on global climate change and food security. *Science* 304:1623–1627.

Lal, R., J. M. Kimble, R. F. Follett, and B. A. Stewart, eds. 1998. *Management of Carbon Sequestration in Soil*. CRC Press/Lewis Publishers, Boca Raton, FL, pp. 37–51.

Lal, R., M. Griffin, J. Apt, L. Lave, and M. G. Morgan. 2004. Managing soil carbon. *Science* 304:393.

Laouina, A. 1992. Recherches actuelles sur les processus d'érosion au Maroc. *Bulletin du Réseau Erosion* 12:292–299.

Laouina, A, R. Nafaa, C. Coelho, M. Chaker, T. Carvalho, A. K. Boulet, and A. Ferreira. 2000. Gestion des eaux et des terres et phénomènes de dégradation dans les collines de Ksar El Kebir, Maroc. *Bulletin du Réseau Erosion* 20:256–274.

Le Bissonnais, Y. 1996. Aggregate stability and assessment of soil crustability and erodibility. I. Theory and methodology. *European Journal of Soil Science* 47:425–437.

Masson, J. M. 1971. *L'érosion des sols par léau en climat Méditerranéen. Méthodes expérimentales pour l'étude des quantités érodées à l'échelle du champ*. Ph.D. Dissertation, Université de Montpellier, France.

Mazour, M. 1992. Les facteurs de risque de l'érosion et du ruissellement dans le bassin versant de l'Isser — Tlemcen — Algérie. *Bulletin du Réseau Erosion* 12:300–313.

Mazour, M. and M. Benmansour. 2002. Effet de l'exposition des versants sur la production de biomasse et l'efficacité anti-érosive dans le Nord-Ouest algérien. Communication at the International Colloquium Land Use, Erosion and Carbon Sequestration, Montpellier, France, 23–28 September 2002.

Mazour, M. and E. Roose. 2002. Influence de la couverture végétale sur le ruissellement et l'érosion des sols sur parcelles d'érosion dans des bassins versants du Nord-Ouest de l'Algérie. *Bulletin du Réseau Erosion* 21:320–330.

Meddi, M. and B. Morsli. 2001. Etude de l'érosion et du ruissellement sur bassins versants expérimentaux dans les Monts de Beni-Chougrane (Ouest d'Algérie). *Zeitschrift für Geomorphologie N. F.* 45:443–452.

Morsli, B. 1996. *Caractérisation, distribution et susceptibilité des sols à l'erosion (cas des montagnes de Beni-Chougrane)*. Magister Thesis, INA, Algiers.

Morsli, B., M. Meddi, and A. Boukhari. 2002. Etude du ruissellement et du transport solide sur parcelles expérimentales. Utilisation de la simulation de pluies, in *Actes du Séminaire sur la Gestion de l'Eau*. Université de Mascara, Algeria, pp. 80–88.

Morsli, B., M. Mazour, N. Mededjel, A. Hamoudi, and E. Roose. 2004. Influence de l'utilisation des terres sur les risques de ruissellement et d'érosion sur les versants semi-arides du nord-ouest de l'Algérie. *Sécheresse* 15:96–104.

Moufaddal, K. 2002. Les premiers résultats des parcelles de mesure des pertes en terre dans le bassin versant de Oued Nakhla dans le Rif occidental (Nord du Maroc). *Bulletin du Réseau Erosion* 21:244–254.

Moukhchane, M. 2002. Différentes méthodes d'estimation de l'érosion dans le bassin versant de Nakhla (Rif occidental, Maroc). *Bulletin du Réseau Erosion* 21:255–265.

Nelson, D. W. and L. E. Sommers. 1996. Total carbon, organic carbon, and organic matter, in D. L. Sparks, A. L. Page, P. A. Helmke, R. H. Loeppert, P. N. Soltanpour, M. A. Tabatabai, C. T. Johnson, and M. E. Sumner, eds., *Methods of Soil Analysis: Part 3 — Chemical Methods*, Soil Science Society of America and American Society of Agronomy, Madison, WI, pp. 961–1010.

Pla Sentis, I. 2002. Soil organic matter effects on runoff and erosion in dryland vineyards of NE Spain. Communication at the International Colloquium Land Use, Erosion and Carbon Sequestration, Montpellier, France, 23–28 September 2002.

Remini, B. 2000. L'envasement des barrages: quelques exemples algériens. *Bulletin du Réseau Erosion* 20:165–171.

Robert, M. 2006. Global change and carbon cycle: the position of soils and agriculture, in E. Roose, R. Lal, C. Feller, B. Barthès, and B. A. Stewart, eds., *Soil Erosion and Carbon Dynamics*, Taylor & Francis, Boca Raton, FL.

Roose, E. 1981. *Dynamique actuelle de sols ferrallitiques et ferrugineux tropicaux d'Afrique Occidentale. Etude expérimentale des transferts hydrologiques et biologiques de matières sous végétations naturelles ou cultivées.* Travaux et Documents ORSTOM 130, ORSTOM, Paris.

Roose, E. 1991. Conservation des sols en zones méditerranéennes. Synthèse et proposition d'une nouvelle stratégie de lutte antiérosive: la GCES. *Cahiers ORSTOM, série Pédologie* 26:145–181.

Roose, E. 1993. Water and soil fertility management. A new approach to fight erosion and improve land productivity, in E. Baum, P. Wolff, and M. Zoebisch, eds., *Topics in Applied Resource Management in the Tropics, Vol. 3, Acceptance of Soil and Water Conservation Strategies and Technologies*, Deutsches Institut für Tropische und Subtropische Landwirtschaft, Witzenhausen, Germany, pp. 129–164.

Roose, E. 1996. Land husbandry: Components and strategy. *FAO Soils Bulletin* 70, FAO, Rome.

Roose, E., M. Arabi, K. Brahamia, R. Chebbani, M. Mazour, and B. Morsli. 1993. Erosion en nappe et ruissellement en montagne Méditerranéenne Algérienne. Réduction des risques érosifs et intensification de la production agricole par la GCES: synthèse des campagnes 1984–1995 sur un réseau de 50 parcelles d'érosion. *Cahiers ORSTOM, série Pédologie* 27:289–307.

Roose, E., R. Chebbani, and L. Bourougaa. 2000. Ravinement en Algérie: typologie, facteurs de contrôle, quantification et réhabilitation. *Sécheresse* 11:317–326.

Roose, E. and B. Barthès. 2001. Organic matter management for soil conservation and productivity restoration in Africa: a contribution from Francophone research. *Nutrient Cycling in Agroecosystems* 61:159–170.

Roose E. and B. Barthès. 2006. Soil carbon erosion and its selectivity at the plot scale in tropical and Mediterranean regions, in E. Roose, R. Lal, C. Feller, B. Barthès, and B. A. Stewart, eds., *Soil Erosion and Carbon Dynamics*, Taylor & Francis, Boca Raton, FL.

Sabir, M., A. Merzouk, and O. Berkat. 1994. Impacts du pâturage sur les propriétés hydriques du sol dans un milieu pastoral aride: Aarid, Haute Moulouya, Maroc. *Bulletin du Réseau Erosion* 14:444–462.

Sabir, M., M. Maddi, A. Naouri, B. Barthès, and E. Roose. 2002. Runoff and erosion risks indicators on the main soils of the Mediterranean mountains of occidental Rif area (Morocco), in *Proceedings of the 12th ISCO (International Soil Conservation Organization) Conference, Sustainable Utilization of Global Soil and Water Resources, Vol. 2*, Beijing, 26–31 May 2002.

Sabir, M., B. Barthès, and E. Roose. 2004. Recherche d'indicateurs de risques de ruissellement et d'érosion sur les principaux sols des montagnes Méditerranéennes du Rif occidental (Maroc). *Sécheresse* 15:105–110.

Tribak, A. 1988. *L'érosion du Prérif Oriental. Contribution à l'étude de la dynamique actuelle dans quelques bassins au Nord de Taza (Maroc).* Ph.D. Dissertation, Université de Grenoble, France.

Wischmeier, W. H. and D. D. Smith. 1978. *Predicting rainfall erosion losses: a guide to erosion planning.* Agriculture Handbook 537, United States Department of Agriculture, Washington D.C.

CHAPTER 9

Carbon, Nitrogen, and Fine Particles Removed by Water Erosion on Crops, Fallows, and Mixed Plots in Sudanese Savannas (Burkina Faso)

A. Bilgo, Georges Serpantié, Dominique Masse, Jacques Fournier, and Victor Hien

CONTENTS

9.1 Introduction 126
    9.1.1 Homogeneous Land Use 126
    9.1.2 Heterogeneous Land Use 127
    9.1.3 Adaptation of the Method 127
9.2 Materials and Methods 128
    9.2.1 Site of Study 128
    9.2.2 Experimental Device 128
    9.2.3 Cropping Systems 128
    9.2.4 Field Measurements 129
    9.2.5 Analyses of the Samples 129
9.3 Results 130
    9.3.1 Physical Properties of the Experimental Soils 130
    9.3.2 Effects of Rainfall Erosivity and Soil Moisture on Runoff and Soil Losses 130
    9.3.3 Effects of Treatments on Runoff and Erosion, Quantitative Approach 131
    9.3.4 Erosion, Qualitative Approach 132
        9.3.4.1 Coarse Sediment Load or "Carriage" 132
        9.3.4.2 Suspension (Collected with Filters) 132
        9.3.4.3 C and N Dissolved in Rains and in Micro-Filtrated Water Samples 134
    9.3.5 Carbon, Nitrogen, and Fine Particle Losses by Erosion 135
9.4 Discussion 135
    9.4.1 Validation of Experimental Data on Erosion 135
    9.4.2 Physical Composition of the Eroded Soils and Consequences 137
    9.4.3 Carbon Losses and Carbon Selectivity 138
    9.4.4 Possible Reasons for Constant Erosion over Years 138
    9.4.5 "Fallow" Effect 138
    9.4.6 Impact of Soil Erosion 139

9.5 Conclusion 139
Acknowledgments 140
References 140

9.1 INTRODUCTION

In tropical zones, and particularly in subhumid areas such as Sudanese savannas, erosion and runoff are important factors of degradation of the cultivated soils and cause declines in crop yields (Lal, 1975, 1983; Roose, 1981; Pieri, 1991). In Sudanese areas of West Africa, many studies have analyzed the process of erosion and characterized the soil losses (Roose, 1981; Pontanier et al., 1986; Collinet, 1988; Mietton, 1988; Casenave and Valentin, 1989; Fournier et al., 2000). On a bare plot, the "splashing" effect of an intense rain on an uncovered soil dissociates soil skeleton and soil plasma, causing a waterproof surface crust. A diffuse runoff evacuates released materials. Downstream, a sheet flood occurs, developing both sheet erosion and regressive erosion like small stairs. Furthermore, linear erosion and gullies can appear when runoff concentrates. Even on low slopes (< 1%), the soil losses are high, around 10 t ha^{-1} yr^{-1}. Little research has evaluated the losses in fine particles, nutrients, and carbon (C) at the plot level (Roose, 1981; Bep A Ziem et al., 2002). However these elements represent the principal losses for the farming systems and for the environment. Moreover, erosion is potentially a cause of C loss, to be taken into account in the strategies of environmental management on a global scale.

In the soils of the Sudanese savannas, C, nitrogen (N), and fine particles are highly related, physically or not. Soil organic C fractions of slow turnover and N soil reserves are aggregated to fine particles (Feller, 1995). Consequently, losses of stabilized soil organic C will be explained largely by fine particle losses and will be parallel to N losses. Yet, N and fine particles are highly subjected to fertility status of soils. Consequently, managing soil organic C should be related to management of N and fine particles. Hence, studying C losses of savanna-cropped soils should be related to a study of N and fine particle losses.

The existing data have to be brought up to date, taking into account the dynamics of the farming systems in the expanding cotton belts of Sudanese savannas. Here there is evidence on the best nonsloping soils of the lowlands that the old cereal farming systems in shifting cultivation (pearl millet, *Pennisetum glaucum,* and sorghum, *Sorghum bicolor*) have made way for ploughed, ridged, and fertilized cotton, sorghum, and maize through permanent cropping. Temporary cropping and fallowing remains only on the lands of secondary interest like the sandy or gravelous upland soils (Serpantié, 2003). These soils are most susceptible to erosion. Two principal ways of management of land against erosion exist, depending on whether land use is homogeneous or not.

9.1.1 Homogeneous Land Use

This chapter does not consider contour hydraulic treatments (micro-dams, stone lines, stone walls, terraces), which are rare in this zone because they require too heavy investments for peasants without external aid (Van Campen and Kebe, 1986) and show little adaptation to the large surfaces that are cropped in cotton zones. Therefore this chapter focuses on agricultural treatments. Tillage reduces runoff temporally but increases its turbidity, maintaining high soil losses (Roose, 1973; Collinet, 1988; Maass et al., 1988). Comparing crops of tilled sorghum and fallows in Saria (Burkina Faso, 0.7% slope), Roose (1993) found 7.3 t ha^{-1} yr^{-1} soil losses under sorghum, 0.51 t ha^{-1} yr^{-1} under herbaceous fallow, and 0.17 t ha^{-1} yr^{-1} under shrubby fallow. The least erosive cropping methods like contour ridges are unfortunately unsuitable. Continuous rectilinear ridges with a light incline (1% on average), frequently used by farmers, are easier to build. They drain excess water, which is common in the Sudanese climate, and reduce the risks of transverse gullies that can appear because of lateral collection of runoff in the case of almost-contour ridges (Nimy, 1999). Mulching-based cropping systems, tested in the wet tropics, strongly reduce erosion (Blancaneaux et al., 1993), but are poorly suited to prevailing Sudanese conditions: freely grazing herds in the dry season, and a short growing season. Thus for the moment, few antierosive agricultural treatments seem to be perfectly appropriate to the Sudanese cotton zones.

9.1.2 Heterogeneous Land Use

A proper study scale is essential for the assessment of soil losses. The farmer may find it beneficial to reduce displacements of nutrients and fine elements, and to preserve them within his own land, but not necessarily to retain them at the same place where erosion occurs. The same conditions occur for C, for which it is important to know the starting and destination places. Local losses are partly recoverable on differently treated contiguous spaces. One can thus assume that variations of cropping practices along a slope will reduce the global erosion of the hill slope. To verify this, the erosion effects of a heterogeneous method of managing a hillside must be characterized, and the balance of erosion must be checked, on a given scale. Such a structured landscape has already been recommended by Wischmeier and Smith (1978) in order to reduce erosion, but in reference to models like buffer strip cropping adapted to mechanized open-fields. Their adaptation to the Sudanese conditions, including different crops, narrow "grass strips" and "hedges," has already been documented (Roose, 1967; Roose and Bertrand, 1971; Boli et al., 1993; Albergel et al., 2000), but not sufficiently concerning fine and organic elements. It would also be advisable to study composite modes of hillside management, such as the crop–fallow mosaics of the natural landscapes of Sudanese uplands, where grass and shrubs represent more cover than in the case of grass strips and hedges.

9.1.3 Adaptation of the Method

Currently, empirical studies on erosion under natural rain using the 'USLE' scale (plot area of 100 m^2 and slope length of 20 m; Wischmeier, 1974), or rainfall simulation on a smaller area (Collinet, 1988) are related to the measurement of a potential of diffuse erosion, which is firstly due to primary rain erosion. However, on a larger area, other erosion processes occur. On the one hand, it is advisable to take into account longer slopes, particularly for the study of grass strips. This more closely accounts for the real conditions of the farmers' fields (Boli et al., 1993). On the other hand the "small water-catchment" scale, carried out on natural water catchments, does not provide a precise account due to the differences in soils and local parameters of rain and catchment geometry between experimental catchments.

In recent experiments conducted in Burkina Faso (Bondoukui), Fournier et al. (2000) studied runoff and erosion during natural rains, under conditions of farming systems of the cotton zones. This original experiment included plots of great length (50 m) and very small width (3.2 m), in opposition to the classical pattern of experimental plots. Ridges inclined in the direction of slope (1% slope) channeled runoff on four interridges of 80 cm each, reproducing farmers' practice. Channeling runoff made it possible to work on experimental plots of very low width since it eliminated "edge-effects." The treatments reproduced the principal land-use classes of the Sudanese uplands (5- to 10-year cropping after clearing, short fallow, long fallow) and principal land-use associations, i.e., crop upstream, grass fallow downstream, and a broad grass strip.

Our research used the same device, though it focused more specifically on the assessment of fine particles, C, and nutrients, in order to evaluate the effects of land use on erosion on a qualitative level. Therefore, the objectives of the present chapter are: (1) to assess erosion of C, N, and fine particles on homogeneous land use modes; and (2) to compare them with results obtained on heterogeneous land-use mode.

9.2 MATERIALS AND METHODS

9.2.1 Site of Study

The experiments were carried out in the area of Bondoukui (11°49'N, 3°49'W; altitude 360 m) in the west of Burkina Faso. In this area, the climate is Sudanese with a wet season of 5 months, a dry season of 7 months, and an average annual rainfall of 850 mm. On deeper soils, the cropping system combines cereals (maize or sorghum) with a commercial crop (cotton) in 2-year rotations, during 10 years after clearing a 10- to 20-year fallow. The average technical system involves: ridged plowing (for cotton and maize), sowing on ridges, low mineral fertilization, manual weeding, and a new ridging of lines around 50 days after sowing.

9.2.2 Experimental Device

Two blocks were built in May 1997 on a sandy plateau covered with ferrugineous tropical eluviated soils (luvisol; FAO) on a 1% slope.

The first experimental block included eight isolated plots, measuring 50 m × 3.2 m, marked by ridges and metal sheet reinforcements. They were laid out on a well-draining soil till 80 cm deep. Three plots were studied for the purposes of this research. The treatments were:

- Plot in homogeneous land use "CROP": annual cotton/corn biennial rotation on ridges. Before the experiment, the plot was cultivated for 8 years since clearing. Another plot, 20-m long, was coupled with this plot in 1999.
- Plot in homogeneous land use "GFAL": grass fallow of 1 or 2 years protected against cattle, excluded in 1999 (cotton on mulch), and on which old ridges remained.
- Plot in composite land use "CROP/GFAL": at the upper part of the plot, a 40-m-long subplot with a biennial rotation of cotton/corn on ridges (like CROP), and at the lower part of the plot, a 10-m-long subplot with permanent meadow, not grazed in the wet season. It contained *Andropogon gayanus* sown in 1997, whose residues remained on the soil in the dry season, without burning (except accidentally in 1999). As compared with classical grass strip treatments, this composite plot had a higher grass area/plot area ratio (1/5 against 1/20 in the classical case).

The second experimental block included three plots, only one of which was retained for this research. They were laid out on a hydromorphic eluviated soil well drained and only tilled to 35-cm depth. The plot in homogeneous land use "SFAL" was a 15-year dense shrubby fallow. The plot dimensions were 25 m × 6.5 m.

9.2.3 Cropping Systems

Previous ridges remained at the beginning of the rainy season. The cultivated plots were ploughed by animal traction (10 to 15 cm deep, ripple relief) in direction of slope at the beginning of the wet season. At that time, only the lateral ridges were manually reconstructed. After sowing cotton or maize on lines and after preemergence chemical weeding, two manual hoe weedings progressively reconstructed the three central ridges on sowing lines, principal ridging occurring 50 days after sowing. Fertilizer was brought twice: 100 kg ha^{-1} of NPK (14-23-14) in holes after the first weeding, and 50 kg ha^{-1} of urea (46% N) in holes after ridging. The plots were protected from fire and grazing. In 1999 however, the strips of *Andropogon gayanus* (CROP/GFAL) burned accidentally, and SFAL burned in 1998 and 1999. GFAL never burned.

9.2.4 Field Measurements

Each block included a recording rain gauge collecting rain at a 1-m height, making it possible to calculate the maximum intensity of rain into 15 or 30 min (I_{15}, I_{30}). Each plot was equipped with a downstream concrete surface for reception of runoff, followed by a tank for the coarse sediment decantation, of a calibrated runoff divisor (thin blade, cut up in multiple "V"), and one or two barrels for the reception of a sample of water downstream from the central V. By gauging from the various containers and maximum level of water measured on divisor blades, runoff amount and maximum runoff intensity were estimated after each rain throughout the years 1998 to 2001.

For the coarse sediment measurement, the tank was allowed to settle, clear water was manually drained, and the bottom was collected, dried, and weighed. The barrels containing samples of runoff including fine sediments mainly were vigorously mixed without any flocculating agent in order not to modify the composition of the aqueous solutions, then sampled (0.66:1 per barrel). Each sample was filtered with the filter paper (about 15 µm pores). Already filtered but still tinted water was again sampled and filtered under aspiration with the micro-filter (0.2 µm pores). The coarse sediments of the tank constituted the "carriage" while the filtered sediments of barrels added to the micro-filtered ones constituted the "suspension" (fine erosion). The final micro-filtrated water was intended to an analysis of soluble minerals.

One of the problems arising from the measurement of erosion under natural rains is the catastrophic nature of the erosive events of rare recurrence. In the present case, light overflows occurred from the barrels of reception during some rainy periods in 1999. Fortunately, the presence of a 20-m-long cultivated plot in which barrels did not overflow and in which runoff was usually identical to that of the 50-m-long cropped plot (R_{50m} = 1.03 R_{20m} + 0.50, with r^2 = 0.93; R_{50m} and R_{20m} in % of rainfall) helped estimate the probable quantities of runoff on the 50-m-long cropped plot. Erosion was appreciated on the barrels of the 50-m-long plot, with a possible slight overestimation due to a decantation effect during overflow.

9.2.5 Analyses of the Samples

The particle-size and chemical analyses of sediment samples were carried out by compartment. The samples of coarse carriage were analyzed for all the events from 1999 to 2001. The small samples of filtered suspensions collected on filter papers in 1998 and 1999 were gathered on each plot according to the amount of soil losses (< 1 t ha^{-1}, 1 to 10 t ha^{-1}, > 10 t ha^{-1}) and were analyzed for C and N. However there was not enough suspension to determine particle-size distribution for each plot and for each type of event. So it was measured for each group of events on gathered samples from several plots, therefore achieving the minimal quantities needed for the laboratory. In order to assess the annual fine particle losses, this average particle-size distribution was affected to soil loss quantity for each event according to its group. Micro-filtrates, presumed to have a content of 100% of fine dispersed particles (fine silt, clay, and organic colloids), were not analyzed. Micro-filtered water and the rainwater were analyzed for N some events during the year 1999.

Organic C and total N in coarse sediments and in soils were determined following the Walkley and Black method (Nelson and Sommers, 1996) and the Kjeldhal method (Bremner, 1996), respectively. Total C and N in fine sediments were determined by dry combustion using a CHN Elemental Analyser (Nelson and Sommers, 1996). Mineral N (NO_3^-, NH_4^+) was also determined in micro-filtered water (Mulvaney, 1996). Particle-size analysis of air-dried samples (< 2 mm) was determined by a combination of dry sieving and sedimentation (pipette method), after destruction of the organic matter and total dispersion (Gee and Bauder, 1986). Physical measurements were carried out *in situ*: bulk density (150-cm^3 cylinder, eight samples); steady state infiltration under a sheet of water was measured with the double ring method of Fournier (1998), using a central ring with 1-m^2 diameter and two replicates per treatment. The sampling of soils for analysis (eight subsamples within the blocks) was made on May 1998 and infiltration measurement on May 2000.

Table 9.1 Topsoil Properties (0 to 10 cm depth) and *in situ* Measurements in the Experimental Plots

	Block 1	Block 2
Clay (< 2 µm; %)	5.9	3.0
Fine silt content (2 to 20 µm; %)	3.8	8.6
Coarse silt content (20 to 50 µm; %)	17.0	20.9
Fine sand content (50 to 200 µm; %)	39.5	31.5
Coarse sand content (200 to 2000 µm; %)	33.8	36.0
Bulk density	1.58	1.58
Total porosity (% in 150 cm^3)	39.0	40.1
Water-stable aggregates (%)	4.9	7.5
Infiltration in CROP (mm h^{-1})	250	—
Infiltration in GFAL (mm h^{-1})	250	—
Infiltration in CROP/GFAL (*A. gayanus*) (mm h^{-1})	500	—
Infiltration in SFAL (mm h^{-1})	—	120
C content (g kg^{-1})	2.3	6.6
N content (g kg^{-1})	0.3	0.6
C/N ratio	8	11
pH in water	5.8	6.5
pH in KCl	4.9	5.3

9.3 RESULTS

9.3.1 Physical Properties of the Experimental Soils

Table 9.1 shows that the soils from the two blocks had comparable particle-size distribution and porosity, except that topsoil clay content was twice lower in Block 2 (15-yr shrubby fallow) than in Block 1 (cultivated for several years). Infiltration was also two times less in the soil of Block 2, though it was better protected by litter, biologically and chemically richer, and better aggregated. This lower infiltration could result from the higher silt/clay ratio, which is an indicator of plugging. Indeed, the hydromorphic pseudo-gley layer was deeper in Block 1 (80 cm) than in Block 2 (35 cm). It could thus be assumed that runoff was caused mainly by soil saturation in Block 2 (in wet years especially), and by surface crusting in Block 1. Within Block 1, where the soil was already very permeable, the perennial grass cover (*Andropogon gayanus*) in CROP/GFAL resulted in a two times greater infiltration (under a sheet of water) than in CROP and GFAL.

9.3.2 Effects of Rainfall Erosivity and Soil Moisture on Runoff and Soil Losses

Table 9.2 presents the annual runoff rate and erosion for each plot over the experimental period. It clearly shows the dramatic effects of high annual rainfall on runoff and erosion: indeed, when compared to averages over the years 1998, 2000, and 2001, the year 1999 had a 50% higher annual rainfall, but runoff was multiplied by 2 to 4, and erosion by 3 to 18, depending on the plot. Years 1998, 2000, and 2001 were characterized by a total rainfall ranging from 680 to 850 mm, lower than or equal to the mean annual rain in Bondoukui (850 mm). The rain intensity was low (only three to four rains with I_{30} higher than 50 mm h^{-1}, one higher than 80 mm h^{-1}), and rains were spaced out so that soils remained dry, producing little runoff and erosion. In contrast, annual rainfall was 1155 mm in 1999 (wettest year of the decade), with seven rains having I_{30} higher than 50 mm h^{-1} (including two higher than 80 mm h^{-1}). Several events occurred when the soils were already very wet, even saturated by a temporary water table in the case of plot SFAL, resulting in great runoff and abundant erosion. The dramatic effect of intense and repeated rains has already been reported (Roose, 1977; Wischmeier and Smith, 1978). Erosion was more affected by wet conditions

Table 9.2 Annual Runoff and Erosion on the Experimental Plots

Period	Plot	Occurrence of Fire	Rain (mm yr^{-1})	Annual Runoff Rate (%)	Maximum Runoff (%)	Coarse Carriage (t ha^{-1} yr^{-1})	Fine Suspension (t ha^{-1} yr^{-1})	Total Erosion (t ha^{-1} yr^{-1})
1998	CROP$_{50m}$		846	20	70	0.9	3.4	4.3
	CROP/GFAL		id	2	23	0.1	0.1	0.2
	GFAL		id	0	0	0.0	0.0	0.0
	SFAL	fire	899	13	51	0.6	0.8	1.4
1999	CROP$_{50m}$		1155	38	90	6.2	30.6	36.7
	CROP$_{20m}$		1155	40	99	9.2	16.4	25.6
	CROP/GFAL	fire	1155	22	83	0.7	8.3	8.9
	GFAL		1155	nd	nd	nd	nd	nd
	SFAL	fire	1098	18	73	0.4	5.1	5.5
2000	CROP$_{50m}$		800	24	68	1.5	8.4	10.0
	CROP/GFAL		800	4	21	0.0	0.8	0.8
	GFAL		800	2	8	0.2	0.0	0.2
	SFAL		666	2	16	0.0	0.0	0.0
2001	CROP$_{50m}$		680	26	52	4.0	11.2	15.3
	CROP/GFAL		680	8	21	0.1	0.5	0.6
	GFAL		680	2	10	0.0	0.0	0.0
	SFAL		652	2	8	0.0	0.0	0.0
Average over 1998, 2000, and 2001[a]	CROP$_{50m}$		775	23 a	63 a	2.1 a	7.7 a	9.9 a
	CROP/GFAL		775	5 b	22 b	0.1 b	0.5 b	0.5 b
	GFAL		775	1 b	6 b	0.1 b	0.0 b	0.1 b
	SFAL		739	6 b	25 b	0.2 b	0.3 b	0.5 b
Weighted average over four years[b]	CROP$_{50m}$		814	25	66	2.5	10.0	12.6
	CROP/GFAL		814	7	28	0.2	1.3	1.3
	GFAL		814	nd	nd	nd	nd	nd
	SFAL		775	7	30	0.2	0.8	1.0

Note: nd: not determined.

[a] Averages over 1998, 2000, and 2001 were compared using a Student t-test; within a column, two different letters mean that the difference was significant at $p < 0.05$.
[b] Weighted averages over four years were calculated as the sum of data relating to 1999 with a 0.1 coefficient (as 1999 was the wettest year of the decade) and averages over 1998, 2000, and 2001 with a 0.9 coefficient.

than runoff: on CROP treatment for example, runoff rate was 1.7 times higher but erosion 3.7 times higher in 1999 than during the three other years, on average.

9.3.3 Effects of Treatments on Runoff and Erosion, Quantitative Approach

The years 1998, 2000, and 2001 were used as experimental replicates to study usual climatic conditions ("modal erosive years"). In such conditions, the effect of complete cropping on runoff and erosion was clear: mean annual runoff rate and erosion were significantly higher on CROP than on the three other plots (23 vs. 1 to 6% and 9.9 vs. 0.1 to 0.5 t ha^{-1} yr^{-1}, respectively; $p < 0.05$). When compared with the three other plots, CROP had a 4 to 18 times higher mean annual runoff rate but a 20 to 150 times greater erosion, thus erosion was much more affected by complete cropping than runoff. Runoff and erosion did not differ significantly between CROP/GFAL, GFAL, and SFAL, though they tended to be higher on CROP/GFAL and SFAL than on GFAL (5 to 6% vs. 1% and 0.5 vs. 0.1 t ha^{-1} yr^{-1}, respectively). Runoff and erosion on CROP/GFAL were thus significantly smaller than on CROP (they were divided by 5 and 20, respectively) but not significantly greater than on GFAL. Therefore, introducing a 10-m-long grass strip at the downstream part of a cultivated plot was very effective for water and soil conservation, the data indicated that it absorbed most of the runoff and trapped most of the sediments produced by the upper part of the plot. The fire that occurred on SFAL in 1998 had noticeable effects: indeed, in the absence of fire (years 2000 and 2001, and GFAL in 1998), runoff and erosion were similar and small in GFAL and SFAL (2% and 0.2 t ha^{-1} yr^{-1}, respectively), whereas fire (SFAL in 1998) resulted in strong increases in runoff and erosion (13% and 1.4 t ha^{-1} yr^{-1}, respectively, i.e., at least seven times more than in the absence of fire).

Runoff and erosion were much higher in the wettest year of the decade (1999). Annual runoff rate reached 20% on SFAL and CROP/GFAL, 40% on CROP, and maximum runoff rate more than 70%, due to topsoil saturation. Erosion reached 36.7 t ha^{-1} yr^{-1} on CROP. Even under such climatic conditions, the grass strip in CROP/GFAL was effective in reducing runoff and soil loss (which were two and four times smaller than in CROP, respectively), though an accidental burning in the dry season of 1999 probably reduced its effectiveness.

9.3.4 Erosion, Qualitative Approach

9.3.4.1 Coarse Sediment Load or "Carriage"

The "carriage" (eroded coarse particles and soil aggregates trapped in the tank) generally represented a low proportion of total erosion ($\leq$ 50%), during the wet year especially (< 20%). It was less than 1 t ha^{-1} yr^{-1}, except on CROP (Table 9.2).

Its contents in C and N were small (3 to 4 g C kg^{-1} and 0.2 to 0.3 g N kg^{-1}), but close to those measured in the topsoil (0 to 10 cm depth). Its C/N ratio ranged from 12 to 16 and was higher than in the topsoils (8 to 11), suggesting that the carried organic matter contained a high proportion of fresh organic residues. These carried materials could thus be considered as a mixture of soil aggregates, washed sands, and litter fragments. Carriage increased at the beginning of the rainy season and more particularly after plowing. In the same way, some cases of strong erosions in C corresponded to erosive rains occurring on tilled soils in the course of the rainy season (weeding, ridging), causing a facilitated mobilization of aggregates. The small runoffs on tilled soils showed high levels of turbidity.

The carriage (more important after plowing) could form a deposit not far from the eroded place. Thus this coarse erosion, of low content in fine particles and in C and N, was not really worth fighting against. However, it increased a lot in the case of highly erosive events and could be exported in lowlands.

9.3.4.2 Suspension (Collected with Filters)

The fine elements in suspension represented the major eroded fraction: it generally accounted for more than 50% of total annual erosion, and more than 70% on CROP. It seemed advisable to separate the solid fractions collected on filter paper (weighed and analyzed) from the fractions collected on micro-filter (weighed but not analyzed).

The filtered residues mainly consisted in fine elements (< 20 µm), except for the most erosive rainfall events: indeed, fine particles represented 76, 79, and 36% of the particles collected on the filter papers for events resulting in < 1, 1 to 10, and > 10 t ha^{-1} erosion, respectively (Figure 9.1). These rates were much higher than those measured in the topsoil (10 to 12%), indicating that suspensions were particularly enriched in fine particles. In the same way, suspension were enriched in C (30 to 55 g kg^{-1}) and N (1.5 to 4 g kg^{-1}) compared to the topsoils (2.3 to 6.6 g C kg^{-1} and 0.3 to 0.6 g N kg^{-1}). These sediment contents in C and N decreased, just as the content in fine particles, with the importance (and the scarcity) of the erosive event. It suggested that runoff produced by the most erosive events could erode soil layers that were deeper and contained less C and N than superficial layers affected by less erosive events. The C and N contents of filtered suspensions were generally lower for CROP than for CROP/GFAL and SFAL, probably due to the presence of litter and enriched upper soil layer on the latter plots. Additionally, considering the

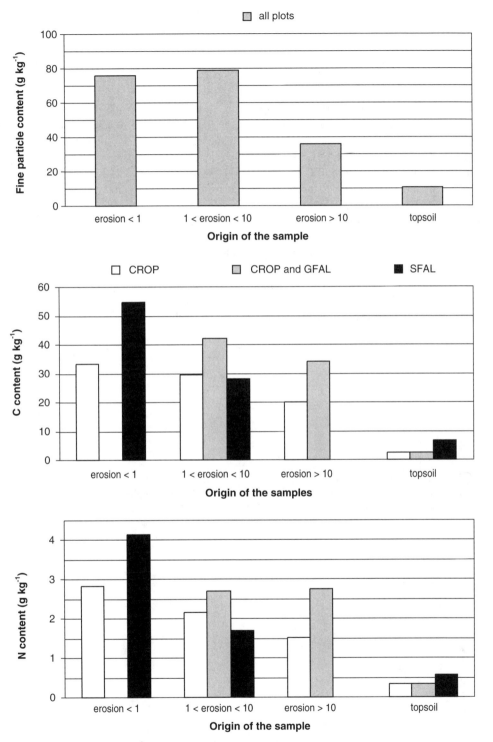

Figure 9.1 Contents in fine particles, carbon, and nitrogen in suspensions filtrated with filter paper, gathered from groups of erosive events (< 1, 1 to 10, and > 10 t ha^{-1} yr^{-1}), and in the topsoil (0 to 10 cm depth).

whole of the analyzed suspensions, the ratios of C content to clay content and of C content to fine particle content were rather constant (ca. 0.090 and 0.035, respectively), highlighting the close relationship between C and fine particles.

The colloidal and very fine eroded fractions, collected by micro-filtration but not analyzed, represented small sediment amounts (0.12 t ha^{-1} in 2000 on CROP). However, these very fine particles, which crossed the pores of the filter paper (15 µm) but were retained by the 0.2-µm pores of micro-filters, must be taken into account in an assessment. Indeed, the very fine particles (fine clay) are rare in the soils under study, but are associated to a fine and processed organic matter. The constant ratio between C and fine particle content allowed the estimation of C content in micro-filtrated suspension. Finally, the amount of C in this fraction remained low, like the C eroded in carriage: on CROP, the mean annual erosion of C for 1998, 2000, and 2001 reached 243 kg C ha^{-1}, of which 11 kg C ha^{-1} (estimate) were related to the micro-filtrated residues and 5 kg C ha^{-1} to carriage (see below).

9.3.4.3 C and N Dissolved in Rains and in Micro-Filtrated Water Samples

Rainwater contents in N-NO$_3^-$ and N-NH$_4^+$ were determined for seven rainfall events in June and July 1999 (Figure 9.2). They were 0.218 (±0.089) mg N-NO$_3^-$ l^{-1} and 0.122 (±0.089) mg N-NH$_4^+$ l^{-1}, equivalent to a gain of 4 kg N ha^{-1} yr^{-1} for the year under study (1155 mm yr^{-1}). In general rainwater N-NO$_3^-$ content was slightly higher than N-NH$_4^+$ content, except for one event where they reached 0.4 and less than 0.05 mg l^{-1}, respectively.

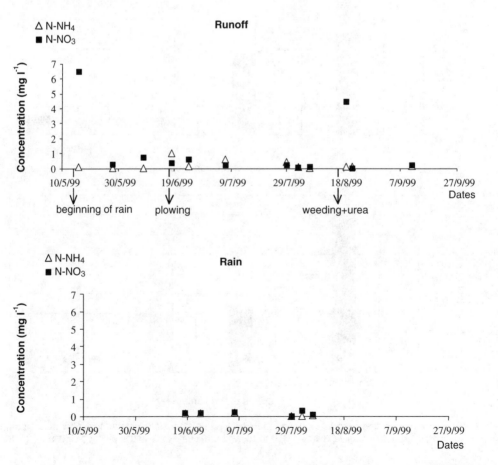

Figure 9.2 Contents in N-NO$_3^-$ (nitrates) and N-NH$_4^+$ (ammonium) of micro-filtrated runoff water (CROP plot, year 1999).

Runoff contents in $N-NO_3^-$ and $N-NH_4^+$ were determined for 12 events from May to September 1999, and varied according to the season (Figure 9.2). Runoff content in $N-NO_3^-$ was high at the beginning of the rainy season (6.5 mg l^{-1} on May 16), then it decreased strongly (< 1 mg l^{-1} from the end of May to mid-August) until fertilization with urea, which corresponded to another peak (4.4 mg l^{-1} on August 19), followed by another strong decrease (< 0.5 mg l^{-1} from August 20 to mid-September). Runoff content in $N-NH_4^+$ was very low at the beginning of the rainy season (< 0.2 mg l^{-1} from May to mid-June), increased just after plowing (1.0 mg l^{-1} on June 18), then decreased and remained low (< 0.5 mg l^{-1} from June 24 to the end of July, and < 0.2 mg l^{-1} in August and September).

Thus runoff was enriched in nitrates at the very beginning of the rainy season and just after fertilization (urea), and in ammonium just after plowing, otherwise its $N-NO_3^-$ and $N-NH_4^+$ contents were slightly richer than concentrations in rainwater. For the year under study, losses in N dissolved in runoff could therefore be roughly estimated at 3 and 6 kg N ha^{-1} yr^{-1} for fallows (20% runoff) and crops (40% runoff), respectively. The soluble C in runoff was not measured due to its small concentration. It could however be estimated from N determinations and data in the literature: in Saria (central Burkina Faso), Roose (1981) reported C/N ratios in runoff water from 0.2 (crops) to 0.6 (fallows). Using these values, losses of soluble C in runoff could be roughly estimated at ca. 1 and 2 kg C ha^{-1} yr^{-1} for crops and fallows in 1999, respectively (Roose reported 5.4- and 1.1-kg ha^{-1} yr^{-1} losses of C dissolved in runoff on cereal and savanna plots, respectively).

9.3.5 Carbon, Nitrogen, and Fine Particle Losses by Erosion

For each event the suspension was multiplied by the average content of the type of erosive event on the considered plot, thus giving an assessment of the global quantity of C, N, and fine-exported elements (Table 9.3). This process, however, limited the precision of the study because the composition of the erosion could have evolved during the season or between years in the same manner as the content of the carriage.

In rather dry years (1998, 2000, and 2001), total annual eroded C ranged from 100 to 350 kg C ha^{-1} yr^{-1} on CROP, and from 0 to 50 kg C ha^{-1} yr^{-1} on the other plots. In 1999 (wet year), it reached 770 kg C ha^{-1} yr^{-1} on CROP, and 160 to 350 kg C ha^{-1} yr^{-1} on the other plots. Suspensions represented the major eroded fraction (> 50% in general) and their C content was much higher than that of coarse sediments (ca. 10 times more). As a consequence, suspensions generally accounted for more than 90% of total eroded C. Eroded C was much greater in 1999 than in drier years (i.e., 3, 10, and 23 times greater on CROP, SFAL, and CROP/GFAL, respectively). In 1998, 2000, and 2001, mean annual eroded C (total) was significantly and much greater on CROP than on the other plots ($p < 0.05$), but did not differ significantly between CROP/GFAL, GFAL, and SFAL. Similar relationships were found for mean carriage C, suspension C, total eroded N, eroded clay, and eroded fine silt. In 1999 (wet year), carriage C, suspension C, eroded C, eroded N, eroded clay, and eroded fine silt were also greater on CROP than on the other plots, but differences between plots were smaller than in 1998, 2000, and 2001 (drier years). The grass strip was effective in reducing C and N erosion, especially in rather dry years: when compared with CROP, annual eroded C and N on CROP/GFAL were 93 to 94% smaller in average in 1998, 2000, and 2001, and 55 to 62% smaller in 1999 (despite an accidental burning in the dry season of 1999).

9.4 DISCUSSION

9.4.1 Validation of Experimental Data on Erosion

Methods of measures used in this study were not the standard ones used in West Africa. Low width of plots is often criticized, even in case of ridging (Boli et al., 1993). However the results of

Table 9.3 Losses in Carbon, Nitrogen, Clay, and Fine Silt on the Experimental Plots

Period	Plot	Carriage C (kg C ha^{-1} yr^{-1})	Suspension C (kg C ha^{-1} yr^{-1})	Total Eroded C (kg C ha^{-1} yr^{-1})	Total Eroded N (kg N ha^{-1} yr^{-1})	Total Eroded Clay (t ha^{-1} yr^{-1})	Total Eroded Fine Silt (t ha^{-1} yr^{-1})
1998	CROP	3	105	108	8.4	1.1	1.4
	CROP/GFAL	0	4	4	0.3	0.1	0.1
	GFAL	0	0	0	0.0	0.0	0.0
	SFAL	3	44	47	3.5	0.1	0.1
1999	CROP	11	760	771	58.0	7.2	10.1
	CROP/GFAL	2	346	348	22.1	2.5	4.1
	GFAL	nd	nd	nd	nd	nd	nd
	SFAL	2	157	159	10.1	1.5	2.4
2000	CROP	5	265	270	20.6	2.9	3.9
	CROP/GFAL	0	25	25	2.1	0.3	0.3
	GFAL	1	0	1	0.1	0.0	0.0
	SFAL	0	0	0	0.0	0.0	0.0
2001	CROP	8	342	350	27.0	3.5	5.2
	CROP/GFAL	0	16	16	1.4	0.2	0.2
	GFAL	0	0	0	0.0	0.0	0.0
	SFAL	0	0	0	0.0	0.0	0.0
Average over 1998, 2000, and 2001[a]	CROP	5[a]	237[a]	243[a]	18.7[a]	2.5[a]	3.5[a]
	CROP/GFAL	0[b]	15[b]	15[b]	1.3[b]	0.2[b]	0.2[b]
	GFAL	0[b]	0[b]	0[b]	0.5[b]	0.0[b]	0.0[b]
	SFAL	1[b]	15[b]	16[b]	1.2[b]	0.1[b]	0.1[b]
Weighted average over four years[b]	CROP	6	289	296	22.6	3.0	4.2
	CROP/GFAL	0	48	48	3.4	0.4	0.6
	GFAL	nd	nd	nd	nd	nd	nd
	SFAL	1	29	30	2.1	0.2	0.3

Note: nd: not determined.

[a] Averages over 1998, 2000, and 2001 were compared using a Student t-test; within a column, two different letters mean that the difference was significant at $p < 0.05$.
[b] Weighted averages over four years were calculated as the sum of data relating to 1999 multiplied by 0.1 (as 1999 was the wettest year of the decade) and averages over 1998, 2000, and 2001 multiplied by 0.9.

the present study were close to those of other studies. On a field cultivated in a prolonged way (CROP), the erosion measured in dry or normal years ranged from 4 to 15 t ha^{-1} yr^{-1}. This was consistent with data reported by Roose (1993) on cultivated ferruginous soils in central Burkina Faso (Saria, 800 mm of annual rain, 0.7% slope), where erosion amounted to 3 to 20 t ha^{-1}·yr^{-1} on 20-m-long plots. In northern Cameroon (1000- to 1500-mm annual rainfall, 1 to 2% slope), Boli-Baboulé et al. (1999) measured erosion between 10 and 15 t ha^{-1} yr^{-1} for cotton-maize rotations involving plowing on recently cleared plots. In similar conditions but in southern Mali, Diallo et al. (2000) reported 18 t ha^{-1} yr^{-1} erosion. This confirmed the interest of plots with low width but great length, which were used in the present study.

In the wettest year of the decade (1999), erosion in CROP increased considerably: 37 t ha^{-1} on a 50-m-long plot, and 26 t ha^{-1} on a 20-m-long plot. Surprisingly, the cumulated runoffs were similar: 464 and 434 mm for the 20- and 50-m-long plots, respectively. The difference in soil losses could be attributed to a bias involving the relative concentration of suspensions by decantation of the overflowed barrels of the 50-m-long plot. Actually several differences appeared between plots, even when barrels did not overflow (Figure 9.3). The runoff events produced similar erosion on both plots when runoff maximal intensity was under a threshold. The difference between the two plots became positive when the peak of runoff exceeded a certain value, then increased with increasing runoff intensity. The erosion was less selective, too. It suggested that linear erosion

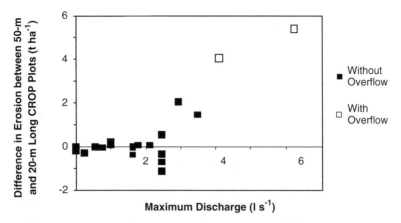

Figure 9.3 Difference in erosion between 50-m- and 20-m-long CROP plots, in relation to maximal discharge on the 50-m-long plot.

be added to sheet erosion on long ridged plots, as observed by Poesen and Bryan (1990) in experimental conditions.

The measurements of erosion were thus coherent. Major erosion in 1999 on the plots of 50 m was due to a combination of several factors: very humid climate, two exceptional erosive events with $I_{30} > 80$ mm h^{-1}, saturation of soils, especially on SFAL. In addition, the length of the slope and ridged relief determined some linear erosion under high rainfalls.

9.4.2 Physical Composition of the Eroded Soils and Consequences

The erosion sediments were separated in two classes: coarse carriage sediments and fine particles in suspension, including micro-particles. Solutes in runoff represented another source of loss. Carriage represents the coarse eroded fractions (litters, sands, and aggregates), which progress by saltation and are deposited in plane anfractuosities or on the zones of deceleration or spreading out of the sheet of runoff, for example on grass strips. This coarse erosion, which accounted only for 20% of the soil losses and 2% of C loss, was measured at the exit of the plots and thus classically considered as a soil loss. It seems also to take part in soil degradation (Collinet, 1988). However, these coarse sediments are poor in nutrients, fine particles, and C, and they settle not far from the eroded zone, in the first anfractuosities or obstacles, reducing the damages due to this type of erosion. This minimal distance covered by coarse sediments undoubtedly explains the relative scarcity of the "alluvial cones" in the current landscapes of cultivated savanna (compared with the Sahel landscapes) and thus gives the impression of small erosion to observers. This carriage is easily retained along bush hedges and narrow grass strips, which gives the illusion of high erosion control effectiveness of these hedges. If there were no strips, the same carriage would be retained in the water-furrows and other hollows.

The coarse particles contrast with suspensions, which accounted for 80% of erosion and primarily consisted in fine particles (75% of fine elements rich in organic matter and nutrients). Though we analyzed gathered samples, the results were consistent with those of Droux et al. (2000), which showed a composition of 50% clay and 40% silt in suspension in the water of Dounfin (Mali) draining into a water catchment of 17.5 km². These materials are a remote exportation and give sedimentation on the zones of spreading rivers, which, with a high flood risk, are seldom cultivated or used as pastures. Considering that eroded soils were sandy, the results confirmed the high selectivity of erosion in Sudanese cropping conditions. The selectivity of this rain erosion has often been observed in savanna regions (Roose, 1973; 1981; 1996; Lal, 1983) as well as in temperate grasslands (Fullen et al., 1998). This selective erosion is particularly detrimental for the farmer and the environment because it results in impoverished soil on the slopes. Organic matter being strongly associated or juxtaposed with fine elements, consequently fine elements and C have the same dynamics.

9.4.3 Carbon Losses and Carbon Selectivity

The C losses by erosion measured by Roose (1981) under sorghum in central Burkina Faso (Saria, 800-mm annual rainfall) amounted to 150 kg C ha^{-1} yr^{-1}, and to 9 kg C ha^{-1} yr^{-1} under savanna. Under cotton and maize, Boli-Baboulé et al. (1999) found 85 to 160 kg C ha^{-1} yr^{-1} in northern Cameroon, and Diallo et al. (2000), 330 kg C ha^{-1} yr^{-1} in southern Mali. In the present study, mean weighed annual eroded C was 296 kg C ha^{-1} yr^{-1} under cotton or maize, and 30 kg C ha^{-1} yr^{-1} under fallow. All these results presented the same order of magnitude.

As regarded C selectivity, the C enrichment ratio (CER, ratio between C content in sediments and in topsoils) was higher in the present experiment (10 for CROP) than in cited experiments (generally 1.5 to 3 for crops). For fine particles and N, enrichment ratios of sediments reached 6 and 6.8 in CROP, respectively, confirming that the selectivity of erosion was very high in the present experiment. The main erosion fraction was suspension sediments. Considering that the soil had been cropped for 7 years at least, it was particularly subjected to disaggregation and dispersion of its small particles, which were rich in C even if the soil was poor. In SFAL, the CER reached 5, which was consistent with values of 3 to 10 under natural vegetation given by cited experiments.

The loss of 1 or 2 kg C ha^{-1} yr^{-1} as C dissolved in runoff was the same order of magnitude as data reported by Roose (1981) in comparable conditions (5.4 and 1.1 kg C ha^{-1} yr^{-1} under cereals and savanna in central Burkina Faso, respectively). This fraction was negligible, however it was likely to be displaced at very long distances, exactly like the micro-particles and colloids fractions difficult to filter and also difficult to settle (less than 11 kg ha^{-1} yr^{-1} according to the results of the present study).

9.4.4 Possible Reasons for Constant Erosion over Years

Yet how does one explain the constant character of the erosion year after year? One could think that, once the fine surface was impoverished by rain erosion, the "sandy mulch" would protect the soil against rain. In the fallow, a cover of grass is apparently enough to fix this sandy mulch, producing a sorted micro-layer on the soil surface, known as "ST3" crust (coarse sands on surface, fine sands in the middle, clay and fine silt film underneath; Casenave and Valentin, 1989). In reality, fine materials extracted and transported by termites in these grasslands lead to an increasing of fine particle content in the topsoil. In the crops, termites in dry season and tillage in wet season (plowing, two or three weedings, ridging) regularly shift and reorganize the soil while bringing on the surface fine particles and associating them with organic matter. Tillage exposes the soil to strong erosion during erosive rainfalls, while the soil is in a fragmentary state. This homogenization results in constant erosion over years, impoverishing the soil gradually. During a cycle of fallow–cropping, a significant variation of the fine particles content has been measured in the 0 to 20 cm depth layer (synchronic measurements; Serpantié, 2003). Between the clearing of fallow and the end of the cropping phase, the fine particle content decreases in the soil (variation of 35 g kg^{-1}). During fallow, it increases. However it would be advantageous to undertake a global assessment of the fine particles because the textural variation remains small compared to the quantity of eroded fine elements. It is also necessary to take into account precisely the sedimentary contributions, which occur in fallow placed downstream from crops.

9.4.5 "Fallow" Effect

In normal or dry years, in unburned and ungrazed fallows, erosion was reduced by a factor of 10 to 20 with respect to cropped plots. The present results thus confirmed those of Roose (1993) and Diallo et al. (2000). Burning the shrubby fallow clearly promoted the runoff and erosion, worsening in the wettest year of the decade. The protective effect of the fallow vegetation increased during the rainy season in direct relationship to the installation and the growth of the herbaceous cover. Runoff and erosion were reduced under fallow, but the fallow also absorbed runoff coming from the upslope. Thus, in normal and dry years, the *Andropogon gayanus* large strip, which represented only 20% of the CROP/GFAL plot, could absorb almost all the runoff and all the solid erosion resulting from the upstream crop. The excess rain and accidental fire in 1999 limited its effectiveness. This confirmed the first results obtained by Fournier et al. (2000). Perennial grass fallow and mulch produced an obstacle and thus reduced the velocity of the runoff sheet and increased its thickness. This surface water flow also infiltrated quickly into the largest pores (in particular the macropores placed in relief on the grass tuft; Planchon and Janeau, 1990). This hypothesis was confirmed by the high infiltration capacity of the perennial grass cover under a sheet of water, on the soil of Block 1 (Table 9.1). What was the origin of such a functional macroporosity under *Andropogon gayanus*? There were no worm casts or fauna pores emerging in the surface. In the present case, the renewal of rooting (thick roots of *Andropogon gayanus*, which fasten the tuft to soil) created the macroporosity. The soil was preserved by its good structural state under tufts (Diallo et al., 1998). In the present experiment (Block 1), this effect of overinfiltration was perfectly expressed because the soil was well drained but easily encrusted in surface and packed under the plowed layer. Only the surface layer was thus opposed to the infiltration, from which came the great "*Andropogon*" effect. On a poorly drained soil, such as the soil of Block 2 (plot SFAL), or on the silty soils of the plains, the "*Andropogon*" effect would be less obvious as suggested by the tests of infiltration of the present study. On these soils, a rate of absorbing fallows of 20% of land should be insufficient. Absorption effect was not at its optimum in 1999 because of an accidental burning of the herbaceous cover and mulch and also due to the existence of runoff resulting from soil saturation by water.

9.4.6 Impact of Soil Erosion

The assessment of erosion in terms of economical and environmental impact is still greatly lacking. Forgetting the scaling factor is a classical mistake. It is often thought that erosion plays a significant role lower production. The N losses are indeed of the same order and magnitude as the deficit of N of the mineral crop balance, which is 25 kg N ha^{-1} yr^{-1} for a cotton-sorghum rotation in Bondoukui (Serpantié, 2003). It is thus a fact that erosion reduces locally the fertility. However this can be compensated by the participation of sedimentation and regeneration of the downstream fallow land, which will be later put into cropping. With annual losses averaging 250 kg C ha^{-1} and 20 kg N ha^{-1}, ordinary erosion locally appears like a dead loss, benefiting downstream sites. However, its impact on the farming system and environment is lower in a cropping system where crops are mixed with fallows, as compared to a completely cropped landscape.

A different conclusion is reached in wet years, because C and N losses are much more severe (770 kg C ha^{-1} and 58 kg N ha^{-1}), even in landscapes including fallows. In such climatic conditions, cultivated soils are brutally impoverished if they are recently tilled and very incompletely covered. It is a dramatic loss because this C is soil organic matter, therefore is a stable organic matter compared with fresh organic residues.

9.5 CONCLUSION

In the Sudanese cotton zones, the selective erosion of the surface layers removes fine particles, C, and nutrients from cultivated plots to downstream areas. Protected fallows (unburned and ungrazed) located on the same slope are potential reception areas, if the soil is well drained. Consequently, according to the type of landscape and to possible burning and grazing of these fallows, erosion can be strongly modified as compared with measurements made on plots managed in a homogeneous way. This process of spatial redistribution of surface water and "fertility" by erosion and sedimentation has only been briefly described in studies (Vallet, 1999). In the process of restoration of the soil fertility by fallows, sedimentation is thus an important process to take into account, though generally forgotten.

In regards to the erosion control, organic matter and nutrient losses over long distances can be reduced by an adequate distribution of fallows or artificial meadows within the cultivated fields or hillsides. Modeling should take into account this process of absorption on various types of soil, and the pattern of land use (in chess-work, in strips). Classic grass strips, which are insufficient to control the runoff losses and the erosion of fine elements (Boli et al., 1993), should also be considered again with larger strips and according to draining capacity of soils. In case of less-filtering soils, and taking into account wet years, erosion could be controlled by a higher proportion of areas covered by perennial grass fallows (Serpantié and Madibaye, 1999; Fournier et al., 2000), by covering practices (mulch, cover crops, associated crops, Mucuna, Cucurbitacea, etc.), reduction of tillage by the employment of weed-killer, hydraulic micro-dams (stone lines). Further research should also consider other linked questions: water-catchment scale (in order to appreciate the complex effects of such associations of crops and fallow), farm or social organization for efficient erosion control, and consequences of strong infiltration on the chemical evolution of the soils.

ACKNOWLEDGMENTS

This work was a shared research between the Institut de l'Environnement et de Recherches Agricoles (INERA, Burkina Faso), the Institut de Recherche pour le Développement (IRD, France), and the Ecole Inter-Etats des Techniciens de l'Hydraulique et de l'Equipement Rural (ETSHER, Ouagadougou) (D.S.O. project number BF 002702 founded by the Netherlands).

The authors thank all contributors, especially M. Da Sewa Silveira and M. Sako Tahirou for topographical studies, plot implementation, and after-rain measurements. They also thank M. P. Zahonero (ETSHER) for his help, and M. E. Roose and M. B. Barthès for their comments on the manuscript.

REFERENCES

Albergel, J., M. Diatta, and Y. Pépin. 2000. Aménagement hydraulique et bocage dans le bassin arachidier du Sénégal, in C. Floret and R. Pontanier, eds., *La Jachère en Afrique Tropicale. Rôles, Aménagement, Alternatives*. John Libbey Eurotext, Paris, pp. 741–751.

Bep A Ziem, B., Z. B. Boli, and E. Roose. 2002. Pertes de carbone par érosion hydrique et évolution des stocks de carbone sous rotation intensive coton/maïs sur des sols ferrugineux sableux du Nord Cameroun. Communication at the International Colloquium Land Use, Erosion and Carbon Sequestration, Montpellier, France, 23–28 September 2002.

Blancaneaux, P., P. L. De Freitas, R. F. Amabile, and A. M. De Carvalho. 1993. Le semis direct comme pratique de conservation des sols des cerrados du Brésil central. *Cahiers ORSTOM, série Pédologie* 28:253–275.

Boli, Z., E. Roose, B. Bep A Ziem, K. Sanon, and F. Waechter.1993. Effet des techniques culturales sur le ruissellement, l'érosion et la production de coton et maïs sur un sol ferrugineux tropical sableux. Recherche de systèmes de cultures intensifs et durables en région soudanienne du Nord-Cameroun. *Cahiers ORSTOM, série Pédologie* 28:309–326.

Boli-Baboulé, Z., E. Roose, and B. Bep A Ziem. 1999. Effets du labour et de la couverture du sol sur le ruissellement et les pertes en terre sur un sol ferrugineux sableux en zone soudanienne du Nord-Cameroun. *Bulletin du Réseau Erosion* 19:372–378.

Bremner, J. M. 1996. Nitrogen — Total, in D. L. Sparks, A. L. Page, P. A. Helmke, R. H. Loeppert, P. N. Soltanpour, M. A. Tabatabai, C. T. Johnson, and M. E. Sumner, eds., *Methods of Soil Analysis: Part 3 — Chemical Methods*, Soil Science Society of America and American Society of Agronomy, Madison, WI, pp. 1085–1121.

Casenave, A. and C. Valentin. 1989. *Les etats de surface de la zone Sahélienne. Influence sur l'infiltration*. Editions de l'ORSTOM, Paris.

Collinet, J. 1988. *Comportement hydrodynamique et erosif des sols de l'Afrique de l'Ouest*. Ph.D. Dissertation, Université de Strasbourg, France.

Diallo, D., E. Roose, B. Barthès, B. Khamsouk, and J. Asseline. 1998. Recherche d'indicateurs d'érodibilité des sols dans le bassin versant de Djitiko (haut bassin du Niger au Sud Mali). *Bulletin du Réseau Erosion* 18:336–347.

Diallo, D., D. Orange, E. Roose, and A. Morel. 2000. Potentiel de production de sédiments dans le bassin versant de Djitiko (103 km^2). Zone soudanienne du Mali Sud. *Bulletin du Réseau Erosion* 20:54–66.

Droux, J. P., M. Mietton, and J. C. Olivry. 2000. Une mesure de l'érosion hydrique mécanique et chimique sur un petit bassin versant gréseux en zone tropicale soudanienne: Le cas du Dounfing, Mali. *Bulletin du Réseau Erosion* 20:68–88.

Feller, C. 1995. *La matière organique dans les sols tropicaux à Argile 1:1. Recherche de compartiments fonctionnels. Une approche granulométrique*. Editions de l'ORSTOM, Paris.

Fournier, J. 1998. Expérimentation d'une nouvelle méthode d'infiltrométrie. *Sud Sciences et Technologies* 1:15–20.

Fournier, J., G. Serpantié, J. P. Delhoume, and R. Gathelier. 2000. Rôle des jachères sur les écoulements de surface et l'érosion en zone soudanienne du Burkina Faso. Application à l'aménagement des versants, in C. Floret and R. Pontanier, eds., *La Jachère en Afrique Tropicale. Rôles, Aménagement, Alternatives*. John Libbey Eurotext, Paris, pp. 179–188.

Fullen, M. A., Zhi Bo Wu, and R. T. Brandsma. 1998. A comparison of the texture of grassland and eroded sandy soils from Shropshire, UK. *Soil and Tillage Research* 46:301–305.

Gee, G. W. and J. W. Bauder. 1986. Particle-size analysis, in A. Klute, ed., *Methods of Soil Analysis: Part 1 - Physical and Mineralogical Methods*, 2nd ed., American Society of Agronomy, Madison, WI, pp. 383–411.

Lal, R. 1975. Role of mulching techniques in tropical soil and water management. Technical Bulletin No. 1, IITA, Ibadan, Nigeria.

Lal, R., 1983. Soil erosion and its relation to productivity in tropical soils, in, S. A. El-Swaïfy, W. C. Moldenhauer, and A. Lo, eds., *Soil Erosion and Conservation*. Soil Conservation Society of America, Ankeny, IA, pp. 237–247.

Maass, J. M., C. F. Jordan, and J. Sarukhan. 1988. Soil erosion and nutrient losses in seasonal tropical agroecosystems under various management techniques. *Journal of Applied Ecology* 25:595–607.

Mietton, M., 1988. *Dynamique de l'Interface Lithosphère-Atmosphère au Burkina Faso. L'Erosion en Zone de Savane*. Ph.D. Dissertation, Université de Grenoble, France.

Mulvaney, R. L. 1996. Nitrogen — Inorganic Forms, in D. L. Sparks, A. L. Page, P. A. Helmke, R. H. Loeppert, P. N. Soltanpour, M. A. Tabatabai, C. T. Johnson, and M. E. Sumner, eds., *Methods of Soil Analysis: Part 3 — Chemical Methods*. Soil Science Society of America and American Society of Agronomy, Madison, WI, pp. 1123–1184.

Nelson, D. W. and L. E. Sommers. 1996. Total carbon, organic carbon, and organic matter, in D. L. Sparks, A. L. Page, P. A. Helmke, R. H. Loeppert, P. N. Soltanpour, M. A. Tabatabai, C. T. Johnson, and M. E. Sumner, eds., *Methods of Soil Analysis: Part 3 — Chemical Methods*, Soil Science Society of America and American Society of Agronomy, Madison, WI, pp. 961–1010.

Nimy, V. 1999. *Etude des techniques palliatives à la raréfaction des jachères. Cas de la fonction "Conservation des eaux et des sols."* IDR Thesis, Bobo-Dioulasso, Burkina Faso, pp. 961–1010.

Pieri, C. 1991. *Fertility of Soils: A Future for Farming in the West African Savannah*. Springer-Verlag, Berlin.

Planchon, O. and J. L. Janeau. 1990. Le fonctionnement hydrodynamique à l'échelle du versant, in Equipe HYPERBAV, ed., *Structure et Fonctionnement Hydropédologique d'un Petit Bassin Versant de Savane Humide*. Editions de l'ORSTOM, Paris, pp. 165–183.

Poesen, J. W. A. and R. B. Bryan. 1990. Influence de la longueur de pente sur le ruissellement: Rôle de la formation de rigoles et de croûtes de sédimentation. *Cahiers ORSTOM, série Pédologie* 28:71–80.

Pontanier, R., H. Moukouri-Kuoh, R. Sayol, L. Seyni-Boukar, and B. Thébé. 1986. Apport de l'infiltromètre à aspersion pour l'évaluation des ressources en sol des zones soudano-sahéliennes du Cameroun, in *Deuxièmes Journées Hydrologiques de l'ORSTOM*, Montpellier, 16–17 October 1986, Editions de l'ORSTOM, Paris.

Roose, E. 1967. Dix années de mesure de l'érosion et du ruissellement au Sénégal. *L'Agronomie Tropicale* 22:123–152.

Roose, E. 1973. *Dix-Sept années de mesures expérimentales de l'erosion et du ruissellement sur un sol ferrallitique sableaux de Basse Côte d'Ivoire. Contribution à l'etude de l'erosion hydrique en milieu intertropical*. Ph.D. Dissertation, Université d'Abidjan.

Roose, E. 1977. Erosion et ruissellement en Afrique de l'Ouest: Vingt années de mesures en petites parcelles Expérimentales. *Travaux et Documents ORSTOM* 78, ORSTOM, Paris.

Roose, E. 1981. *Dynamique actuelle de sols ferrallitiques et ferrugineux tropicaux d'Afrique Occidentale. Etude expérimentale des transferts hydrologiques et biologiques de matières sous végétations naturelles ou cultivées*. Editions de l'ORSTOM, Paris.

Roose, E. 1993. Capacité des jachères à restaurer la fertilité des sols pauvres en zone soudano-sahélienne d'Afrique occidentale, in C. Floret and G. Serpantié, eds., *La Jachère en Afrique de l'Ouest*. Editions de l'ORSTOM, Paris, pp. 223–244.

Roose, E. 1996. Land husbandry: Components and strategy. *FAO Soils Bulletin* 70, FAO, Rome.

Roose, E. and R. Bertrand. 1971. Contribution à l'étude de la méthode des bandes d'arrêt pour lutter contre l'érosion hydrique en Afrique de l'Ouest. Résultats expérimentaux et observations sur le terrain. *L'Agronomie Tropicale* 26:1270–1283.

Serpantié, G. 2003. *Persistance de la culture temporaire dans les savanes cotonnières d'Afrique de l'Ouest. Etude de cas au Burkina Faso*. Ph.D. Dissertation, INA Paris-Grignon.

Serpantié, G. and D. Madibaye. 1999. Recherches participatives sur la culture de *Andropogon gayanus* Kunth var. *tridentatus* Hack en zone soudanienne. Essais participatifs d'installation de peuplements (Bondoukui et Béréba, Burkina Faso), in *Actes de l'Atelier Régional (Proceedings of the Regional Workshop)* Cultures Fourragères et Développement Durable en Zone Sub-Humide, Korhogo, Côte d'Ivoire, 26–29 May 1997. CIRDES, Bobo Dioulasso, Burkina Faso, IDESSA, Bouaké, Côte d'Ivoire, and CIRAD, Montpellier, France, pp. 191–204.

Valet, S. 1999. L'aménagement traditionnel des versants et le maintien des cultures associées: cas de l'Ouest-Cameroun. *Bulletin du Réseau Erosion* 19:37–68.

Van Campen, W. and D. Kebe. 1986. Lutte anti-érosive dans la zone cotonnière au Mali-Sud, in Séminaire (Workshop) Aménagement Hydro-Agricole et Systèmes de Production, CIRAD-DSA, Montpellier, France, 16–19 November 1986, pp. 67–77.

Wischmeier, W. H. 1974. New developments in estimating water erosion, in Proceedings of the 29th Annual Meeting of the Soil Conservation Society of America, Syracuse, NY, pp. 179–186.

Wischmeier, W.H. and D. D. Smith. 1978. *Predicting Rainfall Erosion Losses: A Guide to Erosion Planning*. Agriculture Handbook 537, United States Department of Agriculture, Washington D.C.

CHAPTER 10

Effect of a Legume Cover Crop on Carbon Storage and Erosion in an Ultisol under Maize Cultivation in Southern Benin

Bernard Barthès, Anastase Azontonde, Eric Blanchart, Cyril Girardin, Cécile Villenave, Robert Oliver, and Christian Feller

CONTENTS

10.1 Introduction 143
10.2 Materials and Methods 144
    10.2.1 Description of the Site and Treatments 144
    10.2.2 Soil and Plant Sampling 145
    10.2.3 Carbon and Nitrogen Determination, and Other Analyses 145
    10.2.4 Determinations of Runoff, Soil Losses, and Eroded Carbon 145
    10.2.5 Statistical Analyses 146
10.3 Results 146
    10.3.1 General Properties of Soils 146
    10.3.2 Soil Carbon 146
    10.3.3 Residue Biomass 147
    10.3.4 Runoff, Soil Losses, and Eroded Carbon 149
10.4 Discussion 149
    10.4.1 Changes in Soil Carbon 149
    10.4.2 Residue Biomass 149
    10.4.3 Nitrous Oxide Emissions 150
    10.4.4 Runoff, Soil Losses, and Eroded Carbon 151
10.5 Conclusion 153
Acknowledgments 154
References 154

10.1 INTRODUCTION

Soil organic matter (SOM) management is recognized as a cornerstone for successful farming in most tropical areas, with or without the application of mineral fertilizers (Merckx et al., 2001). Several experiments have demonstrated the direct or indirect positive effects of SOM on chemical, physical and biological properties of soil related to plant response (Sanchez, 1976; Pieri, 1991). Moreover, SOM is an essential reservoir of carbon (C), and SOM management can have significant implications on the global C balance and thus on climate change (Craswell and Lefroy, 2001). In many rural areas of the tropics, the environmental challenge consists of reducing deforestation, increasing organic matter storage in cultivated soils, and reducing soil erosion. Therefore, under the economical conditions prevailing in developing countries, maintaining soil fertility and meeting the environmental challenge require land-use practices that include high levels of organic inputs and soil organic C sequestration (Feller et al., 2001).

Natural fallowing has long been the main practice to maintain soil fertility in tropical areas. However, as its effects only become significant after a period of at least 5 years, natural fallowing is no longer possible in the context of increasing population. Such is precisely the case in southern Benin, where the population density is 300 to 400 inhabitants km^{-2} (Azontonde, 1993). The benefits of legume-based cover crops in Africa (in regions with annual rainfall > 800 mm) as an alternative to natural fallow, to control weeds and soil erosion, and enrich soil organic matter and N are widely recognized (Voelkner, 1979; Raunet et al., 1999; Carsky et al., 2001). In southwestern Nigeria, higher maize (*Zea mays*) yields were obtained in live mulch plots under *Centrosema pubescens* or *Psophocarpus palustris* than in conventionally tilled and no-till plots for four consecutive seasons (Akobundu, 1980).

The effect of relay-cropping maize through *Mucuna pruriens* (var. *utilis*) was assessed in southern Benin from 1988 to 1999 in terms of plant productivity and soil fertility (Azontonde, 1993; Azontonde et al., 1998). The relay-cropping system (M) was compared with traditional maize cropping system without any input (T), and with a maize cropping system with mineral fertilizers (NPK). This chapter focuses on changes in soil C during the period of the experiment in relation to residue biomass C returned to the soil, runoff and soil erosion losses, and loss of C with erosion.

10.2 MATERIALS AND METHODS

10.2.1 Description of the Site and Treatments

The experiment was conducted from 1988 to 1999 at an experimental farm at Agonkanmey (6°24'N, 2°20' E), near Cotonou in southern Benin in an area of low plateaus. The climate is subhumid-tropical with two rainy seasons (March to July and September to November). Mean annual rainfall is 1200 mm and mean annual temperature is 27°C. The soils are classified as Typic Kandiustult (Soil Survey Staff, 1994) or Dystric Nitisols (FAO-ISRIC-ISSS, 1998), and have a sandy loam surface layer overlying a sandy clay loam layer at about 50 cm depth. Most of the land is cultivated to maize (*Zea mays*), beans (*Vigna* sp.), cassava (*Manihot esculenta*), or peanuts (*Arachis hypogea*), often associated with oil palm (*Elaeis guineensis*).

The study was conducted on three 30 × 8 m plots on a 4% slope. These demonstration plots were not replicated, as it is usually difficult in long-term experiments (Shang and Tiessen, 2000), especially when these include runoff plots. Three cropping systems were compared: T (traditional), maize without any input; NPK, maize with mineral fertilizers (200 kg ha^{-1} of NPK 15-15-15, and 100 kg ha^{-1} of urea); M, relay-cropping of maize and a legume cover crop, *Mucuna pruriens* var. *utilis*, with no fertilizer. Maize (var. *DMR*) was cropped during the first rainy season with shallow hoe tillage by hand (hoeing depth was about 5 cm). In M plot, maize was sown through the mucuna mulch from the previous year. Mucuna was sown one month later, and once maize had been harvested, its growth as a relay-crop continued until the end of the second (short) rainy season. During this short rainy season, the T and NPK treatments were maintained as natural fallow. Additional information on the site and soil properties has been provided by Azontonde (1993) and Azontonde et al. (1998).

10.2.2 Soil and Plant Sampling

Undisturbed soil profile samples were collected: (1) in March, June, August, and October 1988 and 1995, at 18 locations per plot for 0 to 10, 10 to 20, and 20 to 40 cm depths, using 0.2-dm^3 soil cores, and (2) in November 1999 at three locations per plot for 0 to 10 and 10 to 20 cm depths in two replicates, and for 20 to 30, 30 to 40, and 50 to 60 cm depths in one replicate, using 0.5-dm^3 soil cores. Soil samples were also obtained with a knife for different depths along the profile walls. Soil bulk density (Db) was determined after oven-drying core samples, whereas the other samples were air-dried, sieved (2 mm) and ground (< 0.2 mm) for C and N analyses.

Aboveground biomass of maize and mucuna was determined every year from five replicates (1 × 1 m) at maize harvest (August) and at mucuna maximum growth (October), respectively. In 1995, following the same pattern, roots of maize and mucuna were collected for 0 to 10, 10 to 20, and 20 to 40 cm depths, and hand-sorted (Azontonde et al., 1998). Annual root biomass was calculated using the ratio of below- to aboveground biomass determined in 1995, and the annual aboveground biomass. Sampling of the aboveground biomass of weeds was done in November 1999 at nine locations per plot, using a 0.25 × 0.25-m frame. Litter was simultaneously and similarly sampled. Root sampling was also carried out in November 1999 on six 0.25 × 0.25 × 0.30-m monoliths per plot: monoliths were cut into three layers (corresponding to 0 to 10, 10 to 20, and 20 to 30 cm depths), and visible roots were hand sorted. With respect to the vegetation cover, we assumed that roots and litter sampled in T and NPK originated from weeds, whereas those sampled in M originated from mucuna. All plant samples were dried at 70°C, weighed for biomass measurement, and finely ground for C determination.

10.2.3 Carbon and Nitrogen Determination, and Other Analyses

Total C content (Ct) of soil samples collected in 1988 and 1995 was determined by the Walkley and Black method (WB), and total N content (Nt) by the Kjehldahl method. Both Ct and Nt of soil samples collected in 1999 were determined by the dry combustion method (DC) using an Elemental Analyser (Carlo Erba NA, 1500). The Ct was analyzed on 60 samples using both WB and DC methods, leading to a relationship (r = 0.971) that was used to convert WB data into DC data. All Ct data are thereafter expressed on a DC basis. The C content of plant samples was determined by dry combustion using an Elemental Analyser (CHN LECO, 600).

Particle-size analysis was performed by the pipette method after removal of organic matter with H_2O_2 and dispersion by Na-hexametaphosphate. Soil pH in water was determined using a 1:2.5 volumetric soil:solution ratio.

10.2.4 Determinations of Runoff, Soil Losses, and Eroded Carbon

Each plot was surrounded by half-buried metal sheets and fitted out with a collector draining runoff and sediments toward two covered tanks set up in series. When the first tank was full, additional flow moved through a divisor into the second tank, both with a capacity of 3-m^3. Runoff and soil loss data were collected from 1993 to 1997.

Runoff amount (m^3) was assessed on every plot after each rainfall event or sequence of events, by measuring the volume of water in each tank and multiplying it by a coefficient depending on divisors. This runoff amount was converted to depth on the basis of the plot area. Annual runoff rate (mm mm^{-1}) was defined as the ratio of annual runoff depth to annual rainfall, and mean annual runoff rate as the ratio of runoff depth to rainfall over 5 years.

The amount of dry coarse sediments (Mg) was deduced by weighing wet coarse sediments collected in the first tank, and oven-drying the aliquots. The quantity of suspended sediments (Mg) was assessed by flocculation and oven-drying of aliquots collected from each tank. Annual soil losses (Mg ha^{-1} yr^{-1}) were computed as the sum of dry-coarse and -suspended sediments over 1 year, and averaged over 5 years to calculate mean annual soil losses.

Annual eroded C (Mg C ha^{-1} yr^{-1}) was calculated as the product of annual soil losses by C content of sediments. Sediment C content was not measured, but was estimated as the product of soil Ct (at 0 to 10 cm depth, for the year under consideration) by an enrichment ratio (Starr et al. 2000). Soil Ct for the year under consideration was interpolated from soil Ct measurements carried out in 1988, 1995, and 1999. The enrichment ratio, defined as the ratio of Ct in sediments to that in the soil (0 to 10 cm depth), was estimated from the data in the literature: on light-textured Ultisols and Oxisols under maize cultivation (with mineral fertilizers) in southern and northern Ivory Coast, with 2100- and 1350-mm annual rainfall, respectively, C enrichment ratios measured in runoff plots by Roose (1980a, 1980b) were 1.9 (7% slope) and 1.4 (3% slope), respectively. Thus, C enrichment ratio of 1.6 was assumed for maize plots (T and NPK). In the absence of literature data regarding cover crops, C enrichment ratio under maize-mucuna (M) was found similar to those measured in runoff plots having comparable soil cover conditions: for two light-textured Oxisols under bush savannas in northern Ivory Coast, with 1200- and 1350-mm annual rainfall, respectively, C enrichment ratio measured by Roose and Bertrand (1972) and Roose (1980b) was 2.6 (4% slope) and 3.4 (3% slope), respectively; and for a sandy Ultisol under banana plantation in southern Ivory Coast (14% slope, 1800-mm annual rainfall), C enrichment ratio was 3.0 (Roose and Godefroy, 1977). Averaging these data, C enrichment ratio of 3.0 was assumed for maize-mucuna (M) rotation. Dissolved C in runoff was neither measured nor taken into consideration.

10.2.5 Statistical Analyses

Differences in mean Ct and Ct stocks were tested by a Student unpaired t-test. Differences in mean annual runoff rates, soil losses, and eroded C were tested by a paired t-test. In both cases, no assumptions were made on normality and variance equality (Dagnélie, 1975).

10.3 RESULTS

10.3.1 General Properties of Soils (Table 10.1)

The clay (< 2 μm) content of the soil ranged between 110 and 150 g kg^{-1} for 0 to 10 cm depth in 1988, and it increased between 1988 and 1999 in T (50%) but not in NPK and M treatments (increase < 15%). The clay content also increased with depth. Moreover, clay content in 1999 was higher at 0 to 10 cm in T than at 10 to 20 cm in NPK and M treatments. The sand (> 50 μm) content was between 600 and 800 g kg^{-1} to 20 cm depth, mainly in the form of coarse sand (> 200 μm) (data not shown). Soil pH was acidic (< 6) and decreased between 1988 and 1999, especially in T and NPK treatments (–0.5 over a decade).

10.3.2 Soil Carbon

Total soil carbon content Ct (g C kg^{-1} soil) was determined through 18- and three-replicate sampling in March 1988 and November 1999, respectively (Table 10.1). The validity of the latter was assessed using 18-replicate sampling done in October 1995 as a reference: following Dagnélie (1975) and Shang and Tiessen (2000), at 95% confidence level, irrespective of the plot and the depth, three-replicate sampling in 1995 would have led to a less than 8% relative error in Ct estimation. Thus, Ct determined in 1999 by three-replicate sampling was representative of the mean value of the plot. Similarly, Ct stock (Mg C ha^{-1}) estimated in November 1999 was representative of the large area.

Table 10.1 Soil Clay Content, pH in Water, Total Carbon Content Ct, C:N Ratio, and Total Carbon Stock in 1988 and 1999 (Mean ± Standard Deviation when Available)

	Depth (cm)	T 1988	T 1999	NPK 1988	NPK 1999	M 1988	M 1999
Clay (g kg^{-1})	0–10	147 ± 1	216	111 ± 6	128	127 ± 6	136
	10–20	nd	339	nd	198	nd	179
pH	0–10	5.6 ± 0.1	5.1	5.6 ± 0.1	5.2	5.2 ± 0.1	5.0
	10–20	5.4 ± 0.2	4.7	5.4 ± 0.2	5.0	5.1 ± 0.2	5.0
Ct (g kg^{-1})	0–10	5.5 ± 0.2	5.3 ± 0.1	5.4 ± 0.1	6.7 ± 1.8	5.2 ± 0.1	11.5 ± 2.0
	10–20	4.6 ± 0.3	4.0 ± 0.7	4.8 ± 0.4	3.8 ± 1.2	4.8 ± 0.4	7.3 ± 0.9
	20–30[a]	4.1 ± 0.2	3.5 ± 0.5	4.0 ± 0.4	3.6 ± 1.1	4.6 ± 0.3	4.4 ± 0.1
	30–40[a]		3.2 ± 0.1		4.1 ± 0.7		4.2 ± 0.2
	50–60	nd	2.4 ± 0.1	nd	3.5 ± 1.8	nd	3.3 ± 0.5
C:N	0–10	10.2 ± 1.0	12.2 ± 0.4	10.8 ± 0.5	11.3 ± 0.1	11.5 ± 0.5	11.9 ± 0.8
	10–20	10.9 ± 1.4	10.1 ± 0.6	10.7 ± 1.8	9.9 ± 0.7	12.0 ± 1.8	11.6 ± 0.8
	20–30[a]	11.4 ± 1.2	8.7 ± 0.5	10.6 ± 1.9	9.3 ± 1.0	12.8 ± 1.7	10.0 ± 1.2
	30–40[a]		8.2 ± 0.8		8.8 ± 1.4		8.9 ± 1.3
	50–60	nd	7.0 ± 0.4	nd	8.8 ± 3.2	nd	8.1 ± 1.4
Ct stock (Mg C ha^{-1})	0–10	7.7 ± 0.7	8.4 ± 0.3	7.3 ± 0.5	10.6 ± 3.4	6.8 ± 0.3	17.4 ± 3.3
	0–20	13.6 ± 0.9	14.5 ± 0.4	14.6 ± 1.0	17.0 ± 3.9	13.8 ± 0.8	28.7 ± 3.9
	0–40	25.9 ± 1.5	24.2 ± 0.5	27.0 ± 1.8	28.8 ± 5.7	27.7 ± 1.7	41.4 ± 4.9
	0–60	nd	32.0 ± 0.3	nd	39.7 ± 3.6	nd	51.7 ± 4.1

Note: nd: not determined.
[a] 20 to 40 cm in 1988.

Differences in Ct between plots were negligible (< 2% at 0 to 20 cm) in March 1988. Between March 1988 and November 1999, Ct increased considerably at 0 to 20 cm depth in M (90%, $p < 0.01$) but changed slightly in T (–8%) and NPK (3%), and for 20 to 40 cm depth (changes < 20%). In November 1999, and as a consequence, Ct at 0 to 20 cm depth was much greater in M than in T (100%, $p < 0.01$) and NPK (80%, $p < 0.05$) treatments. Differences between plots were rather small below this depth, as were differences between NPK and T (< 30% in general) treatments.

Changes in Ct stock (Mg C ha^{-1}) for 0 to 40 cm depth were similar showing small initial differences between plots (< 7%); between March 1988 and November 1999, slight changes in T and NPK (< 15%) treatments but a considerable increase in M (50%, $p < 0.01$); higher final Ct stock in M than in T (70%, $p < 0.01$) and NPK (45%, $p < 0.05$) treatments. Stock of Ct for 0 to 40 cm depth finally attained the value of 24, 29, and 41 Mg C ha^{-1} in T, NPK, and M, respectively. Between 1988 and 1999, mean (± standard deviation) annual changes in Ct stock were 0.1 (±0.1), 0.2 (±0.4), and 1.4 (±0.4) Mg C ha^{-1} yr^{-1} in T, NPK, and M, respectively, for 0 to 20 cm depth; and –0.2 (±0.1), 0.2 (±0.5), and 1.3 (±0.5) Mg C ha^{-1} yr^{-1}, respectively, for 0 to 40 cm depth.

10.3.3 Residue Biomass

Average annual residue biomass (dry matter) returned to the soil in T, NPK, and M was 8.0, 13.0, and 19.9 Mg ha^{-1} yr^{-1}, with 35, 72, and 82% of aboveground biomass, respectively (Table 10.2). Mean annual residue C added was 3.5, 6.4, and 10.0 Mg C ha^{-1} yr^{-1}, with 39, 74, and 84% as aboveground biomass, respectively (aboveground biomass had a slightly more C content than roots). Returned C mainly originated from weeds in T (55% as roots and 17% as aboveground biomass), which represented 44 and 92% of aboveground and belowground residue C, respectively. In contrast, returned C in NPK was mainly from maize (61% as aboveground biomass and 14% as roots). In M, maize and mucuna accounted for similar amounts of residue C, either as aboveground biomass (about 40% each) or roots (8% each). Moreover, maize residue biomass C was of the same order of magnitude in NPK and M (ca. 5 Mg C ha^{-1} yr^{-1}) treatments.

Table 10.2 Residue Biomass Returned to the Soil (Mean ± Standard Deviation)

	Residue Biomass (Dry Matter) (Mg ha^{-1} yr^{-1})			C Content of Residues (g C kg^{-1})			Residue Biomass C (Mg C ha^{-1} yr^{-1})			C:N of Residues		
Origin	T	NPK	M	T	NPK	M	T	NPK	M	T	NPK	M
Maize												
Aboveground	1.44 ± 0.06	7.46 ± 0.19	8.05 ± 0.20	533 ± 26	524 ± 16	538 ± 10	0.77 ± 0.03	3.91 ± 0.12	4.33 ± 0.12	118	75	84
Roots	0.42 ± 0.01	1.76 ± 0.05	1.67 ± 0.07	474 ± 20	508 ± 20	456 ± 21	0.20 ± 0.01	0.89 ± 0.04	0.76 ± 0.03	121	78	91
Subtotal	1.86 ± 0.07	9.22 ± 0.19	9.72 ± 0.21	519 ± 33	521 ± 26	524 ± 23	0.97 ± 0.03	4.80 ± 0.12	5.10 ± 0.13	118	75	86
Mucuna												
Aboveground	0.00	0.00	8.34 ± 0.24	—	—	488 ± 25	0.00	0.00	4.07 ± 0.12	—	—	18
Roots	0.00	0.00	1.88 ± 0.06	—	—	455 ± 20	0.00	0.00	0.85 ± 0.03	—	—	22
Subtotal	0.00	0.00	10.22 ± 0.25	—	—	482 ± 32	0.00	0.00	4.93 ± 0.13	—	—	19
Weeds												
Aboveground	1.36 ± 0.39	1.89 ± 0.29	0.00	440 ± 12	430 ± 10	—	0.60 ± 0.18	0.82 ± 0.14	0.00	35	25	—
Roots[a]	4.77 ± 1.81	1.89 ± 0.92	0.00	400	400	—	1.91 ± 0.72	0.75 ± 0.37	0.00	nd	nd	—
Subtotal	6.13 ± 1.85	3.78 ± 0.96	0.00	409	415	—	2.51 ± 0.74	1.57 ± 0.40	0.00	nd	nd	—
Subtotal												
Aboveground	2.80 ± 0.40	9.35 ± 0.34	16.39 ± 0.32	488 ± 29	505 ± 19	513 ± 27	1.37 ± 0.18	4.73 ± 0.18	8.41 ± 0.17	78	65	51
Roots	5.19 ± 1.81	3.65 ± 0.92	3.55 ± 0.09	406	452	455 ± 29	2.11 ± 0.72	1.64 ± 0.37	1.61 ± 0.04	nd	nd	55
Total	7.99 ± 1.85	13.00 ± 0.98	19.94 ± 0.33	435	490	503 ± 40	3.48 ± 0.74	6.37 ± 0.41	10.02 ± 0.18	nd	nd	52

Note: nd: not determined.

[a] 0 to 30 cm; data resulting from sampling carried out in November 1999, assuming that roots collected in T and NPK were weed roots only and had a C content of 400 g C kg^{-1}.

Table 10.3 Annual Runoff Rates, Soil Losses, and C Erosion

Year	Rainfall (mm yr^{-1})	Runoff Rate (mm mm^{-1})			Soil Losses (Mg ha^{-1} yr^{-1})			C Erosion (Mg C ha^{-1} yr^{-1})		
		T	NPK	M	T	NPK	M	T	NPK	M
1993	1288	0.30	0.13	0.09	41.5	9.8	3.1	0.4	0.1	0.1
1994	1027	0.20	0.10	0.06	31.2	8.2	2.2	0.3	0.1	0.1
1995	1000	0.16	0.08	0.04	10.6	3.8	1.3	0.1	0.0	0.0
1996	1126	0.25	0.12	0.08	40.4	8.9	2.5	0.3	0.1	0.1
1997	1558	0.40	0.15	0.11	46.3	15.6	5.5	0.4	0.2	0.2
Mean	1200	0.28	0.12	0.08	34.0	9.3	2.9	0.3	0.1	0.1
SD[a]	230	0.09	0.03	0.03	14.2	4.2	1.6	0.1	0.0	0.1

[a] SD: standard deviation.

10.3.4 Runoff, Soil Losses, and Eroded Carbon

Annual rainfall ranged between 1000 and 1558 mm, and averaged 1200 mm between 1993 and 1997 (Table 10.3). Mean annual runoff rate in T, NPK, and M treatments was 0.28, 0.12, and 0.08 mm mm^{-1}, and mean annual soil losses were 34.0, 9.3, and 2.9 Mg ha^{-1} yr^{-1}, respectively. Using C enrichment ratios of sediments determined in similar soil and climate conditions (Roose, 1980a, 1980b), mean eroded C was estimated at 0.3, 0.1, and 0.1 Mg C ha^{-1} yr^{-1} in T, NPK, and M treatments, respectively. In plots vulnerable to erosion, eroded C was thus of the same order of magnitude as changes in Ct stock for 0 to 40 cm depth: –0.3 vs. –0.2 Mg C ha^{-1} yr^{-1} in T, and –0.1 vs. 0.2 Mg C ha^{-1} yr^{-1} in NPK. In contrast, eroded C in M was negligible compared with changes in Ct stock: –0.1 vs. 1.3 Mg C ha^{-1} yr^{-1}. Moreover, mean annual runoff rate and soil losses were significantly more in T than in NPK and more in NPK than in M; and eroded C was more in T than in NPK and M ($p < 0.01$) treatments. Additionally, mean annual runoff rate, soil losses, and eroded C increased with the increase in annual rainfall.

10.4 DISCUSSION

10.4.1 Changes in Soil Carbon

At the end of our experiment, Ct stock for 0 to 40 cm depth was 24 Mg C ha^{-1} under unfertilized maize, 29 Mg C ha^{-1} under fertilized maize, and 41 Mg C ha^{-1} under maize-mucuna rotation. Elsewhere in southern Benin and in similar soil conditions, Djegui et al. (1992) reported Ct stocks for 0 to 35 cm depth at 27 Mg C ha^{-1} under oil palm plantation, 30 Mg C ha^{-1} under food crops (with fallow), and 48 Mg C ha^{-1} under forest.

The data on change in Ct stock presented herein are consistent with other published data (Table 10.4). For an Alfisol in southwestern Nigeria, rates of 0.2 Mg C ha^{-1} yr^{-1} were recorded for 0 to 10 cm depth for fertilized maize (Lal, 2000), vs. 0.3 Mg C ha^{-1} yr^{-1} for NPK; in Brazilian Ultisols and Oxisols, rates of around 1 Mg C ha^{-1} yr^{-1} were measured for 0 to 20 cm depth under long-term no-till cropping systems (Bayer et al., 2001; Sá et al., 2001), vs. 1.4 Mg C ha^{-1} in M; in a Nigerian Alfisol, rates beyond 2 Mg C ha^{-1} yr^{-1} have even been measured for 0 to 20 cm depth under a 2-year *Pueraria* cover (Lal, 1998). These data confirm that residue mulching increases Ct stock in tropical soils, especially in cropping systems including legume cover crops.

10.4.2 Residue Biomass

The high rates of Ct increase in M resulted first from high residue biomass returned to the soil, which averaged 20 Mg ha^{-1} yr^{-1} (dry matter). The aboveground biomass of mucuna was 8 Mg ha^{-1}

Table 10.4 Compared Values of Annual Changes in Ct Stock under Various Tropical Cropping Systems Including Reduced or No Tillage

Country and Soil Type	Cropping System (and duration, in yr)	Change in Ct Stock (Mg C ha^{-1} yr^{-1})	Reference
For 0 to 20 cm Depth			
Nigeria, Alfisol	*Pueraria sp.* (2)	+2.1	Lal (1998)
Benin, Ultisol	maize-mucuna (11)	+1.4	this chapter
Brazil, clayey Oxisol	cereals and soybean (10, 22)	+1.0	Sá et al. (2001)
Brazil, clay loam Ultisol	*Cajanus cajan*-maize (12)	+0.9[a]	Bayer et al. (2001)
Nigeria, Alfisol	*Stylosanthes sp.* (2)	+0.4	Lal (1998)
Benin, Ultisol	fertilized maize (11)	+0.2	this chapter
Nigeria, Alfisol	*Centrosema sp.* (2)	+0.1	Lal (1998)
Benin, Ultisol	nonfertilized maize (11)	+0.1	this chapter
For 0 to 10 cm Depth			
Benin, Ultisol	maize-mucuna (11)	+1.0	this chapter
Nigeria, Alfisol	*Cajanus cajan*-maize (3)	+0.7	Lal (2000)
Honduras, various soils	mucuna-maize (1 to 15)	+0.5[b]	Triomphe (1996a)
Benin, Ultisol	fertilized maize (11)	+0.3	this chapter
Nigeria, Alfisol	fertilized maize (3)	+0.2	Lal (2000)
Benin, Ultisol	nonfertilized maize (11)	+0.1	this chapter

[a] For 0 to 17.5 cm depth.
[b] From +0.2 to +1.4 Mg C ha^{-1} yr^{-1}, depending on the site.

yr^{-1}, within the range of published data: 6 to 7 Mg ha^{-1} yr^{-1} in 1-year mucuna fallows in Nigeria (Vanlauwe et al., 2000) and an average of 11 Mg ha^{-1} yr^{-1} in mucuna-maize systems in Honduras (> 2000-mm annual rainfall; Triomphe, 1996b). The ratio of change in Ct stock to residue C measured in these plots also agreed with data in the literature: in a 12-year no-till maize-legume rotations on a sandy clay loam Ultisol in Brazil, Ct stock increase for 0 to 17.5 cm depth represented 11 to 15% of aboveground residue C (Bayer et al., 2001), vs. 15% in M (and 5% in NPK). In contrast, in long-term no-till cereal–legume rotations on clayey Oxisols also in Brazil, the increase in Ct stock for 0 to 40 cm depth represented 22 to 25% of total residue C (Sá et al., 2001), vs. 12% in M (and 3% in NPK). This difference confirms the role of clay content for C sequestration through the development of stable aggregates and hence organic matter protection (Feller and Beare, 1997).

In plots that were left under natural fallow during the short rainy season, weeds represented an important proportion of residue biomass, i.e., 77% in T and 29% in NPK. Weeds represented about 50% of the aboveground residue biomass in T, as was also the case in nonfertilized maize plots studied in Nigeria (Kirchhof and Salako, 2000). These data underline the need for systematic measurements of weed biomass when it represents a noticeable proportion of biomass returned to the soil. In our experiment, weeds were sampled on one day only, and it is likely that it led to some uncertainties. Weed biomass was negligible in M: proportions of aboveground residue biomass for maize, mucuna, and weeds were 49, 51, and 0%, respectively. Similarly, these proportions were 49, 42, and 9%, respectively, in 1–year maize-mucuna plots studied in Nigeria (Kirchhof and Salako, 2000). Indeed, Carsky et al. (2001) reported that weed suppression was often cited as the reason for the adoption of mucuna fallow systems in Africa.

10.4.3 Nitrous Oxide Emissions

Use of nitrogenous fertilizers also impacts nitrous oxide (N_2O) emissions, which can be roughly estimated using Equation 10.1 (Bouwman, 1996):

$$\text{N-}N_2O \text{ emissions (kg ha}^{-1}\text{ yr}^{-1}\text{)} = 1 + [0.0125 \times \text{N-fertilizer (kg ha}^{-1}\text{ yr}^{-1}\text{)}] \quad (10.1)$$

In NPK, N fertilizer was used at the rate of 76 kg N ha^{-1} yr^{-1} (Azontonde et al., 1998). Following Equation 10.1, it resulted in 2-kg N-N$_2$O ha^{-1} yr^{-1} emissions. As the global warming potential of N$_2$O is about 300 times that of CO$_2$ (IPCC, 2001), these N$_2$O emissions were equivalent to more than 0.2-Mg C-CO$_2$ ha^{-1} yr^{-1} emissions, and thus offset Ct increase (0.2 Mg C ha^{-1} yr^{-1}).

In M, mucuna residues supplied the soil with more than 250 kg N ha^{-1} yr^{-1} (Azontonde et al., 1998). In this case, Equation 10.1 led to an overestimation of N$_2$O emissions, as it was established from a set of experiments excluding legume cover crops, which provide N that is less directly available than mineral fertilizers. However, it may give an order of magnitude: by following Equation 10.1, N supply by mucuna residues resulted in 4-kg N-N$_2$O ha^{-1} yr^{-1} emissions, equivalent to 0.5-Mg C-CO$_2$ ha^{-1} yr^{-1} emissions (vs. 1.3 Mg C ha^{-1} yr^{-1} as Ct increases). Though overestimated, these data indicate that from an environmental point of view, Ct increase in soils under legume cover crops could be partly offset by N$_2$O emissions.

10.4.4 Runoff, Soil Losses, and Eroded Carbon

As compared with T, mean annual runoff rate and soil losses were 57 and 73% less in NPK, respectively, and were 71 and 91% less in M, respectively. Protection of the soil surface by vegetation and residues dissipates kinetic energy of rainfall and has an important influence on the reduction of runoff and erosion (Wischmeier and Smith, 1978). Thus, groundcover by mucuna mulch was probably the main reason for less runoff and soil losses in M than in T and NPK treatments. Similarly but to a lesser extent, it is likely that due to large biomass, fertilized maize provided a better groundcover than unfertilized maize. Additionally, residue return determines an increase in SOM that favors aggregate stability (Feller et al., 1996), thus preventing detachment of easily transportable particles, and thereby reducing surface clogging, runoff, and erosion (Le Bissonnais, 1996) Therefore, higher Ct also resulted in less runoff and erosion in M than in NPK, and also less in NPK than in T treatment.

With respect to runoff plots from tropical areas cropped with maize (or sorghum), comparisons with published data show that annual runoff rate was high (> 0.25 mm mm^{-1}) in T and under humid conditions (2100-mm annual rainfall); soil losses were high (> 20 Mg ha^{-1} yr^{-1}) under humid or semiarid conditions (500-mm annual rainfall) and in nonfertilized plots (Table 10.5). In contrast, runoff rate was low (< 0.10 mm mm^{-1}) on steep slopes with clayey soils (Kenya) and under maize-mucuna (M); soil losses were low (< 5 Mg ha^{-1} yr^{-1}) in M treatment. Thus, runoff and erosion increased with increase in annual rainfall and with a decrease in soil surface cover (absence of mulch, nonfertilized plots, semiarid conditions), in accordance with usual observations (Wischmeier and Smith, 1978; Roose, 1996). Under nonfertilized maize in Kenya, low runoff rates (0.02 mm mm^{-1}) resulted in high soil losses (29 Mg ha^{-1} yr^{-1}); assuming that the clayey Alfisol in this study had a stable structure with a high infiltration rate, steep slopes (30%) probably determined the nonselective transport of aggregates in the absence of adequate groundcover.

Mean annual C erosion was estimated at 0.3, 0.1, and 0.1 Mg C ha^{-1} yr^{-1} in T, NPK, and M treatments, respectively. Though mean soil losses were three times more in NPK than in M, eroded C was similar in both treatments probably because of high C content in surface soil (which supplies sediments) and higher C enrichment ratio of sediments in M than in NPK treatments. Indeed, several experiments have indicated that C enrichment ratio increases with decrease in soil losses (Roose, 1980a; 1980b). Thus, mucuna mulch was less effective in reducing the amount of C erosion than in reducing runoff and soil losses; but it was very effective in reducing the proportion of topsoil C that was eroded, which was much less in M than in NPK treatments. This underlines the interest of referring C erosion to topsoil C (enrichment ratio), and to temporal changes in topsoil C.

These data are consistent with those reported in the literature, which showed that C erosion significantly increased with increase in the product of soil losses and soil Ct stock (r = 0.932, p < 0.01; Figure 10.1, drawn up from Table 10.5). The data reported herein show that either soil Ct

Table 10.5 Compared Values of Annual Runoff Rate, Soil Losses, and C Erosion from Runoff Plots Cropped with Maize (or Sorghum) in Tropical Areas

Country	Rainfall (mm yr^{-1})	Slope (%)	Soil Type	Ct Stock[a] (Mg C ha^{-1})	Runoff Rate (mm mm^{-1})	Soil Losses (Mg ha^{-1} yr^{-1})	C Erosion (Mg C ha^{-1} yr^{-1})	Reference
				Nonfertilized Maize				
Kenya	1000	30	Clayey Alfisol	80	0.02	29.0	2.4	Gachene et al. (1997)
Benin	1200	4	Sandy loam Ultisol	20	0.28	34.0	0.3	this chapter
				Fertilized Maize (or Sorghum)				
Ivory Coast	2100	7	Sandy loam Ultisol	34	0.27	89.4	1.8	Roose (1980a)
Kenya	1000	30	Clayey Alfisol	80	0.01	8.4	0.7	Gachene et al. (1997)
Burkina Faso	800	1	Sandy Alfisol	13	0.25	7.3	0.2	Roose (1978)
Zimbabwe	500	5	Sandy Alfisol	15?	0.17	20.6	0.2	Moyo (1998)
Ivory Coast	1350	3	Sandy (gravely) Oxisol	21	0.20	5.5	0.1	Roose (1980b)
Benin	1200	4	Sandy loam Ultisol	22	0.12	9.3	0.1	this chapter
				Maize-Mucuna				
Benin	1200	4	Sandy loam Ultisol	35	0.08	2.9	0.1	this chapter

[a] At 0 to 30 cm.

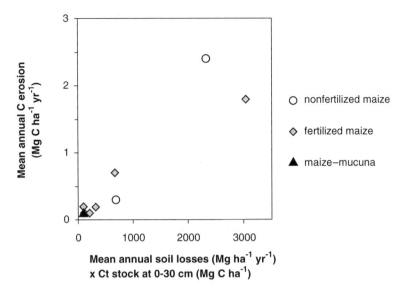

Figure 10.1 Relationship between mean annual C erosion (Mg C ha^{-1} yr^{-1}) and the product of mean annual soil losses (Mg ha^{-1} yr^{-1}) into Ct stock of bulk soil at 0 to 30 cm (Mg C ha^{-1}) (data from Table 10.5).

stock (T and NPK) or soil losses (M) were rather small, thus C erosion was much smaller than in studies from Kenya (high soil Ct stocks on steep slopes) and Ivory Coast (humid conditions), where it ranged from 0.7 to 2.4 Mg C ha^{-1} yr^{-1}.

10.5 CONCLUSION

For this sandy loam Ultisol, relay-cropping of maize and mucuna (M) was very effective in enhancing C sequestration: change in Ct stock for 0 to 40 cm depth was 1.3 Mg C ha^{-1} yr^{-1} over the 12-year period of the experiment, ranging among the highest rates recorded for the eco-region. This increase resulted first from the high amount of residue biomass provided by mucuna, which amounted to 10 Mg DM ha^{-1} yr^{-1} (83% aboveground). Mucuna residues, supplying the soil with N, also favored the production of maize biomass, and total mucuna plus maize residue biomass returned to the soil was about 20 Mg ha^{-1} yr^{-1}. These results indicate the usefulness of mucuna for SOM management. In contrast, nonfertilized (T) and fertilized continuous maize cultivation (NPK) resulted in –0.2- and 0.2-Mg C ha^{-1} yr^{-1} change in Ct stock for 0 to 40 cm depth, respectively. Total residue biomass was 8 and 13 Mg ha^{-1} yr^{-1}, including 77 and 29% by weeds, respectively. These contributions demonstrate the need for weed biomass sampling, especially when noticeable rainfall occurs beside the cropping season. Weed biomass was negligible in M, underlining the potential of mucuna for weed control.

Moreover, the thick mulch produced by mucuna decreased losses by runoff and erosion, which were 0.28, 0.12, and 0.08 mm mm^{-1}, and 34, 9, and 3 Mg ha^{-1} yr^{-1} in T, NPK, and M treatments, respectively. Eroded C was estimated at 0.3, 0.1, and 0.1 Mg C ha^{-1} yr^{-1} in T, NPK, and M, respectively. Thus, C erosion was of the same order of magnitude as changes in soil Ct stock in treatments vulnerable to erosion (T and NPK). In contrast, C erosion under maize-mucuna was negligible as compared to changes in soil Ct stock.

Through its benefits on SOM management, weed suppression, and erosion control, cropping systems including a legume cover may have an adverse impact from a global change standpoint. Indeed, rough estimates show that N_2O emissions resulting from N supply by mucuna may partly offset soil C storage in M treatment. In NPK, N_2O fluxes consecutive to mineral N supply could even offset soil C storage completely. In order to characterize these adverse effects and establish greenhouse gas balances precisely, there is an urgent need for accurate field measurements of N_2O fluxes, especially in cropping systems including legumes.

ACKNOWLEDGMENTS

We thank Barthélémy Ahossi, Maurice Dakpogan, Charles de Gaulle Gbehi, Nestor Gbehi, Branco Sadoyetin, and André Zossou for fieldwork, Jean-Yves Laurent and Jean-Claude Marcourel for technical assistance, and Gérard Bourgeon for his comments on a previous version of the chapter. This work was financially supported by a French PROSE grant (CNRS-ORSTOM program on soil and erosion).

REFERENCES

Akobundu, I. O. 1980. Live mulch: A new approach to weed control and crop production in the tropics, in Proceedings of the British Crop Protection Conference — Weeds, Brighton, pp. 377–382.

Azontonde, A. 1993. Dégradation et restauration des terres de barre (sols ferrallitiques faiblement désaturés argilo-sableux) au Bénin. *Cahiers ORSTOM Série Pédologie* 28:217–226.

Azontonde, A., C. Feller, F. Ganry, and J. C. Rémy. 1998. Le mucuna et la restauration des propriétés d'un sol ferrallitique au sud du Bénin. *Agriculture et Développement* 18:55–62.

Bayer, C., L. Martin-Neto, J. Mielniczuk, C. N. Pillon, and L. Sangoi. 2001. Changes in organic matter fractions under subtropical no-till cropping systems. *Soil Science Society of America Journal* 65:1473–1478.

Bouwman, A. F. 1996. Direct emission of nitrous oxide from agricultural soils. *Nutrient Cycling in Agroecosystems* 46:53–70.

Carsky, R. J., M. Becker, and S. Hauser. 2001. Mucuna cover crop fallow systems: potential and limitations, in G. Tian, F. Ishida, and D. Keatinge, eds., *Sustaining Soil Fertility in West Africa*, Soil Science Society of America Special Publication No 58, Madison, WI, pp. 111–135.

Craswell, E. T. and R. D. B. Lefroy. 2001. The role and function of organic matter in tropical soils. *Nutrient Cycling in Agroecosystems* 61:7–18.

Dagnélie, P. 1975. *Théorie et méthodes statistiques. Applications agronomiques*, 2nd ed. Presses Agronomiques de Gembloux, Belgium.

Djegui, N., P. de Boissezon, and E. Gavinelli. 1992. Statut organique d'un sol ferrallitique du Sud-Bénin sous forêt et différents systèmes de culture. *Cahiers ORSTOM Série Pédologie* 27:5–22.

FAO-ISRIC-ISSS (Food and Agriculture Organization of the United Nations, International Soil Reference and Information Centre, International Society for Soil Science). 1998. *World Reference Base for Soil Resources*. FAO, Rome.

Feller, C., A. Albrecht, and D. Tessier. 1996. Aggregation and organic matter storage in kaolinitic and smectitic tropical soils, in M. R. Carter and B. A. Stewart, eds., *Structure and Organic Matter Storage in Agricultural Soils*. Lewis Publishers, Boca Raton, FL, pp. 309–359.

Feller, C. and M. H. Beare. 1997. Physical control of soil organic matter dynamics in the tropics. *Geoderma* 79:69–116.

Feller, C., A. Albrecht, E. Blanchart, Y. M. Cabidoche, T. Chevallier, C. Hartmann, V. Eschenbrenner, M. C. Larré-Larrouy, and J. F. Ndandou. 2001. Soil organic carbon sequestration in tropical areas. General considerations and analysis of some edaphic determinants for Lesser Antilles soils. *Nutrient Cycling in Agroecosystems* 61:19–31.

Gachene, C. K. K., N. J. Jarvis, H. Linner, and J. P. Mbuvi. 1997. Soil erosion effects on soil properties in a highland area of Central Kenya. *Soil Science Society of America Journal* 61:559–564.

IPCC (Intergovernmental Panel on Climate Change). 2001. *Climate change 2001: The scientific basis. Contribution of working group I to the third assessment report of the IPCC*. Cambridge University Press, U.K.

Kirchhof, G. and F. K. Salako. 2000. Residual tillage and bush-fallow effects on soil properties and maize intercropped with legumes on a tropical Alfisol. *Soil Use and Management* 16:183–188.

Lal, R. 1998. Land use and soil management effects on soil organic matter dynamics on Alfisols in western Nigeria, in R. Lal, J. M. Kimble, R. F. Follett, and B. A. Stewart (eds.), *Soil Processes and the Carbon Cycle*. CRC Press, Boca Raton, FL, pp. 109–126.

Lal, R. 2000. Land use and cropping systems effects on restoring soil carbon pools of degraded Alfisols in Western Nigeria, in R. Lal, J. M. Kimble, and B. A. Stewart, eds., *Global Climate Change and Tropical Ecosystems*, CRC Press, Boca Raton, FL, pp. 157–165.

Le Bissonnais, Y. 1996. Aggregate stability and assessment of soil crustability and erodibility. 1. Theory and methodology. *European Journal of Soil Science* 47:425–437.

Merckx, R., J. Diels, B. Vanlauwe, N. Sanginga, K. Denef, and K. Oorts. 2001. Soil organic matter and soil fertility, in G. Tian, F. Ishida, and D. Keatinge, eds., *Sustaining Soil Fertility in West Africa*. Soil Science Society of America Special Publication No 58, Madison, WI, pp. 69–89.

Moyo, A. 1998. The effect of soil erosion on soil productivity as influenced by tillage with special reference to clay and organic matter losses. *Advances in GeoEcology* 31:363–368.

Pieri, C. 1991. *Fertility of Soils: A Future for Farming in the West African Savannah*. Springer-Verlag, Berlin.

Raunet, M., L. Séguy, and C. Fovet-Rabot. 1999. Semis direct sur couverture végétale permanente du sol: De la technique au concept, in F. Rasolo and M. Raunet, eds., *Gestion Agrobiologique des Sols et des Systèmes de Culture*. CIRAD, Montpellier, France, pp. 41–52.

Roose, E. J. 1978. *Dynamique actuelle de deux sols ferrugineux tropicaux indurés sous sorgho et sous savane Soudano-Sahélienne — Saria (Haute-Volta): Synthèse des campagnes 1971–1974*. ORSTOM, Paris.

Roose, E. J. 1980a. *Dynamique actuelle d'un sol ferrallitique sablo-argileux très désaturé sous cultures et sous forêt dense humide sub-équatoriale du Sud de la Côte d'Ivoire — Adiopodoumé: 1964–1975*. ORSTOM, Paris.

Roose, E. J. 1980b. *Dynamique actuelle d'un sol ferrallitique gravillonnaire issu de granite sous culture et sous savane arbustive soudanienne du Nord de la Côte d'Ivoire — Korhogo: 1967–1975*. ORSTOM, Paris.

Roose, E. J. 1996. Land husbandry, components and strategy. *FAO Soils Bulletin* 70, Rome.

Roose, E. J. and R. Bertrand. 1972. *Importance relative de l'érosion, du ruissellement et du drainage oblique et vertical sous une savane arbustive de moyenne Côte d'Ivoire — Bouaké: 1967–1971*. ORSTOM-IRAT, Abidjan.

Roose, E. J. and J. Godefroy. 1977. *Pédogenèse actuelle d'un d'un sol ferrallitique remanié sur schiste sous forêt et sous bananeraie fertilisée de basse Côte d'Ivoire. Synthèse de huit années d'observations de l'érosion, du drainage et de l'activité des vers de terre à la station IRFA d'Azaguié et la Forêt du Téké*. ORSTOM-IRFA, Abidjan.

Sá, J. C. M., C. C. Cerri, W. A. Dick, R. Lal, S. P. Venske Filho, M. C. Piccolo, and B. E. Feigl. 2001. Organic matter dynamics and carbon sequestration rates for a tillage chronosequence in a Brazilian Oxisol. *Soil Science Society of America Journal* 65:1486–1499.

Sanchez, P. 1976. *Properties and Management of Soils in the Tropics*. John Wiley & Sons, New York.

Shang, C. and H. Tiessen. 2000. Carbon turnover and carbon-13 natural abundance in organo-mineral fractions of a tropical dry forest soil under cultivation. *Soil Science Society of America Journal* 64:2149–2155.

Soil Survey Staff. 1994. *Keys to Soil Taxonomy*, 6th ed. USDA — Soil Conservation Service, Washington, D.C.

Starr, G. C., R. Lal, R. Malone, D. Hothem, L. Owens, and J. Kimble. 2000. Modelling soil carbon transported by water erosion processes. *Land Degradation and Development* 11:83–91.

Triomphe, B. L. 1996a. *Seasonal Nitrogen Dynamics and Long-Term Changes in Soil Properties under the Mucuna/Maize Cropping System on the Hillsides of Northern Honduras*. Ph.D. Thesis, Cornell University, New York.

Triomphe, B. 1996b. Un système de culture original et performant dans une zone de montagne du tropique humide: La rotation maïs/mucuna au Nord-Honduras, in J. Pichot, N. Sibelet, and J. J. Lacoeuilhe, eds., *Fertilité du Milieu et Stratégies Paysannes sous les Tropiques Humides*. CIRAD — Ministère de la Coopération, Paris, pp. 318–328.

Vanlauwe, B., O. C. Nwoke, J. Diels, N. Sanginga, R. J. Carsky, J. Deckers, and R. Merckx. 2000. Utilization of rock phosphate by crops on a representative toposequence in the Northern Guinea savanna zone of Nigeria: Response by *Mucuna pruriens*, *Lablab purpureus* and maize. *Soil Biology and Biochemistry* 32:2063–2077.

Voelkner, H. 1979. Urgent needed: An ideal green mulch crop for the tropics. *World Crops* 31:76–77.

Wischmeier, W. H. and D. D. Smith. 1978. *Predicting Rainfall Erosion Losses: A Guide to Erosion Planning*. USDA, Washington, D.C.

CHAPTER 11

Organic Carbon Associated with Eroded Sediments from Micro-Plots under Natural Rainfall from Cultivated Pastures on a Clayey Ferralsol in the Cerrados (Brazil)

Didier Brunet, Michel Brossard, and Maria Inês Lopes de Oliveira

CONTENTS

11.1 Introduction 157
11.2 Material and Methods 158
11.3 Results and Discussion 160
    11.3.1 Rainfall, Runoff, and Soil Losses 160
    11.3.2 Organic Carbon Concentration in Sediment 161
    11.3.3 Organic Carbon Mass Associated with Sediments and in the Topsoil 163
    11.3.4 The Enrichment Ratio 164
11.4 Conclusion 165
Acknowledgment 165
References 165

11.1 INTRODUCTION

Transformations in land use since the 1950s have affected large areas of the tropics. Conversion of native vegetation to cropping and pastures has been widespread in all biomes of the tropics (humid forests, savannas, volcanic areas, etc.). Among these conversions, monocultures of cultivated pastures with exotic species cover large areas, especially in tropical South America, where pastures cover more than 120 million hectares (Mha). The introduction of pastures has two main objectives: to provide resources for extensive cattle production, and to secure ownership of the land.

In Brazil, the Cerrado region (the savanna biome) covers 22% of the territory, and cultivated pastures with exotic grass species represent 49.5 Mha (Sano et al., 2000). A high proportion of these plant-soil systems is relatively unproductive or in decline. Low productivity pastures are characterized by low liveweight gains during the wet season (from 1000 to 1200 mm) and liveweight losses during the dry season (4 to 6 months) (Rolón and Primo, 1979). Decrease in productivity is due to inadequate cattle and pasture management, and soil factors (Balbino et al., 2002). Soil carbon content (‰) and total carbon stocks are among the main factors that integrate the effects of management when vegetation and tillage are changed. Hence soil carbon is one of the most important indicators affecting soil quality (Doran et al., 1994). Since the 1970s, many experiments on major tropical soil types have indicated that yield decline is caused by soil loss due to erosion (Stocking, 2003). Conversely, Gitz and Ciais (2003) have shown through modeling that changes in land use can cause emission of CO_2 into the atmosphere. Reduction in pasture productivity is generally caused by chemical alterations in the soil (Gijsman and Thomas, 1996), and by the adverse effects of animal trampling on soil physical properties (McCalla et al., 1984; Holt et al., 1996) and especially soil compaction (Willatt and Pullar, 1983; Chanasyk and Naeth, 1995; Greenwood et al., 1997).

The Ferralsols (Latossolos, according to the Brazilian classification) represent 46% of the Cerrado area. The top few cm of soil of low productivity pastures in these soils have extremely low porosity. Soil structure is a strong platy from the surface to 3 cm depth, followed by a combination of compact clods of 1 to 5 cm in size, and clods organized in very porous agglomerated micro-aggregates (Balbino et al., 2002). In addition to research on conservation tillage and cropping systems, there is a need for information on soil loss under a wide range of soil–plant systems. One of the hypotheses concerning pasture decline processes relates to the loss of water and soil caused by runoff, and consequently, loss of organic carbon (C). The infiltration capacity of pasture is, in general, higher than that of arable land. Experimental data on runoff and soil losses exist in the region at scales ranging from 10-m^2 plots (Dedecek et al., 1986; Leprun, 1994; Santos et al., 1998) to a whole watershed (Silva and Oliveira, 1999). However, losses of organic carbon associated with water erosion have rarely been assessed in this region.

This paper presents the results of an experiment carried out under natural rainfall on 1 m^2 erosion micro-plots on a clayey Ferralsol in the Brazilian central plateau, in order to assess runoff, soil losses, and carbon losses under *Brachiaria* pastures. The short-term effects of renewed pastures on runoff, erosion, and eroded C are also discussed.

11.2 MATERIAL AND METHODS

The experimental site was located on a farm at 1000 m above sea level on the Brazilian central plateau (15°13'S, 47°41'W) in an EMBRAPA Cerrados–IRD–Fazenda Rio de Janeiro station. The soil is a homogeneous dark red, clayey Ferralsol with 55 to 65% clay in the upper layers (< 35 cm) and more than 70% in the lower layers (> 55 cm; Table 11.1). The topsoil layer (0 to 0.02 m) has a mean bulk density of 0.9 Mgm^{-3} and a mean C content of 24.2 mg g^{-1}.

The mean annual rainfall is 1200 mm, and the rainy season lasts 7 months from the end of September to the beginning of April. The mean annual temperature is 22°C and those of the coldest and warmest month are 20°C (July) and 23°C (October), respectively.

Table 11.1 Particle Size Distribution, Bulk Density, and Organic Carbon Content (SOC) of the Ferralsol under Study (Mean Data of Experimental Site)

Depth m	Clay g 100 g^{-1}		Coarse Sand g 100 g^{-1}		Fine Sand g 100 g^{-1}		Silt g 100 g^{-1}		Bulk Density g cm^{-3}		SOC mg C g^{-1}	
	Mean	SD	Mean	SD	Mean	SD	Mean	SD	Mean	SD	Mean	SD
0–0.02	53.4	3.7	3.7	0.8	25.2	5.4	13.4	4.9	0.90	0.09	24.2	4.0
0.05–0.15	62.1	4.3	3.5	1.0	13.9	3.4	14.8	3.3	1.13	0.09	19.3	2.7
0.25–0.35	65.8	4.9	3.8	0.8	14.2	3.6	11.8	4.4	1.13	0.15	12.7	2.1
0.55–0.65	70.7	4.1	2.6	0.5	13.3	3.0	9.7	3.1	1.10	0.15	8.4	1.6
0.85–0.95	71.3	2.9	2.6	1.2	14.1	3.8	8.8	3.4	1.06	0.14	6.7	1.1
1.15–1.25	71.0	3.3	2.1	1.0	14.6	2.9	9.2	3.1	1.02	0.09	5.4	0.8

Note: SD: standard deviation. n = 9 for particle size distribution, n = 54 for bulk density for the 0.0 to 0.02 m depth layer, and n = 9 for the other layers.

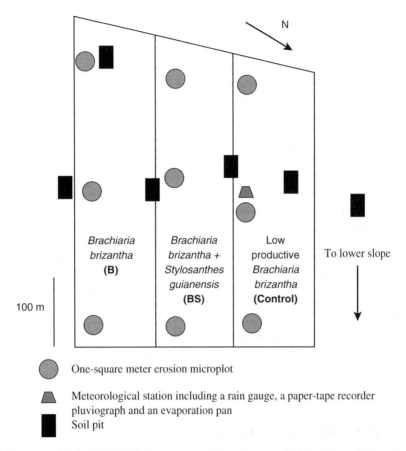

Figure 11.1 Experimental site EMBRAPA Cerrados — IRD — Fazenda Rio de Janeiro (Goiás, Brazil).

Three treatments compared during the first rainy season (Figure 11.1) included:

- Control: a 10-year-old pasture of *Brachiaria brizantha* cv *Marandu*, characterized by a low productivity
- Treatment B: a 21-month-old pasture of the same *Brachiaria*, and restored
- Treatment BSG: a 21-month-old pasture involving the same *Brachiaria* along with a legume *Stylosanthes guianensis* var. *vulgaris,* cv *Mineirão*

The last two treatments, restored pastures, applied to the same pasture as the control, were established in 1999. Soil tillage included disking to 15-cm depth (twice at right angles) to improve germination of *Brachiaria* seeds present in the soil, sowing of the legume, and spreading fertilizer. The fertilizer application was at 40 kg ha^{-1} of phosphorus as triple superphosphate and 74 kg ha^{-1} of sulfur as flower of sulfur. Due to the large plot size (5 ha) and the relative homogeneity of soil and pastures, treatments were not replicated.

Each plot was 400- to 500-m long and had a 3.5% slope. In March 1999, three 1-m^2 erosion micro-plots were demarcated within each plot, at low, middle, and upper positions. These micro-plots were delineated by a frame buried approximately 5 cm into the soil. Runoff generated within the frame border was routed through a pipe into belowground collection tanks. At this 1-m^2 scale, the beginning of runoff process can be observed. Moreover, under the pastures cover, any gully erosion was determined by the nature of nonaggressive rainfalls. However, slope runoff coefficients extrapolated to the watershed scale overestimated watershed runoff (Harms and Chanasyk, 2000). The research on the scale effect on runoff has shown that the sheet flow decreases with an increase in surface area for a given rainfall amount (Molinier et al., 1989). The biomass in micro-plots was cut regularly when it reached 30 cm height.

During the 2000–2001 season, all three treatments (B, BSG, and Control) were managed uniformly so that their effects could be compared. During the 2001–2002 season, only two treatments (B and BSG) could be compared because the control was managed differently: complete cutting (but without disturbing the soil surface and the roots), and leaving soil surface bare throughout the rainy season in order to evaluate the effects of the absence of plant cover (bare soil under control).

Runoff and sediments were collected twice a week between October and May during two successive rainy seasons, from 2000 to 2002. After filtration through 0.2 μm membranes, the sediments were dried at 65°C and weighed. In addition, topsoil in micro-plots was sampled from 0 to 0.20 m in January 2001 and from 0 to 0.02 m at the end of the rainy season of the same year. Particle size distribution was done on the soil of these two layers and on the sediments from the bare soil, but not on the sediments from the pastures due to their small weight. It was measured after dispersion using NaOH, following routine procedures for Ferralsols in Brazil (EMBRAPA, 1997).

Soil and sediment carbon contents were determined by the wet oxidation method (Walkley and Black, 1934, modified by EMBRAPA, 1997). When a runoff event did not produce enough sediment for a C analysis, we assumed that sediment C content was similar to that of the preceding event. The rainfall was recorded weekly on the site by a recording rain gauge.

Statistical analysis was done by Student unpaired t-tests where differences in mean runoff coefficient (%), soil losses (g m^{-2}), sediment C (mg C g^{-1}), soil organic carbon (SOC) content, (mg C g^{-1}), SOC stock of the layer 0 to 0.02 m (g C m^{-2}) and SOC losses (g C m^{-2} yr^{-1}) between plots were tested. No assumptions were made on normality and variance equality (Dagnélie, 1975).

11.3 RESULTS AND DISCUSSION

11.3.1 Rainfall, Runoff, and Soil Losses

Total rainfall was 1101 mm during the 2000–2001 season, and 1304 mm during the 2001–2002 season (Figure 11.2). Compared to other tropical areas, rainfall was not very erosive, since only six daily rainfalls exceeded 50 mm during these two seasons, and the intensity of only seven rains exceeded 50 mm h^{-1} in 30 min.

For a given treatment and landscape position, during the first season, the annual runoff coefficient (RC: annual runoff/annual rainfall, in %) ranged between 0.1 and 0.5% in B and BSG and between 0.8 and 1.9% in the control (Table 11.2). The significant difference between the control and B and BSG may be explained by the ground cover, which was 70 to 80% for B and BSG and 50 to 55% for the control. During the second season, the RC in B and BSG ranged between approximately the same values (0.1 to 0.6%) with slight variations among the micro-plots. The RC was 9% on bare soil micro-plots in the absence of grass cover. Averaged over the two seasons, the RC, which ranged from 0.1 to 0.3%, was not significantly different among the two restored treatments.

Annual soil losses ranged between 43 and 119 g m^{-2} yr^{-1} under pasture during the first rainy season (with 1100-mm annual rainfall), but did not differ significantly among the three treatments though they tended to be more in the control (Table 11.2). During the following season (with 1300-mm annual rainfall), soil losses in B and BSG were similar, but higher than the first season. They were often maximum on the low and minimum on the middle position of the slope (except B). No significant difference was observed between the B and BS treatments. Soil losses were 2073 g m^{-2} yr^{-1} on bare soil.

At this scale, under pasture, few data exist on runoff and soil losses for this region. However Castro et al. (1999) observed a RC of 6%, and 20 g m^{-2} of soil loss on no-till and mulched 1-m^2 plots, under natural rainfall on a clayey Ferralsol in southern Brazil. At another scale, Dedecek et al. (1986) measured soil losses up to 53 Mg ha^{-1} yr^{-1}, in 77-m^2 bare soil plots of a Ferralsol in the same region. Thus, the

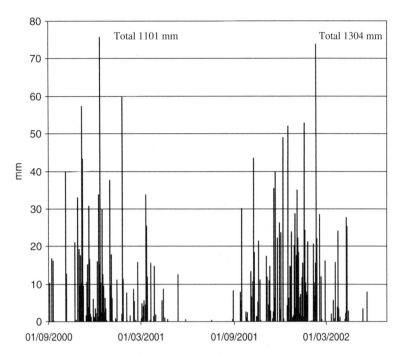

Figure 11.2 Daily rainfall (mm) during the two rainy seasons from 2000 to 2002.

Table 11.2 Runoff Coefficient (RC) in % and Soil Losses in g m^{-2} on 1-m^{-2} Plots

	2000 to 2001				2001 to 2002			
	RC		Soil Losses		RC		Soil Losses	
Plot	Mean	SD	Mean	SD	Mean	SD	Mean	SD
B	0.1 a	0.1	43 a	6	0.3 a	0.3	124 a	40
BS	0.2 a	0.2	72 a	63	0.2 a	0.2	123 a	74
Control	1.3 b	0.6	119 a	92	—	—	—	—
Bare soil under control	—	—	—	—	9.0 b	0.4	2073 b	475

Note: SD: standard deviation. Within a column, data followed by the same letter are not significantly different (p < 0.05).

present data confirmed the major effect of plant cover, with the two kinds of pasture rehabilitation, on runoff generation on these Ferralsols, which are clearly sensitive to erosion as shown by the control bare soil. The present results also indicated that legume introduction in the renewed pasture did not significantly affect soil erosion.

11.3.2 Organic Carbon Concentration in Sediment

For a given treatment, landscape position, and year, sediment C content ranged between 21 and 27 mg C g^{-1}, except in the middle slope position of the control plot where it was 31 mg C g^{-1} during the 2000–2001 season (under pasture) and 40 mg C g^{-1} during the 2001–2002 season (bare soil). But the treatment means, which ranged between 23 and 30 mg C g^{-1}, did not differ significantly (Table 11.3). In comparison, for small pastured watersheds, Owens et al. (2002) observed values ranging from 52 to 72 mg C g^{-1}.

However, sediment C content tended to be somewhat more during the 2001–2002 season than during the 2000–2001 season (+5% on average). This may be explained by the fact that there was an incomplete amount of litter/waste during the first season, part of it having been accidentally

Table 11.3 Organic Carbon Contents in the Sediments (mg C g⁻¹) and the Soil Layers (SOC) 0 to 0.02 m and 0 to 0.10 m, and Carbon Enrichment Ratio

Plot	Sediment C mg C g⁻¹				SOC 0 to 0.10 m mg C g⁻¹		C Enrichment Ratio Reference Layer 0 to 0.10 m		SOC 0 to 0.02 m mg C g⁻¹		C Enrichment Ratio Reference Layer 0 to 0.02 m	
	2000 to 2001		2001 to 2002				2000 to 2001	2001 to 2002			2000 to 2001	2001 to 2002
	Mean	SD	Mean	SD	Mean	SD			Mean	SD		
B	23.6 a	0.9	24.1 a	2.7	18.2 a	2.7	1.3	1.3	22.1 a	0.8	1.1	1.1
BS	23.0 a	1.8	24.7 a	2.1	20.9 a	2.2	1.1	1.2	23.6 a	1.6	1.0	1.0
Control	25.4 a	5.0	—	—	21.8 a	1.7	1.2	—	26.9 a	3.1	0.9	—
Bare soil under control	—	—	30.4 a	8.1	21.8 a	1.7	—	1.4	26.9 a	3.1	—	1.1

Note: SD: standard deviation. Within a column, data followed by the same letter are not significantly different ($p < 0.05$).

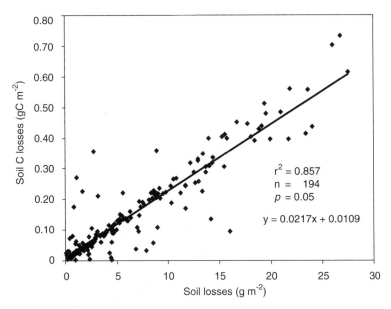

Figure 11.3 Relationship between the soil carbon mass associated to sediments and soil losses under pastures (2000 to 2002).

removed during the first samplings. The SOC content in the micro-plot topsoil (0 to 0.02 m) ranged from 21 to 29 mg C g^{-1}, and was thus similar to sediment C contents. Correlation between topsoil and sediment C contents was stronger in the 2001–2002 season ($r^2 = 0.64$, n = 9) than in the 2000–2001 season ($r^2 = 0.47$).

11.3.3 Organic Carbon Mass Associated with Sediments and in the Topsoil

Soil carbon losses depended strictly on soil losses because C content in the topsoil (0 to 0.02 m) varied little. Considering each runoff event under pasture during these two rainy seasons, the mass of C associated with the sediment (g C m^{-2}) was strongly correlated with the mass of sediment (g soil m^{-2}) ($r^2 = 0.86$, n = 194; Figure 11.3).

Two years after pasture restoration, in May 2001, the mean SOC stock in the 0 to 0.02 m layer was 384, 432, and 499 g C m^{-2} in B, BS, and the Control, respectively (Table 11.4). Although the control had a C stock 30% and 15% higher than B and BSG, respectively, these differences were not significant. Nevertheless, differences may be explained partly by the lower SOC content in B and BSG treatment. The disk plowing in 1999 to 15 cm depth in B and BS treatments may have

Table 11.4 Soil Organic Carbon (SOC) Losses of Micro-Plots (g C m^{-2} yr^{-1}) and SOC Stock of Layer 0 to 0.02 m (g C m^{-2})

	SOC Stock 0 to 0.02 m		SOC Losses			
			2000 to 2001		2001 to 2002	
	Mean	SD	Mean	SD	Mean	SD
B	384 a	19	1.0 a	0.2	3.0 a	1.1
BS	432 b	18	1.7 a	1.5	3.1 a	2.1
Control	499 ab	97	2.7 a	2.0	—	—
Bare soil under control	499 ab	97	—	—	61.8 b	9.4

Note: SD: standard deviation. Within a column, data followed by the same letter are not significantly different (p < 0.05).

accentuated mineralization of organic matter, a dilution of the surface organic matter by the mechanical effect of disking, and also on the soil bulk density, which was lower in treatment B. Additionally, the SOC stock at 0 to 0.02 m depth was 13% more in B than in BSG, and this difference was significant (due to small standard deviations).

Considering the 0 to 0.20 m layer, these differences were smaller or nonexistent, the C stocks being 3.9, 4.6, and 4.5 kg C m^{-2} in B, BS, and Control, respectively (data not shown). Under 12-year-old pasture on Brazilian Ferralsols of the same type, Chapuis-Lardy et al. (2002) observed a C stock of 5.4 kg C m^{-2} for the 0 to 0.2 m layer.

For a given treatment and year, C losses ranged between 1.0 and 3.1 g eroded C m^{-2} yr^{-1} under pastures, and were more in the 2001–2002 than in the 2000–2001 season in B and BSG treatments (Table 11.4). In contrast, on bare soil, C losses were twenty times greater (61.8 g eroded C m^{-2} yr^{-1}) than under pasture during the second rainy season (Table 11.4). Moreover, if we consider separately the runoff events on bare soil for lower and upper slope positions, where mean sediment C content was 25.8 mg C g^{-1}, C losses were strongly correlated with soil losses ($r^2 = 0.99$, n = 86). For the middle slope position, where sediment C content was 39.6 mg C g^{-1} in bare soil, C losses were also strongly correlated with soil losses ($r^2 = 0.995$, n = 42). On this last landscape position, soil losses were the lowest and sediment C content was the highest of all measurements.

Finally, on the 1-m^2 scale, mean annual C losses under pastures represented 0.5% of the SOC stock in the 0 to 0.02 m soil layer, whereas under bare soil they represented 12% of the SOC stock. These data provide some new insights into C fluxes in managed tropical grassland. Representing relatively small quantities of the SOC, the C associated with the eroded sediment may be of some importance in the organic matter redistribution over the landscape down slope. For example, at another scale, on small watersheds in Ohio (< 0.8 ha), Owens et al. (2002) reported mean eroded C losses of 12.7 to 24.0 kg eroded C ha^{-1} yr^{-1} depending on tillage practices (13.8 kg eroded C ha^{-1} yr^{-1} for no-tillage). On 100-m^2 runoff plots from tropical and Mediterranean regions, Roose (2004) reported C losses ranging from 0.1 to 50 kg eroded C ha^{-1} yr^{-1} in well-covered plots (forest, savanna, etc.), 50 to 350 kg eroded C ha^{-1} yr^{-1} under row crops, and up to 3000 kg eroded C ha^{-1} yr^{-1} for bare fallows on steep slopes in very humid regions. In Kenya, Zöbisch et al. (1995) arrived at similar conclusions; on 23-m^2 plots, they observed during a rainy season 773 and 53 kg eroded C ha^{-1} lost on bare fallow and maize–beans rotation, respectively.

11.3.4 The Enrichment Ratio

The enrichment ratio is defined as the ratio of the concentration of any given component in the eroded materials to that in the contributing soils. It is greater than one when the sedimentary materials are enriched. For the two rainy seasons under study, the organic carbon enrichment ratio was more than one for all treatments, considering the 0–0.10 m soil depth layer as a reference (Table 11.3). Then, it ranged between 1.1 and 1.3 under pastures, and was 1.4 in the control bare soil. Considering the 0 to 0.02 m soil depth, the enrichment ratio was close to one. In the control bare soil, the particle size distribution showed that the amounts of clay and silt in sediments ranged from 500 to 690 and 120 to 230 g kg^{-1}, respectively, depending on the micro-plot considered, whereas they were 560 and 110 g kg^{-1} soil in the 0 to 0.02 m layer and 630 and 160 g kg^{-1} in the 0 to 0.10 m layer (data not shown). The small depletion of clay particles in this Ferralsol upper layer was a consequence of sheet erosion over time. De Jong and Kachanovski (1988) reported that about 50% of SOC losses in Canadian grassland sites were due to erosion. But in other cases, for well-managed pastures, the SOC content is generally conserved. For example, Fisher et al. (1994) observed an increase in soil C content to 1-m depth under *Brachiaria humidicola* on a "Llanos" soil in Colombia. Chapuis-Lardy et al. (2002) reported that pastures increased the storage of C in the topsoil of Ferralsols compared to the native Cerrado ecosystems. But under low productivity cultivated pastures, C storage may be lower than in native fields (Da Silva et al., 2004).

11.4 CONCLUSION

At the 1-m^2 scale of this study, runoff under pasture was small, and was significantly smaller in the restored pastures than in the degraded control pasture. Similarly, soil and carbon losses under pasture were small, and were smaller in the restored pastures than in the control, but the differences were not significant. In contrast, runoff, soil, and carbon losses were much greater in control bare soil than under pasture. Considering events individually, eroded C was strongly correlated with soil losses. The C enrichment ratio was about one considering the 0 to 0.02 m soil depth layer, and ranged between 1.1 and 1.4 considering the 0 to 0.10 m soil layer. However, there was no significant differences in C stock between restored and control pastures in the 0.02 m soil layer. The level of eroded C by sheet erosion under experimental pasture conditions was small (0.5% of the initial C stock in the 0.02 m soil layer), but was larger (12%) in the bare soil.

The plant cover in the restored pastures was very efficient in reducing runoff and sediment loss for most rainfall events compared to the control, a 10-year-old *Brachiaria brizantha* pasture. However, differences were small and generally not significant among both renewed pastures with regard to runoff, soil losses, and eroded C.

The erosion of C was a selective process because it was limited to the top soil. There were no rills. The eroded C may be of some importance in the redistribution of soil organic matter over the landscape. However, this process is probably not the main factor responsible for pasture decline in this region.

The data presented allows the assessment of the effect of the pasture restoration on runoff and carbon associated with sediment losses under natural rainfall and runoff conditions. Further investigations are needed to study the sustainability of the restored pastures beyond two successive rainy seasons for numerous runoff and erosion events.

ACKNOWLEDGMENT

Financial and logistic support was received from EMBRAPA Cerrados (Brazil) and IRD (France) during the Pastures/Soil Project 1997 to 2002.

REFERENCES

Balbino, L. C., M. Brossard, J. C. Leprun, and A. Bruand. 2002. Mise en valeur des Ferralsols de la région du Cerrado (Brésil) et évolution de leurs propriétés physiques: une étude bibliographique. *Étude et Gestion des Sols* 9:83–104.

Castro, N. M. dos R., A. V. Auzet, P. Chevallier, and J. C. Leprun. 1999. Land use change effects on runoff and erosion from plot to catchment scale on the basaltic plateau of southern Brazil. *Hydrological Processes* 13:1621–1628.

Chanasyk, D. S. and M. A. Naeth. 1995. Grazing impacts on bulk density and penetrometer resistance in the foothills fescue grasslands of Alberta, Canada. *Canadian Journal of Soil Science* 75:551–557.

Chapuis Lardy, L., M. Brossard, M. L. Lopes Assad, and J. Y. Laurent. 2002. Carbon and phosphorus stocks of clayey Ferralsols in Cerrado native and agroecosystems, Brazil. *Agriculture Ecosystems and Environment* 92:147–158.

Da Silva, J. E., D. V. S. Resck, E. J. Corazza, and L. Vivaldi. 2004. Carbon storage in clayey Oxisol cultivated pastures in the "Cerrado" region, Brazil. *Agriculture Ecosystems and Environment* 103:357–363.

Dagnélie, P. 1975. *Théorie et méthodes statistiques. Applications agronomiques*, 2nd ed. Presses Agronomiques de Gembloux. Belgium.

Dedecek, R. A., D. V. S. Resck, and E. de Freitas, Jr. 1986. Perdas de solo, água e nutrientes por erosão em latossolo vermelho-escuro dos cerrados em diferentes cultivos sob chuva natural. *Revista Brasileira de Ciência do Solo* 10:265–272.

De Jong, E. and R. G. Kachanoski. 1988. The importance of erosion in the carbon balance of prairies soils. *Canadian Journal of Soil Science* 68:111–119.

Doran, J. W., D. C. Coleman, D. F. Bezdicek, and B. A. Stewart, eds. 1994. Defining Soil Quality for a Sustainable Environment. *Soil Science Society of America Journal Special Publication* 35, Madison, WI, U.S.

EMBRAPA. 1997. *Manual de métodos de análise de solo. Centro cacional de pesquisa de solos*, 2nd ed., Rio de Janeiro.

Fisher, M. J., I. M. Rao, M. A. Ayarza, C. E. Lascano, J. I. Sanz, R. J. Thomas, and R. R. Vera. 1994. Carbon storage by introduced deep-rooted grasses in the South American savannas. *Nature* 371:236–238.

Gijsman, A. J. and R. J. Thomas. 1996. Evaluation of some physical properties of an Oxisol after conversion of native savanna into legume-based or pure grass pastures. *Tropical Grasslands* 30:237–248.

Gitz, V. and P. Ciais. 2003. Amplification effect of changes in land use and concentration of atmospheric CO_2. *Comptes Rendus Geoscience* 335:1179–1198.

Greenwood, K. L., D. A. McLeod, and K. J. Hutchinson. 1997. Long-term stocking rate effects on soil physical properties. *Australian Journal of Experimental Agriculture* 37:413–419.

Harms, T. E. and D. S. Chanasyk. 2000. Plot and small-watershed scale runoff from two reclaimed surface-mined watersheds in Alberta. *Hydrological processes* 14:1327–1339.

Holt, J. A., K. L. Bristow, and J. G. McIvor. 1996. The effects of grazing pressure on soil animals and hydraulic properties of two soils in semiarid tropical Queensland. *Australian Journal of Soil Research* 34:69–79.

Leprun, J. C. 1994. Effets de la mise en valeur sur la dégradation physique des sols. Bilan du ruissellement et de l'érosion de quelques grands écosystèmes brésiliens. *Étude et Gestion des Sols* 1:45–65.

McCalla, G. R., W. H. Blackburn, and L. B. Merrill. 1984. Effects of livestock grazing on infiltration rates, Edwards Plateau of Texas. *Journal of Range Management* 37:265–269.

Molinier, M., P. Audry, P., J. C. Desconnets, and J. C. Leprun. 1989. Dynamique de l'eau et des matières dans un écosystème représentatif du Nordeste brésilien. Conditions d'extrapolation spatiale à l'échelle régionale. Rapport final ATP-PIREN, multigr. ORSTOM, Recife, 1989.

Owens, L. B., R. W. Malone, D. L. Hothem, G. C. Starr, and R. Lal. 2002. Sediment carbon concentration and transport from small watersheds under various conservation tillage practices. *Soil and Tillage Research* 67:65–73.

Rolón, J. D. and A. T. Primo. 1979. Experiences in regional demonstration trials in Brazil, in P. A. Sanchez and L. E. Tergas, eds., *Pasture Production in Acid Soils of the Tropics*. Centro International de Agricultura Tropical, Cali, Colombia, pp. 417–430.

Roose, E. J. 2004. Carbon losses by erosion and enrichment ratio: Influence of green cover, soils and erosion processes on slopes of tropical countries. Proceedings of the International Colloquium Land Use, Erosion and Carbon Sequestration, 23–28 September 2002, Montpellier, France.

Sano, E. E., A. de O. Barcellos, and H. S. Bezerra. 2000. Assessing the spatial distribution of cultivated pastures in the Brazilian savanna. *Pasturas Tropicales* 23:2–15.

Santos, D., N. Curi, M. M. Ferreira, A. R. Evangelista, A. B. da Cruz Filho, and W. G. Teixeira. 1998. Perdas de solo e produtividade de pastagens nativas melhoradas sob diferentes práticas de manejo. *Pesquisa Agropecuária Brasileira* 33:183–189.

Silva, C. L. and C. A. S. Oliveira. 1999. Runoff measurement and prediction for a watershed under natural vegetation in central Brazil. *Revista Brasileira de Ciência do Solo* 23:695–701.

Stocking, M. A. 2003. Tropical soils and food security: the next 50 years. *Science* 302:1356–1359.

Walkley, A. and I. A. Black. 1934. An examination of the Degtjareff method for determining soil organic matter and a proposed modification of the chromic acid titration method. *Soil Science* 37:29–38.

Willatt, S. T. and D. M. Pullar. 1983. Changes in soil physical properties under grazed pastures. *Australian Journal of Soil Research* 22:343–348.

Zöbisch, M. A., C. Richter, B. Heiligtag, and R. Schlott. 1995. Nutrient losses from cropland in the Central Highlands of Kenya due to surface runoff and soil erosion. *Soil and Tillage Research* 33:109–116.

CHAPTER 12

Runoff, Soil, and Soil Organic Carbon Losses within a Small Sloping-Land Catchment of Laos under Shifting Cultivation

Vincent Chaplot, Yves Le Bissonnais, and J. Bernadou

CONTENTS

12.1 Introduction 167
12.2 Materials and Methods 168
    12.2.1 Site 168
    12.2.2 Land Use and Land Management 170
    12.2.3 Evaluation of Water Erosion 171
    12.2.4 Soil Sampling and Soil and Sediment Analyses 171
    12.2.5 Evaluation of the Soil Structural Stability 171
12.3 Results 172
    12.3.1 The Soil Characteristics 172
    12.3.2 The Runoff, Sediment, and Carbon Losses during the 2002 Rainy Season 173
    12.3.3 The Evaluation of the Soil Structural Stability 176
12.4 Discussion 176
    12.4.1 The Scale Effect for Water Erosion 176
    12.4.2 The Impact of the Fallow Period on Water Erosion 177
12.5 Conclusion 178
Acknowledgments 178
References 179

12.1 INTRODUCTION

An understanding of the impact of human activities on bio-geochemical cycling is important to determine, for instance, the implications of changing soil properties on crop yields and food security. The extent of human-induced soil degradation is so alarming that the acquisition of new scientific knowledge is of paramount importance. The effect of rapid changes in land use on water and soil resources in the tropics caused by demographic, economic, political, and cultural factors is well documented. Land-use conversion, such as deforestation of natural ecosystems and inappropriate land use, has notable effects on water supply, water quality, soil erosion, and soil degradation (Ingram et al., 1996). One of the consequences of the conversion of the tropical rainforests to pastures or cultivation is a decrease of the porosity of the topsoil where organic matter and nutrients concentrate. Such densification leads to a decrease in infiltration (Husain et al., 2002) inducing more runoff, nutrient leaching, and erosion responsible for an important reduction of onsite fertility (i.e., soil degradation) as well as offsite consequences (flooding, decrease in groundwater recharge, eutrophication hazards, water pollution by heavy metals and pesticides, sedimentation in valleys and reservoirs). All these on- and offsite effects may jeopardize the future of natural ecosystems and the economic development of societies.

In sloping lands of the tropics, the principal traditional agricultural practice consists of shifting cultivation. In this system, the land management leads to successive periods of cropping and fallow. This nonintensive practice preserves soil fertility in the long term (Sanchez and Hailu, 1996) through improvement of nutrient cycling, as has been documented in northern Vietnam by Fagerstrom et al. (2002). Nowadays, in many sloping lands of the tropics, the shifting cultivation cycle (i.e., the time period between two successive clearing/cropping phases on the same site) has been shortened to 3 to 5 years whereas ecological sustainability may require a minimum fallow period of 10 years (Sanchez and Hailu, 1996). Such dramatic reduction or suppression of fallows may exacerbate water erosion at the catchment level.

The direct impacts of fallow on the reduction of soil water erosion over catchments are well documented. Gafur et al. (2003) in Bangladesh indicated that the sediment loss from a catchment under fallow was about 0.3 kg m^{-2} yr^{-1}, that is, six times lower than that under cultivation. The median peak discharge under successive crops increased by a factor of seven and annual runoff increased by 16%. In Western Africa, results from runoff plots and lysimeters revealed C losses by erosion and leaching between 10 and 1900 kg C ha^{-1} yr^{-1}, depending on vegetal cover and annual rainfall (e.g., Roose and Barthès, 2001). In Cameroon, soil losses amounting to 0.12 kg m^{-2} yr^{-1} under fallow and 10.9 kg m^{-2} yr^{-1} under cultivation have been reported (Ambassa-Kiki and Nill 1999). The reasons for lower erosion under fallow than under cultivation may be due to the decrease in detachment rate (by 64%; Mamo and Bubenzer, 2001) and the increase in water infiltration (Husain et al., 2002).

Yet, there is a strong need for quantitative data on the impact of the reduction of fallow duration on soil erodibility and C losses during the cropping period of the shifting cultivation cycle.

Objectives of the project were: (1) to evaluate runoff, soil, and soil organic carbon (SOC) losses within a small sloping-land catchment of Laos under shifting cultivation, and (2) to assess the impact of the fallow period on water erosion. The study was conducted in the mountainous areas of northern Laos where shifting cultivation covers one third of agricultural land (Dufumier and Weige,l 1996). It involved simultaneous evaluations of water, sediment, and C erosion at the outlet of a 0.6-ha catchment and on 1-m^2 microplots under upland rice cultivation following a 4-year fallow period (RF4) or under 3-year continuous upland rice cultivation (RF0). Measurements were also made for soil aggregate stability (e.g., Le Bissonnais and Arrouays, 1997; Barthès and Roose, 2002).

12.2 MATERIALS AND METHODS

12.2.1 Site

The study site, a 0.6 ha-catchment, is located within a hillslope in northern Laos (Luang Prabang province, Figure 12.1). The average annual rainfall over the last 30 years is 1403 mm, and the mean annual temperature is 25°C. Two distinct seasons characterize the study site: a wet season from April to October, and a dry season from November to March.

The specific hillslope is 170 m long with a convexo-concave or complex slope. The altitude ranges from 505 m in the streambed to 584 m at the summit, and the mean slope gradient is 46%. At downslope position, average slope gradient is 30% with a range of 5 to 33%. The slope gradient

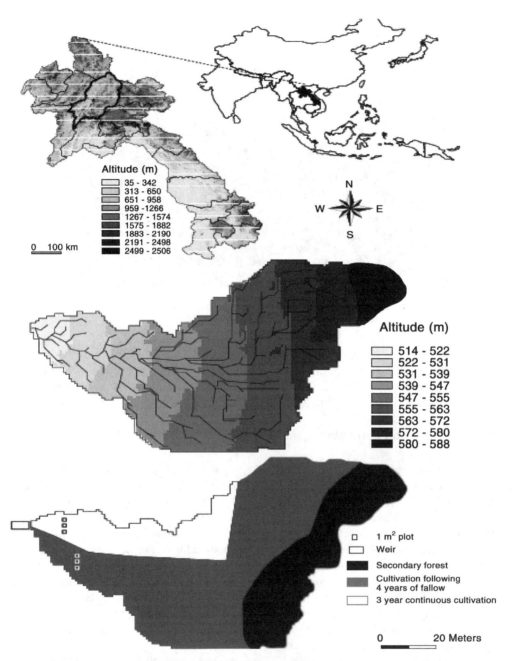

Figure 12.1 Location of the study catchment within the Luang Prabang province (Laos). Topographic conditions and estimated flow pathways from the digital elevation model with a 5-m mesh. Land-use type and history and position of micro-plots for the evaluation of runoff, sediment, and carbon losses.

increases in the upslope direction to reach a mean of 54% at midslope, and decreases afterward to 25% at the summit. Within this hillslope, a permanent gully stretches perpendicularly to the contour from the footslope to the midslope position. A weir constructed toward backslope of this gully defines the catchment outlet (Figure 12.1 and Figure 12.2). The geological substratum is mainly constituted by argilites, siltstones, and fine-grained sandstone from Permian to Upper Carbonifer. The spatial distribution of soils developed from these bedrocks is controlled mainly by the topography. Alfisols are the most representative soils within the catchments. They comprise all the surface

Figure 12.2 Picture of the study 0.6-ha catchment from the opposite hillslope on April 11, after clear cut and burning. Position of the weir at the outlet of the catchment.

area under crops (Figure 12.1). Alfisols are deep (from 2.5 to 4.5 m) clayey soils marked by a typical argillic B horizon with clay films (thickness ≥ 1 mm). Due to truncation by erosion, this argillic horizon frequently outcrops to the soil surface. With increasing erosion rates, especially in the upslope direction, the argillic horizon disappears and soils are predominately Inceptisols.

12.2.2 Land Use and Land Management

This catchment is representative of the slash-and-burn systems of Southeast Asia without inputs. In particular, it shows the effect of the gradual reduction of the fallow period from 10 to 15 years during the 1970s to 2 to 5 years in 2000s, and the gradual encroachment of continuous cultivation on the whole catchment area (Figure 12.1 and Figure 12.2). The cropping system is predominantly comprised of 80% of rotations between crops (mainly rice, *Oryza sativa*, and possibly maize, *Zea mays*, and Job's tear, *Coix lacryma Jobi*) and grassy/woody fallows. Forest covers less than 20% of the whole catchment area. Upland rice and Job's tear are the most common crops. On hillslopes, crops are generally located at backslope and midslope positions whereas the slope summits are under forest and the bottomlands under tree plantations. The first management operation generally occurring from March to April comprises slash and burn. The soil preparation for sowing consists in a manual shallow tillage (0 to 2 cm) using a hoe. Crops are seeded manually and weeded throughout the growing season, until the harvest.

Experiments were conducted during 2002. There were two contrasting historic land uses within the catchment: 4-year natural wooded/grassy fallow period (RF4) and 3-year continuous upland rice cultivation (RF0). It was thus possible to evaluate the impact of the fallow period on the runoff, sediment, and C losses since these two land uses occurred on a similar geological substrate (schist), slope angle (45%), slope position (backslope) and aspect (west), soil type (Alfisols), cropping system (upland rice), and land management. These two land uses have been under a slash-and-burn system since 1960 using similar land use and management alternating between upland rice cultivation (2 to 3 years) and periods of natural fallow (5 to 6 years). The vegetation in RF4 treatment was cut on March 10, both treatments were burned on March 22, and seeded on May 15. Each plot was weeded on June 19, and August 1 and 27 by shallow (0–2 cm) cultivation using a hand hoe.

12.2.3 Evaluation of Water Erosion

During the 2002 rainy season, runoff rate and amount, sediment and C losses were measured both at the catchment outlet and on 1-m² plots. At the catchment outlet, a weir was constructed and automatic water level recorder and water sampler were installed for the estimation of water, sediment and C losses (Figure 12.1). For each land-use history, three bounded 1-m² microplots, each separated by 2 m, were installed. Metal sheets were inserted in the soil to 0.1 m depth, just after the burn. Field measurements were carried out during the rainy season from May 15, 2002 to November 3, 2002. After June 5, measurements were considered to occur under conditions of steady-state soil loss because no significant soil cracking and rills were observed within the plots. For each rainfall event, rainfall characteristics such as rainfall amount and maximum or average rainfall intensity were computed using an automatic rain gauge with a 6-min time interval. After each rainfall event, the runoff amount from each micro-plot was measured and an aliquot was collected and oven-dried to estimate sediment concentration and sediment discharge. In addition, soil surface features including crusting and soil surface cover were evaluated visually every week. Soil roughness was quantified monthly using a laser device at each plot according to a 5-cm regular grid. A total of 210 samples were collected for 35 rainstorm events.

12.2.4 Soil Sampling and Soil and Sediment Analyses

At the end of the rainy season and after the harvest of rice, a soil profile was described for each treatment. The following parameters were measured: (1) number, type (Soil Survey Staff, 1999), and thickness of horizons (including loose saprolite); (2) moist Munsell chroma and value; (3) structure and main features; (4) texture; (5) bulk density; and (6) organic C content. The bulk densities were estimated by the volumetric method using 250-ml volume cylinders. Soil organic carbon content was measured using wet oxidation techniques of Heanes (1984). This method is a modification of the original Walkley and Black method where oxidization of organic C is achieved using only heat of reaction and dilution. However, the Walkley and Black method underestimates organic C content (especially when high proportion of black C from slash-and-burn practices occurs). The Heanes technique was chosen because it involves a hot plate digestion stage. Before the determination of organic C, soil samples were oven-dried at 105°C for 24 h and subsequently sieved through a 2-mm sieve.

To estimate the annual eroded C at the catchment outlet and from the 1-m² plots, a determination of organic C content of sediments was performed for the main rainfall events of 2002 and for an additional set of four events randomly selected over the range of 35 events. The mean organic C content was used to compute the annual organic C loss.

12.2.5 Evaluation of the Soil Structural Stability

Soil structural stability of RF0 and RF4 treatments was assessed on soil aggregates collected on May 5 (before the rainy season and just after the burn) at a systematic location over the plot boundary. A large quantity of soil (around 5 kg) was collected from the 0 to 5 cm layer and aggregates 3 to 5 mm in size were obtained by dry sieving. These aggregates represented 91% and 93% of the total soil weight for RF0 and RF4, respectively.

Before measurements, the aggregates were oven-dried at 40°C for 24 hours. An evaluation of the C content of the soil samples was performed using the Heanes (1984) method.

The Le Bissonnais laboratory test (Le Bissonnais, 1996) is based on a combination of three treatments, each corresponding to different wetting conditions and energy inputs: fast-wetting, slow-wetting, and mechanical breakdown. The fast-wetting test consisted of the gentle immersion of 5-g subsamples of dried aggregates in 50-ml deionized water. The second operation, 10 min later, consisted of the collection of the aggregates after the water had been removed by a pipette. For slow-wetting, a 5-g subsample of dried aggregates was put on a filter paper and then wetted by capillarity on a suction table at a pressure of 0.3 kPa. Residual aggregates were collected after 30 min. For mechanical breakdown, a 5-g subsample, immersed first in ethanol for 10 min, was placed in a flask with 50 ml of deionized water and made up to 250 ml with water and afterward rotated end-over-end ten times. After 30 min, the excess water plus the suspended particles were removed using a pipette. The residual material was finally collected.

Subsequently, the size distribution of the aggregates was evaluated. To do so, aggregates were put on a 50-µm sieve immersed in ethanol and agitated five times. The fraction with a diameter lower than 50 µm was weighed after a 48-h oven-drying at 105°C. The remaining fraction was collected, oven-dried at 105°C for 48 h, and then submitted to dry-sieving through a column of six sieves with mesh sizes of 2, 1, 0.5, 0.2, 0.1, and 0.05-mm. Weights of aggregates collected on each sieve were measured and expressed as the percentage of the sample dry mass. The aggregate stability was expressed by the mean weight diameter MWD calculated as follows:

$$\text{MWD} = \frac{\sum (x_i \times w_i)}{100} \qquad (12.1)$$

with x_i the mean intersieve size and w_i the percentage of particles retained on each i sieve. Additional measurements of texture and organic C were performed for each plot on the 3 to 5 mm aggregates.

12.3 RESULTS

12.3.1 The Soil Characteristics

Soil profiles of the two treatments showed a surface organo-mineral horizon with a thickness of 2 to 10 cm. The topsoil texture was clayey, clay fraction estimated from six replicates averaged 53% (Table 12.1). There were no strong differences in topsoil texture between the two treatments. The clay content in RF0 ranged from 47 to 59%, and differences among replicates were slightly lower for RF4 with a clay content of from 53 to 57%. For both RF4 and RF0, fine silts represented 27% of total soil. The mean sand content was less than 12%. On average, A horizon exhibited similar organic C contents and densities for the 0 to 5 cm layer. The mean organic C content from the three replicates was 22.5 g C kg^{-1} for RF0 and 22.7 g C kg^{-1} for RF4 (Table 12.1). No significant

Table 12.1 Main Soil Characteristics (Texture in Five Classes; Organic C; Bulk Density; C Stock) of the 0 to 5 cm Layer for Three Replicates of the Two Treatments[a]

	RF0				RF4			
	1	2	3	Mean	1	2	3	Mean
Soil texture (g kg^{-1})								
Clay (< 2 µm)	499	594	470	521	575	534	539	549
Fine silts (2 to 50 µm)	273	255	274	267	275	267	269	270
Coarse silts (20 to 50 µm)	94	66	107	89	57	85	84	75
Fine sands (50 to 200 µm)	62	50	76	63	42	43	42	42
Coarse sands (200–2000 µm)	72	35	73	60	51	71	66	63
Organic C content (g C kg^{-1})	24.3	15.3	28.1	22.5	21.8	24.3	21.9	22.7
Bulk density (Mg m^{-3})	0.98	0.88	0.92	0.93	0.91	0.97	0.91	0.93
Organic C stock (g C m^{-2})	1191	673	1293	1046	992	1179	996	1056

[a] Treatments: continuous cultivation (RF0); cultivation following a 4-year fallow (RF4). Houay Pano catchment (Luang Prabang province, Laos)

differences were observed between these two treatments for black C and mineral-bound C (data not presented). The mean bulk density was 0.93 Mg m^{-3} in both cases with values ranging between 0.88 and 0.98 Mg m^{-3}. As a consequence, the mean organic C stocks computed for the 0 to 5 cm layers were around 1050 g C m^{-2} for both RF0 and RF4. Areas where A horizons were the thickest were marked by a homogeneous reddish brown matrix (5YR4/3) whereas some more reddish spots (5YR4/4) were observed in soils with shallow surface layer. The structure was blocky subangular. Numerous fine, dead and nondeviated roots characterized the RF0, and medium to coarse roots were observed in the RF4 treatment. Just after sowing, a greater proportion of free aggregates on the soil surface was observed at RF4 than at RF0 (40% in average against 15%). In both cases, stable aggregates were relatively coarse (6.5 mm in mean) but slightly coarser at RF0 than at RF4 (6.9 vs. 5.3 mm in average). Some crusted surface remained even after sowing, 30% in RF4 and 10% in RF0. No particle sorting was observed in either treatment. The crust surface in RF4 was partly covered by an algae (from 5 to 40% of the whole crust surface area) and occasionally by mosses (5% in one replicate). In both treatments, several stumps persisted after slash and burn on the soil surface. Stumps were on average 8 m apart. Some compacted earthworm casts embedded in the crust were observed only in RF4. The boundaries of the underlying mineral clayey horizon were diffuse. This horizon showed a reddish matrix (5YR4/3) with numerous vertical and elongated browner spots, probably resulting from the mixing of surface organo-mineral material by bioturbation. In both treatments, the subsurface horizon with a blocky to columnar structure showed numerous clay films on both vertical and horizontal surfaces of peds but little porosity. Underlying horizons were reddish (10YR4/3) and red (5R4/6) to a depth of 1.5 m, becoming structureless with increasing depth.

12.3.2 The Runoff, Sediment, and Carbon Losses during the 2002 Rainy Season

The 2002 rainy season, from May 25 to October 25, was characterized by a total of 35 rainfall events with an amount of 1023 mm. Minimum and maximum rainfall amounts were 4.5 and 162 mm, respectively, with a median of 17 mm. These events showed a median maximum rainfall intensity in 6 min of 40 mm h^{-1} with values ranging between 5 and 135 mm h^{-1}. The most extreme event occurred on July 20 (Figure 12.3). It produced a total rainfall amount of 132 mm with a maximum intensity of 100 mm h^{-1} in 6 min. Four main rainfall events occurred toward the end of the growing cycle, on September 9, October 2 and 6, and November 3.

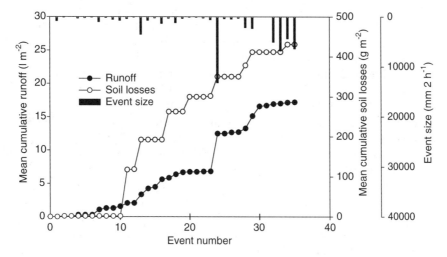

Figure 12.3 Mean cumulative runoff and soil losses at the 0.6-ha catchment outlet during the 2002 rainy season as a function of the rainfall event size.

Table 12.2 General Statistics for Runoff (R), Sediment Concentration (SC), and Soil Losses (SL) for the 35 Rainfall Events of 2002[a]

	R Rate mm mm^{-1}	R Amount l m^2 yr^{-1}	SC g l^{-1}	SL g m^2	EOC g C kg^{-1}	EOC g C m^2
Catchment Level						
Min.	0.001		0.1	1	17.4	0.0
Max.	0.062		133.0	116	32.3	3.1
Av.	0.017		25.1	12	26.1	0.3
SD.	0.018		29.6	27	5.6	1.2
Sum		17.2		431		11.2
RF0 Micro-plots						
Min.	0.035		0.3	5	21.6	0.0
Max.	0.731		36.5	1383	44.7	48.5
Av.	0.218		9.4	60	35.6	2.2
SD.	0.168		7.4	249	12.3	16.2
Sum		223.1		2114		75.2
RF4 Micro-plots						
Min.	0.039		0.2	5	31.3	0.0
Max.	0.727		28.6	422	33.0	18.6
Av.	0.197		4.8	28	32.1	0.9
SD.	0.142		4.7	76	0.8	9.8
Sum		202.1		975		31.2

Abbrev.: minimum, Min.; maximum, Max.; average, Av.; standard deviation, SD.

[a] Two treatments (continuous cultivation, RF0; cultivation following a 4-year fallow, RF4); and three replicates are considered for microplots. General statistics on eroded organic carbon (EOC) at the microplot and catchment levels computed from five randomly selected events, and total EOC computed over 2002. Houay-Pano catchment (Luang Prabang province, Laos)

Results of water runoff, sediment, and C losses for these 35 rainfall events are presented in Table 12.2 and Figure 12.3 through Figure 12.5. At the catchment level, the mean annual runoff rate (R) was 0.017 mm mm^{-1}, with values for individual events ranging from 0.001 to 0.062 mm mm^{-1} (Table 12.2). During the five first events, the runoff coefficient was very low and only a slight increase of the cumulative amount occurred (Figure 12.3). No sediment and C erosion occurred during this period. Runoff and soil losses progressively increased up to event number 30 with the exception of event number 23. At the end of the rainy season, the total runoff amount was 17.2 l m^2 yr^{-1} and the total sediment losses were 431 g m^2 yr^{-1} (Figure 12.3). The events under study were characterized by a mean annual sediment concentration of 25.1 g l^{-1}, with a maximum value of 133 g l^{-1} on July 20. The mean annual C concentration in sediments, computed from the five selected events, was 26.1 g C kg^{-1} and the computed C losses were 11.2 g C m^2 yr^{-1}.

On the microplots, mean annual runoff rates and total runoff amounts were slightly higher for RF0 than for RF4 (0.218 vs. 0.197 mm mm^{-1}, and 223 vs. 202 l m^{-2} yr^{-1}, respectively; Table 12.2 and Figure 12.4). However, mean sediment concentration was twice as much for RF0 than for RF4 (9.4 vs. 4.8 g l^{-1}, $p < 0.05$). Consequently, total soil losses were twice as high for RF0 than for RF4 (2.1 vs. 1.0 kg m^{-2} yr^{-1}, $p < 0.05$; Figure 12.5). Slightly mean sediment C content for RF0 than for RF4 (35.6 vs. 32.1 g C kg^{-1}) also resulted in about 2.5 times higher total C losses for RF0 than for RF4 (75.2 vs. 31.2 g C m^{-2} yr^{-1}, $p < 0.05$). Thus fallow prior to cultivation reduced mean runoff and mean sediment C content by 10%, but reduced mean sediment concentration and total soil and C losses by 50 to 60%.

The preceding fallow period significantly decreased not only the overall soil erosion but also the preferential erosion of C since the mean organic C enrichment ratio of sediments decreased

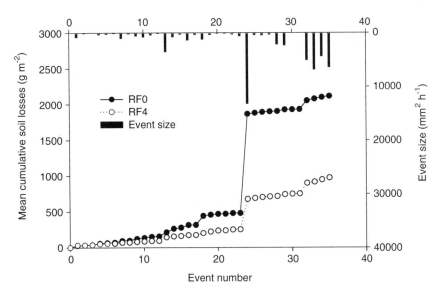

Figure 12.4 Mean cumulative runoff amount during the 2002 rainy season estimated from three 1-m² micro-plot replicates under continuous cultivation (RF0) and under cultivation following a 4-year fallow (RF4).

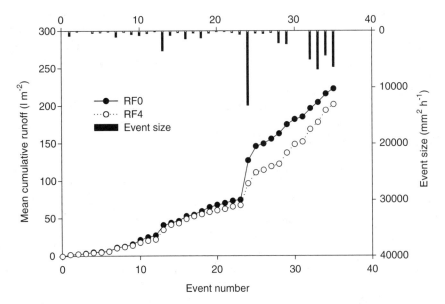

Figure 12.5 Mean cumulative sediment losses during the 2002 rainy season estimated from three 1-m² micro-plot replicates under continuous cultivation (RF0) and under cultivation following a 4-year fallow (RF4).

from 1.6 (RF0) to 1.4 (RF4) with reference to the 0 to 5 cm soil layer, and from 1.7 (RF0) to 1.5 (RF4) with reference to the 0 to 10 cm soil layer (data not presented). During 2002, the amount of eroded C represented 7.2 and 3% of organic C stocks for the 0 to 5 cm soil layer in RF0 and RF4, respectively. No significant differences in soil surface cover were observed between the two treatments during the study period (Figure 12.6).

The mean cumulative runoff and sediment losses for the two treatments studied during the rainy season are presented in Figure 12.4 and Figure 12.5. At the very onset of the rainy season (from May 25 to June 6) and under conditions of bare soil (soil surface cover of less than 10% at both situations, Figure 12.6) and with low-sized events, few differences existed among the two

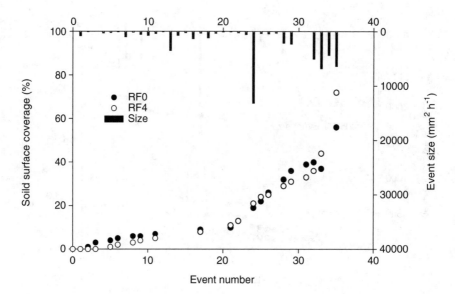

Figure 12.6 Soil surface coverage by vegetation including crop and weed under upland rice cultivation (RF0) and cultivation following a 4-year fallow (RF4). Mean of three field replicates.

treatments. Differences became significant from event ten. Most of the erosion produced during 2002 occurred in the middle of the rainy season, especially during the most extreme event of July 20. This event accounted for 65 and 43%, respectively, of the total annual soil losses on RF0 and RF4. In addition, during this major event for which the average soil surface cover was 20% in both treatments, more soil erosion was observed on RF0 than on RF4 (Figure 12.5). Afterward, the water erosion occurred even with greater proportion of soil surface cover, which increased from 20% on July 20, 30% at mid-August, 40% at mid-September, and to 56% for RF0 and 72% for RF4 for the last rainfall event on November 3. There were no significant differences in soil roughness between the two treatments.

12.3.3 The Evaluation of the Soil Structural Stability

The mean weight diameter (MWD) estimated for the three replicates within each treatment (RF0 and RF4) is presented in Figure 12.7. The MWD ranged from 2.36 to 3.19 mm, which is rather high even for soils of the temperate region (Le Bissonnais and Arrouyas, 1997). The average MWD was slightly higher for RF4 (3.13 mm) than for RF0 (2.94 mm). Significant differences among treatments were observed for mechanical breakdown, in which more slaking and dispersion occurred in RF0. These results demonstrated that there was lower dispersion after 4 years of fallow than with continuous cultivation. Furthermore, the low dispersion was due to the aggregate protection against raindrop impact caused by 4 years of fallow. However, fallow provided few benefits in these Alfisols in terms of the aggregate slaking caused by increased air compression, and the breakdown due to swelling tensions or physico-chemical dispersion.

12.4 DISCUSSION

12.4.1 The Scale Effect for Water Erosion

Results at the catchment level showed a low annual runoff rate (< 0.02 mm mm^{-1}) and moderate soil losses (0.4 kg m^{-2} yr^{-1}) compared to the results of Gafur et al. (2003) in Bangladesh showing

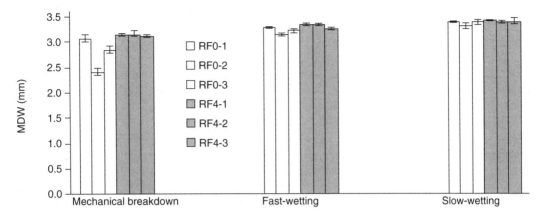

Figure 12.7 Mean weight diameter (MWD) of stable aggregates in 0 to 5 cm soil samples under continuous cultivation (RF0) and cultivation following a 4-year fallow (RF4) including three field replicates (1, 2, 3); the aggregation test included three treatments (mechanical breakdown, fast- and slow-wetting) and was carried out on three laboratory replicates for each sample (mean and standard deviation).

runoff from 1-ha catchments of about 0.2-mm mm^{-1} and sediment losses in catchments under cultivation of about 18 kg m^{-2} yr^{-1}. In Cameroon, soil losses of 10.9 kg m^{-2} yr^{-1} at the small catchment scale have been reported under cultivation (Ambassa-Kiki and Nill, 1999). The lower water erosion in northern Laos may be explained by the high infiltration rate in Alfisols. High infiltration rates may occur along slopes in local depressions or as a result of biological features such as tree stumps, root networks, and macropores. In addition, a high infiltration rate may also occur within the central gully as observed in other sites by Bryan and Poesen (1989) and Slattery and Bryan (1992), and confirmed by our visual observations. Such a high infiltration may limit the transport of sediments by reducing detachment and transport processes on longer distances as proposed by Kinnell (2000) and Chaplot and Le Bissonnais (2003). As a consequence, C losses of 11 g C m^2 yr^{-1}, which corresponded to 0.11 mg C ha^{-1} yr^{-1}, were slightly lower than those evaluated in other catchments (e.g., Gregorich et al., 1998: from 13 to 49 g C m^{-1} yr^{-1} in the Canadian prairies). A similar hypothesis then for low water and soil losses at the catchment level may explain the small amount of eroded C.

The runoff amount was 12 times more and the sediment losses were 2 to 5 times more at the microplot than at the catchment level. These trends may be explained by the higher infiltration and sedimentation possibilities at the catchment level as reported by Le Bissonnais et al. (1998). Longer slope at the catchment level than in microplots resulted in more runoff velocity and greater transport capacity, and as a consequence, in three to five times more sediment concentration.

12.4.2 The Impact of the Fallow Period on Water Erosion

Within the catchment, a comparison between water erosion on plots differing only in the duration of a fallow period revealed a greater erosion under continuous cultivation than under cultivation following a fallow period. The preceding fallow period decreased the total yearly runoff amounts by 10% and total soil erosion by 50%. Such differences between plots with similar soil characteristics and properties, topographic conditions, orientation, and soil surface cover mainly occurred for the intense rainstorm events. Subsequently, runoff was surprisingly more for RF4 than for RF0, but this did not result in more soil erosion in RF4 than in RF0. Further investigations must be conducted to understand such a temporal behavior.

The lower soil erodibility after a fallow period may first be explained by a greater resistance of the soil aggregates to mechanical breakdown. Such a greater resistance was not well explained by differences in soil structure, clay, and soil C content of aggregates as previously observed at other sites (Le Bissonnais and Singer, 1993; Le Bissonnais and Arrouays, 1997; Chenu et al., 2000; Malam Issa et al., 2001). No differences existed among RF0 and RF4 treatments regarding the soil C content and stock, though the RF4 treatment was expected to have more C content and stock due to enhanced possibilities of physical C sequestration under fallow (Feller and Beare 1997; Larré-Larrouy et al. 2004). The higher aggregate resistance after fallow may be due to increased binding and gluing of aggregates or the organic C quality (Chenu et al., 2000; Malam Issa et al., 2001), a result that needs further studies.

The low soil erodibility after a fallow period may be partly due to the presence of algal and moss crusts protecting soil surface against raindrop impact. This hypothesis is also supported by the observation on differences in water erosion that occurred during the first half of the rainy season and especially during its strongest events when algae and mosses were present on the soil surface, and no differences were observed between two treatments during the second half of the rainy season after the removal of these biological features by the rainfall. However, additional research needs to be conducted to establish a direct relation between the fallow period, biological crusts, and the overall water erosion.

Finally, these results indicated that the stability of aggregates in surface soil is an important predictor of water, sediment, and C losses by water erosion. Similar conclusions were made by Barthès et al. (2000) in Benin, Cameroon, and Mexico under different climate (400- to 1600-mm annual rainfall), soil type (sandy clay loam Nitosol, loamy sand Ferralsol, loamy Regosol), and management (from savanna to long-duration mouldboard plowing) systems.

12.5 CONCLUSION

This chapter quantified runoff, soil, and SOC losses within a small sloping-land catchment of Laos under shifting cultivation. In addition, some of the processes involved in interrill erosion were identified by using a combination of laboratory and field surveys. Soil erosion and especially eroded C were relatively low at the small catchment level due to the high infiltration rate limiting the transport of sediments. At this level, erosion was transport-limited. Although at the micro-plot level soil detachment and runoff production were high, this was not apparent at the catchment outlet, suggesting the existence of high infiltration at specific locations: gullies and biological features such as roots and stumps remaining after slash and burn. Furthermore, these biological features provide habitats and nutrients for a range of living organisms, and potentiating infiltration pathways. Thus fallow periods reduce soil erosion by both limiting the detachment capacity and increasing infiltration at specific locations.

In addition, the fallow period within a shifting cultivation cycle, protects soil against erosional processes, both locally and at the micro-plot level by limiting soil detachment due to raindrop impact through an enhanced aggregate stability and by biological crusts by algae and mosses.

Further studies are necessary in order to confirm these results and to define an optimal duration of the fallow period in relation to soil type. Additional research is also needed to better understand the processes and mechanisms involved in water erosion of clayey sloping lands of the tropics, and to identify management options for sustainable use.

ACKNOWLEDGMENTS

This research is part of the Management Soil Erosion Consortium (MSEC) aiming in Southeast Asia at a better understanding of water erosion within catchments. Authors gratefully acknowledge the Asian Development Bank (ADB) and the International Water Management Institute (IWMI) for funding this project and the Soil Survey and Land Classification Center for hosting foreign research in Laos. Special thanks are due to Khampaseuth Xayyathip, Julien Tessier, Khambay

Phonmisa, and Norbert Silvera for help with field work, H. Gaillard and O. Duval for their skilled technical assistance and analysis of the data. Finally the authors acknowledge C. Valentin, the director of the IRD research unit, and helpful suggestions from Bernard Barthès, Christian Feller, and Eric Roose in reviewing this chapter.

REFERENCES

Ambassa-Kiki, R. and D. Nill. 1999. Effects of different land management techniques on selected topsoil properties of a forest Ferralsol. *Soil and Tillage Research* 52:259–264.

Barthès, B., A. Azontonde, B. Z. Boli, C. Prat, and E. Roose. 2000. Field-scale run-off and erosion in relation to topsoil aggregate stability in three tropical regions (Benin, Cameroon, Mexico). *European Journal of Soil Science* 51:485–495.

Barthès, B. and E. Roose. 2002. Aggregate stability as an indicator of soil susceptibility to runoff and erosion; validation at several levels. *Catena* 47:133–149.

Bryan, R. B. and J. Poesen. 1989. Laboratory experiments on the influence of slope length on runoff, percolation, and rill development. *Earth Surface Processes and Landforms* 4:211–231.

Chaplot, V. and Y. Le Bissonnais. 2003. Runoff features for interrill erosion at different rainfall intensities, slope lengths and gradients in an agricultural loessial hillslope. *Soil Science Society of America Journal* 67:844–851.

Chenu, C., Y. Le Bissonnais, and D. Arrouays. 2000. Organic matter influence on clay wettability and soil aggregate stability. *Soil Science Society of America Journal* 64:1479–1486.

Dufumier, M. and J. Y. Weigel. 1996. Minorités ethniques et agriculture d'abattis-brûlis au Laos. *Cahiers des Sciences Humaines* 32:195–208.

Fagerstrom, M., S. Nilsson, M. van-Noordwijk, T. Phien, M. Olsson, A. Hansson, and C. Svensson. 2002. Does *Tephrosia candida* as fallow species, hedgerow or mulch improve nutrient cycling and prevent nutrient losses by erosion on slopes in northern Viet Nam? *Agriculture Ecosystems and Environment* 90:291–304.

Feller, C. and M. H. Beare. 1997. Physical control of soil organic matter dynamics in the tropics. *Geoderma* 79:69–116.

Gafur, A., J.R. Jensen, O. K. Borggaard, and L. Petersen. 2003. Runoff and losses of soil and nutrients from small watersheds under shifting cultivation (Jhum) in the Chittagong Hill Tracts of Bangladesh. *Journal of Hydrology* 279:292–309.

Gregorich, E. G., K. J. Geer, D. W. Anderson, and B. C. Liang. 1998. Carbon distribution and losses: Erosion and deposition effects. *Soil and Tillage Research* 47:291–302.

Heanes, D. L. 1984. Determination of total organic-C in soils by an improved chromic acid digestion and spectrophotometric procedure. *Communications in Soil Science and Plant Analysis* 15:1191–1213.

Husain J., H. H. Gerke, and R. F. Huttl. 2002. Infiltration measurements for determining effects of land use change on soil hydraulic properties in Indonesia. *Advances in GeoEcology* 35:229–236.

Ingram, J., J. Lee, and C. Valentin. 1996. The GCTE soil erosion network: A multidisciplinary research program. *Journal of Soil and Water Conservation* 51:377–380.

Kinnell, P. I. A. 2000. Effect of slope length on sediment concentrations associated with side-slope erosion. *Soil Science Society of America Journal* 64:1004–1008.

Larré-Larrouy, M. C., Blanchart E., A. Albrecht, and C. Feller. 2004. Carbon and monosaccharides of a tropical Vertisol under pasture and market-gardening: Distribution in secondary organomineral separates. *Geoderma* 119:163–178.

Le Bissonnais, Y. and M. J. Singer. 1993. Seal formation, runoff and interrill erosion from seventeen California soils. *Soil Science Society of America Journal* 57:244–249.

Le Bissonnais, Y. 1996. Aggregate stability and assessment of soil crustability and erodibility: I. Theory and methodology. *European Journal of Soil Science* 47:425–437.

Le Bissonnais, Y. and D. Arrouays. 1997. Aggregate stability and assessment of soil crustability and erodibility: II. Application to humic loamy soils with various organic carbon contents. *European Journal of Soil Science* 48:39–48.

Le Bissonnais, Y., H. Benkhadra, V. Chaplot, D. Fox, D. King, and J. Daroussin. 1998. Crusting and sheet erosion on silty loamy soils at various scales from m^2 to small catchments. *Soil Technology* 46:69–80.

Malam Issa, O., Y. Le Bissonnais, C. Defarge, and J. Trichet. 2001. Role of a cyanobacterial cover on structural stability of sandy soils in the Sahelian part of western Niger. *Geoderma* 101:15–30.

Mamo, M. and G. D. Bubenzer. 2001. Detachment rate, soil erodibility, and soil strength as influenced by living plant roots part I: Laboratory study. *Transactions of the ASAE* 44:1167–1174.

Roose, E. and B. Barthès. 2001. Organic matter management for soil conservation and productivity restoration in Africa: A contribution from Francophone research. *Nutrient Cycling in Agroecosystems* 61:159–170.

Sanchez, P. A. and M. Hailu, eds. 1996. Alternatives to slash-and-burn agriculture. *Agriculture, Ecosystems and Environment* 58:1–2.

Slattery, M. and R. B. Bryan. 1992. Hydraulic conditions for rill incision under simulated rainfall: A laboratory experiment. *Earth Surface Processes and Landforms* 17:127–146.

Soil Survey Staff. 1999. *Soil taxonomy. A basic system of soil classification for making and interpreting soil surveys.* 2nd ed. USDA, Washington D.C.

CHAPTER 13

Soil Erodibility Control and Soil Carbon Losses under Short-Term Tree Fallows in Western Kenya

Anja Boye and Alain Albrecht

CONTENTS

13.1 Introduction 181
13.2 Materials and Methods 182
    13.2.1 Site Description 182
    13.2.2 Experimental Design and Management 183
    13.2.3 Rainfall Simulation 184
    13.2.4 Soil Sampling and Analyses 184
    13.2.5 Data Analyses 185
13.3 Results 185
    13.3.1 Soil C Content, Bulk Density, C/N Ratio, and C Stocks 185
    13.3.2 Water Stable Aggregates and Soil Strength 186
    13.3.3 Runoff, Sediment Concentration, and Soil Loss 188
    13.3.4 C Content of Sediments, Enrichment Ratio, and Soil C Losses 189
    13.3.5 Principal Component Analysis 190
13.4 Discussion 192
    13.4.1 Impact of Improved Fallows on Runoff 192
    13.4.2 Control of Soil Loss by Improved Fallows 192
    13.4.3 Effect of No-Tillage on Soil Properties, Runoff, and Soil Loss 193
    13.4.4 C Content of Sediments and Enrichment Ratio 193
    13.4.5 Effect of Land Management on Soil C Losses and Soil C Stocks 193
13.5 Conclusion 194
Acknowledgments 194
References 195

13.1 INTRODUCTION

Extensive research has been undertaken to understand the processes of soil erosion by water for different climates and soil types. While soil erosion is a natural process, anthropogenic influence through cultivation has exacerbated the rate of soil erosion. Accelerated soil erosion is the major land degradation process in Africa (Cooper et al., 1996). Soil erosion by water is a three-phase process: (1) detachment of soil particles by rain drops, (2) transport of detached particles by runoff, and (3) deposition of detached and transported particles. Cultivation makes the land more susceptible to runoff and soil erosion by removal of the permanent plant cover. Several studies have reported close relationships between soil erodibility, soil organic carbon (SOC), and macro-aggregation (Le Bissonnais, 1996; Barthès et al., 2000; Barthès and Roose, 2002). SOC is widely acknowledged as one of the most important soil parameters to maintain good soil health (Doran et al., 1996).

However, a considerable challenge exists in maintaining adequate SOC levels in cultivated soils, especially in the tropics, where SOC losses through cultivation, decomposition, and erosion often exceed inputs. The main sources of SOC input in the tropics are biomass (above and below ground biomass) and manures, which are often less than required to maintain adequate SOC levels (Nandwa, 2001). Agroforestry is a good management option to produce sufficient biomass and to maintain or increase SOC. In western Kenya, agroforestry practices such as planted fallows produce 20 ton biomass per hectare in 8 to 18 months, which, when returned to the soil, increases SOC and soil macro-aggregation (Niang et al., 1998; IMPALA, 2001; 2002; Mutuo, 2004). Similar findings have been reported by Ingram (1990). The potential for soils to store SOC has received much attention, and several studies have shown the potential of agroforestry to sequester C above and below ground and in soil (Kursten and Burschel, 1993; Dixon, 1995; Ingram and Fernandes, 2001; Albrecht and Kandji, 2003). However, much less is known about the specific potential of planted/improved fallows to sequester C and reduce erosion-induced C losses. Several studies have reported selective detachment and transport of SOC and fine particles, resulting in depletion of SOC for *in situ* soil and enhanced SOC for depositional areas (Watung et al., 1996; Wan and El-Swaify, 1997; Jacinthe et al., 2002; Owens et al., 2002; Lal, 2003). Some studies have reported eroded C to be subjected to accelerated mineralization and thereby to contribute to CO_2 emissions from soils. In contrast, other studies suggest that deep burial of deposited sediments promotes C sequestration (Jacinthe et al., 2002; McCarty and Ritchie, 2002). Reducing runoff and soil erosion remains crucial for controlling erosion-induced C losses and more research is needed to fully understand the fate of eroded C. In Kenya, agroforestry is widely practiced to control runoff and soil erosion (Cooper et al., 1996). Van Roode (2000) reported that contour strips and hedges in association with terracing increase infiltration under the vegetative strips. However, focus has mainly been on sloping hillsides and catchment scales, and little attention has been given to the role agroforestry can play in controlling interrill erosion.

Minimum and no-till (NT) reduce runoff through accumulation of SOC and enhanced soil aggregation (Arshad et al., 1999; Franzluebbers, 2002). There is a negative correlation between enhanced SOC under NT and runoff. Several studies have found SOC to accumulate in the near surface soil layers for soils under NT compared to conventionally tilled soils (Ingram and Fernande,s 2001). However, the accumulation of SOC under NT has generally been assessed several years after conversion because time is a crucial factor in SOC accumulation. Few studies have focused on SOC accumulation under NT shortly after conversion and in association with agroforestry and planted fallows.

Thus, the aim of this study was to assess runoff and soil loss from long-term cultivated Ferralic Arenosol and Ferralsol under simulated rainfall from planted fallows (improved fallows). Specific objectives were to assess the effects of: (1) short-term improved fallows on runoff, soil, and carbon (C) losses, and (2) no-till on runoff, soil, and C losses. The study was conducted at harvest of the first maize crop after fallowing.

13.2 MATERIALS AND METHODS

13.2.1 Site Description

The study was conducted on two farms in western Kenya, Masai farm and Luero farm in July 2001. Masai farm (sandy loam) is located in Busia District (00°34.407'N, 034°11.554'E) at an

Table 13.1 Topsoil Characteristics (0 to 15 cm Depth) at the Beginning of the Experiment for the Two Study Sites, Masai and Luero (Kenya)

Soil Type	Sand (%)	Silt (%)	Clay (%)	SOC (g kg^{-1})	N (g kg^{-1})	C/N	pH$_{H_2O}$	Total P (g kg^{-1})	Exch. Ca (cmol$_c$ kg^{-1})	Exch. Mg (cmol$_c$ kg^{-1})
Sandy loam (Masai)	71	12	17	7.8	0.48	16.3	5.4	0.18	2.27	0.68
Clay (Luero)	35	25	40	16.9	1.40	12.1	5.3	0.47	3.94	1.23

Sand: 50 to 2000 µm; silt: 2 to 50 µm; clay: 0 to 2 µm; SOC: soil organic carbon; N: total nitrogen; P: phosphorus; exch.: exchangeable; Ca: calcium; Mg: magnesium.

altitude of 1290 m. Rainfall is bimodal with an annual mean of 1200 mm. Mean annual temperature is 21°C. The soil is a coarse Ferralic Arenosol (FAO) with 17% clay, 12% silt, and 71% sand. The slope gradient is 6%. The Luero farm (clay soil) is located in the highlands of western Kenya in Vihiga District (00°06.818'N, 034°31.488'E) at an altitude of 1620 m. Rainfall is bimodal with an annual mean of 1800 mm. Mean annual temperature is 22°C. The soil is a fine mixed nito-humic Ferralsol (FAO) with 40% clay, 24% silt, and 35% sand. The slope gradient is 7%. Table 13.1 lists important properties of the surface soil for the two sites.

13.2.2 Experimental Design and Management

The experimental design for each farm was a randomized block design with three replicates, each plot measuring 18 × 16 m. The objective was to compare continuous cultivation (CC) of maize (*Zea mays*) intercropped with beans (*Phaseolus vulgaris*), with intercropping of maize and beans preceded by a 18-month improved fallow of two legumes: *Crotalaria grahamiana* (IF-Cg) or *Tephrosia candida* (IF-Tc). The experiment was established in July 1999 (Figure 13.1). Improved fallows were planted at the end of the cropping season in the former bean rows, which had been harvested in late June. The fallows were allowed to grow until February 2001, when they were slashed and the land prepared for the following maize crop. Maize and beans were harvested every season (December 1999, July 2000, December 2000, August 2000, December 2001, and August 2001) for the control plots. During the first season (short rainy season 1999), all treatments were manually weeded between September and October 1999. Only the control plots (CC) were weeded (IMPALA, 2001; 2002) during the following two seasons (long rainy season 2000 and short rainy season 2000). In February 2001, the fallows (IF-Cg and IF-Tc) were slashed by cutting the stem about 10 cm above the ground level.

After the 18-month fallow phase, maize and beans were planted by splitting each plot into two parts. One part was tilled with a hand hoe, disturbing the soils to 10 cm depth (CT). The other part was left undisturbed (NT) except for planting operations (direct sowing). The returned biomass was incorporated into the soil for the CT plots, whereas it was left on the soil surface for the NT plots. The residue biomass returned (from the improved fallow and the weeds) was 1.7 t ha^{-1} for IF-Cg and 2.1 t ha^{-1} for IF-Tc for the sandy soil, and 3.9 t ha^{-1} for IF-Cg and 7.8 t ha^{-1} for IF-Tc for the clayey soil. The woody stems were removed from the system and used by the farmers. The

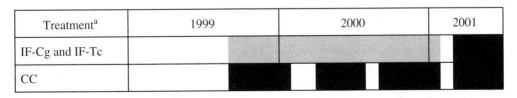

Figure 13.1 Cropping sequences for the cropping systems under study. Black represents the cropping phase and grey the fallow period. CC is continuous cultivation; IF-Cg improved fallow treatment with *Crotalaria grahamiana;* and IF-Tc improved fallow treatment with *Tephrosia candida*.

removed biomass amounted to 8.7 t ha^{-1} for IF-Cg and 10.6 t ha^{-1} for IF-Tc for the sandy loam, and 19.7 t ha^{-1} for IF-Tc for the clayey soil. Maize and fallows were planted at a spacing of 75 cm between rows and 25 cm within row. All plots were weeded twice between April and June 2001, and were harvested in August 2001.

13.2.3 Rainfall Simulation

A field rainfall simulator (ORSTOM type) was used to simulate rainfall. This simulator simulates rainfall over 4 m^2 but measures runoff from a 1-m^2 plot. Rainfall is produced by a single nozzle, which sprinkles water downward. The nozzle is placed at a height of 4 m, enabling the raindrops to reach terminal velocity. More detailed information on the ORSTOM rainfall simulator is given by Asseline and Valentin (1978).

Two rainfall intensities were chosen, 50 and 90 mm hr^{-1}, to simulate low and high intensity rainstorms. Rainfall simulations were carried out at crop harvest (maize intercropped with beans) in August 2001. The simulation regime consisted of three events. The objective was to simulate rain on dry, wet, and very wet soils. The first event was carried out on dry soils with medium rainfall intensity (50 mm hr^{-1}), and it was continued until steady runoff occurred (maximum 90 min). The following day two simulations were conducted on wet soils. The first event had an intensity of 50 mm hr^{-1} and the following one of 90 mm hr^{-1}. The duration was 30 min. There was a break of 15 min between the two rainfall simulations to allow runoff to cease. There were three replicates for each treatment and subtreatment. Runoff was measured on a 1-min interval and sediment samples were collected for every 2 min until steady runoff was achieved. At steady state, runoff sediment samples were collected at 5-min intervals. Before each simulation the loose biomass was removed by hand. Thus, simulations were done on a bare soil surface when soil cover was less than 5%.

The data for only the very wet run are presented in this chapter. The objective is to discuss runoff, soil erosion, and soil C losses for different land use systems and soil types. Majority of soil loss occurs at the on-set of the rainy season during high intensity storms. Thus, the very wet run simulated such scenarios, and provide an indication of soil C losses during natural rainfall.

13.2.4 Soil Sampling and Analyses

Soil samples were collected to determine bulk density, soil aggregate stability, soil C and nitrogen (N) contents, and C/N ratio. Soil was sampled in June 2001 (before harvest of beans) at 0 to 5 and 5 to 10 cm depths using 98-cm^3 cores with three replicates for each plot. Water stable aggregates (WSA) were determined by wet sieving after shaking, with three replicates for each soil sample. Fifty g of air-dried soil was passed through a 2-mm sieve and shaken in 300 ml of water for 1 hour in a tumbler shaker at 50 revolutions per minute. The sample was then sieved through 212- and 20-μm sieves. For both soil types, WSA larger than 212 μm were expressed on a coarse sand-free basis (Albrecht et al., 1992; Feller et al., 1996). These samples were then bulked to form one sample per replicate (n = 3). Soil resistance to penetration and shear stress were measured onsite using a penetrometer CL 700A (kg cm^{-2}) and a torvane CL 600 (kg cm^{-2}), respectively. Soil resistance to penetration and shear stress were measured at the soil surface after each rainfall simulation next to the 1-m^2 plot. Each measurement was replicated six times.

Total C and N contents of soil and sediment samples were determined by the CNS Carlo Erba micro-analyzer method. In the absence of carbonates, all C was considered organic. Soil C stocks were calculated for equivalent depth (0 to 10 cm) and for equivalent mass (this mass was 87.5 kg m^{-2} and corresponded to the smallest mass of the 0 to 10 cm soil layers under study, which was in the clay soil under IF-Tc and NT) basis (Ellert and Bettany, 1995).

13.2.5 Data Analyses

The data were statistically analyzed using ANOVA for a completely randomized block design with final runoff rate, runoff depth, sediment concentration, and soil loss as variables in the first analysis. In the second analysis, the variables were percentage water stable aggregates, soil C content, bulk density, C/N ratio, and soil resistance to penetration and shear stress. The third analysis had enrichment ratio, C losses, and soil C stocks as variables. Statistical significance was determined at the 95% confidence level with Tukey's test. Sediment C content was averaged for the three replicates, thus no statistical analyses were done for this variable and for sediment C/N ratio.

A principal component analysis (PCA) was done with the ADE4 statistical package (Thioulouse et al., 1997), in order to identify the dominant factors explaining eroded C losses for the three land-use systems and two subtreatments. The variables were percentage water stable aggregates, soil resistance to penetration and shear, soil C content, and soil C losses.

13.3 RESULTS

13.3.1 Soil C Content, Bulk Density, C/N Ratio, and C Stocks (Table 13.2)

Soil C content, bulk density (BD), and C stocks were influenced by site ($p \leq 0.001$). The clay soil had significantly higher soil C content for both depth increments regardless of the treatment. Soil C content was more than double for the clay soil: 23.6 vs. 10.4 g C kg^{-1} at 0 to 5 cm depth and 20.5 vs. 8.8 g C kg^{-1} at 5 to 10 cm depth. Soil C content decreased with depth for both sites

Table 13.2 Effect of Cropping System on Soil Organic Carbon (SOC), C/N Ratio, Bulk Density, and Soil C Stock for the Sandy Loam and the Clay Soil

Soil Type and Treatment[a]	SOC (g C kg^{-1})		C/N	Bulk Density (g cm^{-3})		SOC Stock (g C m^{-2})	
	0 to 5 cm	5 to 10 cm	0 to 5 cm	0 to 5 cm	5 to 10 cm	0 to 10 cm	First 87.5 kg m^{-2}
Sandy loam, CC	8.6Aa[bc]	8.6Aa	13.9ABa	1.32Ca	1.37Ba	1157Aa	753Aa
Sandy loam, IF-Cg	11.3Aa	8.7Aa	13.5ABa	1.36Ca	1.49Cb	1408Aa	875Aa
Sandy loam, IF-Tc	11.2Aa	9.0Aa	14.5Ba	1.33Ca	1.56Cb	1466Aa	885Aa
Clay, CC	21.0Ba	19.6Ba	13.2Aa	1.08Bb	1.09Aa	2202Ba	1742Ba
Clay, IF-Cg	22.8Ba	20.0Ba	12.9Aa	1.04Bb	1.07Aa	2259Ba	1871Ba
Clay, IF-Tc	27.1Cb	21.6Ba	13.3Aa	0.90Aa	1.03Aa	2382Ba	2131Cb
LSD[d] for the sandy loam	3.7	2.8	1.1	0.16	0.12*	396	260
LSD for the clay	3.3**	2.4	0.8	0.06***	0.09	206	159***
LSD for site effect	3.1***	2.3***	1.2	0.11***	0.11***	284***	193***
Sandy loam, CT	10.4	9.3	14.1	1.30	1.45	1359	861
Sandy loam, NT	10.4	8.3	13.8	1.37	1.49	1328	814
Clay, CT	20.8	20.7	13.2	1.01	1.06	2142	1808
Clay, NT	26.4	20.2	13.1	1.00	1.06	2421	2021
LSD for the sandy loam	2.6	2.0	1.1	0.11	0.08	282	184
LSD for the clay	2.4***	1.8	0.6	0.04	0.09	204**	131**

[a] CC is continuous cultivation, IF-Cg improved fallow treatment with *Crotalaria grahamiana*, and IF-Tc improved fallow treatment with *Tephrosia candida*.
[b] Means followed by the same upper case letter in the same column are not statistically different at $p \leq 0.05$.
[c] Means followed by the same lower case letter for each site are not statistically different at $p \leq 0.05$.
[d] LSD at $p \leq 0.05$.
*, **, *** significant at 0.05, 0.01, and 0.001, respectively.

(15% in the sandy loam and 13% in the clay soil). Bulk density was lower in the clay soil than in the sandy loam for both depths, 25% at 0 to 5 cm depth (1.01 vs. 1.33) and 38% at 5 to 10 cm depth (1.06 vs. 1.47). Bulk density increased with depth for both sites (10% and 5% for the sandy loam and clay soil, respectively). The C/N ratio (0 to 5 cm depth) ranged from 13.2 to 14.5 and was not affected by soil type. However C/N ratio was significantly higher (ca. 10%) in the sandy loam under IF-Tc than in the clay soil under CC and IF. The C stocks were significantly more in the clay soil than in the sandy loam: 70% greater considering the 0 to 10 cm depth layer (2280 vs. 1340 g C m^{-2}), and 130% greater considering an equivalent soil mass (87.5 kg m^{-2}, which was the smallest mass of the 0 to 10 cm soil layers under study: 1920 vs. 840 g C m^{-2}).

Treatment (CC vs. IF) and tillage practices (CT vs. NT) had stronger effect on soil properties in the clay soil than in the sandy loam. For the former, soil C content at 0 to 5 cm depth increased by 29% under IF-Tc (27 vs. 21 g C kg^{-1}, $p \leq 0.01$), and was intermediate under IF-Cg (not significant). In contrast, soil C content did not differ significantly among treatments at 5 to 10 cm depth (it ranged from 19.6 to 21.6 g C kg^{-1}). In the sandy loam, IF did not significantly increase soil C content at 0 to 5 cm and 5 to 10 cm depths (it ranged from 8.6 to 11.3 g C kg^{-1}). NT increased soil C content in the clay soil by 27% at 0 to 5 cm depth, but tillage did not influence soil C content in the sandy loam and at 5 to 10 cm depth in the clay soil. In the clay soil, IF-Tc reduced BD at 0 to 5 cm depth by 17% (0.90 vs. 1.08 g cm^{-3}, $p \leq 0.001$), but no significant effect was observed under IF-Cg and at 5 to 10 cm depth. In the sandy loam, differences in BD were not significant at 0 to 5 cm depth (BD ranged from 1.32 to 1.36 g cm^{-3}), but increased under IF-Tc and IF-Cg at 5 to 10 cm depth (14 and 9%, respectively, $p \leq 0.02$). Tillage did not affect BD for the two sites for both depths. Soil C stocks at 0 to 10 cm depth did not differ significantly among treatments though they were more under IF treatments than under CC (22 to 27% in the sandy loam, 3 to 8% in the clay soil). Considering C stocks on equivalent soil mass basis (the upper 87.5 kg m^{-2}), there were reduced differences among treatments in the sandy loam (16 to 18%) but increased differences among treatments occurred in the clay soil (7 to 22%) so that difference between IF-Tc and CC was significant ($p \leq 0.001$). Both calculations (0 to 10 cm depth and 87.5 kg m^{-2}) indicated that C stock was 12 to 13% and significantly greater under NT than under CT in the clay soil ($p \leq 0.015$), but was not affected by tillage in the sandy loam. Large variations were observed in C stocks depending on the method of calculation. For the sandy loam, C stocks calculated for the 0 to 10 cm depth and equivalent soil mass differed considerably (1157 to 1466 vs. 753 to 885 g C m^{-2}), which was not the case for the clay soil (2202 to 2382 vs. 1742 to 2131 g C m^{-2}). The larger differences in C stocks for the sandy loam can be explained by larger BD. Indeed, soil mass for 0 to 10 cm depth was 1400 Mg for the sandy loam and 1050 Mg for the clay soil.

In short, soil C was more and BD lower in the clay soil than in the sandy loam and at 0 to 5 than at 5 to 10 cm depth. As compared with CC and conventional tillage, soil C content generally increased in fallow treatments or under NT at 0 to 5 cm depth in the clay soil, but neither at 5 to 10 cm depth nor in the sandy loam. The C/N ratio (0 to 5 cm depth) tended to be greater in the clay soil, but was not affected by treatment or tillage. The BD tended to decrease under fallow treatments in the clay soil but not in the sandy loam (it increased at 5 to 10 cm depth), and was not affected by tillage. The C stocks were more in the clay soil than in the sandy loam at 0 to 10 cm depth and on equivalent soil mass. Soil C stocks at 0 to 10 cm depth were not significantly affected by treatment, whereas C stocks on equivalent soil mass basis were more after IF (IF-Tc) in the clay soil. Soil C stocks were not significantly affected by tillage methods.

13.3.2 Water Stable Aggregates and Soil Strength (Table 13.3)

Water stable aggregates (WSA) and soil strength were influenced by site ($p < 0.001$). Soil strength was measured *in situ* as soil resistance to penetration (RP) and soil resistance to shear (RS). The WSA were more in clay soil for both depths (350 to 420 vs. 50 to 60 g kg^{-1} at 0 to 5 cm depth and 350 to 410 vs. 40 to 50 g kg^{-1} at 5 to 10 cm depth). WSA generally decreased with

Table 13.3 Effect of Cropping System on Water Stable Aggregates (WSA) and Soil Strength for the Sandy Loam and the Clay Soil

Soil Type and Treatment[a]	WSA (g kg^{-1})		Soil Resistance to penetration (kg cm^{-2})	Soil Resistance to shear (kg cm^{-2})
	0 to 5 cm	5 to 10 cm	0 to 10 cm	0 to 2 cm
Sandy loam, CC	49.2Aa[bc]	38.6Aa	1.02Db	1.95Bb
Sandy loam, IF-Cg	42.2Aa	52.7Aa	0.62Aa	1.65Aa
Sandy loam, IF-Tc	60.1Aa	49.7Aa	0.68ABa	1.48Aa
Clay, CC	348.3Ba	353.2Ba	0.87CDab	2.38Ca
Clay, IF-Cg	403.1Cb	391.7Cab	1.05Db	2.23Ca
Clay, IF-Tc	421.1Cb	406.1Cb	0.82BCa	2.68Db
LSD[d] for the sandy loam	27.8	17.2	0.17**	0.25**
LSD for the clay	33.9**	45.6*	0.21*	0.18***
LSD for site effect	29.9***	29.6***	0.19***	0.21***
Sandy loam, CT	49.9	42.2	0.73	1.59
Sandy loam, NT	51.1	51.8	0.81	1.80
Clay, CT	368.0	374.4	0.99	2.57
Clay, NT	413.6	393.0	0.83	2.30
LSD for the sandy loam	21.3	12.5	0.13	0.18*
LSD for the clay	24.2**	33.8	0.15*	0.15**

[a] CC is continuous cultivation, IF-Cg improved fallow treatment with *Crotalaria grahamiana*, and IF-Tc improved fallow treatment with *Tephrosia candida*.
[b] Means followed by the same upper case letter in the same column are not statistically different at p ≤ 0.05.
[c] Means followed by the same lower case letter for each site are not statistically different at p ≤ 0.05.
[d] LSD at p ≤ 0.05.
*, **, *** significant at 0.05, 0.01, and 0.001, respectively.

depth, except for sandy loam IF-Cg and clay soil CC. Soil resistance to shear (RS) was 20 to 80% more in the clay soil than in the sandy loam for all treatments (2.2 to 2.7 vs. 1.5 to 2.0 kg cm^{-2}), but the effect of soil type on RP did not observe a clear trend.

Treatment and tillage methods influenced WSA and soil strength. At 0 to 5 cm depth, WSA increased significantly (p = 0.003) under both IF treatments in the clay soil (16% for IF-Cg and 21% for IF-Tc), but the effect of IF was not significant in the sandy loam. At 5 to 10 cm depth, WSA also increased significantly under both IF treatments in the clay soil (11% for IF-Cg and 15% for IF-Tc, only the latter being significant), but the increase was not significant in the sandy loam (though it reached 30 to 40%). The NT increased WSA by 12% (p = 0.002) in the clay soil at 0 to 5 cm depth, but tillage did not affect WSA at 5 to 10 cm depth or in the sandy loam. Soil strength was higher under CC than under IF treatments in the sandy loam: RP was 50 to 65% higher, and RS 18 to 32% higher. The relatively high soil strength under CC may be attributed to soil crusting. Under CC, the soil surface crusted within the first few minutes of the simulated rainfall event. There was no crusting on the clay soil. For the clay soil, IF-Tc increased RS by 13% but no increase was observed for IF-Cg. Conversely, IF-Cg significantly increased RP by 21%, but the effect of IF-Tc was not significant. No clear trends were observed for RP and RS in relation to tillage: NT increased RP and RS by 11 and 13 % in the sandy loam (significant for RS only) but decreased RP and RS by 16 and 11% in the clay soil, respectively.

In summary, WSA increased under IF treatments and NT in the clay soil at 0 to 5 cm depth and under IF-Tc at 5 to 10 cm depth, but was not affected by treatment and tillage in the sandy loam. Resistance to penetration was not clearly affected by soil type. For the sandy loam RP was 50 to 65% more under CC than under IF treatments but was not affected by tillage, whereas for the clay soil it was not clearly affected by fallowing but was 16% smaller under NT than under CT. Resistance to shear was 20 to 80% more in the clay soil than in the sandy loam, and was 20 to 30% more under CC than IF treatments in the sandy loam, whereas fallow effect was not clear in the clay soil. As compared with CT, resistance to shear under NT was 13% more in the sandy loam but 11% lower in the clay soil.

13.3.3 Runoff, Sediment Concentration, and Soil Loss (Table 13.4)

Final runoff rate (FRR) and runoff depth (RD) were highly influenced by site ($p \leq 0.001$). Generally, FRR and RD were significantly lower on the clay soil than on the sandy loam. When comparing treatment across sites, FRR was 36, 50, and 74% lower; and RD was 32, 78, and 60% smaller under CC, IF-Cg and IF-Tc in the clay soil than in the sandy loam, respectively (however differences in FRR and RD between sites were not significant for IF-Cg). Sediment concentration (SC) was less clearly influenced by site ($p = 0.004$). Under CC, it was twice higher on the clay soil than on the sandy loam, but under IF treatments it tended to be lower on the clay soil. Soil loss was highly affected by site ($p \leq 0.001$) and was two and six times greater on the sandy loam than on the clay soil for CC and IF treatments, respectively.

Treatment significantly affected FRR at both sites ($p \leq 0.001$ and $p = 0.004$ for the sandy loam and the clay soil, respectively), with greater reductions under IF treatments on the clay soil. As compared with CC, FRR was reduced by 71 to 73% for IF treatments on the clay soil and by 66% for IF-Cg and 29% for IF-Tc on the sandy loam. A similar trend was observed for RD: IF-Cg and IF-Tc significantly reduced RD by 89 and 58% on the clay soil, and by 68 and 29% on the sandy loam, respectively. Tillage did not influence the runoff barometers for the two sites. On the clay soil, IF significantly reduced SC ($p = 0.009$), which was three times lower than under CC (0.7 vs. 2.1 g l^{-1}). On the sandy loam, in contrast, the differences in SC between treatments were small (< 10%) and not significant (SC ranged from 0.88 to 0.96 g l^{-1}). Additionally, SC was 45% smaller under NT than under CT on the clay soil, but 50% greater under NT than CT on the sandy loam. However, these differences were not significant. Treatments clearly influenced SL on the clay soil,

Table 13.4 Effect of Cropping System on Runoff and Soil Loss for the Sandy Loam and the Clay Soil

Soil Type and Treatment[a]	Final Runoff Rate[b] (mm hr^{-1})	Runoff Depth[b] (mm)	Sediment Concentration[b] (g l^{-1})	Soil Loss[b] (g m^{-2})
Sandy loam, CC	70Cc[cd]	28Cc	0.96Aa	28.5Cb
Sandy loam, IF-Cg	24Aa	9Aa	0.88Aa	12.7Ba
Sandy loam, IF-Tc	50Bb	20Bb	0.94Aa	25.5Cb
Clay, CC	45Bb	19Bb	2.12Bb	13.1Bb
Clay, IF-Cg	12Aa	2Aa	0.72Aa	2.1Aa
Clay, IF-Tc	13Aa	8Aa	0.65Aa	4.2Aa
LSD[e] for the sandy loam	18***	8**	0.64	10.5**
LSD for the clay	18**	11*	0.90**	2.9***
LSD for site effect	15***	8***	0.71**	8.0***
Sandy loam, CT	51	21	0.74	20.3
Sandy loam, NT	45	17	1.11	24.1
Clay, CT	26	11	1.50	7.0
Clay, NT	20	9	0.82	5.9
LSD for the sandy loam	15	6	0.49	8.5
LSD for the clay	13	8	0.70	2.3

[a] CC is continuous cultivation, IF-Cg improved fallow treatment with *Crotalaria grahamiana*, and IF-Tc improved fallow treatment with *Tephrosia candida*.
[b] Final runoff rate, runoff depth, sediment concentration, and soil loss were measured over a 30 min period.
[c] Means followed by the same upper case letter in the same column are not statistically different at $p \leq 0.05$.
[d] Means followed by the same lower case letter for each site are not statistically different at $p \leq 0.05$.
[e] LSD at $p \leq 0.05$.
*, **, *** significant at 0.05, 0.01, and 0.001, respectively.

where it was three and six times smaller in IF-Tc and IF-Cg than in CC, respectively (4 and 2 vs. 13 g m^{-2}, p = 0.020). On the sandy loam, SL was twice as small in IF-Cg than in CC (13 vs. 29 g l^{-1}, p = 0.020), but did not differ significantly between IF-Tc and CC (26 vs. 29 g m^{-2}). Tillage did not influence SL for either of the two sites.

In short, runoff was smaller on the clay soil than on the sandy loam (30 to 80%) and for IF treatments than for CC (30 to 90%), but was not influenced by tillage. Sediment concentration was not clearly affected by soil type (for CC it was higher on the clay soil, for IF treatments it tended to be higher on the sandy loam). It was three times lower under IF treatments than under CC on the clay soil, but did not differ significantly between IF treatments and CC on the sandy loam. The influence of tillage on SC neither followed a clear trend nor was it significant. Soil loss was more on the sandy loam than on the clay soil, and was lower under IF treatments than under CC (on the clay soil especially), but was not significantly influenced by tillage methods.

13.3.4 C Content of Sediments, Enrichment Ratio, and Soil C Losses

The effect of soil type on sediment C content was not clearly defined (for CC and IF-Tc it was 63 and 94% higher on the sandy loam than on the clay soil, but for IF-Cg it was 67% lower on the sandy loam; Table 13.5). The C/N ratio of sediments was 26 to 46% higher on the sandy loam than on the clay soil across treatments (14 to 15 vs. 10 to 12). The C enrichment ratio of sediments (ER) and C losses were strongly influenced by site (p ≤ 0.001). The ER was higher on the sandy loam (3.4 to 6.5) than on the clay soil (1.4 to 2.7), but the difference was significant for CC and IF-Tc only (6.1 vs. 1.5 and 6.5 vs. 1.4, respectively, p ≤ 0.01). The C losses were 3.4, 3.6, and 14 times more on the sandy loam than on the clay soil for CC, IF-Cg, and IF-Tc, respectively (however the difference was not significant for IF-Cg). They were maximum for IF-Tc on sandy loam (1.95 g C m^{-2}) and minimum for IF-Cg and IF-Tc on clay soil (0.12 to 0.14 g C m^{-2}).

On the sandy loam, C content of sediments was 40% lower for IF-Cg but 40% higher for IF-Tc than for CC (37 and 73 vs. 52 g kg^{-1}). On the clay soil, C content was 95 and 20% higher for

Table 13.5 Effect of Cropping System on Sediment C Content, Sediment C/N, Ratio of Sediment Enrichment in C, and C Losses for the Sandy Loam and the Clay Soil

Soil Type and Treatment[a]	Sediment C (g C kg^{-1})	Sediment C/N	Enrichment Ratio	C Losses (g C m^{-2})
Sandy loam, CC	52.1	14.3	6.1Cb[bc]	1.43Bab
Sandy loam, IF-Cg	37.4	15.1	3.4Ba	0.43Aa
Sandy loam, IF-Tc	73.3	13.9	6.5Cb	1.95Bb
Clay, CC	31.9	11.1	1.5Aa	0.42Ab
Clay, IF-Cg	62.3	12.0	2.7Bb	0.12Aa
Clay, IF-Tc	37.8	9.5	1.4Aa	0.14Aa
LSD[d] for the sandy loam	—	—	1.6**	1.14*
LSD for the clay	—	—	0.2***	0.10***
LSD for site effect	—	—	1.0***	0.77***
Sandy loam, CT	38.5	13.5	3.9	0.76
Sandy loam, NT	70.0	15.4	6.7	1.77
Clay, CT	37.2	10.9	1.8	0.24
Clay, NT	50.9	10.8	2.0	0.21
LSD for the sandy loam	—	—	1.1***	0.92*
LSD for the clay	—	—	0.1	0.08

[a] CC is continuous cultivation, IF-Cg improved fallow treatment with *Crotalaria grahamiana*, and IF-Tc improved fallow treatment with *Tephrosia candida*.
[b] Means followed by the same upper case letter in the same column are not statistically different at p ≤ 0.05.
[c] Means followed by the same lower case letter for each site are not statistically different at p ≤ 0.05.
[d] LSD at p ≤ 0.05.
*, **, *** significant at 0.05, 0.01, and 0.001, respectively.

IF-Cg and IF-Tc than for CC, respectively (62 and 38 vs. 32 g C kg^{-1}). Additionally, sediment C content was 40% (clay) to 80% higher (sandy loam) under NT than under CT. Sediment C/N ratio was slightly lower for IF-Tc than for CC (3% on the sandy loam and 14% on the clay soil), but slightly higher for IF-Cg than for CC (6 and 8%, respectively). On the sandy loam, ER was similar in CC and IF-Tc but was twice lower in IF-Cg (3.4 vs. 6.1 to 6.5, p = 0.010). On the clay soil, it was also similar in CC and IF-Tc but was twice as high in IF-Cg (2.7 vs. 1.4 to 1.5, p = 0.010). Additionally, ER was 70% higher for NT than for CT on the sandy loam (6.7 vs. 3.9, p ≤ 0.001), but was not significantly influenced by tillage on the clay soil (though 11% higher for NT). As compared with CC, both IF treatments reduced C losses by 70% on the clay soil (0.12 to 0.14 vs. 0.42 g C m^{-2}, p ≤ 0.001). On the sandy loam, differences in C losses between CC and IF treatments were not significant though C losses were 70% more for IF-Tc and 40% lower for IF-Cg than for CC (1.95, 0.43, and 1.43 g C m^{-2}, respectively; C losses were 4.5 times more for IF-Tc than for IF-Cg, p ≤ 0.05). In contrast, C losses were not influenced by tillage on the clay soil, but were 2.3 times more for NT than for CT on the sandy loam (p = 0.035).

C losses represented 0.03 to 0.13% of soil C stock at 0 to 10 cm depth in the sandy loam, but 0.01 to 0.02% only in the clay soil. Eroded C as a proportion of C stock (0 to 10 cm) was thus six to 20 times more for the sandy loam than for the clay soil, whereas soil loss was only two to six times more on the former than on the latter. The amount of eroded C as a proportion of soil C stock at 0 to 10 cm depth was not clearly influenced by treatment or tillage. However it was four times more for CC than for IF-Cg on both soil types, and twice more for NT than for CT on the sandy loam.

In summary, sediment C content was not clearly affected by site (it was more on the sandy loam for two out the three treatments) but was generally more for IF treatments than for CC (except for IF-Cg on the sandy loam). Sediment C/N ratio was more on the sandy loam than on the clay soil, with IF-Cg > CC > IF-Tc. The CER of the sediments was more on the sandy loam than on the clay soil, was similar for CC and IF-Tc, and was twice more for NT than for CT on the sandy loam. C losses were more on the sandy loam, generally lower for fallow treatments (except IF-Tc on the sandy loam), and were twice more for NT on the sandy loam. The proportion of soil C stock lost with sediments was much higher for the sandy loam than for the clay, but was not clearly affected by fallow or tillage treatments.

13.3.5 Principal Component Analysis (Figure 13.2)

The eigen values of the principal component analysis (PCA) showed that the first factor accounted for 59% of the total inertia. On the correlation circle, this factor was represented by the horizontal axis (F1), which opposed water stable aggregates (WSA), soil C content, soil resistance to shear (RS), on the one hand, and C losses, on the other (Figure 13.2a). The second factor accounted for 23% of the total variation (F2), and was mainly explained by soil RP. The first two axes accounted for 82% of the inertia.

The factorial map of treatments (Figure 13.2b) showed the effects of soil type and treatment on soil C losses. The points, which represented the plots, clustered into two main groups: the first group, on the right part of the map, included the plots located on sandy loam, whereas the second group, on the left part of the map, included the plots on clay soil. Thus the projection on the F1 axis led to a contrast between the clay soil, which had more soil C content, WSA, and RS but lower C losses, and the sandy loam, where soil C content, WSA, and RS were lower and C losses higher. This projection also showed that on clay soil, the plots representing IF-Tc CT, IF-Tc NT, and IF-Cg NT (Numbers 9, 10, and 12), on the left, had lower C losses than CC CT, CC NT and IF-Cg CT (Numbers 7, 8, and 11). This was interpreted as resulting from more WSA after fallowing, which however was not achieved for IF-Cg CT (Number 11).

The projection on the F2 axis allowed to distinguish among plots on sandy loam between those under CC, toward the top of the map and having more RP, and those under IF treatments, which

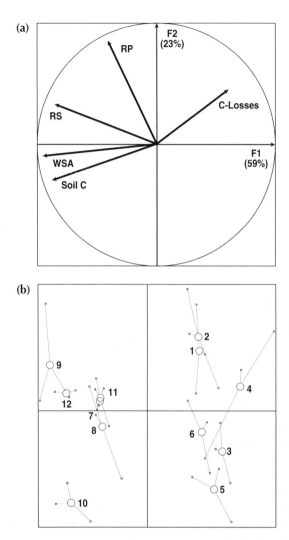

Figure 13.2 Results from principal component analysis (PCA) on soil carbon losses: (a) F1-F2 correlation circle of variables (RP and RS are soil resistance to penetration and shear, respectively, and WSA water stable aggregates). (b) Factorial map of treatments. No.'s 1–6 sandy loam: (1) CC-CT, (2) CC-NT, (3) IF-Tc-CT, (4) IF-Tc-NT, (5) IF-Cg-CT, (6) IF-Cg-NT; No.'s 7–12 clay soil: (7) CC-CT, (8) CC-NT, (9) IF-Tc-CT, (10) IF-Tc-NT, (11) IF-Cg-CT, (12) IF-Cg-NT.

had lower RP (this separation was not possible on clay soil). The projection on the F2 axis also separated the plots according to tillage. On the sandy loam, plots under NT had more RP than their counterparts (same treatment) under CT. On the clay soil, in contrast, plots under NT had lower RP than their counterparts under CT (except for IF-Cg, Numbers 11 and 12). Additionally, IF-Tc NT on clay soil (Number 10) was at the bottom of the map, clearly below the other IF plots (Numbers 9, 11, and 12), and this was related with a lesser RP on the former than on the latter.

In short, the PCA analysis indicated that C losses were negatively related to the soil C content and WSA, which were lower: (1) on sandy loam than on clay soil and (2) among plots on clay soil, under continuous cultivation than after fallow (except on IF-Cg CT). On the clay soil, increases in soil C content and WSA through fallowing resulted in reduction in C loss. In contrast, C losses on sandy loam cannot be easily explained using the PCA, as the plots mainly ranged according to RP, which was perpendicular to C losses.

13.4 DISCUSSION

13.4.1 Impact of Improved Fallows on Runoff

Results of runoff amount and rate indicated that fallowing had a significant effect on reducing runoff. Indeed, runoff was lower for plots previously under improved fallow than for plots under continuous cultivation (30 to 90% reduction). Lower runoff amount and rate may be attributed to improvement in soil structure after the fallow phase. When land is taken out of cultivation (natural fallow or planted fallow for several years, protected from fire and grazing), there is a combined build-up of SOC and soil aggregation (Ingram, 1990; Niang et al., 1996; IMPALA, 2001; 2002; Mutuo, 2004). Several studies have reported close relationship between runoff and soil aggregation (Le Bissonnais, 1996; Barthès et al., 2000; Barthès and Roose, 2002). In this study, improvement in WSA through fallowing was more important for the clay soil than for the sandy loam (18% vs. 4% in average at 0 to 5 cm depth), resulting in a greater reduction in runoff (−70 vs. −50% in average). For the sandy loam, the reduction in runoff after fallowing was mainly caused by a lower susceptibility to crusting. Indeed, the soil surface of the sandy loam crusted quickly (after 10 min), which strongly hindered infiltration. It is well established that surface sealing promotes overland flow (Bryan and De Ploey, 1983; Le Bissonnais, 1996; Rao et al., 1998) and is often prevailing on degraded soils. Values of RP and RS indicated that cropping IF reduced crusting on the sandy loam (on average, RP and RS were reduced by 36 and 20%, respectively). Reduced RS and RP after biomass return were also reported by Zeleke et al. (2004). However, for the clay soil in the present study, there was no significant difference in RP among treatments, and RS increased under IF-Tc (13%). Biomass return also reduces BD, and this was the case for the clay soil in the present study. Differences in BD among treatments were not significant for the sandy loam soil at 0 to 5 cm depth. This is contrary to the findings of Zeleke et al. (2004), who reported that biomass return decreased in BD in a sandy soil but not in a clay soil. Decrease in BD with biomass return has been related to increased infiltration rates, which is also confirmed in the present study.

13.4.2 Control of Soil Loss by Improved Fallows

Plots previously under IF had less soil loss. Improved fallows reduced soil loss by 68 to 84% on the clay soil and by 11 to 55% on the sandy loam, indicating that improvement in soil structure due to IF was highly dependent on clay and soil C contents. On the sandy loam, reduction in soil loss after fallowing depended mainly on reduction in runoff and transportability of detached particles, since topsoil properties such as C content, WSA, and BD were not significantly affected. On the clay soil, reduction in soil loss after fallowing was more clearly associated with increases in C content and WSA and decrease in BD. A close relationship between topsoil WSA and soil susceptibility to runoff and erosion were also reported by Barthès and Roose (2002). The strong impact of fallowing on soil loss reduction was expected for the clay soil, due to its potential to form stable aggregates through the association between organic matter and clay particles. However, this study also showed that soil loss may be significantly reduced (−55%) on sandy loam when *Crotalaria grahamiana* was used as improved fallow.

The soil losses measured in this study were in the same range as those reported in other studies with simulated rainfall (Merzouk and Blake, 1991; Meyers and Wagger, 1996). Soil loss measured from 1-m^2 plots primarily results from splash detachment by rain drops, and, therefore, is an indication of interrill erosion. Scaling up soil loss from 1-m^2 to slope and catchment scales has been widely reviewed. Merzouk and Blake (1991) reported agreement between values of soil erodibility measured under simulated rainfall and the magnitude of soil erosion observed in the field. Other studies reported that soil loss is underestimated on microplots due to the short slope length (Le Bissonnais et al., 1998). However, simulated rainfall on microplots in the field enables detailed investigations on splash detachment and soil erodibility. Additionally, it provides reliable indication of runoff and soil loss for different soil types and land-use systems.

13.4.3 Effect of No-Tillage on Soil Properties, Runoff, and Soil Loss

The NT improves soil structure, due to the stabilization of the soil surface by increased SOC content and the accumulation of crop residues (Ingram and Fernandes, 2001; VandenBygaar et al., 2002), and by the lack of mechanical disturbance and its consequences on biological activity (Beare et al., 1994). In this study, changes in soil physical properties under NT depended on soil texture. For the clay soil, topsoil C content and WSA were significantly more under NT than under CT, but the increase was limited (27% for C content, 12% for WSA). For the sandy loam, topsoil C content and WSA did not differ significantly among NT and CT. Recent conversion from CT to NT probably explained the limited effects of tillage practices on soil properties. Indeed, measurements were made at the end of the first cropping season under NT, after many years under CT. Improvement in soil physical properties under NT is a slow process, especially for degraded soils (Ingram and Fernandes, 2001). Rhoton et al. (2002) observed that topsoil SOC and WSA under NT increased by 17% after 4 years and by 70% after 14 years. Thus, more increase in SOC and WSA under NT may be possible in the soils under this study, but for longer durations. Soil BD did not differ either among tillage systems, which was contrary to the results of Rhoton et al., (2002). These authors found BD to increase with the conversion from CT to NT. In the present study, BD varied according to soil type and depth, as was also reported by Arshad et al., (1999). Moreover, runoff and soil loss were not influenced by tillage in the present study, probably due to the recent conversion from CT to NT. Indeed, several studies have reported that runoff and soil losses were reduced under NT, and are related to the accumulation of SOC under NT (Arshad et al., 1999; Franzluebbers, 2002; Rhoton et al., 2002). Bradford and Huang (1994) reported similar results from experiments under simulated rainfall. In the present study, significant increases in topsoil C and WSA in the clay soil under NT indicated that runoff and soil loss may be reduced over time. Long-term experiments are needed to confirm this hypothesis.

13.4.4 C Content of Sediments and Enrichment Ratio

Soil erosion decreases SOC by selective detachment and transport of fine particles (Watung et al., 1996; Wan and El-Swaify, 1997; Jacinthe et al., 2002; Lal, 2003), resulting in an enrichment of sediments with SOC relative to the *in situ* soil (Wan and El-Swaify, 1997; Owens et al., 2002). This study also showed that sediments were enriched in SOC. The ER was higher for the sandy loam than for the clay soil (3.4 to 6.5 vs. 1.4 to 2.7), indicating that erosion was more selective on the former than on the latter. Sediments had a higher C/N ratio on the sandy loam (14 to 15, close to that of the topsoil and of the plants) than on the clay soil (10 to 12, lower than that of the topsoil), indicating that eroded C was less processed and less protected in the former than in the latter. These results suggested that eroded C in the clay soil was mainly in the form of processed organic matter protected within aggregates and removed along with them, whereas eroded C in the sandy loam was mainly in the form of particulate organic matter.

13.4.5 Effect of Land Management on Soil C Losses and Soil C Stocks

The SOC losses ranged from 0.12 to 1.95 g C m^{-2}, which corresponded with C losses measured by Jacinthe et al. (2002) under simulated rainfall on long-term NT plots. The present study demonstrated the potential of improved fallows in reducing runoff and soil losses (on 1-m^2 plots). A PCA showed that C losses may be explained by topsoil SOC content, WSA, and RS, and were thus influenced by soil type and land use. Indeed, topsoil C content, WSA, and RS were lower and

C losses more in the sandy loam than in the clay soil. For the clay soil, increases in topsoil SOC content and WSA after improved fallows similarly resulted in smaller C losses. For the sandy loam, topsoil SOC content and WSA, as well as C losses, were less clearly affected by improved fallows, but increase in RS under NT was associated with an increase in C losses. Thus, increases in C losses were associated with decreases in topsoil SOC content and WSA for the clay soil (after fallow), but with increase in RS for the sandy loam (under NT).

Topsoil C stocks in the sandy loam and clay soil were less than those reported by Wilson (1997) and Nandwa (2001) for intensively cultivated soils in Kenya. The present study showed that improved fallows increased topsoil C stocks, especially for the clay soil. The only significant increase in topsoil C stock resulting from fallow, which reached 22%, was for IF-Tc on clay soil, when stocks were calculated on an equivalent soil mass basis. The increase was not significant when stocks were calculated for the 0 to 10 cm depth layer (equivalent depth), indicating the importance of calculation on equivalent soil mass when discussing management-induced changes in SOC and nutrient storage, as was recommended by Ellert and Bettany (1995). Additionally, NT resulted in an increase in topsoil SOC stock (on equivalent soil mass or depth) in the clay soil but not in the sandy loam, which also confirmed the findings of Arshad et al. (1999). Their results showed greater SOC stocks under NT for a silty loam but no increase for a sandy loam soil.

13.5 CONCLUSION

The objectives of this chapter were to evaluate the effects of improved fallows (with *Crotalaria grahamiana* or *Tephrosia candida*) and no-tillage on runoff, soil, and C losses for a sandy loam and a clay soil under maize-beans cultivation. The results showed that runoff, soil, and C losses were lower on the clay soil than on the sandy loam. The data also showed that short-term improved fallows reduced and controlled runoff, soil, and C losses during the following cropping phase on both soil types, but that the reduction was more on the clay soil than on the sandy loam. These trends were attributed to a build-up of topsoil C and WSA during the fallow phase, which was less important in general for the sandy loam than for the clay soil. Nevertheless, improved fallow with *Crotalaria grahamiana* was very effective in reducing runoff, soil, and C losses on the sandy loam, mainly due to a reduction in crusting.

Soil C stocks were more in the clay soil and were more clearly increased by improved fallows than in the sandy loam. The C enrichment ratio of sediments was significantly higher for the sandy loam, indicating that higher proportions of topsoil C were removed than on the clay soil. Sediment enrichment was not affected by treatments. Moreover, the proportion of topsoil C stock lost with sediments was much higher for the sandy loam than for the clay soil, but was not clearly affected by treatments.

No-tillage did not significantly influence runoff and soil losses. However, no-tillage increased topsoil WSA, C content, and C stock in the clay soil, and increased sediment enrichment ratio and C losses for the sandy loam. No definite trends were observed for soil strength: under NT, soil resistance to shear and penetration decreased in the clay soil but increased in the sandy loam. However, all the results regarding tillage practices must be confirmed by long-term experiments. Indeed, measurements were made at the end of the first cropping season under NT, following many years under CT. As improvement in soil properties under NT is considered a slow process in general, long-term experiments are needed to further examine the effects of no-till on water, soil, and C conservation.

ACKNOWLEDGMENTS

The authors thank the European Commission (Project INCO-DEV n° ICA4-2000-30011), the Institut de Recherche pour le Développement (IRD), and the World Agroforestry Centre (ICRAF) for financing this research work.

REFERENCES

Albrecht, A. and S. T. Kandji. 2003. C sequestration in tropical agroforestry systems. *Agriculture, Ecosystems, and Environment* 99:15–27.

Albrecht, A., L. Rangon, and P. Barret. 1992. Effets de la matière organique sur la stabilité structurale et la détachabilité d'un vertisol et d'un ferrisol (Martinique). *Cahiers ORSTOM, séroe Pédologie* 27:121–133.

Arshad, M. A., A. J. Franzluebbers, and R. H. Azooz. 1999. Components of surface soil structure under conventional and no-tillage in northwestern Canada. *Soil and Tillage Research* 53:41–47.

Asseline, J. and C. Valentin. 1978. Construction et mise au point d'un infiltromètre à aspersion. *Cahiers ORSTOM, série Hydrologie* 15:321–349.

Barthès, B., A. Azontonde, B. Z. Boli, C. Prat, and E. Roose. 2000. Field-scale runoff and erosion in relation to topsoil aggregate stability in three tropical regions (Benin, Cameroon, Mexico). *European Journal of Soil Science* 51:485–495.

Barthès, B. and E. Roose. 2002. Aggregate stability as an indicator of soil susceptibility to runoff and erosion; validation at several levels. *Catena* 47:133–149.

Beare, M. H., P. F. Hendrix, and D. C. Coleman. 1994. Water-stable aggregates and organic matter fractions in conventional- and no-tillage soils. *Soil Science Society of America Journal* 58:777–786.

Bradford, J. M. and C. Huang. 1994. Interrill soil erosion as affected by tillage and residue cover. *Soil and Tillage Research* 31:353–361.

Bryan, R. B. and J. De Ploey. 1983. Comparability of soil erosion measurements with different laboratory rainfall simulators. *Catena* 4:33–56.

Cooper, P. J. M., R. R. B. Leaky, M. R. Rao, and L. Reynolds. 1996. Agroforestry and the mitigation of land degradation in the humid tropics and sub-humid tropics of Africa. *Experimental Agriculture* 32:235–290.

Dixon, R. K. 1995. Agroforestry systems: Sources or sinks of greenhouse gases? *Agroforestry Systems* 31:99–116.

Doran, J. W., M. Sarrantonio, and M. A. Liebig. 1996. Soil health and sustainability. *Advances in Agronomy* 56:1–54.

Ellert, B. H., and J. R. Bettany. 1995. Calculation of organic matter and nutrients stored in soils under contrasting management regimes. *Canadian Journal of Soil Sciences* 75:529–538.

Feller, C., A. Albrecht, and D. Tessier. 1996. Aggregation and organic carbon storage in kaolinitic and smectitic tropical soils, in M. R. Carter and B. A. Stewart, eds., *Structure and Organic Matter Storage in Agricultural Soils*, CRC Press, Boca Raton, FL, pp. 309–360.

Franzluebbers, A. J. 2002. Water infiltration and soil structure related to organic matter and its stratification with depth. *Soil and Tillage Research* 75:1–9.

IMPALA. 2001. *Project Report Year 1 (INCO-DEV Project No ICA4-CT-2000-30011)*. World Agroforestry Centre, Nairobi, Kenya.

IMPALA. 2002. *Project Report Year 2 (INCO-DEV Project No ICA4-CT-2000-30011)*. World Agroforestry Centre, Nairobi, Kenya.

Ingram, J. 1990. The role of trees in maintaining and improving soil productivity — A review of the literature, in Prinsley R. T., ed., *Agroforestry for Sustainable Production, Economic Implications*. Commonwealth Science Council, London, pp. 243–303.

Ingram, J. S. I. and E. C. M. Fernandes. 2001. Managing carbon sequestration in soils: Concepts and terminology. *Agriculture, Ecosystems and Environment* 87:111–117.

Jacinthe, P. A., R. Lal, and J. M. Kimble. 2002. Carbon dioxide evolution in runoff from simulated rainfall on long-term no-till and plowed soils in southwestern Ohio. *Soil and Tillage Research* 66:23–33.

Kursten, E. and P. Burschel. 1993. CO_2-mitigation by agroforestry. *Water, Air, and Soil Pollution* 70:533–544.

Lal, R. 2003. Soil erosion and the global carbon budget. *Environment International* 29:437–450.

Le Bissonnais, Y. 1996. Aggregate stability and assessment of soil crustability and erodibility: I. Theory and methodology. *European Journal of Soil Science* 47:425–437.

Le Bissonnais, Y., H. Benkhadra, V. Chaplot, D. Fox, D. King, and J. Daroussin. 1998. Crusting, runoff and sheet erosion on silty loamy soils at various scales and upscaling from m^2 to small catchment. *Soil and Tillage Research* 46:69–80.

McCarty, G. W. and J. C. Ritchie. 2002. Impact of soil movement on carbon sequestration in agricultural ecosystems. *Environmental Pollution* 116:423–430.

Merzouk, A. and G. R. Blake. 1991. Indices for the estimation of interrill erodibility of Moroccan soils. *Catena* 18:537–559.

Meyers, J. L. and M. G. Wagger. 1996. Runoff and sediment loss from three tillage systems under simulated rainfall. *Soil and Tillage Research* 39:115–129.

Mutuo, P. 2004. *Potential of improved tropical legume fallows and zero tillage practices for soil organic carbon sequestration.* Ph.D. Dissertation, Imperial College, University of London.

Nandwa, S. M. 2001. Soil organic carbon (SOC) management for sustainable productivity of cropping and agroforestry systems in Eastern and Southern Africa. *Nutrient Cycling in Agroecosystems* 61:143–158.

Niang, A., J. De Wolf, M. Nyasimi, T. S. Hansen, R. Rommelse, and K. Mwendwa. 1998. *Soil fertility recapitalisation and replenishment project in Western Kenya. Progress Report February 1997–July 1998. Pilot Project Report No. 9.* Regional Agroforestry Research Centre, Maseno, Kenya.

Niang, A., S. Gathumbi, and B. Amadalo. 1996. The potential of short duration improved fallow for crop production enhancement in the highlands of western Kenya. *East African Agriculture and Forestry Journal* 62:103–114.

Owens, L. B., R. W. Malone, D. L. Hothem, G. C. Starr, and R. Lal. 2002. Sediment carbon concentration and transport from small watersheds under various conservation tillage practices. *Soil and Tillage Research* 67:65–73.

Rao, K. P. C., T. S. Steenhais, A. L. Cogle, S. T. Srinivasan, D. F. Yule, and G. D. Smith. 1998. Rainfall infiltration and runoff from an Alfisol in semi-arid tropical India. I. No-till systems. *Soil and Tillage Research* 48:51–59.

Rhoton, F. E., M. J. Shipitalo, and D. L. Lindbo. 2002. Runoff and soil loss from midwestern and southeastern U.S. silt loam soils as affected by tillage practice and soil organic matter content. *Soil and Tillage Research* 66:1–11.

Thioulouse, J., D. Chessel, S. Doledec, and J. M. Olivier. 1997. ADE4: A multivariate analysis and geographical display software. *Statistics and Computing* 7:75–83.

Vanden Bygaart, A. J., X. M. Yang, B. D. Kay, and J. D. Aspinall. 2002. Variability in carbon sequestration potential in no-till soil landscapes of southern Ontario. *Soil and Tillage Research* 65:231–241.

Van Roode, M. 2000. *The effects of vegetative barrier strips on surface runoff and soil erosion in Machakos, Kenya.* Ph.D. Dissertation, Universiteit Utrecht.

Wan, Y. and S. A. El-Swaify. 1997. Flow induced transport and enrichment of erosional sediment from a well-aggregated and uniformly-textured Oxisol. *Geoderma* 75:251–265.

Watung, R. L., R. A. Sutherland, and S. A. El-Swaify. 1996. Influence of rainfall energy flux density and antecedent soil moisture content on splash transport and aggregate enrichment ratios for a Hawaiian Oxisol. *Soil Technology* 9:251–272.

Zeleke, T. B., M. C. J. Grevers, B. C. Si, A. R. Mermut, and S. Beyene. 2004. Effect of residue incorporation on physical properties of the surface soil in the South Central Rift Valley of Ethiopia. *Soil and Tillage Research* 77:35–46.

CHAPTER 14

Soil and Carbon Losses under Rainfall Simulation from Two Contrasting Soils under Maize-Improved Fallows Rotation in Eastern Zambia

G. Nyamadzawo, P. Nyamugafata, R. Chikowo, T. Chirwa, and P. L. Mafongoya

CONTENTS

14.1 Introduction 197
14.2 Materials and Methods 198
    14.2.1 Site Description 198
    14.2.2 Rainfall Simulations 199
    14.2.3 Data Analyses 199
14.3 Results 200
    14.3.1 Kalunga Site 200
    14.3.2 Msekera Site 200
    14.3.3 Comparison between Kalunga and Msekera Sites 202
14.4 Discussion 204
14.5 Conclusion 205
Acknowledgments 205
References 205

14.1 INTRODUCTION

Soil loss through erosion has been associated with losses in soil organic carbon (SOC). Studies on watersheds have shown that SOC is easily lost because it is usually sediment bound. Nutrient enrichment ratios of eroded sediments of up to 5.7 have been reported (Coleman et al., 1990). The close association of organic matter with plant nutrients in the soil makes erosion of soil organic matter a strong indicator of the overall plant nutrient loss resulting from erosion (Follet et al., 1987).

Runoff losses have been determined using different methods (Sharpley et al., 1994; Zhang and Miller, 1996), but in spite of these there is no widely accepted procedure. Runoff experiments have often used the whole watershed (Lowrance, 1992; Goss et al., 1993), or use simulated rainfall on microplots. Simulated rainfall has the advantage that many parameters can be controlled compared to natural rainfall for watershed measurements (Barthès et al., 1999). However microplots used with simulated rainfall (mesocosm study) only give information on the soil detachability and erodibility and not on total erosion that takes place in the field as the complete hydrodynamic process of soil detachment, transport, deposition, and resuspension cannot be represented. Therefore the use of microplots in mesocosm studies has been retained in order to evaluate the potential of soil erosion and sediment transport.

Improved fallow systems increase soil organic matter (SOM) and improve soil physical and chemical properties (Juo and Lal, 1977). However, there is little work that has looked at how long these benefits can last after fallowing and how SOC losses are affected by soil type in the Southern Africa region. Most work done to date on improved fallows has concentrated on effect of nutrient mineralization and cycling (Hartemink et al., 1996; Mafongoya et al., 1997; Chikowo et al., 2003). The objective of this study was to determine the potential erodibility of two different soils and losses in SOC at fallow termination and after one maize-cropping season using simulated rainfall in a maize-improved fallow rotation. We hypothesized that soil and SOC losses are lower under improved fallows and increase during the cropping phase.

14.2 MATERIALS AND METHODS

14.2.1 Site Description

The study was conducted at Msekera Research Station and Kalunga Farmers Training Centre in Eastern province of Zambia from the 1996 through 2001 growing seasons. Msekera Research Station is located 32°34'S and 13°34'E at an altitude of 1032 m above sea level with a slope of about 4 to 5%, and has a mean annual rainfall of 1092 mm and a mean annual temperature of 25°C. The soil is classified as Haplic Luvisol (FAO-ISSS-ISRIC, 1998). Kalunga Farmers Training Centre is located 32°33'S and 13°50'E at an altitude of 1015 m above sea level with a slope of about 2%, with a mean annual rainfall of 1000 mm and a mean annual temperature of 28°C. The soil is classified as Ferric Luvisol (FAO-ISSS-ISRIC 1998). Selected soil properties from the 0 to 20 cm depth, as in November 1996, are shown in Table 14.1. Both sites were established in the 1996 and 1997 seasons and had been under natural vegetation prior to trial establishment. The plots were under planted fallows in the 1996 and 1997 seasons before they were put under maize in 1997 and 1998 seasons. They were again under planted fallows for 2 years from 1998 to 1999 and 1999 to 2000 seasons. Planted fallows were cut at the beginning of the 2000 and 2001 seasons after growing for 24 months, and maize was grown in the 2000 and 2001 season. The fallow treatments were planted to *Sesbania sesban* and *Tephrosia vogelii* (two legumes), natural fallow (NF) along with continuous maize control. Grass in NF plots was burnt before tillage at fallow termination. The plots were tilled using an ox-drawn plow and ridges were then made using hoes. The maize crop was planted on top of the ridges. The treatments were arranged in a complete randomized block design (RCBD), using slope as a blocking factor, with three replicates. The gross

Table 14.1 Selected Properties of the Soils under Study from the 0 to 20 cm Depth, November 1996

	Msekera	Kalunga
Clay (g kg^{-1})	150	70
Sand (g kg^{-1})	660	770
Silt (g kg^{-1})	190	160
pH (0.01 M CaCl$_2$)	4.3	4.5
Organic carbon (g kg^{-1})	10	4
Total nitrogen (g kg^{-1})	1.0	0.4
Phosphorus (ppm)	36	7
Exchangeable K (mmol kg^{-1})	2	1

plot size was 10 × 10 m, and a spacing of 1 × 1 m and 1 m × 0.5 m was used for *S. sesban* and *T. vogelii*, respectively.

14.2.2 Rainfall Simulations

Rainfall simulations were conducted at a rainfall intensity of 35 mm h^{-1} on 1-m^2 plots surrounded by a 50-cm buffer zone. Simulations were carried out soon after cutting the fallows, before land preparation in October 2000, and just before land preparation for the second maize crop in October 2001. At fallow termination, the twigs and litter were not removed from the plots. In October 2001, the maize stover was removed from the plots before simulations as these are normally fed to livestock. At fallow termination, rainfall simulations in the natural fallow plots were carried out without removing the grasses and weeds and their litter and these were burnt just before land preparation for the first maize crop. During the simulation events the microtopography of the soil was not tampered with. A portable rainfall simulator based on a single full cone nozzle principle, calibrated after Panini et al. (1993), was used. The plots, which had an average slope of 2% at Kalunga and 4% at Msekera, were demarcated and hydrologically confined using aluminum sheets installed on all sides leaving about 7 cm of the sheets above the ground. A metal flume was anchored at the outlet, leading into a small trench to collect runoff. Borehole water with a pH of 6.9 and an electrical conductivity of 190 mS cm^{-1} (Standard error Se = 3.2) was used for simulations. Rainfall simulations involved dry and wet runs. Dry runs were conducted on a dry soil and wet runs were carried out at the same spot used for dry runs the following day. In October 2000, rainfall simulations continued for 3 hours or until steady state runoff was attained. The antecedent soil moisture content (weight basis) at 0 to 10 cm depth for dry and wet runs at Kalunga site were 6% (Se = 0.6) and 10% (Se = 0.2), respectively, for both October 2000 and 2001. At Msekera the mean initial soil moisture content at 0 to 10 cm was 9% (Se = 0.2) at fallow termination and 6% (Se = 0.9) in October 2001 dry runs, and 12% (Se = 1.2) for wet runs for both October 2000 and 2001. In 2000, runoff intensity (mm h^{-1}) was periodically measured by sampling water flowing from each plot. The sediments in the collected runoff were dried and bulked for analysis. In October 2000, total runoff losses were estimated by plotting runoff against time and the area under the curve was calculated and expressed as a percentage of the total rainfall discharged from the plots. In 2001, a container was anchored at the base of the outlet to collect all the runoff and sediments. Runoff was then estimated by summing up runoff collected from the container and that collected during periodic sampling. The sediment collected in the container was weighed before being mixed with the solids separated from runoff collected during the simulations. Solids were separated from water through centrifugation, dried at 60°C for 12 hours, weighed and analyzed for carbon. The SOC was determined using the modified Walkley-Black procedure (Nelson and Sommers, 1982) and a correction factor of 1.3 was used in the calculations.

14.2.3 Data Analyses

For estimating infiltration rate, the empirical Horton-type model was used. The balance of rain minus runoff estimated infiltration, Infiltration [(I) = precipitation (P) − runoff (Q).] A modified version of the Horton-type equation proposed by Morin and Benjamin (1977) was fitted to the infiltration data:

$$i = i_f + (i_o - i_f) e^{-R/K} \qquad (14.1)$$

where: i is estimated instantaneous infiltration rate (mm h^{-1}); i_f is final infiltration rate (mm h^{-1}); i_o is initial infiltration rate (mm h^{-1}); R is cumulative rainfall (i.e., intensity × time, in mm); K is the infiltration rate decay coefficient, which expresses infiltration dynamics as affected by soil properties (mm) and was calculated when infiltration data was fitted to the equation. Data on time

Table 14.2 Runoff and Steady-State Infiltration Rates under Rainfall Simulations at Kalunga in October 2000 (00) and October 2001 (01)

Treatment	Time to Runoff (min)			Steady-State Infiltration Rate (mm hr^{-1})			Runoff Coefficient (%)		
	00 dry	01 dry	01 wet	00 dry	01 dry	01 wet	00 dry	01 dry	01 wet
C. maize	10.0	9.0	5.5	7.0	7.0	7.0	36.0	71.0	75.0
S. sesban	60.0	21.0	7.8	nd	21.0	12.0	0.0	12.0	51.0
T. vogelii	104.0	14.0	7.0	nd	16.0	12.0	0.0	27.0	66.0
Natural F.	116.0	14.0	10.0	nd	20.0	14.0	0.0	11.0	65.0
Lsd	30.6	11.4	1.1	—	12.3	2.1	—	4.6	7.5

Note: C. maize: continuous maize; Natural F.: natural fallow; Lsd at $P < 0.05$; nd: not determined.

to ponding, time to runoff, amount of runoff and quantity of soil losses, nutrient losses, and steady state infiltration rates were subjected to analysis of variance using Genstat Statistical package.

14.3 RESULTS

14.3.1 Kalunga Site

At fallow termination (October 2000), SOC content in the topsoil (0–10 cm) was 30% to 70% greater under fallow treatments than under continuous maize, and among fallow treatments, was 14 to 25% greater under natural than under improved fallow treatments ($P < 0.05$; Table 14.3). Time to runoff and steady state infiltration rate decreased while runoff, soil loss, sediment OC content, and OC loss generally increased from October 2000 to October 2001 (dry runs), and from dry to wet runs in October 2001 (Table 14.2 and Table 14.3). However, under continuous maize, steady state infiltration rate and sediment OC content were constant over the two seasons (sediment OC content was also constant over 2001 runs for S. sesban and T. vogelii treatments). When compared to fallow treatments, plots under continuous maize generally showed large differences: smaller time to runoff and steady state infiltration rate, and greater runoff. Soil loss was twice greater under continuous maize than in fallow treatments for both 2001 runs ($P < 0.05$). Loss of OC under continuous maize was also twice greater than in fallow treatments during dry runs (October 2001; $P < 0.05$), but was only 10 to 30% greater during wet runs. When compared to topsoil (0 to 10 cm), sediments had 2.6 to 4.5 times more OC, but the effect of treatments was unclear (enrichment ratio was higher for S. sesban treatment). In short, susceptibility to runoff, erosion, and OC loss increased after fallow termination and from dry to wet initial conditions, and was greater under continuous maize than in fallow treatments. Differences between fallow treatments were small in general. Additionally, the effect of treatments on sediment OC enrichment was unclear.

14.3.2 Msekera Site

At fallow termination (October 2000), topsoil OC content (0 to 10 cm) was 20 to 50% greater in fallow treatments than for continuous maize ($P < 0.05$; Table 14.5). From October 2000 to October 2001 (dry runs), time to runoff generally increased, but it decreased sharply in NF treatment (Table 14.4). In October 2001, time to runoff generally decreased by ca. 50% from dry to wet runs. Steady state infiltration rate decreased from October 2000 to October 2001 and from dry to wet runs, except under continuous maize where it was constant. Runoff rate increased markedly from October 2000 to October 2001, then slightly from dry to wet runs in October 2001. Susceptibility to runoff was generally higher (i.e., smaller time to runoff and steady state infiltration rate, and

Table 14.3 Soil and OC Losses at Kalunga Site in October 2000 (00) and 2001 (01)

Treatment	Total Soil Loss (g m^{-2})			Sediment OC Content (mg C g^{-1})			Topsoil OC* (mg C g^{-1})	OC Enrichment Ratio**			OC loss (g C m^{-2})	
	00 dry	01 dry	01 wet	00 dry	01 dry	01 wet	00	00 dry	01 dry	01 wet	00 dry	01 dry	01 wet
C. maize	nd	41.0	71.0	10.0	11.0	10.0	3.0	3.3	3.7	3.3	nd	0.45	0.71
S. sesban	0.0	15.0	36.0	—	17.0	18.0	4.0	—	4.3	4.5	0.00	0.26	0.65
T. vogelii	0.0	16.0	40.0	—	15.0	16.0	4.4	—	3.4	3.6	0.00	0.24	0.64
Natural F.	0.0	20.0	26.0	—	13.0	21.0	5.0	—	2.6	4.2	0.00	0.26	0.55
Lsd	—	15.0	34.0	—	3.4	4.0	0.2	—	—	—	-	0.11	0.06

Note: C. maize: continuous maize; Natural F.: natural fallow; Lsd at $P < 0.05$.

* Soil OC content at 0 to 10 cm in October 2000.
** Sediment OC content/topsoil OC content (0 to 10 cm) in October 2000.

Table 14.4 Runoff and Steady State Infiltration Rates under Rainfall Simulations at Msekera in October 2000 (00) and October 2001 (01)

	Time to Runoff (min)			Steady State Infiltration Rate (mm hr^{-1})			Runoff Coefficient (%)		
	00 dry	01 dry	01 wet	00 dry	01 dry	01 wet	00 dry	01 dry	01 wet
C. maize	3.0	5.0	2.9	5.0	5.0	4.8	55.0	70.0	82.0
S. sesban	7.0	8.0	4.0	18.0	8.0	5.0	37.0	58.0	69.0
T. vogelii	7.0	9.0	3.5	8.0	7.1	4.8	24.0	61.0	68.0
Natural F.	27.0	5.8	3.5	nd	6.1	4.8	2.0	68.0	71.0
Lsd	17.8	4.8	0.9	7.2	2.8	1.2	12.0	11.1	3.2

Note: C. maize: continuous maize; Natural F.: natural fallow; Lsd at $P < 0.05$. nd: not determined.

greater runoff loss) under continuous maize than under fallow treatments in 2000 (dry runs) and 2001 (both runs). However, in October 2001, steady state infiltration rate in *T. vogelii* and NF treatments did not differ significantly from that under continuous maize. Among the fallow treatments, NF had a smaller susceptibility to runoff than improved fallow treatments in 2000 (dry runs) but not in 2001. In October 2001, soil loss was 1.4 to 2.3 times greater in wet than in dry runs. When compared to fallow treatments, plots under continuous maize lost 1.7 to 3.1 times more sediments during dry runs, and 2.5 to 4.3 times more during wet runs (October 2001). Among the fallow treatments, soil loss was 1.5 to 1.9 times greater under NF than in improved fallow treatments, but the difference was significant in wet runs only ($P < 0.05$). Sediment OC content as well as sediment enrichment in OC did not vary much between years and between runs. Sediment OC content was 40 to 70% smaller under continuous maize than in fallow treatments ($P < 0.05$ in general), and among the fallow treatments, was 20 to 40% smaller under NF than for improved fallow treatments. Sediment enrichment in OC was the smallest under continuous maize (1.3 to 1.6) and the greatest under *T. vogelii* treatment (3.6 to 4.1). From dry to wet runs (October 2001), total OC loss increased by 40 to 70% under fallow treatments, but by 180% under continuous maize. Total OC loss was 50 to 80% greater under continuous maize than under fallow treatments during wet runs ($P < 0.05$), whereas differences between treatments were unclear during dry runs (nevertheless OC loss was 50% greater under NF than under *S. sesban* ($P < 0.05$)). In short, runoff and soil losses were greater in wet than in dry runs, and higher under continuous maize than in fallow treatments. Sediment OC enrichment tended to be greater in fallow treatments than in continuous maize, and OC loss was greater under continuous maize than in fallow treatments during wet runs only. Differences between fallow treatments were small in general.

14.3.3 Comparison between Kalunga and Msekera Sites

There was a significant site and treatment interaction on time to runoff ($P < 0.05$) except for continuous maize. There was a significant site and treatment interaction on soil loss for *S. sesban* and on OC loss for *T. vogelii* only ($P < 0.05$). If we compare the two sites, time to runoff and steady state infiltration rate were smaller and runoff losses generally higher at Msekera than at Kalunga (15 vs. 7% clay at 0 to 20 cm, and 7 to 11 vs. 3 to 5 mg C g^{-1} at 0 to 10 cm in October 2000, respectively; Table 14.1, Table 14.3, and Table 14.5). Differences in soil loss between sites were less clear. Sediment OC content and OC loss were generally greater at Msekera than at Kalunga due to greater topsoil OC content at Msekera. In contrast, sediment enrichment in OC was generally smaller at Msekera than at Kalunga. However, at both sites continuous maize had the lowest time to runoff, steady state infiltration rates and the highest runoff and soil losses relative to fallow treatments. Susceptibility to runoff and OC losses was thus greater at Msekera than at Kalunga.

Table 14.5 Soil and OC Losses at Msekera Site in October 2000 (00) and 2001 (01)

Treatment	Total Soil Loss (g m^{-2})			Sediment OC Content (mg C g^{-1})			Topsoil OC* (mg C g^{-1})	OC Enrichment Ratio**			OC Loss (g C m^{-2})		
	00 dry	01 dry	01 wet	00 dry	01 dry	01 wet	00	00 dry	01 dry	01 wet	00 dry	01 dry to check	01 wet to check
C. maize	nd	43.0	99.0	12.0	10.0	12.0	7.2	1.7	1.4	1.7	nd	0.43	1.19
S. sesban	nd	14.0	24.0	29.0	28.0	27.0	11.0	2.6	2.5	2.5	nd	0.39	0.65
T. vogelii	nd	17.0	23.0	31.0	34.0	35.0	8.5	3.6	4.0	4.1	nd	0.58	0.81
Natural F.	0.0	26.0	40.0	—	22.0	21.0	11.0	—	2.0	1.9	0	0.57	0.84
Lsd	—	22.0	23.0	13.0	22.0	5.5	0.5	—	—	—	—	0.13	0.21

Note: C. maize: continuous maize; Natural F.: natural fallow; Lsd at $P < 0.05$.

* Soil OC content at 0 to 10 cm in October 2000.
** Sediment OC content/topsoil OC content (0 to 10 cm) in October 2000.

14.4 DISCUSSION

The decrease in time to runoff and steady state infiltration rate from fallow termination (October 2000) to the beginning of the second cropping season (October 2001) in fallow treatments at both sites could be attributed to introduction of tillage and lack of leaf/grass litter on the surface. The leaf/grass litter protected the soil surface resulting in higher infiltration rates at fallow termination and this had disappeared through burning (natural fallow) and soil incorporation (improved fallows) as a result of tillage after the first cropping. Reduced steady state infiltration rates then resulted in increased runoff and soil losses (Nyamadzawo et al., 2003). High rates of soil loss have been attributed to tillage (Havelin et al., 1990), as conventional tillage using an ox drawn plow destroyed soil structure. Total soil losses increased in fallow treatments after one cropping season showing that the benefits of fallowing decreased with the introduction of cropping and tillage.

Although soil loss increased after fallow termination, runoff and soil loss in fallow treatments generally remained lower relative to continuous maize at both sites, due to the destruction of structure resulting from continuous cultivation. At Msekera, after one cropping season, steady state infiltration rates in *T. vogelii* and NF treatments decreased and did not differ from that under continuous maize. This shows that at Msekera the benefits of fallowing using these species were short-lived and there was need to return to fallowing earlier. Two years of cropping before fallowing again may be optimal, as was also suggested by Mafongoya and Dzowela (1999) although they were evaluating the length of cropping phase from a soil fertility perspective. This rapid decline in steady state infiltration rates after the first crop seems to be accompanied by a decline in soil fertility as fertile topsoil gets eroded (Mafongoya and Dzowela, 1999).

Sediment bound OC was highest in fallow treatments relative to continuous maize, in relation with greater topsoil OC content. However, due to much greater soil losses under continuous maize than in fallow treatments at both sites, OC loss was generally greater under continuous maize. The OC enrichment ratio of sediments was higher in fallow treatments than in continuous maize at Msekera, indicating selective detachment and transport of fine particles (including small aggregates), which include more OC. This is also because fallow treatments had more SOC than continuous maize in the topsoil. This observation is supported also by the findings of Wan and El-Swaify (1997). Palis et al. (1997) also found out that carbon enrichment ratios were high for aggregated soils with high clay contents. Soil OC losses can be used as indicators of plant nutrient losses because most nutrients are associated with OC, which is washed away with the soil (Follet et al., 1987).

Greater runoff at Msekera than at Kalunga was perhaps related to greater topsoil content in fine particles (15 vs. 7% clay, and 34 vs. 23% clay + silt, respectively), resulting in a greater amount of particles available for dispersion and surface clogging. The relatively low time to runoff obtained at fallow termination at Msekera was due to relatively higher initial topsoil moisture content during simulations in October 2000 (9% compared to 6% in October 2001).

Fallow treatments were less susceptible to runoff and erosion than continuous maize, but improved fallows were largely comparable to natural fallow. Thus, planting leguminous fallows did not result in clear benefit, especially at Kalunga (at Msekera, soil loss was smaller under NF than under improved fallow treatments). Therefore, improved fallows could mainly be promoted for soil fertility benefits through N fixation, as they neither enhance soil hydraulic properties nor reduce erodibility when compared with natural fallow. The increase in susceptibility to runoff, soil and OC loss from dry to wet runs was due to reduced time to runoff and, therefore, more runoff as the rainfall simulations were conducted for a fixed duration (30 min).

The results from this work indicate the potential OC erosion and erodibility under improved fallow systems. Although estimates of erodibility made from small plots cannot be readily extrapolated to the field or landscape scale (Stomph et al., 2002; van de Giesen et al., 2000), they are indicative of the differences between treatments and management. Extrapolations are difficult using this kind of data, plots are often too small and may not be representative of soil variability and some of the processes occurring in large fields, as for example in sediment deposition (De Boer and Campbell, 1989; Govers, 1991; Mathier and Roy, 1996).

14.5 CONCLUSION

This study showed that fallowing increased topsoil OC content and steady state infiltration rates, and reduced runoff, erosion, and loss of OC relative to continuous maize. However these benefits decreased from fallow termination to the beginning of the second cropping season. At Kalunga, there were no differences in runoff and soil loss among improved fallow and NF treatments. Thus the interest in fallow planting was not clear in regard to soil and water conservation. At Msekera improved fallow treatments had less soil erosion from the plots relative to NF, but had similar runoff and OC losses. Runoff, soil and OC losses were higher at Msekera relative to Kalunga because of differences in initial soil moisture content, slope and soil type (texture especially).

ACKNOWLEDGMENTS

We thank the ICRAF Zambian team for allowing us to use their sites for the work reported in this chapter. This work is an output of IMPALA project funded through the European Union Project No. ICA4-CT2000-30011.

REFERENCES

Barthès, B., A. Albrecht, J. Asseline, G. De Noni, and E. Roose. 1999. Relationship between soil erodibility and topsoil aggregate stability or carbon content in a cultivated Mediterranean highland (Aveyron, France). *Communications in Soil Science and Plant Analysis* 30:1929–1938.

Coleman, T. L., A. U. Eke, U. R. Bishnoi, and C. Sabota. 1990. Nutrient losses in eroded sediments from limited resource farm. *Field Crop Research* 24:105–117.

Chikowo, R., P. Mapfumo, P. Nyamugafata, G. Nyamadzawo, and K. E. Giller. 2003. Nitrate-N dynamics following improved fallows and maize root development in a Zimbabwean sandy clay loam. *Agroforestry Systems* 59:1897–195.

De Boer, D. H. and I. A. Campbell. 1989. Spatial scale dependence of sediment dynamics in a semiarid badland drainage basin. *Catena* 16:277–290.

FAO-ISSS-ISRIC. 1998. World Reference Base for Soil Resources. *World Soil Resources Report* 84, FAO, Rome.

Follet, R. F., S. C. Gupta, and P. G. Hunt. 1987. Conservation practices: Relations to the management of plant nutrients for crop production, in R. F. Follet, ed., *Soil Fertility and Organic Matter as Critical Components of Production Systems*. SSSA Special Publication 19, Soil Science Society of America, Madison, WI, p. 19–52.

Goss, M. J., K. R. Howse, P. W. Lane, D. G. Christian, and G. L. Harris. 1993. Loss of nitrate-nitrogen in water draining from autumn sawn crops established by direct drilling and mouldboard ploughing. *Soil Science* 44:35–48.

Govers, G. 1991. A field study on topographical and topsoil effects on runoff generation. *Catena* 18:91–111.

Hartemink, A. E., R. J. Buresh, B. Jama, and B. H. Janssen. 1996. Soil nitrate and water dynamics in *Sesbania sesban* fallows, weed fallows and maize. *Soil Science Society of America Journal* 60:568–574.

Havelin, J. L., D. E. Kissel, L. D. Maddaux, M. M. Cleassen, and J. H. Long. 1990. Crop rotation and tillage effects on soil organic carbon and nitrogen. *Soil Science Society of America Journal* 34:448–452.

Juo, A. S. and R. Lal. 1977. The effects of fallow and continuous cultivation on the chemical and physical properties of an Alfisol in Western Nigeria. *Plant and Soil* 47:567–584.

Lowrance, R. 1992. Nitrogen outputs from a field-size agricultural watershed. *Journal of Environmental Quality* 21:602–607.

Mafongoya, P. L., P. K. R. Nair, and B. H. Dzowela. 1997. Multipurpose tree prunings as a source of nitrogen to maize under semiarid conditions in Zimbabwe: Part 3. Interactions of pruning quality and time and method of application on nitrogen recovery by maize in two soil types. *Agroforestry Systems* 35:57–70.

Mafongoya, P. L. and B. H. Dzowela. 1999. Biomass production of tree fallow and their residual effects on maize in Zimbabwe. *Agroforestry Systems* 47:139–151.

Mathier, L. and A. G. Roy. 1996. A study of the effects of spartial scale on parameters of sediment transport equation for sheet wash. *Catena* 26:161–169.

Morin, J. and Y. Benjamin. 1977. Rainfall infiltration into bare soils. *Water Resources Research* 13:813–817.

Nelson, D. W. and L. E. Sommers. 1982. Total carbon, organic carbon and organic matter, in A. L. Page, D. R. Keeney, D. E. Baker, R. H. Miller, E. Roscoe, Jr., and J. D. Rhoades, eds., *Methods of Soil Analysis 2*. American Society of Agronomy, Madison, WI, pp. 539–579.

Nyamadzawo, G., P. Nyamugafata, R. Chikowo, and K. E. Giller. 2003. Partitioning of simulated rainfall under maize-improved fallow rotations in Kaolinitic Soils. *Agroforestry Systems* 59:207–214.

Palis, R. G., H. Ghandiri, C. W. Rose, and P. G. Saffigna. 1997. Soil erosion and nutrient loss: 3. Changes in the enrichment ratio of total nitrogen and organic carbon under rainfall detachment and entrainment. *Australian Journal of Soil Research* 35:891–905.

Panini, T., M. P. Salvador Sanchis, and D. Torri. 1993. A portable rain simulator for rough and smooth morphologies. *Quaderni di Scienza del Suolo* 5:47–58.

Sharpley, A. N., S. C. Chapa, R. Wedephahl, J. T. Sims, T. C. Daniel, and K. R Reddy. 1994. Managing agricultural P for protection of surface water. Issues and Options. *Journal of Environmental Quality* 23:437–451.

Stomph, T. J., N. de Ridder, T. S. Steenhuis, and C. van de Giesen. 2002. Scale effects on Hortonian overland flow and rainfall-runoff dynamics: Laboratory validation of a process-based model. *Earth Surface Processes and Landforms* 27:847–855.

Van de Giesen, N. C., T. J. Stomph, and N. de Ridder. 2000. Scale effects of Hortonian overland flow and rainfall-runoff dynamics in West African catena landscape. *Hydrological Processes* 14:165–175.

Wan, Y. and S. A. El-Swaify. 1997. Flow-induced transport and enrichment of erosional sediment from a well-aggregated and uniformly-textured oxisol. *Geoderma* 75:251–265.

Zhang, X. C. and W. P. Miller. 1996. Physical and chemical crusting processes after runoff and erosion in furrows. *Soil Science Society of America Journal* 60:860–865.

SECTION 3

Carbon Transfer in Rivers

CHAPTER **15**

Origins and Behaviors of Carbon Species in World Rivers

Michel Meybeck

CONTENTS

15.1 Introduction 209
15.2 Origins, Ages, and Levels of Riverine Carbon 210
15.3 Carbon Behavior in River Systems 214
    15.3.1 Carbon-Specific Discharge Relations 215
    15.3.2 Influence of Wetlands, Lakes, and Reservoirs 217
    15.3.3 Particulate Organic Carbon and Riverine Total Suspended Solids (TSS) 217
    15.3.4 Autochthonous Fluvial Carbon 220
    15.3.5 Anthropogenic Impacts on Fluvial Carbon and Long-Term Trends 220
    15.3.6 The Fossil POC Controversy 222
15.4 Global Control and Budgets of Riverine Carbon Export 224
    15.4.1 Variability of Carbon Yields 224
    15.4.2 Global Budget of River Atmospheric Carbon Carried by Rivers 229
15.5 Perspectives and Propositions 231
    15.5.1 Improving Our Knowledge of Holocene River Carbon (Pristine Conditions) 231
    15.5.2 Future Evolution of Riverine Carbon at the Anthropocene 233
Acknowledgments 235
References 235

15.1 INTRODUCTION

It is only since the 1980s that the interest of the scientific community has increased in riverine carbon species, including particulate inorganic and organic carbon (PIC and POC) and dissolved inorganic and organic carbon (DIC and DOC). Before the 1980s, 99% of carbon analyses performed in surface waters were concerning HCO_3^-, the dominant species in most waters for pH < 8.2 and, in few of them, CO_3^{2-}, the dominant species for more alkaline waters, both of them constitute DIC when expressed in mg C.L^{-1}. The development of the organic carbon analyses in the 1970s gradually permitted the analysis of DOC and POC, after a filtration on glass-fiber filters at 0.5 or 0.7 µm porosity as in U.S. rivers (Malcom and Durum, 1976) or in the Amazon (Williams, 1968). In the early 1960s, the classic compilation in world rivers by Livingstone (1963) considered mainly HCO_3^-, the dominant anion in most surface waters. During the 1980s, the systematic analysis of carbon species in world major rivers was the principal target of the SCOPE-Carbon project initiated and directed by E. T. Degens, regularly published in the proceedings of his institute (*Mitt. Geol. Paleont. Inst. Univ. Hamburg*) and synthesized in 1991 (Degens et al., 1991). Concurrently with the SCOPE-Carbon project, two studies concerned themselves for the first time with river organic carbon at the global scale: Schlesinger and Melack (1980) and Meybeck (1981, 1982). The studies triggered a general interest for riverine carbon at the regional to global scales (Mulholland and Elwood, 1982; Kempe, 1982; 1984; Artemyev, 1993; Ittekot, 1988; Meybeck, 1993a; 1993b; Amiotte-Suchet and Probst, 1993; Probst et al., 1994; Ludwig et al., 1996; 1998; Amiotte-Suchet et al., 2003).

In addition to studies that focused on river carbon fluxes to oceans, the sources, sinks, and pathways of river carbon have been studied through riverine and estuarine case studies (Degens et al., 1991). Such studies have attempted to differentiate natural processes from anthropogenic sources of carbon, particularly from organic species resulting from organic wastes releases — domestic and industrial — and from eutrophication. In France, for instance, numerous studies on riverine carbon have been done on major river systems and estuaries including the Loire (Meybeck et al., 1988), the Gironde and Garonne (Cauwet and Martin, 1982; Veissy, 1988; Etcheber, 1986), and the Rhone (Sempere et al., 2000). The organic carbon cycle and its coupling with nitrogen and phosphorus cycles has also been studied and modeled since 1989 within a multidisciplinary program, the PIREN-Seine (Meybeck et al., 1998; Servais et al., 1998; Billen et al., 1994).

The increasing interest in riverine carbon is now linked to the global carbon budget and cycle (Kempe, 1982; 1984; Aumont et al., 2001; Andrews and Schlesinger, 1999; Smith et al., 2001) with regards to the past, present, and future linkage between river carbon atmospheric CO_2 and climate and to their coupling with other major biogeochemical cycles as nitrogen, phosphorus, and sulfur (Meybeck and Vörösmarty, 1999; Rabouille et al., 2000). The past climate, over the geologic time scale, may have been very much influenced, even regulated, by the atmospheric CO_2 uptake during weathering of surficial noncarbonated minerals — as fresh volcanic rocks — and its subsequent transfer from continents to oceans as riverine DIC (Berner et al., 1983; Probst et al., 1994; Amiotte-Suchet et al., 2003).

During the 1970s interest increased in DOC and POC analyses, particularly their most labile forms, which control the river's oxygen levels, these analyses have gradually replaced the BOD_5 and COD measurement in regular surveys. Finally organic carbon exports by rivers can be compared to the terrestrial primary production of their basin: although these fluxes are only of the order of 1% of the primary production (Meybeck, 1982), they provide important information on the continental cycle of carbon.

It is not possible to address all these questions here: global carbon budgets, their geographic distributions, and their evolution during geological times will not be presented here and the reader can refer to Meybeck (1982, 1993a, 1993b), Probst et al. (1994), Ludwig et al. (1996), and Amiotte-Suchet et al. (2003). This chapter focuses instead on the origins and behavior of river carbon species and their alteration by human activities. These processes are discussed at the global scale particularly on the basis of the GEMS-GLORI database developed by Meybeck and Ragu (1996, 1997). Most examples presented herein are taken from French river basins, which are very diverse, particularly concerning human impacts, and are also well studied.

15.2 ORIGINS, AGES, AND LEVELS OF RIVERINE CARBON

Riverine carbon dated since the original atmospheric CO_2 fixation ranges from hundreds of millions years (carbonated rocks) to a few days (autochthonous POC and PIC). River carbon origins and ages are detailed in Table 15.1 in two broad categories: (1) Old particulate carbon resulting

Table 15.1 Origins and Ages — Since Original Atmospheric CO_2 Fixation — of Carbon Species in Rivers

	Sources	Age (y)	Flux # 10^{12} g C yr^{-1}	A	B	C	D	E	F
PIC	Geologic	10^4–10^8	170	•					•
DIC	Geologic	10^4–10^8	140		•	•			•
	Atmospheric	0–10^2	245		•	•			•
DOC	Soils	10^0–10^3	200			•			•
	Pollution	10^{-2}–10^{-1}	(15?)					•	
CO_2	Atmospheric	0	(20 to 80)			•	•		
POC	Soil	10^0–10^3	(100)	•					•
	Algal	10^{-2}	(< 10)				•		•
	Pollution	10^{-2}–10^0	(15)					•	
	Geologic	10^4–10^8	(80)						•

Note: A: land erosion, B: chemical weathering, C: global warming and UV changes, D: eutrophication, E: organic pollution, F: basin management damming.

Data from Meybeck, M. 1993a. C, N, P, and S in rivers: From sources to global inputs, in R. Wollast, F. T. Mackenzie, and L. Chou, eds., *Interaction of C, N, P, and S Biogeochemical Cycles and Global Change*. Springer-Verlag, pp. 163–193; Meybeck, M. 1993b. Riverine transport of atmospheric carbon: sources, global typology and budget. *Water, Air, Soil Pollution* 70:443–464; Meybeck, M. and C. Vörösmarty. 1999. Global transfer of carbon by rivers. *Global Change Newsletters* 37:12–14.

from the mechanical erosion of carbonate rocks, this carbon is mostly PIC but growing evidence shows old recycled sedimentary POC (Kao and Liu, 1997; Di Giovanni et al., 2002) and old dissolved inorganic carbon (DIC) resulting from the dissolution of carbonate rocks by acids during weathering reactions. (2) Recent carbon, organic carbon originating from erosion and leaching of soils (DOC and POC) and atmospheric CO_2 implied in most weathering reactions:

- Reaction 1: noncarbonated minerals + CO_2 + H_2O → HCO_3^- + cations + weathered minerals + dissolved silica
- Reaction 2: $(Ca, Mg)CO_3$ + CO_2 + H_2O → 2 HCO_3^- + Ca^{2+} and Mg^{2+}

In Reaction 1, all riverine DIC originates from atmospheric CO_2 and from soil organic acids. In Reaction 2, only half of it originates from CO_2.

Other natural origins of riverine carbon are autochtonous POC resulting from recent debris of algae and macrophytes, particularly in eutrophied rivers. Another autochtonous source of PIC is calcite precipitation when pH exceeds 8.2, which is commonly the case for eutrophied rivers in carbonated basins. The Loire River is a typical example of such behavior (Meybeck et al., 1988). Anthropogenic sources of organic carbon are mostly found in organic wastes from agro-industries and cities.

Organic carbon ages are now determined by ^{14}C dating on river DOC and POC. Average POC ages range from few hundred years to few millenniums: 1260 years in the Amazon river and 4600 years in Hudson river (Raymond and Bauer, 2001) and in Siberian rivers the organic matter may be some 6000 to 8000 years old and originate from peat deposits accumulated since the last glacial period (Artemyev, 1993). Raymond and Bauer also dated DOC in northwestern Atlantic rivers from modern age to 1380 years. Evidence of fossil POC, millions of years old, in some sedimentary river basins exposed to high erosion rates is growing, this topic is fully developed in Section 15.3.6. Throughout this chapter the term *fossil POC* will be applied for ages exceeding 10,000 years. Organic wastes have carbon ages ranging from few days to few years.

Riverine carbon concentration is classically reported in mg C/L. Particulate forms of river carbon can be expressed in mg C/L, yet there are also reported as % C of total suspended solids (TSS), termed here PIC% and POC%. The sum of river carbon species originating from recent CO_2 is termed total atmospheric carbon (Meybeck, 1993a):

$$TAC = DOC + \text{nonfossil POC} + 100\% \text{ silicate weathering DIC} +$$
$$50\% \text{ carbonate weathering DIC} \qquad (15.1)$$

Carbon levels and sources in world rivers are highly variable (Meybeck, 1993a; 1993b; 2003b), their ranges are reported in Table 15.2 for major rivers (area > 10,000 km²) discharging to oceans from the GEMS-GLORI register (Meybeck and Ragu, 1996; 1997). The TAC is minimum in noncarbonated river basins where the terrestrial vegetation production is low as in the African savanna where it barely exceeds 10 mg/L. The Gambia river (West Africa) (Lesack et al., 1984; Meybeck et al., 1986) is a good example of such river with a DOC dominance in TAC. The maximum levels of TAC reach 50 mg/L in easily erodible and weathered carbonated basins where it is transported mostly as DIC. In most limestone regions as in the Jura Mountains DIC is by far the dominant species of TAC.

The occurrence of DIC, and more particularly of HCO_3^-, has been reviewed by Meybeck (2003b) from a set of 1200 pristine rivers worldwide. In most rivers, for which the anions sums, (an expression of the total salt content), ranges from 200 to 6 000 μeq/L, bicarbonate is the dominant anion. Below 200 μeq/L, the pH is usually < 5.5 and CO_2 and organic acids are the dominant carbon species while HCO_3^- is close to zero. Above 6,000 μeq/L, SO_4^{2-} then Cl^- are dominating, and $CaCO_3$ is gradually precipitated in soils and river waters as mineralization increases. When

Table 15.2 Range of Average Carbon Species Concentrations in World Rivers and Global Discharge Weighted Average (mg C/L)

	Minimum		Maximum		Average
Carbon Derived from Soil and Atmosphere					
DIC [1]	< 1	Silicate rocks; acid black waters (Rio Negro, Bra.)	18	Silicate rocks (La Réunion)	6.4
			35	Carbonate rocks	
DOC	< 1	Karst springs	40	Peat bog outlets	5.0
POC [2]	< 0.1	Lake outlets	10	Highly erodible soils (Missouri)	2.35
Carbon Derived from Rock Erosion and Weathering					
DIC	0	Silicate rocks	35	Carbonate rocks	3.15
PIC	0	Silicate rocks; lake outlets	600	Huang He (Yellow R.)	2
POC [3]	0	Nonsedimentary basins	100	Sedimentary rocks, shales, and loess	1.3
Autochtonous Carbon					
PIC [4]	0	Carbonated basins	> 10?	Huang He; Loire; some lakes, outlets	?
POC [5]	0	Oligotrophic river basins	3	Eutrophic rivers, basins and lake outlets	< 0.2
Carbonaceous Pollution					
TOC	0	Unpolluted basins	10?	Polluted basins; lower Seine: 7 mg/L	?
All Carbon Sources					
TOC	1.5		45		8.8
TIC	0		635		11.55
TAC	8		50		13.75

Note: [1] excluding dissolved CO_2; [2] soil POC; [3] fossil POC if age > 10,000 y; [4] precipitated calcite; [5] algal POC.

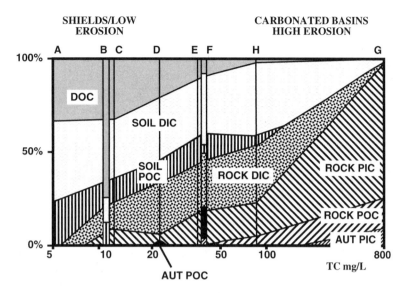

Figure 15.1 Origins of carbon species in selected rivers with increasing total carbon (TC) concentrations (annual averages). Soil-derived carbon: DOC, soil DIC, soil POC. Rock-derived carbon: rock DIC, rock PIC, rock POC. River-derived carbon: autochthonous (algal) POC, autochthonous PIC(in black). A = Gambia (West Africa), B = Rio Negro (Brazil), C = Solimoes (Brazil), D = Loire (winter average), E = Mackenzie, F = Loire (summer average), G = Huang He, H = Dranse (French Alps). (Data from Meybeck, M. 1993a. C, N, P, and S in rivers: From sources to global inputs, in R. Wollast, F. T. Mackenzie, and L. Chou, eds., *Interaction of C, N, P, and S Biogeochemical Cycles and Global Change*. Springer-Verlag, pp. 163–193.)

considering 700 world basins (exorheic and endorheic), between 3,200 and 200,000 km², HCO_3^- is the dominant anion — i.e., the HCO_3^- proportion exceeds 40% of the anion sum — for more than 90% of basins.

Examples of the various origins of river carbon are presented in Figure 15.1 within a gradient of total carbon (TC) concentrations from Gambia to Huang He (Yellow River). When TC exceeds 50 mg C/L, the proportion of TAC:TC drops rapidly below 50% and in the Huang He more than 90% of TC is of fossil origin, both as PIC and POC, mostly from the loess deposits (50,000 to 200,000 years old). In the Brazilian Rio Negro basin, devoid of any carbonate rocks and poorly drained, DIC is near zero due to the very acidic pH (4.2) and most TC and TAC are under the DOC form (Moreira-Turcq et al., 2003).

When the total organic carbon (TOC = DOC + POC) of rivers is plotted against the total inorganic carbon (TIC = DIC + PIC) for the few rivers in which all carbon species are analyzed, there is no correlation and the carbon concentration ranges are relatively limited (Figure 15.2a), from 5 to 40 mg/L, for all rivers. The Huang He TIC reaches hundreds of mg C/L, which indicates the exceptional nature of this river in which the average TSS level exceeds 10,000 mg/L. Other major rivers such as the Amazon, the Mackenzie, or the Danube are relatively well clustered in Figure 15.2. The TOC/TIC ratio ranges from 0.2 (Huang He) to nearly 10 depending on lithology, climate, runoff pathways, and terrestrial primary production. All carbon species are still rarely documented in rivers and it is difficult to infer these influences on TOC and TIC distribution without setting up an appropriate database at the global scale.

Both DOC and DIC are not directly correlated but their ratio DOC/DIC is probably a major control of pH or the proton concentration of river waters with limited eutrophication (Figure 15.2b). In very dilute waters, as those found in Central Amazon, common pH values are < 5.5 and bicarbonates are close to zero while organic acids are important with DOC values exceeding 10 mg/L. As a result, the DOC/DIC ratio exceeds 50. In rivers draining carbonate rocks, DOC may be quite limited (it is often less than 1 mg/L in karstic springs) while DIC is very high (Table 15.2):

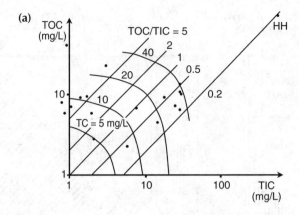
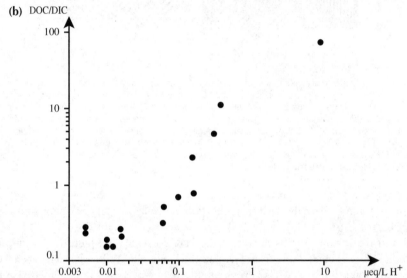

Figure 15.2 Organic carbon vs. inorganic carbon in world rivers. (a) Total organic carbon (TOC) vs. total inorganic carbon (TIC) (Data from Meybeck, M. and A. Ragu. 1996. Environment Information and Assessment Rpt. UNEP, Nairobi, 250 pp.) Diagonal lines represent the equal TOC/TIC ratios and curves the total carbon (TC = TOC + TIC). Note the specific position of the Huang He (HH). (b) DOC/DIC ratio vs. proton concentrations (μeq/l) in some world rivers, annual averages (Data set from Meybeck, M. and A. Ragu. 1996. Environment Information and Assessment Rpt. UNEP, Nairobi, 250 pp.; and others.)

as a result, the DOC/DIC ratio can be < 0.2 for pH values close to 8.2. If all world river waters were fully mixed, their pH would be around 7.2 for discharge-weighted averages of 6.4 mg/L for DIC and 5 mg/L for DOC. This average reflects the dominance of waters influenced by noncarbonated wet tropics at the global scale.

15.3 CARBON BEHAVIOR IN RIVER SYSTEMS

Carbon behavior can be studied through multiple aspects as concentrations vs. river specific discharge, seasonal variations, lake inlet-outlet comparisons, longitudinal profiles in river main course, trends, and relationship with TSS. In eutrophied and polluted river systems such as the Seine River the organic carbon can also be modeled through a multi-process approach considering both carbon production (P) and respiration (R) within the hydrological network.

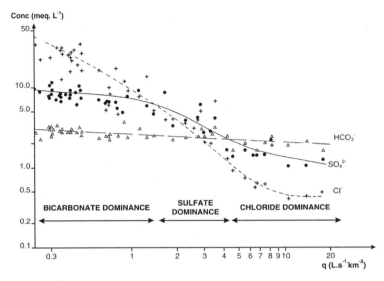

Figure 15.3 Compared relations of HCO_3^-, Cl^-, and SO_4^{2-} (meq/L) with river discharge in the Dolores River at Cisco, Utah (U.S. Geol. Survey analyses). (From Meybeck, M. 1985. Variabilité dans le temps de la composition chimique des rivières et de leurs transports en solution et en suspension. *Rev. Fr. Sci. Eau*, 4, 93–121. With permission.)

15.3.1 Carbon-Specific Discharge Relations

The DIC is generally not very variable with the specific discharge (q in L s^{-1} km^{-2}) compared to Cl^- and SO_4^{2-}, which may be much more variable. Chloride and sulfate often originate from either natural or anthropogenic point sources, such as saline springs, mining, industrial, and urban wastes waters, which are diluted during floods. The pristine Dolores River, a Utah tributary of the Colorado characterized by saline springs inputs, is a good example of such contrasted evolution (Figure 15.3). While Cl^- and SO_4^{2-} are diluted during high waters, HCO_3^- remains relatively stable. In this peculiar river, waters may be dominated by Cl^- at low flows, then SO_4^{2-}, then HCO_3^-.

DIC behaviors in three monolithologic stream basins are compared in Figure 15.3: in carbonated basins, as the karstic Brevon stream, a tributary to Lake Geneva (French Alps), DIC is relatively stable, while for the Diege, a crystalline-rock stream basin in Central France, and for the basalt-draining Rhue stream (Auvergne, France), the DIC vs. q slope is well marked (Figure 15.4). It is interpreted as a marked DIC difference between surface runoff water and groundwater, and is even more contrasted for basalt basins.

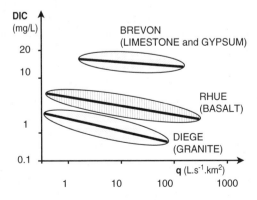

Figure 15.4 Compared evolution of DIC with river runoff in three small monolithologic French streams: Brevon (Haute Savoie), Rhue at Egliseneuve (Auvergne), and Diège (Corrèze).

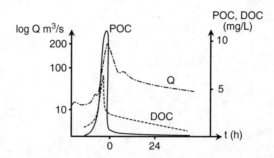

Figure 15.5 Evolution of dissolved (DOC) and particulate (POC) organic carbon concentration during an exceptional flood event in the Cannone mountainstream (Corsica), Mediterranean river regime. (Data from Löye-Pilot, M. D. 1985. Les variations des teneurs en carbone organique (dissous et particulaire) d'un torrent de montagne: La Solenzana, Corse. *Verh. Internat. Verein. Limnol.* 22:2087–2093. With permission.)

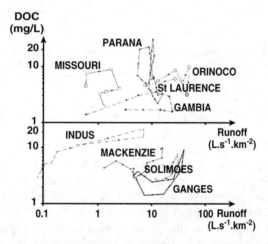

Figure 15.6 Cyclic relations between dissolved organic carbon (DOC) and river specific discharge (q) during annual floods for some major rivers. Clockwise cycles: Missouri, Ganges, Solimoes, Mackenzie. Counterclockwise cycle: Saint Lawrence. No definite cycle: Parana and Indus. (Data from the SCOPE carbon program [Degens et al., 1982] and *Mitt. Geol. Pal. Inst. Univ. Hamburg*, 52, 55, 58, and 64.)

The DOC evolution with q is very illustrative of the pedogenic origin of this carbon species: it generally increases during the rising stage of big floods then decreases during the falling stage as for the Cannone mountain stream (33 ha) in the Solenzara basin in Corsica, on crystalline rocks (Figure 15.5) (Löye-Pilot, 1985). As a result a clockwise cycle of DOC vs. q is generally observed in streams. This cycle is also common in large rivers such as in the Mackenzie River in Canada (Figure 15.6) or in the Congo River (Seyler et al., this volume).

The DOC vs. q relationships are very specific. In the Moisie basins (Quebec), Naiman (1982) observed a counterclockwise cycle with a maximum DOC level in the falling stage of high waters after the snowmelt. Other counterclockwise relationships have been observed. In two New Zealand paired basins, the Mamaï and the Larry, opposite DOC behavior is described by Moore (1987): the Mamaï cycle is clockwise while the Larry cycle is counterclockwise, due to the DOC rich peat-draining waters in this basin that are diluted during floods by less concentrated waters. For the lower Saint Lawrence, a counterclockwise cycle is also observed, it is probably generated by the dilution of DOC-rich Ottawa River, that drains Canadian shield wetlands, by the DOC-poor upper Saint Lawrence that originates from the Great Lakes system, in which DOC levels are ten times lower than for the Ottawa river.

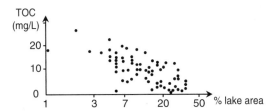

Figure 15.7 Inverse relationship between average TOC concentration and the proportion of lake area in a set of 78 small Finish basins. (Redrawn from Kortelainen, P. and J. Mannio, 1988. Natural and anthropogenic acidity sources for Finnish lakes. *Water Air Soil Poll.* 42:341–352.)

15.3.2 Influence of Wetlands, Lakes, and Reservoirs

Water bodies play a complex role in carbon transfers particularly for organic carbon. Shallow wetlands and peat lands are prime sources of DOC, which may exceed 40 mg/L (Moore, 1987). Deeper water bodies such as lakes and reservoirs are, on the contrary, prime sinks of DOC, which is degraded by UV action and by bacterial activity at such time scales. In the U.S., for instance, Mulholland and Watts (1982) demonstrated the importance of the wetlands on the TOC export rate in regionalized organic carbon budgets.

Conversely, in Finland, Kortelainen, and Mannio (1988) have found an inverse relationship between TOC levels (mostly in the DOC form) and the occurrence of lakes expressed in % of basin area (Figure 15.7). In such formerly glacierized regions, DOC is very high when lakes are absent, probably resulting from multiple wetlands.

All lakes and reservoirs are also very efficient sinks of particulate detrital carbon, either eroded PIC or soil POC. The trapping efficiency of particulates exceeds 90% when the water residence time exceeds 6 months, a figure common in lakes and in the largest reservoirs where it may exceed 100 years and 2 years, respectively. The occurrence of reservoirs should be carefully considered in any model of particulate carbon transfer as done in the French Dordogne River (Veissy, 1988). In mesotrophic to eutrophic temperate lakes occurring in carbonated basins, there is a seasonal calcite precipitation when algal development increases pH values above 8.2. Such autochthonous PIC production is known as lake milking and some lake sediments may contain up to 90% of autochthonous $CaCO_3$ as in Annecy Lake, a subalpine lake in France (Crouzet and Meybeck, 1971) in shallow waters far from detrital inputs.

Waterbodies influence on carbon levels can be contradictory and must be carefully established in all basins.

15.3.3 Particulate Organic Carbon and Riverine Total Suspended Solids (TSS)

At a given station, DOC, POC, and TSS concentrations are increasing with specific discharge q during floods, as for the Solenzara (Figure 15.5). Such common behavior relates to preferential soil leaching and erosion during these events. However TSS variations commonly range by two or three orders of magnitude (Meybeck et al., 2003) while POC ranges by one order of magnitude and DOC by less than one order of magnitude. Since both DOC and TSS increase with q, limited positive correlations between individual DOC and TSS can be expected at most stations. But such relation is not validated on annual averages at the global scale (Figure 15.8). Very high TSS levels, as those found in the Huang He (TSS > 10,000 mg/L), do not influence average DOC levels, which are at 2.25 mg/L in this river, a value also common in the low-turbidity rivers.

The positive correlation between DOC and POC concentrations and the river specific discharge is well illustrated by the Rhône River study made by Sempere et al. (2001), which included a major flood reaching 100 L s^{-1} km^{-2} (Figure 15.9). Since the slope of these relations are different, the TOC-q relation in log-log scale is not linear. Such positive relationship found between POC

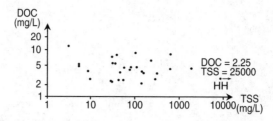

Figure 15.8 Dissolved organic carbon (DOC) variations vs. total suspended solids (annual averages). Note the specific position of the Huang He (HH).

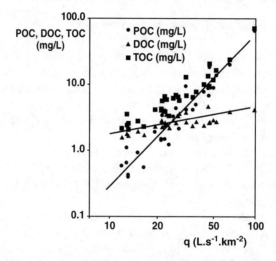

Figure 15.9 Dissolved (DOC), particulate (POC), and total (TOC) organic carbon concentrations on individual samples (mg/L) in the Rhone River at mouth vs. specific discharge q during an exceptional wet year (log scales). (Data from Sempere, R., B. Charriere, F. van Wambeke, and G. Cauwet. 2000. Carbon inputs of the Rhône River to the Mediterranean Sea: Biochemical implications. *Global Biogeochem Cycles* 14:669–681.)

concentrations — in mg C/L — and TSS concentrations — in mg/L — is known since the pioneering study conducted by Malcom and Durum (1976), and has been verified in almost all rivers (Meybeck, 1982; 1993b). The POC-TSS relationship at the station level is such that during floods surficial erosion of organic-rich soil layer — as sheet erosion — is relatively less important than the deeper soil erosion — as rill erosion and river bank erosion both characterized by lower POC contents. Also, at low flows river autochthonous carbon as algal and macrophytes detritus may be important.

In addition to this relation established at given stations on individual samples, a global relation between annual average POC concentration and TSS is also observed. When annual POC and TSS are plotted on a log-log scale, the diagonal lines correspond to equal POC contents lines (Figure 15.10). Meanwhile the POC concentrations (mg/L) are increasing with TSS, the POC contents in % of TSS decrease with increasing average TSS levels from more than 20% for very low TSS (< 10 mg/L) to 1% or less for high TSS (> 1000 mg/L) (Figure 15.11). This relationship, which is more often described by the inverse correlation between POC% and TSS (Figure 15.11 and Figure 15.12) is universally verified at this time (Meybeck, 1982; Ittekot, 1988, Figure 15.8) and has been used for modeling the river organic export at the global scale by Ludwig et al. (1996) (see Section 15.4.2).

As seen previously, DOC export is not related to TSS levels or yields but POC export is directly linked to the sediment yield. As a result the POC/TOC ratio gradually increases from about 10% in the least erosive basins (sediment yields < 20 t km^{-2} yr^{-1}) to more than 90% for the most erosive

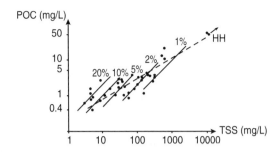

Figure 15.10 Relation between annual average particulate organic carbon concentrations (POC in mg/L) and total suspended solids (TSS in mg/L) in world rivers. Diagonals represent lines of equal POC contents in TSS (POC %TSS). The Huang He (HH) is outside the graph at POC = 0.5% for 25,000 mg/L. (From Meybeck, M. 1993a. C, N, P and S in rivers: From sources to global inputs, in R. Wollast, F. T. Mackenzie, and L. Chou, eds., *Interaction of C, N, P and S Biogeochemical Cycles and Global Change*. Springer-Verlag, pp. 163–193. With permission.)

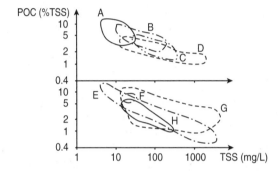

Figure 15.11 Relationship between individual measurements of particulate organic carbon content (POC in %TSS) and total suspended solids (TSS in mg/L): range during one hydrological year. A = Saint Lawrence, B = Niger, C = Mississippi, D = Indus, E = Missouri, F = Zaire, G = Yang Tse, H = Orenoco. (Data from the SCOPE carbon program [Degens et al., 1982] and *Mitt. Geol. Pal. Inst. Univ. Hamburg*, 52, 55, 58, and 64.)

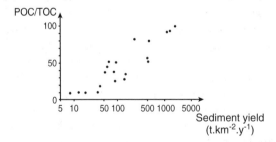

Figure 15.12 Proportions of POC in %TOC vs. river sediment yield in some world rivers. (Data from Meybeck, M. and A. Ragu. 1996. River discharges to the oceans: An assessment of suspended solids, major ions, and nutrients. Environment Information and Assessment Rpt. UNEP, Nairobi, 250 pp.)

basins (yields > 1000 t km^{-2} yr^{-1}) (Figure 15.12). At the global scale the decrease of annual average POC% levels with increasing annual TSS average is complex: the least turbid rivers are generally found in low relief basins in which soil derived particles are dominating; the most turbid rivers are linked to steep basins and semiarid basins with limited vegetal cover, or to very erodible surficial rocks as loess and shales. In all cases the erosion of organic-rich soil layer is minor compared to stream incision. Climate and vegetation are probable second order controls in such relation: for a similar average level of TSS, the annual POC% may vary over a factor two and more between rivers (Figure 15.11). Such influence remains to be studied.

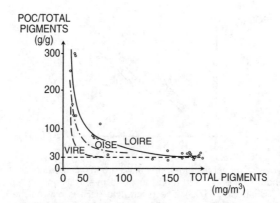

Figure 15.13 Relation between the POC/total pigments ratio (g/g) vs. total pigments in the Loire River compared to two other eutrophied French rivers: Oise (Seine tributary) and Vire (Normandy). (Data from Dessery, S., C. Dulac, J. M. Laurenceau, and M. Meybeck. 1984. *Archiv. für Hydrobiol.* 100, 2:235–260; Meybeck, M., G. Cauwet, S. Dessery, M. Somville, D. Gouleau, and G. Billen. 1988. *Estuarine Coastal Shelf Sci.* 27:595–624.)

15.3.4 Autochthonous Fluvial Carbon

In eutrophic rivers where summer chlorophyll A exceeds 100 µg/L, as the Rhine, Seine, and Loire in Western Europe, the ratio POC/total pigments (in g/g) observed in rivers reaches a limit around 30 ± 5 during algal peaks (Figure 15.13). Such ratio is similar to the one describing algal blooms in lakes and was described by Dessery et al. (1984) on the Oise, a tributary of the Seine, then confirmed in the Seine and Loire basins (Meybeck et al., 1988). This ratio is commonly used to express the algal POC from total pigments levels (as the chlorophyll A + phaeopigment measurements by the Lorenzen method). Most algal POC is a highly labile POC species (Servais et al., 1998) easily degradable in few days; consequently, the production/respiration ratio is well below one, as for instance in turbid estuaries receiving eutrophic river waters (Garnier et al., 1998).

The eutrophic lower Loire River (France), which drains carbonated rocks and has high DIC levels is a good illustration of seasonal variations of autochthonous POC and PIC (Figure 15.14). In this basin the winter algal POC is around 0.8 mg C/L — a value attributed to the erosion of benthic diatoms and macrophytes during the high flow period — and the nonalgal POC (detrital plus anthropogenic) is much higher (2.5 mg C/L). In summer algal POC reaches 5 mg C/L and dominates nonalgal POC (0.5 mg C/L), while the autochthonous PIC reaches 3.6 mg C/L, i.e., precipitated calcite ($CaCO_3$) reaches 30 mg/L (Meybeck et al., 1988). The sum of autochthonous species (autochthonous PIC and algal POC) dominates the sum of detrital and soil-derived species (detrital PIC + detrital POC+ DOC) in summer, however the corresponding fluxes are limited. Summer algal POC is totally degraded in the turbid Loire estuary resulting in a severe dissolved oxygen depletion (Meybeck et al., 1988).

15.3.5 Anthropogenic Impacts on Fluvial Carbon and Long-Term Trends

Each carbon species has its own sensitivity to climate change and to direct human pressures (Meybeck and Vörösmarty, 1999; Table 15.1). Unlike for Cl^- and SO_4^{2-}, the HCO_3^- long-term trends in rivers (> 20 years) under heavy human pressures are relatively stable as observed in the Rhine River (Kempe 1982), in the Saint Lawrence River (Tremblay and Cossa, 1987), or Kura River in Caucasus (Nikanorov and Tsirkunov 1984; Figure 15.15): DIC is not very sensitive to anthropic influence. In the southeastern Baltic drainage basin the bicarbonate concentration has increased from 1946 to 1978 by 5 to 7% for the Neva, 6 to 8% for the Neman, and 7 to 4% for the Western Dvina Rivers (Stakalsky, 1984).

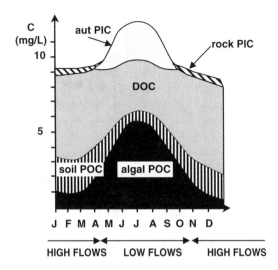

Figure 15.14 Seasonal evolution of minor carbon species in the eutrophied Loire River at mouth. Aut. PIC = summer precipitated calcite, rock PIC = detrital carbonate minerals. (Data from Meybeck, M., G. Cauwet, S. Dessery, M. Somville, D. Gouleau, and G. Billen. 1988. *Estuarine Coastal Shelf Sci.* 27:595–624.) Autochthonous carbon species are dominating in summer low flows (autochthonous PIC and algal POC). (Corrected from Meybeck et al., 1988 and Meybeck, 1993b.) (The dominant C species, DIC around 24 mg/l, is omitted here.)

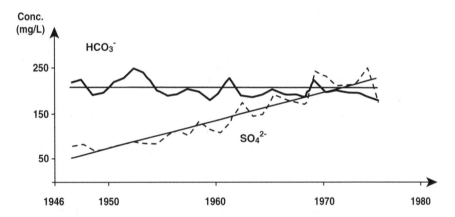

Figure 15.15 Bicarbonate sand sulfate trends in the Kura (Caucasus, 188,000 km²). (From Nikanorov and Tsirkunov, 1984.)

The most complete analysis of DIC trends has been reported by Raymond and Cole (2003) for the alkalinity records in the Mississippi between 1953 and 2001 at St. Francis Ville. They reported a slight increase (23 to 27 mg/L) while the river discharge increased from 450 to 650 km³/y. As a result the DIC flux increased by 45% in half a century. This trend is attributed to increase in rainfall in the basin.

The input of organic matter from domestic and industrial wastewater to rivers is generally known as "organic pollution" but the term carbonaceous pollution is preferred to avoid confusion with organic toxic substances. Such impact has been monitored and modeled through BOD_5 and COD since the 1940s, and is now gradually replaced by direct TOC measurements.

The TOC or DOC have been rarely surveyed for more than 20 years and their trends are still poorly documented. (The longest record of DOC in rivers is reported for peat land headwaters in U.K. since 1962 and is inferred from water color measurement [Worrall et al., 2003], and is discussed in Section 15.5.2.) It cannot be used to assess the global sensitivity of organic carbon to human pressures but the impact of a megacity on the river organic carbon can be considered.

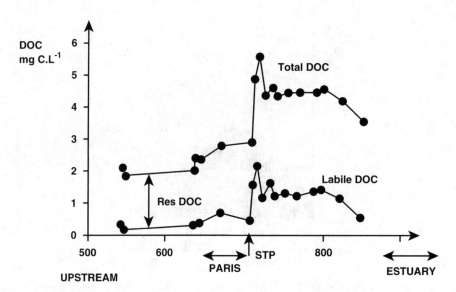

Figure 15.16 Longitudinal profiles of (km) total dissolved organic carbon, easily degradable (lab. DOC) and resistant (res. DOC) in the Seine River across Paris megacity. Most treated sewage (8 M equivalent people) is injected at KP 700 (STP). (From Servais, P., Z. Idlafkih, G. Billen, J. Garnier, J. M. Mouchel, M. Seidl, and M. Meybeck. 1998. Le carbone organique, in M. Meybeck, G. de Marsily, and E. Fustec, eds., *La Seine en son Bassin*, Elsevier, Paris, pp. 483–529. With permission.)

The impact of Paris (10 million population) on the Seine River is a good example of carbonaceous pollution. The DOC increases markedly downstream because of the release of treated domestic sewage from the gigantic Seine-Aval plant (8 million equivalent people) despite a satisfactory efficiency of this plant (80%) for TOC removal (Servais et al., 1998). Then most of the excess DOC is degraded within one week before reaching the estuary (Figure 15.16).

In the more industrialized countries of Western Europe and North America, industrial TOC sources were common until the 1970s in some industrial sectors as pulp and paper, agro-industries, and others. These industrial wastes generally decreased in these regions by at least an order of magnitude between 1960 and 1980.

The equilibrium of inorganic carbon species ($CO_2 - HCO_3^- - CO_3^{2-}$) in rivers is complex and depends on alkalinity, on the algal production vs. bacterial respiration ratio (P/R), and therefore, on eutrophication, and on human sources of labile TOC. Even in well-buffered aquatic systems, such as the Loire River, pronounced nychthemeral cycles of dissolved oxygen and pH are observed (Moatar et al., 1999), with pH values exceeding 9.2 and related calcite precipitation. In the Rhine River, the long-term averages of DIC species, including CO_2, are also regulated by P/R ratio (Kempe, 1982).

The control of carbonaceous pollution in the more industrialized countries from both domestic and industrial sources is one of the major success stories of water quality control and restoration such as in the Rhine, Thames, and Seine rivers. Due to imbalance between waste water treatment and waste collection in megacities, other parts of the world, e.g., India, Brazil, and China, are likely to face a rapid increase in degradable DOC and POC in some river basins (Meybeck, 2002; 2003a).

15.3.6 The Fossil POC Controversy

The occurrence of fossil POC in riverine carbon is debated (Ludwig, 2001). During flood events in turbid rivers when TSS exceeds 1,000 mg/L, POC% contents are generally below 1%. In the most turbid rivers (TSS > 5,000 mg/L) eroding loess (Huang He) or shales (Taïwan and Tunisia

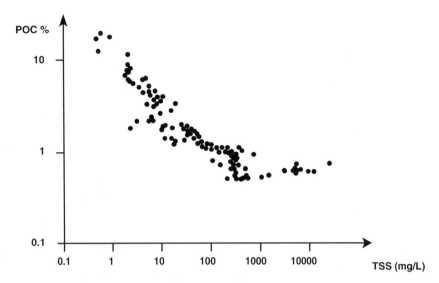

Figure 15.17 Relation between POC contents in suspended matter (%TSS) and TSS in the Lanyang-Hsi River, Taiwan. Note the stabilized POC% above 1000 mg TSS/L. (Data from Kao, S. J. and K. K. Liu. 1997. Fluxes of dissolved and nonfossil organic particulate organic carbon from a small Oceania river (Lanyang Hsi) in Taïwan. *Biochemistry* 39:255, 269.)

rivers), the POC% content reaches 0.5%, a level very close to those found in these types of sedimentary rocks. Meybeck (1993a, 1993b) has interpreted this limit as the transfer by such rivers of a geologic-old POC, as kerogens, without major degradation.

The POC% vs. TSS evolution in the River Lanyang-Hsi in Taïwan (Kao and Liu, 1997) is very illustrative (Figure 15.17). The TSS levels vary over four orders of magnitude in this coastal river basin exposed to heavy rains, rapid land-use change, and active tectonics. During high floods, TSS exceeds 1,000 mg/L and may reach 30,000 mg/L during typhoons. From 1,000 to 30,000 mg TSS/L, the POC% is constant at 0.6 ± 0.2%. The ^{14}C analyses performed by Kao and Liu (1997) have confirmed the fossil nature of such particles which constitute 70% of the annual load for this river, characterized by one of the highest yields of eroded matter (> 1,000 t km^{-2} yr^{-1}). A similar dominant source of geologic POC has also been demonstrated by Huon et al. (this volume) in the headwaters of the Apure, a major Andean tributary of the Orenoco, in another highly eroded basin.

In France, a similar recycling of fossil POC of Toarcian age has also been demonstrated in some small basins (< 100 km^2) of the Terres Noires badlands in the Durance Valley, a lower Rhône tributary, where sediment yields also exceed 1,000 to 5,000 t km^{-2} yr^{-1} (Di Giovanni et al., 2002). This POC input is probably an important source of POC for the whole Rhône river downstream of its confluence with the Durance. The POC (%TSS) vs. TSS relationship presents a clockwise cycle (Figure 15.18) due to two different flood types, the pluvial-oceanic and pluvio-nival floods in winter and spring and the Durance flood in late summer-autumn of this year, characterized by much lower POC contents — as low as 0.8% as can be observed from the 1994 and 1995 study of Sempere et al. (2000).

A reassessment of the POC budget for the Rhone during this exceptional flow year (1994 to 1995) is proposed here from the data of Sempere et al. (2000) assuming that most of the POC carried at very low contents (POC% 1.5) is originating from the Toarcian Marnes Noires erosion (Table 15.3). For this river our estimate of the fossil POC load is 6% compared to 94% for soil POC and the total fossil carbon as PIC, POC, and DIC is estimated to be 44% of the total exported carbon. In this basin the chlorophyl A is quite low due to the high water velocity and turbidity, therefore the algal POC can be neglected.

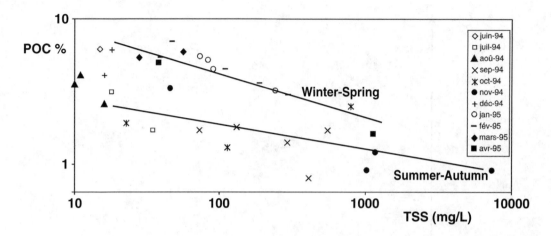

Figure 15.18 Relation between POC contents in suspended matter (%TSS) and TSS in the Rhone River at mouth in 1994 to 1995, an exceptional high flow year. Winter–spring: alpine tributaries dominance; summer–autumn: Mediterranean floods dominance (log scales). (Data from Sempere, R., B. Charriere, F. van Wambeke, and G. Cauwet. 2000. Carbon inputs of the Rhône River to the Mediterranean Sea: Biochemical implications. *Global Biogeochem Cycles* 14:669–681.)

Table 15.3 Example of Decomposition of River Carbon Sources for Carbon Main Species: the Carbonated Rhone River Basin for an Exceptional Flood Year (1994 to 1995)[a]

	Inorganic Carbon			Organic Carbon			Total Carbon	
	DIC	PIC	TIC	DOC	POC	TOC	(mg/L)	(%)
Nonfossil carbon (mg/L)	19.8[b]	—	19.8	2.4	3.4[b]	5.8	25.6 (TAC)	56
Fossil carbon (mg/L)	16.2[b]	4	20.2	—	0.22[b]	0.22	20.4 (TFC)	44
Total (mg/L)	36	4	40	2.4	3.6	6.0	46	
Total (%)	78.3	8.7	87	5.2	7.8	13		100

Note: TAC, total atmospheric carbon; TFC, total fossil carbon.

[a] Reestimated based on data from Sempere, R., B. Charriere, F. van Wambeke, and G. Cauwet. 2000. Carbon inputs of the Rhône River to the Mediterranean Sea: Biochemical implications. *Global Biogeochem Cycles* 14:669–681.
[b] Not separated by authors.

Meybeck (1993) proposed that up to 30% of global riverine POC could be of geologic origin due to the influence of the very highly turbid rivers on the balance of solid transport to the oceans. This estimate was strongly questioned by Ludwig (2001). A revised estimate lowered it to 20 to 25% (see Section 15.4.2). This question is important for the global carbon cycle and will not be resolved until a satisfactory set of ^{14}C ages is available on river particulates from turbid rivers (TSS > 1000 mg/L).

15.4 GLOBAL CONTROL AND BUDGETS OF RIVERINE CARBON EXPORT

15.4.1 Variability of Carbon Yields

The exportation of carbon, or yields (Y_C), is generally expressed in g C m^{-2} yr^{-1}, equivalent to t km^{-2} yr^{-1} a unit more commonly used in river transport, which facilitates the comparison with primary production, either terrestrial or aquatic. The Y_{TOC} is usually two orders of magnitude lower than terrestrial production, i.e., the export of organic carbon is generally between 0.5 and 2% of the production (Meybeck, 1982).

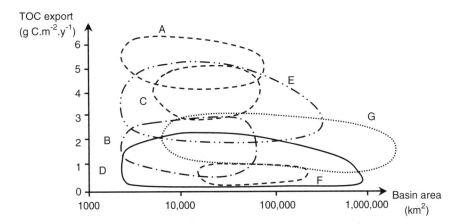

Figure 15.19 Distribution of TOC exportation rates (g C m^{-2} yr^{-1}) vs. river basin area in six North American regions. A = New Brunswick to Maryland (n = 13), B = Maryland to Georgia (n = 13), C = Florida to Louisiana (n = 8), D = Texas and California to Oregon (n = 14), E = Washington and Alaska (n = 7), F = NW Territories — Nelson Basin (n = 4). (Data from Mulholland, P. J. and J. A. Watts. 1982. Transport of organic carbon to the oceans by rivers of North America: A synthesis of existing data. *Tellus* 34:176–186.)

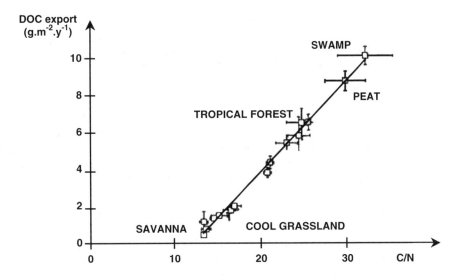

Figure 15.20 DOC export vs. average C/N ratio (g.g^{-1}) in contrasted river basins. (From Aitkenhead, J. A. and W. H. Mac Dowell. 2000. Soil C:N ratio as a predictor of annual riverine DOC flux at local and global scales. *Global Biogeochemical Cycles*. 14:127–138. With permission.)

For a given region the annual DOC yield is independent from the river basin size as shown for the U.S. by Mulholland and Watts (1982) (Figure 15.19). Most of the exported soil DOC is very resistant and is not further degraded during its routing within the river network, typically some weeks to months for the longest rivers.

The DOC yield is directly linked to the C/N ratio in soils of river basins according to Aitkenhead and MacDowell (2000) (Figure 15.20). The highest yields are observed for C/N ≈ 20, typical of wetlands, which confirm the importance of such water bodies in carbon transfers.

The TOC annual yields are increasing faster than water runoff (Meybeck, 1993a) (Figure 15.21), i.e., TOC average concentration, represented as a diagonal in a log-log diagram, increases from about 5 mg/L for dry and warm regions (q < 100 mm/yr), to 15 mg/L for the wet regions (q =

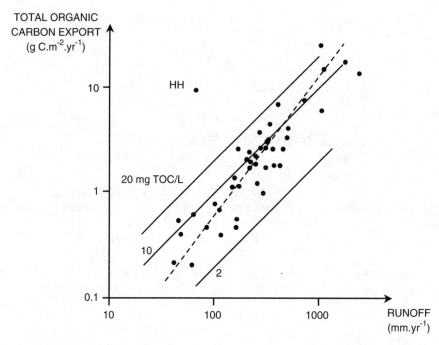

Figure 15.21 Average yields of TOC (g C m^{-2} yr^{-1}) in selected world rivers as a function of river runoff; diagonal lines: equal TOC concentration. Note the exceptional situation of the Huang He River (HH). (Data from Meybeck, M. and A. Ragu. 1996. River discharges to the oceans: An assessment of suspended solids, major ions, and nutrients. Environment Information and Assessment Rpt. UNEP, Nairobi, 250 pp.)

1,000 mm/yr). Such trend could be explained by the growing terrestrial production as q increases and/or rapid processing of organic matter in dry regions.

The Huang He is once more a noted exception in this global scale analysis. Although it is characterized by the world's lowest POC (%TSS) (0.5%), the very high transport of suspended particles from loess erosion results in a world's maximum TOC yield mostly of fossil origin (age > 10,000 years).

The DIC export is dependant on: (1) basin lithology and (2) river runoff (Meybeck, 1986; Amiotte-Suchet et al., 2003). Carbonated basins have the highest DIC levels, about an order of magnitude higher than noncarbonated crystalline basins (e.g., volcanic rocks, gneiss, granite, most metamorphic rocks) and DIC export is generally proportional to river runoff with near constant DIC level for a given rock type. Figure 15.22 illustrates the enormous range of DIC concentration, DIC proportion in anionic sum, and DIC yields or export rates. Bicarbonate is the dominating anion for more than 80% of the world's river waters including those in endorheic regions (Meybeck, 2003).

DOC and POC export rates (Y_{DOC}, Y_{POC}) can be compared on a log-log scale that allows for a better visualization of their relative variation over two orders of magnitude (Figure 15.23). The Y_{DOC} vs. Y_{POC} correlation is weak at the global scale and the Y_{DOC}/Y_{POC} ratio ranges from 0.2 to 10, i.e., each of the forms of organic carbon can be dominant in river systems. The transport is preferentially as DOC in low relief basins with very low solid load and many wetlands as in the Rio Negro (Brazil) and many African rivers (Seyler et al., this volume). POC may dominate in humid high relief regions as in the upper Amazon and in the Asian Monsoon basins. The TOC yields range from 0.25 g C m^{-2} yr^{-1} in lowland dry regions to 16 g C m^{-2} yr^{-1} in humid high relief regions. The exception of the Huang He is once more noted.

The DIC and DOC export rates in our set of documented rivers at the global scale are not correlated as Y_{DIC} is also highly dependant on runoff and on lithology.

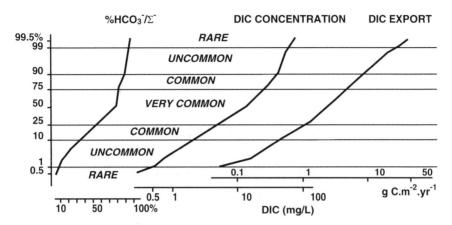

Figure 15.22 Global distribution of DIC concentrations, ionic ratios (HCO$_3^-$/total anions), and DIC yields in medium-size pristine basins (vertical scales: probability scale). (From Meybeck, M. 2003b. Global occurrence of major elements in rivers, in K. K. Turekian and H. D. Holland, eds., *Treatise of Geochemistry*, vol. 5. J. I. Drever, Elsevier, pp. 207–224. With permission.)

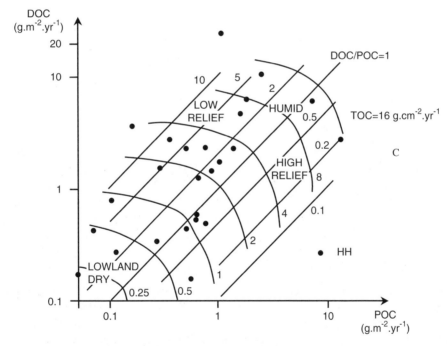

Figure 15.23 Compared exportation rates of DOC vs. POC (g C m^{-2} yr^{-1}) in world rivers. Note the exceptional situation of the Huang He (HH). Diagonals represent lines of equal DOC/POC export; curves represent equal TOC = DOC + POC export. (Data from Meybeck, M. and A. Ragu. 1996. River discharges to the oceans: An assessment of suspended solids, major ions, and nutrients. Environment Information and Assessment Rpt. UNEP, Nairobi, 250 pp.)

When all forms of carbon are adequately measured their exportation (Y_{TC}) can be compared. Figure 15.24a compares these yields in an increasing order of Y_{TC} from the Gambia River in West Africa with Y_{TC} = 0.9 g m^{-2} yr^{-1} to the Dranse River in the French Prealps Y_{TC} = 70 g m^{-2} yr^{-1}. As the carbon export is increasing the proportion of fossil carbon, i.e., "rock DIC" from carbonate rock weathering, "rock PIC" from carbonate rock erosion, "rock POC" from sedimentary rock, increases while the proportion of total atmospheric carbon Y_{TAC} (= Y_{DOC} + $Y_{SOILDIC}$ + $Y_{SOILPOC}$) decreases.

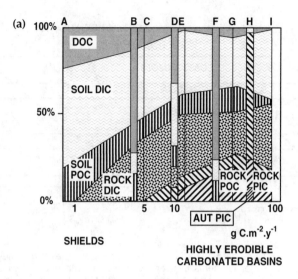

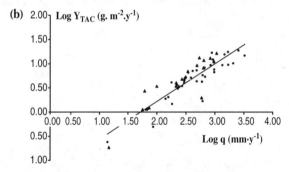

Figure 15.24 Exportation of total carbon by rivers. (a) Decomposition of carbon species exportation in a gradient of total carbon (TC) exportation gCm^{-2}y^{-1}. A = Gambia, B = Zaïre, C = Mississippi, D = Solimoes, E = Chiang Jiang, F = Negro, G = Brahmaputra, H = Huang He, I = Dranse. (Data from Meybeck, M. 1993a. C, N, P, and S in rivers: From sources to global inputs, in R. Wollast, F. T. Mackenzie, and L. Chou, eds., *Interaction of C, N, P and S Biogeochemical Cycles and Global Change*. Springer-Verlag, pp. 163–193.) (b) Yield of riverine total atmospheric carbon (g C m^{-2} yr^{-1}) vs. runoff (mm yr^{-1}) for major rivers. (Data from Meybeck and Ragu 1996, triangles: highly carbonated basins, log-log scale.)

The export of total atmospheric carbon (TAC) can be expressed by its yield per unit basin area (Y_{TAC} in g m^{-2} yr^{-1}). The TAC = atm. DIC + soil POC + POC has been computed on the basis of a first data set from major rivers Meybeck and Ragu (1996), and estimates of the proportions of atmospheric DIC/Total DIC, and of soil POC/Total POC for each basin have been computed. The Y_{TAC} ranges from 0.24 g C m^{-2} yr^{-1} to 18.8 g C m^{-2} yr^{-1} and is directly related to river runoff, which appears to be its essential control factor (Figure 15.24b). In carbonated basins, TAC is mostly exported as DIC while in shields and noncarbonated sedimentary basins, TAC is mostly exported as organic carbon either dissolved, particulate, or a combination of both. The maximum Y_{TAC} could exceed 20 g m^{-2} yr^{-1} on smaller basins, either in humid tropical regions (TOC dominance), or in limestone basins exposed to very heavy rains. In very few regions of the world where these factors are combined as in Borneo or in New Guinea, the Y_{TAC} may exceed 50 g m^{-2} yr^{-1}. If such levels are confirmed, these regions may be regarded as hot spots of carbon transfers with yields reaching more than 10 times the world average Y_{TAC} estimated to 6.0 g m^{-2} yr^{-1} (Meybeck, 1993a; 1993b). Such heterogeneity in the river yields of atmospheric carbon was noted and mapped for the first time by J. L. Probst and his colleague at the 2-degree resolution (Amiotte-Suchet et al., 2003; Ludwig et al., 1996).

15.4.2 Global Budget of River Atmospheric Carbon Carried by Rivers

The global budgets of river organic and inorganic carbon were first addressed by Schlesinger and Melack (1980) and Meybeck (1982, 1993a, 1993b), using simple extrapolations for major biomes. In this approach, the retention and processing of carbon through the river basins by multiple fluvial filters, as deposition of coarse particulates in slopes and piedmonts, lake and reservoir carbon processing and retention, and floodplain and estuary retention and processing are not explicitly addressed.

The global budget of total atmospheric carbon, presented in Table 15.4, still corresponds to this generation of global models (Meybeck, 1993a) based on climatic or morphoclimatic typologies of river basins. When these riverine fluxes are reaggregated into major biomes, following an approach initiated by Schlesinger and Melack (1980), regional discrepancies are very much apparent for DOC and soil POC but not for atmospheric DIC (e.g., cold regions and humid tropics have identical yields) which depends on runoff and lithology, among others.

The second generation of models and budgets concerns estimates of average yields in individual grid cells based on multiple regressions analysis using river flux data from whole documented basins of various sizes in which runoff, temperature, lithology, and soil organic carbon are applied

Table 15.4 Global Flux of River Total Atmospheric Carbon to Oceans (TAC = DOC + atm DIC + soil POC) by Biome

	DOC	atm DIC	atm DC	soil POC	TAC	A (Mkm^2)	Q (km^3 yr^{-1})	q (m yr^{-1})
Total Input to Oceans								
C_{exp} (g m^{-2} yr^{-1})	1.99	2.44	4.43	0.99	5.42	99.9	37,400	0.375
Fc_{atm} (Tg yr^{-1})	199	244	443	99	542			
% carbon species (1)	37	45	82	18	100			
Total Cold Regions								
C_{exp} (g m^{-2} yr^{-1})	1.31	2.5	3.8	0.42	4.25	23.35	5,500	0.235
Fc_{atm} (Tg yr^{-1})	30.5	59.2	89.7	9.9	99.6			
% carbon species (1)	30.6	59.4	90.0	10.0	100			
Total Temperate Regions								
C_{exp} (g m^{-2} yr^{-1})	1.5	4.5	6.0	1.5	7.5	22.0	10,250	0.465
Fc_{atm} (Tg yr^{-1})	32.2	100	132.2	33.7	165.9			
% carbon species (1)	19.4	60.3	79.7	20.3	100			
Total Arid + Savanna Regions								
C_{exp} (g m^{-2} yr^{-1})	0.15	0.6	0.75	0.23	0.98	28.7	2,430	0.085
Fc_{atm} (Tg yr^{-1})	4.45	17.1	21.5	6.75	28.3			
% carbon species (1)	15.7	60.4	76.1	23.9	100			
Total Humid Tropics Regions								
C_{exp} (g m^{-2} yr^{-1})	5.1	2.6	7.7	1.9	9.6	25.8	19,210	0.745
Fc_{atm} (Tg yr^{-1})	131.5	67.6	199.1	48.9	248			
% carbon species (1)	53	27.3	80.3	19.7	100			

Note: Fc_{atm} = Total flux of riverine carbon carried to ocean; A = Exorheic drainage area of climatic zone; Q = Total river discharge; q = Average river runoff; DOC = Dissolved organic carbon; DC = DOC + DIC; POC = Particulate organic carbon; (1) Percentage of carbon species in total atmospheric carbon budget.

From Meybeck, M. 1993b. Riverine transport of atmospheric carbon: sources, global typology and budget. *Water, Air Soil Pollution* 70:443–464. With permission.

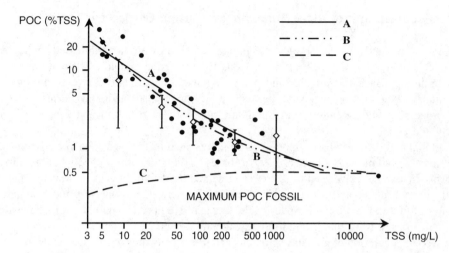

Figure 15.25 Global range of POC (%TSS) and TSS annual averages for major rivers. (For detailed data, see Meybeck, M. and A. Ragu. 1996. River discharges to the oceans: An assessment of suspended solids, major ions, and nutrients. Environment Information and Assessment Rpt. UNEP, Nairobi, 250 pp.). Vertical bars: distribution for given TSS classes from Ittekot (1988). (a) Initial global regressions of Meybeck (1982). (b) Corrected regression of Meybeck (1993b). (c) Maximum fossil POC hypothesis of Meybeck (1993b).

as factors (Bluth and Kump, 1994; Amiotte-Suchet and Probst, 1995; Ludwig and Probst, 1998; Ludwig et al., 1996; Amiotte-Suchet et al., 2003). This generation of budgets allows for the mapping of carbon yields at 1 to 2° resolution, illustrating the striking ranges of yields exceeding two orders of magnitude at the global scale.

A third generation of models and budgets includes some material processing on each cell, then its routing from one cell to the other along the river network. As such the major fluvial filters (Meybeck and Vörösmarty, 2004) can be taken into account as the floodplain processing and the reservoir retention. A first application of this approach has been realized for the total nitrogen budget (Green et al., 2004), yet not for riverine carbon.

The POC-TSS relationship (Figure 15.25) can also be used to partition the world budget of river POC fluxes to ocean into classes of suspended matter (Meybeck, 1982). In the very turbid rivers, the POC is 0.5% of TSS, a value also noted in the parent rock, for instance loess in the Huang He basin. It is postulated that, in world rivers, above 1500 mg/L of TSS, 75% of POC is fossil, and that this proportion reaches 100% for 15,000 mg/L (i.e., the TSS levels of the Huang He), and that fossil POC reaches 10% of total POC from 150 to 500 mg TSS/L. This relationship is combined with a partition of the global TSS budget in TSS classes based on dozens of documented rivers (Meybeck, 1982, revised). The TSS load 15.9×10^9 g/yr estimated as such is on the lower range of most recent estimates, which are around $20 \pm 5 \times 10^9$ g/yr. The TSS budgets and the related POC budgets have been estimated in each class of average yearly TSS (Table 15.5): below 150 mg TSS/L the total TSS load is estimated to be only 7.3% of the global fluxes and the corresponding POC flux to be 26% of the global figure and 100% soil POC; above 1500 mg TSS/L the corresponding TSS load is 34.5 % of the global figure and the POC load is 15.5%, essentially of fossil origin. According to this hypothesis the fossil POC load is around 50×10^{12} g/yr, i.e., 25% of the total POC load to the coastal zone, however this point is still debated (see Section 15.3.6). For very low TSS levels (< 20 mg/L) a similar partition can be made for autochthonous algal carbon. However, the related fluxes can be considered as negligible at the global scale.

Other approaches have been tested: Aitkenhead and MacDowell (2000) used the soil C/N ratio to estimate DOC yields at the global scale. The regional to global sequestration of organic carbon on basins, has been recently addressed by Stallard (1998), Smith et al. (2001), and Einsele et al. (2001).

Table 15.5 Global Model of POC and Fossil POC Budgets by TSS Classes

TSS Class (mg/L)	< 15	15–50	50–150	150–500	500–1500	1500–5000	5000–15,000	> 15,000[a]	Global Mean
TSS load (% of global)	0.0	1.3	6.0	14.2	44	12.6	1.3	20.6	15.9 10^{15} g/yr
Water volume (% of global)	3.8	18.2	32	20.2	23.4	10.8	0.07	0.45	37 400 km³/yr
POC content (% of TSS)	15	6	3.4	1.8	0.9	0.6	0.5	0.45	1.12%
% fossil POC in TSS	0	0	0	10	30	75	90	100	
POC load (% of global)	0.9	6.8	18.3	22.9	35.5	6.7	0.55	8.2	178 10^{12} g/yr
Fossil POC load (% of global POC)	0	0	0	2.3	10.6	5.0	0.5	8.2	47.3 10^{12} g/yr

[a] Huang He only (1980 to 1990 flux).
Adapted from Meybeck, M. 1982. Carbon, nitrogen, and phosphorus transport by world rivers. *American J. Science* 282:401–450.

15.5 PERSPECTIVES AND PROPOSITIONS

15.5.1 Improving Our Knowledge of Holocene River Carbon (Pristine Conditions)

Major improvements in our knowledge of riverine carbon origins and controls have been made during 1990s through: (1) use of global GIS information for flux spatialization (Amiotte-Suchet et al., 2003; Ludwig et al., 1996; Probst et al., 1994), (2) DOC and POC dating (Raymond and Bauer, 2001), (3) use of organic matter stable isotopes to differentiate sources, particularly fossil carbon (see Huon, this volume), (4) new relationship between DOC export and C/N ratio in soils (Aitkenhead and MacDowell, 2000), and (5) direct experiments on CO_2 doubling over DIC export from forest stands (Andrews and Schlesinger, 2001), etc. These developments confirm the different turnover rates of river carbon: 0 to 100 years for atmospheric-derived DIC, more than 100 years for soil DOC in most cases, more than 1000 years for soil POC in many cases, and 10^4 to 10^8 years for fossil carbon as PIC and POC in sedimentary rocks.

These POC sources are schematically presented as a working hypothesis in Figure 15.26 in a gradient of river total suspended matter. This relation can be observed throughout the globe on the basis of yearly averages. In the absence of organic soil cover the POC% should remain stable and reflect the average POC found in the sedimentary rocks of the basin. This origin dominates > 1000 mg TSS/L. From 10 to 1000 mg TSS/L soil POC is added and dominating. Below 10 mg TSS/L the autochthonous POC (algae and macrophytes, debris) gradually replaces soil POC and may dominate < 5 mg/L, which is mostly observed on lake outlets (Meybeck et al., 2003).

Depending on local conditions, particularly the soil and vegetation cover, the points of equal POC species proportions (A1 and A2, B1 and B2) may be observed at various TSS levels.

Another critical question is the proportions of DIC from carbonate minerals weathering found in rivers. The consideration of pristine monolithologic stream basins (n = 250) in France (Meybeck, 1986) and of a set of 1200 pristine rivers (Meybeck, 2003b) allows one to propose a general relationship between the % of atmospheric DIC vs. total river DIC in river basins worldwide (Figure 15.27): (1) there are few examples of very low DIC (< 5 mg/L) in river basins with few limestone outcrops or some trace carbonate minerals (e.g., in shales), (2) in some volcanic terrains the DIC can reach 10 mg/L, entirely originating from atmospheric CO_2, (3) pure carbonated rocks are rarely characterized by river DIC < 25 mg/L in any climate and vegetation condition, and (4) DIC levels > 45 mg/L are observed in < 1% of river water, generally influenced by hydrothermalism and evaporation. For rivers with < 5 mg DIC/L (25 mg HCO_3^-/L or 400 µeq HCO_3^-/L), the DIC may be considered to originate exclusively from the atmosphere.

Yet our knowledge of the river carbon in pristine conditions can be greatly improved if all carbon species (see Table 15.1) were quantified in a data set covering all climate, relief, vegetation, and lithology conditions: since the first papers on this topic (Schlessinger and Melack, 1980;

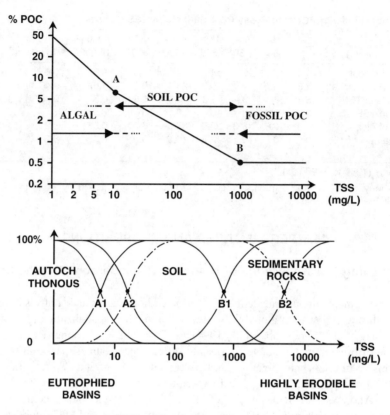

Figure 15.26 Working hypothesis on POC origins in rivers inferred from the POC content (%TSS) vs. total suspended solids relationship. A1 = noneutrophied rivers and lake oulets, A2 = eutrophied rivers (e.g., Loire), B1 = maximum fossil POC hypothesis (e.g., Huang He), B2 = minimum POC hypothesis.

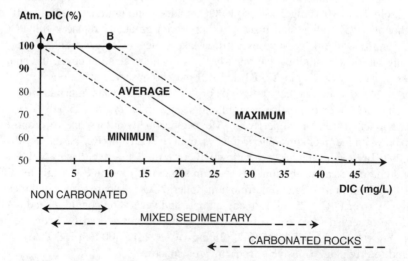

Figure 15.27 Model of DIC origins in river basins with variable proportions of carbonated rocks (% atmospheric DIC vs. total DIC as a function of average DIC level). Range of noncarbonated basins: A = Caroni, B = Reunion and Blue Nile.

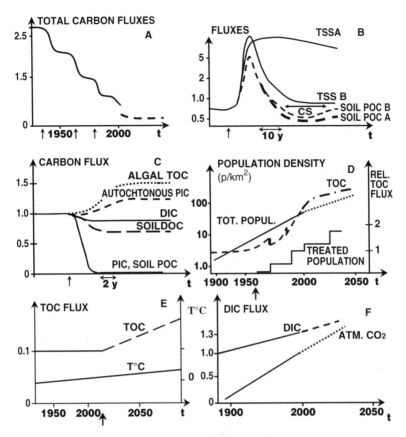

Figure 15.28 Working hypotheses for the contrasted future of carbon fluxes at the Anthropocene. (All fluxes are normalized to pristine average conditions.) (a) Stepwise reduction of fluxes in neoarheic rivers with water consumption periods. (b) Stepwise evolution of TSS and soil POC, without proper soil conservation practices (1) and with soil recovery (2); (CS = soil organic carbon sequestration). (c) Contrasted trends after damming (settling vs. autochthonous carbon production). (d) Nonlinear response of river TOC flux downstream of fast-growing megacity with marked time-lag between population growth and urban sewage collection. (e) Threshold TOC increase due to permafrost melt in arctic basins. (f) Gradual DIC increase in forested area due to CO_2 elevation. (Arrows indicate start of pressure.)

Meybeck, 1981), not much progress has been made in this direction compared to other major fields of river geochemistry.

15.5.2 Future Evolution of Riverine Carbon at the Anthropocene

The anthropogenic forcing of the earth surface system (e.g., climate, land cover) is now equivalent and even greater than the natural one. The term Anthropocene was used by the Russian scientist Vernadskii in the 1920s to qualify this type of evolution and has been recently reused by Crutzen (Crutzen, 2002; Crutzen and Störmer, 2000). The Anthropocene concept can perfectly be applied to river systems and a conventional date at 1950 has been proposed as a starting point (Meybeck, 2002; 2003a; Meybeck and Vörösmarty, 2004). Several hypotheses are presented herein on the future evolution of river carbon transport (Figure 15.28): which show very different patterns including stepwise increase or decrease, nonlinear responses, and thresholds typical of earth system functioning (Steffen et al. 2003).

River systems evolution can be described in a set of major syndromes as flow regulation, river course segmentation, neoarheism, river silting, eutrophication, chemical contamination, acidification, salinization, and others (Meybeck, 2003a). One of the most spectacular evolutions is the gradual stepwise decrease in river flow due to irrigation in dry and semiarid regions as has been observed for the Colorado, Nile, Amu Darya, and, to a lesser extent, for the Orange, Murray, Indus or Huang He rivers. These river systems are no longer connected to the oceans or to internal seas (neo-arheism). Neoarheism lead to the reduction of all river carbon fluxes, whether dissolved or particulate (Figure 15.28a).

Land cover change due to human pressures has numerous impacts on river carbon exports. It is hypothesized (Figure 15.28b) that forest clearing and particularly clearcutting dramatically increases soil erosion in a first phase thus exporting more soil POC. If the soil layer is completely lost, as in semiarid regions (case a) the parent rock may be exposed. If the exposed rock is erodible, a new long-term equilibrium is attained with high erosion rates but with much lower soil POC export. In temperate regions (case b), a gradual recovery after forest cutting is likely: a phase of lower soil POC export by rivers during which organic carbon is gradually accumulating in the soil (carbon sequestration, CS) and a new equilibrium is reached. Such patterns are difficult to quantify on the basis of direct river measurements since they occur over long periods (decades) and other anthropogenic perturbations can be involved.

The impact of river damming is presented in Figure 15.28c: the construction of large reservoirs also leads to the reduction of TSS fluxes, hence of soil POC fluxes, reaching 90% for reservoirs exceeding 6 months of water residence time. At the global scale, reservoir trapping can exceed 30% of the large river sediment fluxes (Vörösmarty et al., 2003). In large reservoirs, soil DOC can be degraded by UV, but there is also a production of autochthonous TOC from algal development and sometimes a precipitation of calcite, which is partially exported in the river outlet during whitening events (Figure 15.28c). Most detrital carbon inputs (as POC and PIC) and part of DIC inputs are retained in lakes or reservoirs.

River courses downstream of megacities are extremely impacted particularly from untreated urban organic wastes (carbonaceous pollution). Figure 15.28d is an attempt to describe the evolution of the TOC fluxes or levels downstream of a fast-growing city for which the exponential population growth rate is faster than the sewage collection and treatment rate, an evolution so far encountered in most capitals of the developing nations. Above a certain population density threshold, which corresponds to the self-assimilating capacity of the river system (i.e., proportional to the ratio waste inputs/river discharge), there is a general increase of TOC with possible stepwise improvement phases related to the installation of waste waters treatment plants. Since the collection/treatment rate is lagging behind population development, the organic carbon excess (i.e., labile TOC) gradually increases until anoxic conditions are met, either in the river itself or in ultimate receiving waters (e.g., estuaries). Due to partial waste treatment and biogeochemical processing of labile TOC in the river, the TOC increase is not proportional to population development.

In addition to human pressures and impacts, which are now widely distributed at the global scale, yet with an heterogeneous distribution, two global impacts may eventually influence river carbon fluxes: (1) temperature changes and (2) CO_2 increase. In the arctic regions characterized by the dominance of permafrost and frozen wetlands as peatbogs (Figure 15.28e), a temperature increase ($T-T_0$) above the melting point threshold of permafrost would lead to a major extrasource of TOC in rivers, mostly as DOC. Another impact of temperature increase on DOC export by peatsoils has been envisaged by Reynolds and Frenner (2001): the activation of phenol oxidase enzyme. This point has been questioned by Evans et al. (2002) who invoke hydrological changes as the key control.

The long-term observation of peat catchments in Northern England since 1962 shows various DOC trends, stationary for most catchments and positive for one, with marked interannual variations possibly linked to pH, river flows, and land-use change in addition to enzymatic latch mechanisms (Worral et al., 2003). Such debate emphasizes the complexity of the organic carbon-export processes.

It has been already postulated that DIC levels were very stable in world rivers, at least compared to other species (e.g., Cl^-, SO_4^{2-}, Na^+, and K^+), which are increasing by factors 2 to 10 in all impacted basins. However this may not be true on the long term. A recent CO_2 doubling experiment at a 1/1 scale in the Duke Forest, U.S. (Andrews and Schlesinger, 2001) demonstrated that groundwater was enriched in DIC during the experiment. The authors suggested that such carbon-enriched water may act as a potential carbon sequestering site, postulating a long residence time for groundwaters ($\gg$ 10 years). It is however more likely that these DIC-enriched surficial ground waters will mostly feed the local river system at low flows, thus increasing DIC fluxes. The confirmation of this trend can only be achieved on carefully selected long-term river surveys: the increase of DIC flux in the Mississippi is much more due to increased rainfall on the basin than to DIC level increase between 1953 and 2001 (Raymond and Cole, 2003). Finally it must be noted that the process of river acidification that resulted in decreased DIC fluxes, once very pronounced in Scandinavia and the northeastern U.S. and Canada, has now gradually decreased.

Due to the combination of enhanced and decreased carbon fluxes under human impacts it is not yet possible to predict the direction of these contrasted river carbon evolutions at the global scale: these multiple impacts will have to be addressed at the local to regional scale for each carbon species, taking into account some new carbon sources and filters and the modifications of existing ones within the Anthropocene.

ACKNOWLEDGMENTS

The data collection and interpretation has been possible through the European Union Carbo-Europe program, and, for the French rivers, through the LITEAU program from the French Ministry of environment. Technical help from Séverine Roussenac is greatly appreciated.

REFERENCES

Aitkenhead, J. A. and W. H. Mac Dowell. 2000. Soil C:N ratio as a predictor of annual riverine DOC flux at local and global scales. *Global Biogeochemical Cycles.* 14:127–138.

Amiotte-Suchet, P. and J. L. Probst. 1993. Flux de CO_2 consommé par altération chimique continentale: Influence du drainage et de la lithologie. *C.R. Acad. Sci. Paris* 317 (II): 615–622.

Amiotte-Suchet, Ph., J. L. Probst, and W. Ludwig. 2003. Worldwide distribution of continental rock lithology: Implications for the atmospheric/soil CO_2 uptake by continental weathering and alkalinity river transport to the oceans. *Global Biogeochem. Cycles* 17 (2):7.1–7.13.

Andrews, J. A. and W. H. Schlesinger. 2001. Soil CO_2 dynamics, acidification and chemical weathering in a temperate forest with experimental CO_2 enrichment. *Global Biogeochem. Cycles* 15:149–162.

Artemyev, V. E. 1993. Geochemistry of Organic Matter in the River-Sea System. Nauka, Moscow, 190 pp.

Aumont, O., J. C. Orr, P. Monfray, W. Ludwig, P. Amiotte-Suchet, and J. L. Probst. 2001. Riverine-drivers interhemispheric transport of carbon. *Global Biogeochem. Cycles* 15:393–405.

Berner, R. A., A. C. Lasaga, and R. M. Garrels. 1983. The carbonate-silicate geochemical cycle and its effect on atmospheric carbon dioxide over the past 100 million years. *Am. J. Sci.* 283:641–683.

Billen, G., J. Garnier, and Ph. Hanset, 1994. Modelling phyto-plankton development in whole drainage networks: The River-Strahler model applied to the Seine river system. *Hydrobiologia* 289:119–137.

Bluth, G. J. S. and L. R. Kump. 1994. Lithologic and climatologic controls of river chemistry. *Geochim. Cosmochim. Ac.* 58:2341–2359.

Cauwet, G. and J. M. Martin. 1982. Organic carbon transported by French river. *Mitt. Geol-Päläont. Inst. Univ. Hambourg* 52:475–481.

Clair, T. A., J. M. Ehrman, and K. Higuchi. 1999. Changes in fresh water carbon exports from Canadian terrestrial basins to lakes and estuaries under a 2*CO_2 atmospheric scenario. *Global Biogeochem Cycles* 13:1091–1097.

Crouzet, E. and M. Meybeck. 1971. Le lac d'Annecy et son bassin versant — Premières données climatologiques, hydrologiques, chimiques et sédimentologiques. *Archives Sciences Genève* 24, 3:437–486.

Crutzen, P. J. and E. F. Störmer. 2000. The Anthropocene. *IGPB Newsletter* 41:17–18.

Crutzen, P. J. 2002. Geology of Mankind — The Anthropocene. *Nature* 415:23.

Degens, E. T., S. Kempe, and J. E. Richey. 1991. Biogeochemistry of major world rivers. John Wiley & Sons, New York.

Dessery, S., C. Dulac, J. M. Laurenceau, and M. Meybeck. 1984. Evolution du carbone organique particulaire "algal" et détritique dans trois rivières du bassin parisien. *Archiv. für Hydrobiol.* 100, 2:235–260.

Di Giovanni, C., J. R. Disnar, and J. J. Macaire. 2002. Estimation of the annual yield of organic carbon released from carbonates and sholes by chemical weathering. *Global Planet. Changes* 32:195–210.

Einsele, G., J. Yan, and M. Hinderer. 2001. Atmospheric carbon burial in modern lakes and its significance for the global carbon budget. *Global Planet. Change.* 30:167–195.

Evans, C. D., C. Freeman, D. T. Monteilh, B. Reynolds, and N. Fenner. 2002. Terrestrial export of organic carbon. *Nature* 415:861–862.

Garnier, J., G. Billen, Ph. Hanset., P. Testard, and M. Coste. 1998. Développement algal et eutrophisation dans le réseau hydrographique de la Seine, in M. Meybeck, G. de Marsily, and E. Fustec, eds., *La Seine en son basin*, Elsevier, Paris, pp. 593–626.

Green, P., C. Vörösmarty, M. Meybeck, J. Galloway, B. Peterson, and E. Boyer. 2004. Preindustrial and contemporary fluxes of nitrogen through rivers: A global assessment based on typology. *Biogeochem.*, in press.

Ittekot, V. 1988. Global trends in the nature of organic matter in river suspensions. *Nature* 332:436–438.

Kao, S. J. and K. K. Liu. 1997. Fluxes of dissolved and non-fossil organic particulate organic carbon from a small Oceania river (Lanyang ttsi) in Taïwan. *Biochemistry* 39:255,269.

Kempe, S. 1982. Long-term record of the CO_2 pressure fluctuations in fresh waters, in E. T. Degens, ed., *Transfer of Carbon and Minerals in Major Wolrd Rivers. Mitt. Geol. Paläont. Inst. Hamburg* 52:91–332.

Kempe, S. 1984. Sinks of anthropogenically enhanced carbon cycle in surface fresh waters. *J. Geophys. Res.* 89:4657–4676.

Kempe, S., M. Pettine, and G. Cauwet. 1991. Biochemistry of European rivers, in E. T. Degens, S. Kempe, J. Richey, eds., *Biogeochemistry of Major World Rivers.* SCOPE 42, John Wiley & Sons, pp. 169–212.

Kortelainen, P. and J. Mannio 1988. Natural and anthropogenic acidity sources for Finnish lakes. *Water Air Soil Poll.* 42:341–352.

Lesack, L. F. W., R. E. Hecky, and J. M. Melack. 1984. Transport of carbon, nitrogen, phosphorus and major solutes in the Gambia river, west Africa. *Limnol. Oceano.* 29:816–830.

Livingstone, D. A. 1963. Chemical composition of rivers and lakes. Data of Geochemistry chapter G. U.S. Geological Survey Prof. Paper 440 G:G1–G64.

Löye-Pilot, M. D. 1985. Les variations des teneurs en carbone organique (dissous et particulaire) d'un torrent de montagne: La Solenzana, Corse. *Verh. Internet.Verein.Limnol.* 22:2087–2093.

Ludwig, W. 2001. The age of river carbon. *Nature* 409:466–467.

Ludwig, W., J. -L. Probst, and S. Kempe. 1996. Predicting the oceanic input of organic carbon by continental erosion. *Global Biogeochem Cycles* 10(1):23–41.

Ludwig, W., P. Amiotte-Suchet, G. Munhoven, and J. -L. Probst. 1998. Atmospheric CO_2 consumption by continental erosion: Present-day controls and implications for the last glacial maximum. *Global and Planetary Change* 16–17(1–4):107–120.

Malcom, R. L., W. H. Durum. 1976. Organic carbon and nitrogen concentration and annual organic load for six selected rivers of the U.S. *U.S. Geol. Survey Water Supply Paper* 1817, 21 pp.

Meybeck M., Lô H. M., Cauwet G., Gac J. V. 1986. Geochemistry of the Gambia River during the 1983 high water stage. *Mitt. Geold Palaeont. Inst. Hamburg* 64:461–474.

Meybeck M. 1986. Composition chimique naturelle des ruisseaux non pollués en France. *Sci. Geol.Bull.* 39:3–77.

Meybeck M. and C. Vörösmarty. 2004. Fluvial filtering of land to ocean fluxes: From natural state to Anthropocene. *C.R. Geosciences* (in press).

Meybeck M., L. Laroche., H. H. Dürr, and J. P. Syvitski. 2003. Global variability of daily total suspended solids and their fluxes. *Global Planet Changes* 39:65–93.

Meybeck, M. 1981. River transport of organic carbon to oceans, in Flux of organic carbon by rivers to oceans. Conf. 800940, U.S. Dept. Energy, Office Energy Research, Washington D.C., pp. 219–249.

Meybeck, M. 1982. Carbon, nitrogen and phosphorus transport by world rivers. *American J. Science* 282:401–450.

Meybeck, M. 1985. Variabilité dans le temps de la composition chimique des rivières et de leurs transports en solution et en suspension. *Rev. Fr. Sci. Eau,* 4, 93–121.

Meybeck, M. 1986. Composition chimique naturelle des ruisseaux non pollués en France. *Sci. Geol.Bull.* 39:3–77.

Meybeck, M. 1993a. C, N, P and S in rivers: From sources to global inputs, in R. Wollast, F. T. Mackenzie, and L. Chou, eds., *Interaction of C, N, P and S Biogeochemical Cycles and Global Change.* Springer-Verlag, pp. 163–193.

Meybeck, M. 1993b. Riverine transport of atmospheric carbon: sources, global typology and budget. *Water, Air Soil Pollution* 70:443–464.

Meybeck, M. 2002. Riverine quality at the Anthropocene: Propositions for global space and time analysis, illustrated by the Seine River. *Aquatic Sciences* 64:376–393.

Meybeck, M. 2003a. Global analysis of river systems: from earth system controls to Anthropocene controls. *Phil. Trans. Royal Acad.* 354:1440.

Meybeck, M. 2003b. Global occurrence of major elements in rivers, in K. K. Turekian and H. D. Holland, eds., *Treatise of Geochemistry,* vol. 5. J. I. Drever, Elsevier, pp. 207–224.

Meybeck, M. and A. Ragu. 1996. River discharges to the oceans: An assessment of suspended solids, major ions, and nutrients. Environment Information and Assessment Rpt. UNEP, Nairobi, 250 pp.

Meybeck, M. and A. Ragu. 1997. Presenting Gems Glori, a compendium of world river discharge to the oceans. *Int. Ass. Hydrol. Sci. Publ.* 243:3–14.

Meybeck, M. and C. Vörösmarty. 1999. Global transfer of carbon by rivers. *Global Change Newsletters* 37:12–14.

Meybeck, M., G. Cauwet, S. Dessery, M. Somville, D. Gouleau, and G. Billen. 1988. Levels, behaviour and tentative budgets of nutrients (organic C, P, N, Si) in the eutrophic Loire estuary. *Estuarine Coastal Shelf Sci.* 27:595–624.

Meybeck, M., G. de Marsily, and E. Fustec. 1998. *La Seine en son Bassin.* Elsevier, Paris, 750 pp.

Moatar, F., A. Poirel, and C. Obled, 1999. Analyse des séries temporelles de mesure de l'oxygène dissous et du pH sur la Loire au niveau du site nucléaire de Dampierre (Loiret). *Hydroecologie Appl.* 11:127–151.

Moore, T. R. 1987. Dissolved organic carbon in forested and cutover drainage basins, Westland, New Zealand. *Int Ass. Hydrol. Sci. Publ.* 167:481–487.

Moreira-Turcq, P., P. Seyler, J. L. Guyot, and H. Etcheber. 2003. Exportation of organic carbon from the Amazon river and its main tributaries. *Hydrol. Processes* 17:1329–1344.

Mulholland, P. J. and J. A. Watts. 1982. Transport of organic carbon to the oceans by rivers of North America: A synthesis of existing data. *Tellus* 34:176–186.

Mulholland, P. J. and Elwood, J. W. 1982. The role of lake and reservoir sediments as sinks in the perturbed global carbon cycle. *Tellus* 34:490–499.

Naiman, R. J. 1982 Characteristics of sediment and organic carbon export from pristine boreal forest watersheds. *Can. J. Fish. Aquatic Sci.* 39:1699–1718.

Nikanorov, A. M. and V. V. Tsirkunov. 1984. Study of the hydrochemical regime and its long-term variations in the case of some rivers in the USSR. *Int. Ass. Hydrol. Sci. Publ.* 150:288–293.

Probst, J. L., J. Mortatti, and Y. Tardi. 1994. Carbon river fluxes and weathering CO_2 consumption in the Congo and Amazon river basins. *Applied Geochemistry* 9:1–13.

Rabouille, C., Machouzie F. T., and Ver L. M. 2001 Influence of human perturbation on carbon nitrogen and oxygen biogeochemical cycles in the global costal ocean. *Geochem. Cosmochim. Acta* 65,3613–3641.

Raymond, P. A. and J. E. Bauer. 2001. Riverine export of aged terrestrial organic matter to the North Atlantic Ocean. *Nature* 407:497–500.

Raymond, P. A. and J. J. Cole. 2003. Increase in the export of alkalinity from North America's largest river. *Science* 301:88–91.

Reynolds, B. and N. Fenner. 2001. Export of organic carbon from peat soils. *Nature* 412:785.

Schlesinger, W. T. T., Melack, J. M. 1980. Transport of organic carbon in the world's river. *Tellus* 33:172–187.

Sempere, R., B. Charriere, F. van Wambeke, and G. Cauwet. 2000. Carbon inputs of the Rhône River to the Mediterranean Sea: Biochemical implications. *Global Biogeochem Cycles* 14:669–681.

Servais, P., Z. Idlafkih, G. Billen, J. Garnier, J. M. Mouchel, M. Seidl, and M. Meybeck. 1998. Le carbone organique, in M. Meybeck, G. de Marsily, and E. Fustec, eds., *La Seine en son bassin,* Elsevier, Paris, pp. 483–529.

Smith, S. V., W. H. Renwick, R. W. Buddemeier, and C. Crossland. 2001. Budgets of soil erosion and deposition for sediments and sedimentary organic carbon across the conterminous U.S. *Global Biogeochem Cycles* 15:697–707.

Stakalsky, B. G. 1984. Study of anthropogenic influence on water quality in some rivers of the Baltic Sea basin. *Int. Ass. Hydrol. Sci. Publ.* 150:295–302.

Stallard, R. F. 1998. Terrestrial sedimentation and the carbon cycle: Coupling weathering and erosion to carbon burial. *Global Biogeochemist. Cy.* 12:231–257.

Steffen, W., A. Sanderson, P. D. Tyson, J. Jäger, P. Matson, B. Moore, III, F. Oldfield, K. Richardson, H. J. Schellenhuber, B. L. Turner, II, and R. J. Wason. 2004. *Global Change and the Earth System*. Springer, Berlin, 336 pp.

Tremblay, G. H. and D. Cossa. 1987. Major ion composition of the St Lawrence river: Variations since the start of industrialization. *Mitt. Geol. Paläont. Inst. Univ. Hamburg* 64:289–293.

Veissy, E. 1998. Transferts de carbone organique, d'azote et de phosphore des bassins versants aux estuaires de la Gironde et de l'Adour. Thèse de Doctorat, Université de Bordeaux I, 281 pp.

Vörösmarty, C. J., M. Meybeck, B. Fekete, K. Sharma, P. Green, and J. Syvitski. 2003. Anthropogenic sediment retention: Major global-scale impact from the population of registered impoundments. *Global Planet Changes* 39:169–190.

Williams, P. J. 1968. Organic carbon and nutrients in the Amazon. *Nature* 218:937–938.

Worral, F., T. Burt, and R. Shedden. 2003. Long-term record of riverine dissolved organic matter. *Biogeochemistry* 64:165–178.

CHAPTER 16

Carbon, Nitrogen, and Stable Carbon Isotope Composition and Land-Use Changes in Rivers of Brazil

Luiz Antonio Martinelli, Plinio Barbosa de Camargo, Marcelo Corrêia Bernardes, and Jean-Pierre Henry Balbaud Ometto

CONTENTS

16.1 Introduction 239
16.2 Study Areas 241
16.3 Methods 242
16.4 Results 243
    16.4.1 Difference between Watersheds 243
    16.4.2 Correlation between Suspended Solids and Compositional Attributes of the Coarse Suspended Solids and Fine Suspended Solids Fractions 245
    16.4.3 Sources of Organic Matter to the Riverine Size Fractions 248
16.5 Discussion 250
16.6 Conclusions 252
References 252

16.1 INTRODUCTION

Particulate organic matter (POM) transported by rivers is generally composed of a mixture of terrestrial materials originated in their watersheds with aquatic material, like phytoplankton and aquatic plants (Angradi, 1993; Barth et al., 1998; Martinelli et al., 1999; Lobbes et al., 2000; Kendall et al., 2001). The relative contribution of each one of these sources is variable among watersheds, and it depends on several factors such as the land use, topography, precipitation, presence of reservoirs, etc. Above all, several rivers face a significant temporal variability in the composition of the POM, suggesting that the relative contribution of both sources is also variable. In general, there is a dominance of terrestrial sources during high water flow, and aquatic sources, mainly phytoplankton, during low flow (Barth et al., 1998; Martinelli et al., 1999; McCusker et al., 1999; Kendall et al., 2001).

During high flows, precipitation is also generally high, consequently the overland flow is also high, generating more soil erosion. The increase in flow and suspended particles is followed by a decrease in the POM attached to the suspended particles (Meybeck, 1982; Zhang et al., 1998). During low flow, relatively low transport of suspended particles allows more light penetration and consequently the increase of primary production. This is especially true in rivers where the presence of reservoirs and lakes increases the contribution of phytoplankton to the POM (Angradi, 1993; Bird et al., 1998; Barth et al., 1998; Martinelli et al., 1999). An exception to this pattern is the Amazon, which has little variation in the composition of the POM during the annual hydrologic cycle (Quay et al., 1992). In this system the fine POM (< 63 µM) is composed mainly of recalcitrant material from soils and the coarse POM (> 63 µM) is composed mainly of plant debris. On the other hand, autochthonous sources, like phytoplankton, appear to be a minor source of POM (Hedges et al., 1986; Devol and Hedges, 2001).

The main terrestrial sources of carbon (C) to river are soil organic matter (SOM) and soil vegetation cover (Hedges et al., 1986; Bird et al., 1998; Lobbes et al., 2000; Onstad et al., 2000). The isotopic composition of the C present in the terrestrial vegetation tissues varies according to their photosynthetic pathways: C3 and C4. Several investigators, using $\delta^{13}C$ values of the POM, have distinguished the presence of C4 organic matter in riverine organic matter in Brazil (Martinelli et al., 1999), Africa (Bird et al., 1998), China (Zhang et al., 1998), Canada (Barth et al., 1998), and the U.S. (Goñi et al., 1998; Onstad et al., 2000). The amount of C4 material transported to coastal zones as riverine organic matter could underestimate the amount of terrestrial C load to these areas due to the similarities of the isotopic signal of the transported POM and the ocean signal. (Goñi et al., 1998). Furthermore, it has been demonstrated that turbid rivers deliver to coastal zones a significant portion of the total nitrogen (N) loading as particulate nitrogen (Mayer et al., 1998). On the other hand, these authors call attention to the fact that in some areas, like in the Mississippi basin, the relative importance of N-delivery by particulate forms has decreased as a consequence of damming and overuse of N fertilizers. Therefore, it is important to understand how human actions (e.g., land-use change, damming, and overuse of fertilizers and other chemicals) alter the composition and quantity of the POM transported by the rivers.

Tropical areas of the world are experiencing high rates of changes in land use and land cover (Houghton, 1994). Therefore, watersheds in these tropical areas are like "natural large scale laboratories," where one can investigate the effects of such changes on the POM transported by rivers. The main land-use changes involve the replacement of original forests (C3) by C4 vegetation such as grasses for pasture or agriculture (corn and sugarcane). For instance, the main land-use change in the Amazon watershed is the replacement of the original forest by pasture (). Analyses of land-use changes made in seven major watersheds in the State of São Paulo (southeast Brazil) revealed that approximately 65% of the land is covered with C4 vegetation (50% pasture and 15% sugarcane).

The main objective of this chapter is to compare changes on the composition of riverine organic matter in watersheds with diverse land uses. Thus, we compared the published data on the characteristics of riverine organic matter of six Brazilian watersheds with distinct land uses: Amazon, Ji-Paraná, Piracicaba, Mogi-Guaçu, Cabras, and Pisca. We compared the C and the N concentrations, and the stable C isotope composition ($\delta^{13}C$) of two size classes of organic matter (coarse and fine) from rivers of these Brazilian watersheds. The main land-use changes in these watersheds were the replacement of the original forest vegetation by pasture or sugarcane plantations, both C_4 plants. The strategy was to test if the signal of these new vegetation covers has already reached the rivers of these watersheds.

The main land cover change in the Amazon watershed has been the replacement of the original forest by pasture. Thus, about 10% (at 560,000 km² in 1998) of the Brazilian Amazon has already been deforested (www.inpe.br). This type of change started 30 to 40 years ago and is mainly concentrated in the southern and eastern regions of the watershed. The Ji-Paraná River is one of the major tributaries of the Madeira River, which is in turn the major tributary of the Amazon River. The Ji-Paraná watershed is located in the southwest region of the Amazon watershed (State of Rondônia), where deforestation is occurring at high rates. Between 1950 and 2000, the population increased from 37,000 to 1.4 million people (www.ibge.gov.br). This increase in the population

CARBON, NITROGEN, AND STABLE CARBON ISOTOPE COMPOSITION AND LAND-USE CHANGES 241

Figure 16.1 Study area showing the location of the Brazilian river watersheds.

was followed by an increase in the deforested area, up to 60,000 km² (25% of the state of Rondônia), over the same period (Pedlowski et al., 1997, www.inpe.gov.br). The Piracicaba, the Mogi-Guaçu, the Cabras, and the Pisca are all located in the State of São Paulo, in the southeast region of Brazil. In this state deforestation started over a century ago, in the end of the nineteenth century. In this case, the original vegetation (mainly Atlantic forest) was replaced by coffee. From 1907 to 1934, approximately 80,000 km² (32% of the State of São Paulo) of the Atlantic Forest was replaced by coffee (Brannstrom, 2000). After its decline, coffee was replaced by pasture and sugarcane in the watersheds located in the State of São Paulo.

16.2 STUDY AREAS

The Amazon watershed encompasses an area of approximately 6 million km² and is mostly composed of tropical rain forests (Figure 16.1; Table 16.1). The south border of this watershed has been severely affected by the conversion of forests into pastures. The total deforested area is nearly 600,000 km². The Ji-Paraná watershed, with drainage area of 75,000 km², is located in the State of Rondônia, in the southwestern Amazon watershed and is one of the regions most severely affected by deforestation (Table 16.1). Intensive land use is concentrated along highway BR-364 within the limits of the Ji-Paraná River watershed, with the most intensive land cover changes in the central part of this watershed. The Piracicaba River watershed is located in the southeastern region of

Table 16.1 Total Area, Final Average Discharge, Average Coarse and Fine Suspended Solids (SS) Concentration, Population Density, and Percent Area Covered with C4 Plants in Each Watershed

	Amazon	Ji-Paraná	Mogi	Piracicaba	Pisca	Cabras
Area (km²)	6,300,000	48,211	13,200	12,400	130	50
Discharge (m³/s)	175,000	1,390	309	174	3.7	0.8
Coarse suspended solids (mg/L)	69.58	2.26	14.17	8.42	14.54	5.55
Fine suspended solids (mg/L)	262.25	20.51	67.49	48.42	29.85	39.34
Population density (hab/km²)	3.4	9.8	96	341	460	200
Area under C4 plants (%)	10	25	48	75	70	83

Brazil in the state of São Paulo and has an area of 12,400 km² (Table 16.1). Only 10% of the original forest in this watershed is still intact, most of the original vegetation has been replaced by pasture and silviculture in the eastern side and by sugarcane in the western side. Rivers of this watershed were severely affected by untreated domestic sewage and industrial effluents especially toward the western side of the watershed. The Mogi-Guaçu watershed (13,200 km²) is also located in the state of São Paulo and its land cover is similar to that of the Piracicaba River watershed (Table 16.1). The main difference is that the population in the Piracicaba watershed is almost 3 times higher than the population in the Mogi-Guaçu watershed; that is approximately 1.2 million people. The Cabras and the Pisca are small subwatersheds of the Piracicaba watershed and their areas are 50 and 130 km², respectively (Table 16.1). The former is almost totally dominated by pasture and the latter by sugarcane. Another important difference between these two subwatersheds is the larger volume of domestic sewage discarded in the Pisca watershed in comparison with the Cabras watershed.

16.3 METHODS

The main channel of the Amazon watershed, the Solimões/Amazon river, was sampled by Hedges et al. (1986) four times during 1982 and 1983. The first time during the early rising water, the second time during the midrising, the third time near the peak of the flood, and the last time in the mid-falling water stage. Water samples were collected in eleven sites along the main channel: six along the Solimões River and five along the Amazon River. The Ji-Paraná River was sampled by Bernardes et al. (in press) seven times covering several stages of the river hydrography between 1999 to 2001. A total of five sites were collected along the Ji-Paraná River, two sites along the Comemoração River, and two sites in the Pimenta Bueno River, both rivers form the Ji-Paraná. The Piracicaba River and its main tributaries (Jaguari and Atibaia) were collected bimonthly between January 1996 and May 1997. A total of two sites were collected in the Piracicaba and Jaguari rivers and three in the Atibaia River (Krusche et al., 2002). The Mogi-Guaçu River was sampled monthly in three sites between January 1997 and June 1998 (Domingues, 2000). During the same period, the Cabras and the Pisca streams were also sampled monthly in three sites each (Ometto, 2000). The sampling in the Amazon watershed was conducted using a winch that allowed vertical integration of the water column. In each sampling site eighteen equidistant vertical sampling profiles were taken (Richey et al., 1990). In other watersheds, river water samples were collected in the middle of the channel at 60% of the total depth using an electric pump. The coarse fraction was separated from the fine fraction by using a sieve (63 µm) The fine suspended solid (FSS) fraction (< 63 mm and > 0.1 mm) and ultra-filtered-dissolved organic matter (UDOM) fraction (< 0.1 mm and > 1,000 Daltons) were isolated in the laboratory with a Millipore tangential flow ultra-filtration system (model Pellicon-2), using membrane cartridges having a nominal 0.1 µm pore size (model Durapore VVPP) and a 1,000 Daltons molecular weight nominal cut off (model PLAC), respectively. (For details see Krusche et al., 2002.)

Several parameters were determined on the coarse (> 63 µm) and fine (3 µm) fractions of riverine suspended organic matter. The weight of suspended solids (SS), concentration (mg/L) of the coarse (CSS) and fine (FSS) fractions were determined for every sampling site. Carbon (%OC) and nitrogen (%N) are reported as percent of total coarse (CSS) or fine suspended solids (FSS). Weight concentrations (in mg C/L or mg N/L) were determined by multiplying particulate organic C or N (grams C or N/100 gram POM) by FSS or CSS concentrations (in mg POM/L). The C and N loads (tC/yr or tN/yr) were obtained by multiplying the C and N weight concentrations by the water discharge. Finally the C and N loads per unity area (t C/km² yr or t N/km² yr) were obtained by dividing the C and N loads by the area of each watershed. The stable C isotopic composition of the fine and coarse POM was also determined (Bernardes et al., 2004). Areas of the watersheds covered with C_4 plants were estimated with

GIS using Landsat images for the Amazon watershed (www.inpe.br), Ji-Paraná (Bernades et al., 2004), Piracicaba (Martinelli et al., 1999), Mogi (Silva, A. M., unpublished), and using aerial photographs for the Pisca and the Cabras watersheds (Ometto et al., 2000). For the Brazilian Amazon region GIS is available for the total deforested area, which is approximately 600,000 km². This represents approximately 12% of the Brazilian Amazon. As forest-to-pasture is the most common conversion in the Amazon region (www.inpe.br), the area covered with C_4 pasture plants was assumed to be about 10%.

Differences between sampling sites were assessed by ANOVA followed by Tukey Honest test for unequal number of data (). Correlation among parameters was assessed by Pearson parametric correlation (). Statistical significant difference is indicated in the text by $P < 0.01$ (1% level of probability).

16.4 RESULTS

The POM was separated into coarse (> 63 μm) and fine fractions (< 63 μm) based on the fact that both fractions have different compositions and different sources (Hedges et al., 1986). Most of the POM is transported as fine particles. At least 80% of the organic matter for the Ji-Paraná, the Piracicaba, the Mogi, and the Pisca rivers and 63% for the Amazon is transported as fine fraction. The exception was the Cabras stream, where only 50% is transported in this fraction.

The percentage of organic carbon (OC) in the coarse fraction ranged from 0.54 to 48.7%, with an average of 7.59 ± 6.87% (n = 376). The average concentration of the fine fraction was significantly smaller than that of the coarse fraction ($P < 0.01$) and equal to 5.46 ± 3.97% (n = 348), ranging from 0.64 to 31.3%. In comparison, the average concentration of %N did not differ between both fractions, with an average of 0.59 ± 0.73% (n = 376) and 0.65 ± 0.55 (n = 348) for the coarse and fine fractions, respectively. The C:N ratios of the coarse fraction ranged from 5.0 to 66.1, with an average of 15.7 ± 5.6 (n = 376). The average C:N ratio of the fine fraction (9.3 ± 2.41, n = 348) was significantly ($P < 0.01$) smaller than the C:N ratio of the coarse fraction, ranging from 3.2 to 18.5. Finally, the $\delta^{13}C$ values of the coarse fraction ranged from –35.2 to –16.2%, with an average of –25.7 ± 2.5% (n = 359). The $\delta^{13}C$ values of the fine fraction ranged from –32.8 to –16.9%, with an average of –25.1 ± 2.9% (n = 348), which is significantly higher than the average for the coarse fraction.

16.4.1 Difference between Watersheds

The Amazon watershed had the lowest values of coarse %OC and coarse %N ($P < 0.01$), while the Pisca stream had the highest values ($P < 0.01$) (Table 16.2). There was no difference between

Table 16.2 Averages of Compositional Parameters of the Riverine Coarse and Fine Particulate Organic Matter of the Basins Investigated in This Study

	Amazon	Ji-Paraná	Piracicaba	Mogi	Cabras	Pisca
Coarse — %OC	1.08a	8.95b	7.45b	6.97b	6.51b	13.75c
Coarse — %N	0.04a	0.53b	0.73b	0.50b	0.52b	1.05c
Coarse — C:N	25a	18a	13b	14b	13b	16b
Coarse — $\delta^{13}C$ (%)	–28.0a	–29.1b	–25.5c	–24.1d	–22.1e	–25.0f
Coarse — kgC/km².yr	940a	86a	511a	415a	817a	755a
Fine — %OC	1.15a	8.04b	5.54c	4.37c	4.23c	7.85b
Fine — %N	0.10a	0.72b	0.74b	0.54b	0.65b	0.92b
Fine — C:N	11a	11a	8b	9b	7b	10b
Fine — $\delta^{13}C$ (%)	–26.9a	–28.1b	–26.1a	–22.8c	–22.0d	–21.2e
Fine — kgC/km² yr	4395a	1180a	2256a	2002a	2196a	2419a

Note: Different letters indicate statistically significant differences.

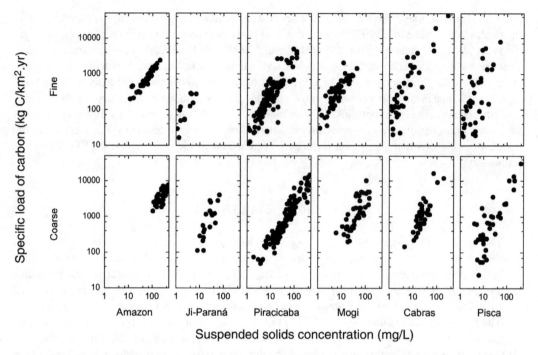

Figure 16.2 Plot of specific load of carbon vs. suspended solids concentrations for the coarse and fine size fractions in the Amazon, Ji-Paraná, Piracicaba, Mogi, Cabras, and Pisca river watersheds.

the concentrations of %OC and %N in other watersheds (Table 16.2). The C:N ratios of the coarse fraction of the Amazon and the Ji-Paraná watersheds were higher (P < 0.01) than the C:N ratios of other watersheds (Table 16.2). The average C:N ratios in the Amazon and in the Ji-Paraná watersheds were 25 and 18, respectively. In other watersheds these values ranged from 13 to 16. The $\delta^{13}C$ values of the coarse fraction was different between watersheds. The Ji-Paraná and the Amazon had the lowest values, –28.0 and –29.1‰, respectively. The average values of other watersheds were at least 3‰ higher than the values in the Ji-Paraná and in the Amazon. The Cabras watershed had the highest average value, approximately –22‰ (Table 16.2). The variability of the specific load of C for the coarse fraction was very high for each sampling site (Figure 16.2). In other words, the standard deviation of each average value was very high (data not shown). This is the reason why the estimates of the specific load of C are not statistically different between watersheds (Table 16.2).

As for the coarse fraction, the average %OC of the fine fraction was lower in the Amazon watershed (P < 0.01). The average concentrations of other watersheds were four to eight times higher than the Amazon's (Table 16.2). The highest average concentrations were observed in the Ji-Paraná (P < 0.01) with values close to 8%. The Piracicaba, the Mogi, and the Cabras watersheds had intermediate average, ranging from 4.2 to 5.5% (Table 16.2). The average fine %N was also the lowest in the Amazon watershed (P < 0.01); other watersheds had values five to nine times higher than that of the Amazon's, but with no significant difference between them (Table 16.3). The average C:N ratio of the fine fraction was 11 for the Amazon and the Ji-Paraná watersheds. This value was approximately one to three units higher (P < 0.01) than other basins, where no difference was observed (Table 16.3). There was a significant difference (P < 0.01) in the $\delta^{13}C$ of the fine fraction between most watersheds. The exceptions were the Amazon and the Piracicaba, with no statistically significant difference between them. The average values for these two watersheds were –26.9 and –26.1‰, respectively (Table 16.2). The average $\delta^{13}C$ values for Ji-Paraná were 1 to 2‰ lower than the Amazon and the Piracicaba watersheds. On the other hand the $\delta^{13}C$ values in other watersheds were 4 to 7‰ higher than for these two watersheds. The specific load of C for the fine fraction, like the coarse fraction, was also considerable variable during the sampling

CARBON, NITROGEN, AND STABLE CARBON ISOTOPE COMPOSITION AND LAND-USE CHANGES   245

Table 16.3 Average, Standard Deviation, and Number of Samples (between Brackets) of Selected Parameters of the Coarse and Fine Suspended Solid Fractions

	Coarse	Fine
%OC	7.59 ± 6.87a	5.46 ± 3.97b
	(376)	(348)
%N	0.59 ± 0.73a	0.65 ± 0.55a
	(376)	(348)
C:N	15.72 ± 5.56a	9.28 ± 2.41b
	(376)	(348)
$\delta^{13}C$ (%)	−25.7 ± 2.53a	−25.1 ± 2.88b
	(359)	(311)

Note: Different letters indicate statistically significant differences.

period for each site (Figure 16.2). Even with such high variability, the averages between the watersheds of the southeast region of Brazil (Piracicaba, Mogi, Cabras, and Pisca) were similar (Table 16.2). The Amazon watershed had the highest specific load of C and the Ji-Paraná the lowest (Table 16.2). However, due to the high standard deviation of such estimates, these values did not statistically differ between watersheds.

16.4.2 Correlation between Suspended Solids and Compositional Attributes of the Coarse Suspended Solids and Fine Suspended Solids Fractions

Seasonal differences between watersheds can be assessed by correlating the suspended solids concentration with compositional attributes. The coarse %OC was inversely correlated with the coarse SS concentration only in the Amazon and the Piracicaba watersheds (Figure 16.3). The coarse %N was not correlated with coarse SS concentration in Cabras and Pisca watersheds (Figure 16.4). Another important difference between watersheds was small variability in %OC and %N of the coarse fraction in the Amazon compared to other watersheds, especially compared to the Piracicaba and the Pisca watersheds. For instance, the coarse %OC ranged from 0.54 to 3.30% in the Amazon, compared to 1% to 49% in the Piracicaba and in the Pisca watersheds (Figure 16.3).The fine %OC and fine %N were inversely correlated with the FSS concentration in all watersheds (Figure 16.3 and Figure 16.4). The range of values was large especially in Piracicaba and Pisca watersheds. The fine %OC ranged from 0.6 to about 30% in these watersheds and from 1 to 1.6% in the Amazon (Figure 16.3).

The C:N ratio of the coarse fraction was inversely correlated with the coarse SS concentration only in the Amazon and in the Cabras basins, and positively correlated in the Ji-Paraná, the Piracicaba, and the Mogi basins (Figure 16.5). Although the C:N ratios of the Amazon and the Ji-Paraná were generally higher than those of the other watersheds, the range of the values was similar between watersheds. For the fine fraction, the fine SS concentration was correlated positively with the C:N ratios in Ji-Paraná, Piracicaba, and Mogi watersheds (Figure 16.5). In this case the range of C:N ratios in the Amazon watershed was again smaller than those in other watersheds. There was a difference of only three units between the lowest and the highest C:N ratio of the fine fraction in these watersheds. In contrast, in the Piracicaba and Pisca watersheds, the difference between the lowest and the highest C:N ratios was nine and twelve units, respectively (Figure 16.5).

Finally, the $\delta^{13}C$ of the coarse fraction was positively correlated with the coarse SS concentration in the Amazon, Cabras, and Pisca watersheds, and inversely correlated in Mogi watershed (Figure 16.6). The $\delta^{13}C$ values were less variable in the Amazon and in the Ji-Paraná in comparison with other watersheds. For instance, while the range of $\delta^{13}C$ values was −28.9 to −27.1% in the Amazon, it was from −29.6 to −20.8% in Piracicaba watershed (Figure 16.6). The $\delta^{13}C$ of the fine fraction

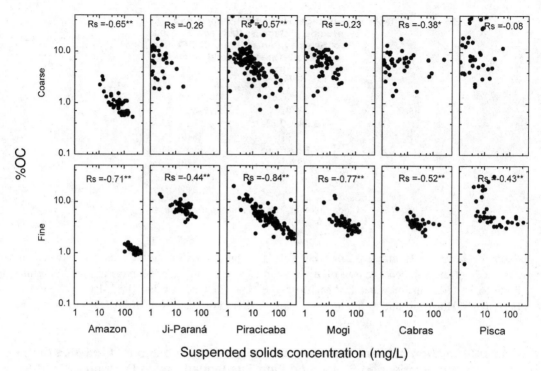

Figure 16.3 Plot of %OC vs. suspended solids concentrations for the coarse and fine size fractions in the Amazon, Ji-Paraná, Piracicaba, Mogi, Cabras, and Pisca river watersheds.

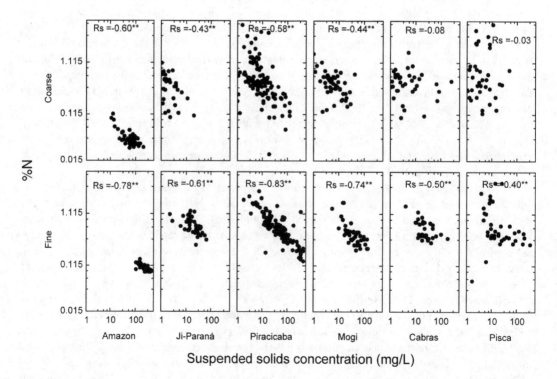

Figure 16.4 Plot of %N vs. suspended solids concentrations for the coarse and fine size fractions in the Amazon, Ji-Paraná, Piracicaba, Mogi, Cabras, and Pisca river watersheds.

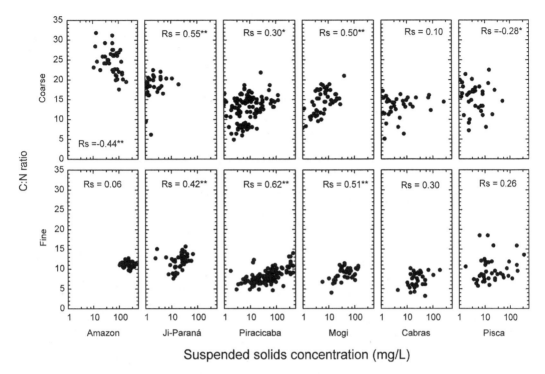

Figure 16.5 Plot of C:N vs. suspended solids concentrations for the coarse and fine size fractions in the Amazon, Ji-Paraná, Piracicaba, Mogi, Cabras, and Pisca river watersheds.

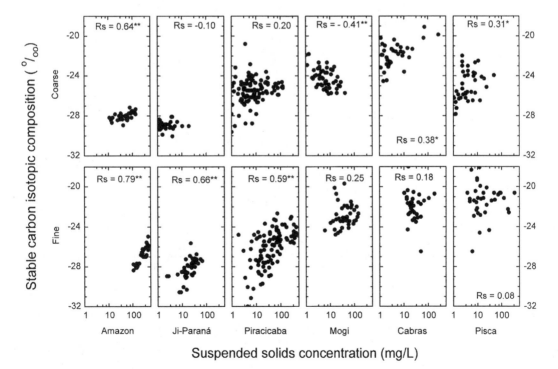

Figure 16.6 Plot of stable carbon isotopic composition ($\delta^{13}C$) vs. suspended solids concentrations for the coarse and fine size fractions in the Amazon, Ji-Paraná, Piracicaba, Mogi, Cabras, and Pisca river watersheds.

was positively correlated with the concentration of fine SS only in the Amazon, Ji-Paraná, and Piracicaba watersheds (Figure 16.6). The range of δ¹³C values was particularly high in the Piracicaba, where the lowest value was equal to –32‰ and the highest –22.7‰, a difference of approximately 9.3‰ between these extreme values.

16.4.3 Sources of Organic Matter to the Riverine Size Fractions

Inverse of the C:N ratio (N:C) was plotted against the δ¹³C values of the riverine coarse and fine suspended solids, and potential end-members as an attempt to constrain the sources of organic matter between watersheds (Figure 16.7). The coarse fraction of the Amazon and Ji-Paraná watershed

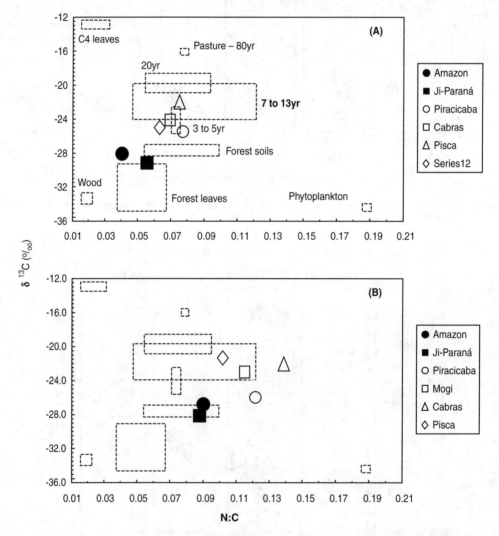

Figure 16.7 Plot of N:C ratio vs. δ¹³C for the coarse (A) and fine (B) size fractions and the following end-members: tree leaves from forests, forest soil, soil covered with a pastures of age between 3 to 5 years, 7 to 13 years, 20 years, and 80 years (rectangles representing the distribution of all points). Tree leaves collected in Ji-Paraná (Martinelli, L. A. and Ehleringer, J. E., unpublished data); forest soil organic matter — soils collected in forests near Manaus and Santarém (Telles, E. V., unpublished data); pasture soils — soils samples collected at several sites in the Rondônia State (Neill et al., 1997).

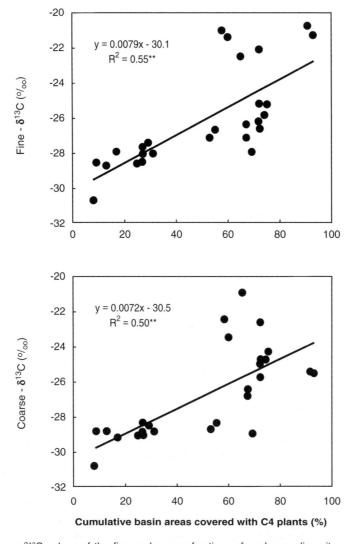

Figure 16.8 Average δ¹³C values of the fine and coarse fractions of each sampling site as a function of the cumulative area of each subwatershed covered with C4 plants (pasture or sugarcane).

were plotted near the field occupied by forest. The other four watersheds had higher δ¹³C values and higher N:C (smaller C:N), and plotted near the pasture soils 3 to 5 and 7 to 13 years of age (Figure 16.7). In general N:C ratios for fine fractions were higher than for coarse fraction. This fact displaced the Amazon and Ji-Paraná from sources of tree leaves to forest soils, and the other watersheds appear to be originated by a mixture of organic matter from pasture soils and some N-enriched source that could be phytoplankton.

The cumulative basin areas covered with C4 for each sampling site of each basin (except for the Amazon) was plotted against the δ¹³C values of the coarse and fine fractions for further confirmation of the influence of C4 plants (pasture and sugarcane) (Figure 16.8). The correlation coefficients were significant and around 0.50 for both fine and coarse fractions (Figure 16.8). This means that a considerable part of the variance in the population can be explained by the area under pasture or sugarcane in each subwatershed.

16.5 DISCUSSION

Results show that the C4-derived organic matter introduced in the landscape by the replacement of the original vegetation by C4 crops has already reached the riverine POM in the watersheds analyzed. One of the evidences was the significant correlation between the cumulative areas of the watersheds covered with C4 plants and the $\delta^{13}C$ values of the coarse and fine fractions of the POM (Figure 16.8). It is surprising a strong correlation existed between five different watersheds. These watersheds encompass a variety of topography, soil types, and vegetation cover. The N:C vs. $\delta^{13}C$ plot of the POM fractions (Figure 16.7) show that the POM in the rivers of the southeast region, plotted near the soils cultivated with C4 plants, either for the coarse or for the fine fraction. The exceptions were the rivers of northern region (Amazon and Ji-Paraná), which had as main source of organic matter leaves from the Amazon rain forest for the coarse fraction, and SOM from forest soils (Figure 16.7). Several authors have also reported the presence of C4 material in the POM of rivers in different parts of the world. For example, most of the POM samples had $\delta^{13}C$ values characteristics of C3 plants in the upper St. Lawrence River in Canada, although Barth et al. (1998) found five samples with $\delta^{13}C$ ranging from –16.3 to –22.4‰. They attribute these high values to the influence of corn grown in the St. Lawrence watershed. Cornfields were also responsible for the high isotopic values found in the POM of the Shuangtaizihe Estuary in North China (Zhang et al., 1998). In the continental U.S., rivers crossing extensive areas of grasslands had the highest $\delta^{13}C$-POM values (Onstad et al., 2000). For example, the $\delta^{13}C$-POM in the lower Colorado River (State of Texas) and in the Brazos River was equal to –19.7‰ and –18.5‰, respectively, indicating that a significant proportion of their POM was derived from C4 plants. POM with ^{13}C-enriched values was also found in the lower Mississippi River by Onstad et al. (2000) and Goñi et al. (1998). The same was true for rivers draining the savannas of the Cameroon, where the $\delta^{13}C$ values were equal to –23.3‰ in the Sanaga River and equal to –21.1‰ in the Mbam River. The highest $\delta^{13}C$ values were especially observed during the rainy season, when the overland flow carries more C4-derived C from the savanna soils (Bird et al. 1998). Finally, high $\delta^{13}C$ values were also found in bottom sediments samples of rivers draining the Australian savannas (Bird and Pousai, 1997).

Although there was a clear signal of C4-derived organic matter in most rivers of this study, there was a large variability in the $\delta^{13}C$ of the POM between rivers and also in a same river, according to the time of the year. For instance, the $\delta^{13}C$ of the fine POM of the Piracicaba River changed from –24 to –25‰ during periods of high suspended solids concentration to values varying –32 to –28‰ during periods of low suspended solids concentration (Figure 16.7). A similar trend was observed in the Sanaga and in the Mbam rivers of the Cameroon, and the explanation for this trend was the higher flow of C4-derived material during the high water period (Bird et al., 1998). The ^{13}C-enrichment observed in the Piracicaba River during high concentrations of suspended solids was followed by an increase of the C:N ratio of the POM. The C:N ratios varied from 10 to 15 at high suspended solids concentrations, decreasing to 5 to 8 at periods of low concentration of suspended solids. These ratios have been interpreted as due to SOM and phytoplankton sources, respectively (Kendall et al., 2001). Following this interpretation, it seems that the main source of organic matter to the fine POM of the Piracicaba River during the low water period is phytoplankton and C4-derived material, brought to the river through soil erosion of sugarcane and pasture fields (Martinelli et al., 1999). The source of phytoplankton to the Piracicaba River is probably a series of reservoirs along it that were built to generate electricity. There are several examples in the literature showing that reservoirs and lakes provide to downstream sectors of rivers a POM with low $\delta^{13}C$ and C:N ratios. For instance, this was the case in four large systems of the U.S. (the Mississippi, the Colorado, the Rio Grande, and the Columbia; Kendall et al., 2001). Although the Piracicaba watershed has a much larger area covered with C4 than the Mogi watershed (Table 16.1), the $\delta^{13}C$ values of the fine POM in the Mogi River is significantly higher than in the Piracicaba River (Figure 16.5). This can be due to the deposition of C4 material induced by soil erosion in the reservoirs, overwhelming the importance of phytoplankton originated POM downstream (Martinelli et al., 1999). A similar trend was observed in the Sanaga River by Bird et al. (1998), who noticed that the Mbakou dam also trapped C4-rich material derived from savanna soils in Cameroon.

The low C:N ratios observed in the fine POM of the Piracicaba River are probably caused by phytoplankton because N-rich remains of phytoplankton tend to concentrate in fine-grain minerals (Hedges and Oades, 1997; Amelung et al., 1999). It is also relevant to consider that the observed N-enrichment could be caused by the attachment of N-rich material provided by wastewater to the suspended solid particles (Martinelli et al., 1999; Krusche et al.,, in press) and by nitrogenous organic matter selectively accumulated in the fine fractions with time by preferential sorption on soil minerals before they are eroded into aquatic systems (Hedges et al., 2000, Aufdenkampe et al., 2001). These mechanisms would also explain the higher N content observed in the fine than in the coarse fraction. The total average C:N ratio was 16 ± 6 for the coarse fraction and was only 9 ± 2 for the fine fraction, a ratio significantly lower than that found in the coarse fraction (Table 16.3).

The Ji-Paraná River had two features in common with the Piracicaba River. First, the C:N ratio of the fine fraction increased with an increase in the fine suspended solids concentration. The C:N ratio of the fine fraction ranged from 8 to 15 between the periods of low and high suspended solids concentration, respectively. Second, the $\delta^{13}C$ values also increased with an increase in the fine suspended solids concentration. However, the increase in the $\delta^{13}C$ values were not so high as was the case in the Piracicaba River (Figure 16.7). During the low water period, the combination of low suspended solids concentration, low C:N ratios, and low $\delta^{13}C$ values suggests that phytoplankton is the main source of POM to the Ji-Paraná (Bernardes et al., in press). In contrast, an increase in the C:N ratio and $\delta^{13}C$ value was observed during the high water period. One possible source for the average bulk POM would be the forest soils that enter the river via overland flow (Figure 16.7). However, the state of Rondônia has one of the largest deforestation rates in the Amazon watershed. The original rain forest is usually replaced by C4 pasture. This replacement started in the mid-1970s, with intensification in the following decades, occurring mainly in the upper region of the Ji-Paraná watershed. Although this replacement started only 30 to 40 years ago, it was noticed that some small streams of the Ji-Paraná watershed are already showing the C4-derived material in their POM composition (Charbel, nonpublished data). In addition, Bernardes et al. (in press) found high $\delta^{13}C$ in the ultra filtered-dissolved organic matter of two tributaries of the Ji-Paraná River, the Rolim-de-Moura, and the Jaru rivers, which have one of the highest areas under pasture. Therefore, it seems that the C4 material from the pastures in Rondônia is already entering in the watershed of the Ji-Paraná River.

A direct relation increase in the $\delta^{13}C$ values of the fine POM fraction with increase in suspended solids concentration, without alteration in the C:N ratio, was observed in the Amazon River (Figure 16.7). The $\delta^{13}C$ increased from –28.5‰ to –25‰ (Figure 16.7), high values observed in the upper reaches of the river (Cai et al., 1988; Quay et al., 1992) where the suspended solids concentration is also high (Bob Meade year). This increase of 2.5‰ may be related to extensive banks of C4 macrophytes living in the Amazon River floodplain, as suggested by Cai et al. (1988), although Martinelli et al. (2003) showed that the C4 macrophytes banks are concentrated in the lower reaches of the river. In addition, using lignin-phenols as a tracer of C4- and C3-derived material, Hedges et al. (1986) found that a maximum of 10% of the Amazon POM may be comprised of C4 material. Therefore, it is more likely that the higher $\delta^{13}C$ values found in the upper river, are caused by the influence of plants growing in the higher altitudes of the Andes that tend be enriched in ^{13}C atoms in relation to plants of the lowlands of the Amazon. In fact, Korner et al. (1988) found that plants of high altitudes have high $\delta^{13}C$ values, and those in the lowland forests of the Amazon tend to have low $\delta^{13}C$ (average –32‰) mainly due to the abundance of water (Martinelli et al., 1998). The lower $\delta^{13}C$ values observed in the lower reaches of the Amazon may be caused by the constant input of these isotopically depleted plants to the river. This hypothesis is supported by the fact that Quay et al. (1992) estimated that at least 35% of the POM exported by the Amazon River is derived from the Terra-Firme forests of the lowland regions of the Amazon basin. Another factor that could decrease the isotopic signal of the POM is the presence of phytoplankton. However, the importance of phytoplankton growth in the Amazon River appears to be minor, mainly due to the high turbidity of the water (Hedges et al., 1986; Richey et al., 1990). In fact, the average C:N ratio of this river was 11 for the fine fraction and 25 for the coarse fraction, both much higher than the typical C:N ratios found for phytoplankton.

16.6 CONCLUSIONS

The riverine coarse and fine POM was already altered by human intervention on the soil cover. Changes in the $\delta^{13}C$ values indicated that the new established C4 vegetation (pasture or sugarcane) was already transported from the terrestrial to the aquatic environment. However, there occurred large seasonal difference between watersheds, suggesting that each watershed has its own functional characteristics. For instance, the presence of reservoirs altered not only the timing but also the proportion in which the C4 material was transferred from the terrestrial to the aquatic ecosystems. The increase of the residence time of the water, together with the increase of N and P loads, enhanced the phytoplankton growth, which became an important component of the fine POM in those watersheds affected by reservoirs. There are only a few, if any, pristine rivers in the world. Therefore, most rivers of the world are likely to be altered similar to the changes observed in the Amazon River system.

REFERENCES

Amelung, W., R. Bol, and C. Friedrich. 1999. Natural C-13 abundance: A tool to trace the incorporation of dung-derived carbon into soil particle-size fractions. *Rapid Communications in Mass Spectrometry* 13:1291–1294.

Angradi, T. R., 1993. Stable carbon and nitrogen isotope analysis of seston in a regulated rocky mountain river, U.S. *Regulated Rivers: Research & Management* 8:251–270.

Aufdenkampe, A. K., J. I. Hedges, J. E. Jeffrey, A. V. Krusche, and C. A. Llerema. 2001. Sorptive fractionation of dissolved organic nitrogen and amino acids onto fine sediments within the Amazon Basin. *Limnology and Oceanography* 46:920–935.

Barth, J. A. C., J. Veizer, and B. Mayer. 1998. Origin of particulate organic carbon in the upper St. Lawrence: Isotopic constraints. *Earth and Planetary Science Letter* 162:111–121

Bernardes, M. C., L. A. Martinelli, A. V. Krusche, J. Gudeman, M. Moreira, R. L. Victoria, J. P. H. B. Ometto, B. V. R. Ballester, A. K. Aufdenkampe, J. E. Richey, J. I. Hedges. 2004. Organic matter composition of rivers of the Ji-Paraná River basin (southwest Amazon) as a function of land use changes. *Ecological Applications,* in press.

Bird, M. I., P. Giresse, and S. Ngos. 1998. A seasonal cycle in the carbon-isotope composition of organic carbon in the Sanaga River, Cameroon. *Limnology and Oceanography* 43:143–146.

Bird, M. I. and P. Pousai. 1997. $\delta^{13}C$ variations in the surface soil organic carbon pool. *Global Biogeochemical Cycles* 11:313–322.

Brannstrom, C. and M. S. Oliveira. 2000. Human modification of stream valleys in the western plateau of São Paulo, Brazil: Implications for environmental narratives and management. *Land Degradation & Development* 11:535–548.

Cai, D. -L., F. C. Tan, and J. M. Edmond. 1988. Sources and Transport of Particulate Organic Carbon in the Amazon River and Estuary. *Estuarine, Coastal and Shelf Science* 26:1–14.

Devol, A. H. and J. I. Hedges. 2001. Organic matter and nutrients in the main stem Amazon River, in M. E. McClain, R. L. Victoria, and J. E. Richey, eds., *The Biogeochemistry of the Amazon Basin.* Oxford University Press, Oxford, pp. 275–306.

Domingures, T. F. 2000. Dinâmica do Carbono e Elementos Relacionados nos Rios Piracicaba e Mogi-Guaçu (SP): Avaliação de Duas Bacias Hidrográficas com Diferentes Graus de Desenvolvimento. *Tese de Mestrado,* Fevereiro, 2000.

Goñi, M. A., K. C. Ruttenberg, and T. I. Eglinton. 1998. A reassessment of the sources and importance of land-derived organic matter in surface sediments from the Gulf of Mexico. *Geochimica et Cosmochimica Acta* 62:3055–3075.

Hedges, J. I., E. Mayorga, E. Tsamarkis, M. E. McClain, A. Aufdenkampe, P. Quay, J. E. Richey, R. Benner, S. Opsahl, B. Black, T. Pimentel, L. Quintanilla, and L. Maurice. 2000. Organic matter in Bolivian tributaries of the Amazon River: A comparison to the lower mainstream. *Limnology and Oceanography* 45: 1449–1466.

Hedges, J. I. and J. M. Oades. 1997. Comparative organic geochemistries of soils and marine sediments. *Organic Geochemistry* 27:319–361.

Hedges, J. I., W. A. Clark, P. D. Quay, J. E. Richey, A. H. Devol, and U. D. Santos. 1986. Compositions and fluxes of particulate organic material in the Amazon River. *Limnology and Oceanography* 31:717–738.

Houghton, R. A. 1994. The worldwide extent of land-use changes. *BioScience* 44:305–313.

Kendall, C., R. S. Steven, and V. J. Kelly. 2001. Carbon and nitrogen isotopic compositions of particulate organic matter in four large river systems across the U.S. *Hydrological Process* 15:1301–1346.

Korner, C., G. D. Farquhar, and Z. Roksandic. 1988. A global survey of carbon isotope discrimination in plants from high-altitude. *Oecologia* 74:623–632.

Krusche, A. V., L. A. Martinelli, R. L. Victoria, M. C. Bernardes, P. B. Camargo, and S. Trumbore. 2002. Composition of particulate organic matter in a disturbed basin of southeast Brazil (Piracicaba River basin). *Water Research* 36 (11):2743–2752.

Krusche, A. V., P. B. de Camargo, C. E. Cerri, M. V. Ballester, L. B. L. S. Lara, R. L. Victoria, and L. A. Martinelli. 2003. Acid rain and nitrogen deposition in a sub-tropical watershed (Piracicaba): Ecosystem consequences. *Environmental Pollution* 121:389–399.

Lobbes, J. M., H. P. Fitznar, and G. Kattner. 2000. Biogeochemical characteristics of dissolved and particulate organic matter in Russian rivers entering the Arctic Ocean. *Geochimica et Cosmochimica Acta* 64 (17):2973–2983.

Martinelli, L. A., S. Almeida, I. F. Brown, M. Z. Moreira, R. L. Victoria, L. S. L. Sternberg, C. A. C. Ferreira, and W. W. Thomas. 1998 Stable carbon isotope ratio of tree leaves, boles and fine litter in a tropical forest in Rondônia, Brazil. *Oecologia* 114:170–179.

Martinelli, L. A., M. V. R. Ballester, A. V. Krusche, R. L. Victoria, P. B. Camargo, M. Bernardes, and J. P. H. B. Ometto. 1999. Landcover changes and 13C composition of riverine particulate organic matter in the Piracicaba River basin (southeast region of Brazil). *Limnology and Oceanography* 44:1826–1833.

Martinelli, L. A., R. L. Victoria, P. B. Camargo, M. C. Piccolo, L. Mertes, J. E. Richey, A. H. Devol, and B. R. Forsberg. 2003. Inland variability of carbon-nitrogen concentrations and $\delta^{13}C$ in Amazon floodplain (várzea) vegetation and sediment. *Hydrological Process* 17:1419–1430.

Mayer, L. M., R. G. Keil, S. A. Macko, S. B. Joye, K. C. Ruttenberg, and R. C. Aller. 1998. Importance of suspended particulates in riverine delivery of bioavailable nitrogen to coastal zones. *Global Biogeochemical Cycles* 12 (4):573–579.

McCusker, E. M, P. H. Ostrom, N. E. Ostrom, J. D. Jeremiason, and J. E. Baker. 1999. Seasonal variation in the biogeochemical cycling of seston in Grand Traverse Bay, Lake Michigan 1999. *Organic Geochemistry* 30:1543–1557.

Meade, R. H., T. Dunne, J. E. Richey, U. M. Santos, and E. Salati. 1985. Storage and Remobilization of Suspended Sediment in the Lower Amazon River of Brazil. *Science* 228:488–490.

Meybeck M. 1982. Carbon, nitrogen and phosphorus transport by world rivers. *American Journal of Science* 282:401–450.

Neill, C., J. M. Melillo, P. A. Steudler, C. C. Cerri, J. F. L. Moraes, M. C. Piccolo, and M. Brito. 1997. Soil carbon and nitrogen stocks following forest clearing for pasture in the southwestern Brazilian Amazon. *Ecological Applications* 7:1216–1225.

Ometto, J. P. H. B., L. A. Martinelli, M. V. Ballester, A. Gessner, A. Krusche, R. L. Victoria, and M. Williams. 2000. Effects of land use on water chemistry and macroinvertebrate in two streams of the Piracicaba River basin, Southeast Brazil. *Freshwater Biology* 44:327–337.

Onstad, G. D., D. E. Canfield, P. D. Quay, and J. I. Hedges. 2000. Sources of particulate organic matter in rivers from the continental U.S.: Lignin phenol and stable carbon isotope compositions. *Geochimica et Cosmochimica Acta* 64:3539–3546.

Pedlowski, M. A., V. H. Dale, E. A. T. Matricardi and E. P. da Silva Filho. 1997. Patterns and impacts of deforestation in Rondônia, Brazil. *Landscape and Urban Planning* 38:149–157.

Quay, P. D., D. O. Wilbur, J. E. Richey, J. I. Hedges, A. H. Devol, and R. Victoria. 1992. Carbon cycling in the Amazon River — Implications from the C-13 compositions of particles and solutes. *Limnology and Oceanography* 37:857–871.

Richey, J. E., J. I. Hedges, A. H. Devol, P. D. Quay, R. Victoria, L. Martinelli, and B. R. Forsberg. 1990. Biogeochemistry of carbon in the Amazon River. *Limnology and Oceanography* 35:352–371.

Zhang, J., S. M. Liu, H. Xu, Z. G. Yu, S. Q. Lai, H. Zhang, G. Y. Geng, and J. F. Chen. 1998. Riverine Sources and Estuarine Fates of Particulate Organic Carbon from North China in Late Summer. *Estuarine, Coastal and Shelf Science* 46:439–448.

CHAPTER **17**

Organic Carbon Transported by the Equatorial Rivers: Example of Congo-Zaire and Amazon Basins

Patrick Seyler, A. Coynel, Patricia Moreira-Turcq, Henri Etcheber, C. Colas, Didier Orange, Jean-Pierre Bricquet, André Laraque, Jean Luc Guyot, Jean-Claude Olivry, and Michel Meybeck

CONTENTS

17.1 Introduction 256
17.2 Methodology 256
    17.2.1 General Characteristics of the Studied Basins 256
        17.2.1.1 The Congo-Zaire Basin 256
        17.2.1.2 The Amazon Basin 258
17.3 Sampling and Analytical Methods 259
    17.3.1 Sampling Frequency 260
    17.3.2 Sampling and Analysis Procedures 260
17.4 Results 260
    17.4.1 Spatial Variations of TSS, POC, and DOC Concentrations in the Congo-Zaire River and Tributaries 260
    17.4.2 Spatial Variations of TSS, POC, and DOC Concentrations in the Amazon River and Tributaries 261
17.5 Discussion 264
    17.5.1 Factors Controlling the TSS, POC, and DOC Concentrations in the Congo-Zaire and Amazon Basins 264
    17.5.2 Organic Carbon Flux from Congo-Zaire and Amazon Basins 265
        17.5.2.1 Congo-Zaire Basin 267
        17.5.2.2 Amazon Basin 268
    17.5.3 Congo-Zaire and Amazon Organic Carbon Yields: Comparisons with World Ranges 269
17.6 Conclusions 269
References 272

17.1 INTRODUCTION

The consequences of deforestation and land management of equatorial rainforests extend beyond the regional issues of conservation and sustainable land use to global issues of climate change. With CO_2 and CH_4 levels continuing to rise in the atmosphere, many scientists consider the tropical forests as one of the key issues in the global carbon (C) budget. Do the tropical forests act as a source or a sink for atmospheric C? The answer to this question warrants evaluating precisely how much C is retained in soils and vegetation and how much is emitted into the atmosphere and transported to oceans by rivers.

The erosion of C from land to sea via rivers represents a major pathway in the global C cycle (Kempe, 1979; Degens et al., 1984). With respect to the total flux of C and that carried by world rivers (G t yr^{-1}), the contribution of organic C is estimated to represent ~40%, with 16% being exported from the tropical rain forest environment (see Meybeck et al., this volume). Investigations conducted by SCOPE/CARBON program since 1980 (Degens, 1982), have substantially improved our knowledge of the fluvial C fluxes (Degens et al., 1984; 1985; 1991; Kempe et al., 1993; Lewis and Saunders, 1989; Richey et al., 1990; 1991; Paolini, 1995). However, even if the global figures are more precise today, important gaps persist, mainly due to the lack of data for some rivers (Meybeck, 1982; 1993b; IGBP, 1995; Billen et al., 1998), and the scarcity of river sampling unsuited to compute realistic riverine C fluxes. This chapter is concerned with the determination of fluxes of organic C in the two largest rivers in the world: the Congo-Zaire River and the Amazon River, which are both responsible for almost 50% of the freshwater inputs into the Atlantic Ocean (Degens et al., 1991; Probst, 1990). Whereas a large amount of data has been published concerning the Amazon, including the cycling and fluxes of bioactive organic compounds (Richey et al., 1980; Junk, 1985; Ertel et al., 1986; Hedges et al., 1986; Quay et al., 1992; Hedges et al., 1994; Mounier et al., 1998; Patel et al., 1999; Moreira-Turcq et al., 2003), almost nothing has been published on the Congo-Zaire basin with a few exceptions (Martins and Probst, 1991; Seyler et al., 1995). Consequently, comparisons between these two vast rivers draining the largest rainforest areas of the world, and — up to now — weakly impacted by anthropogenic perturbations have not been done.

This chapter considers the organic C species distributions in the two mainstreams and their tributaries, to compare levels and yields between the two basins and to present quantitative estimates of organic fluxes from Congo-Zaire and Amazon rivers entering the Atlantic Ocean.

17.2 METHODOLOGY

17.2.1 General Characteristics of the Studied Basins

17.2.1.1 The Congo-Zaire Basin

The Congo-Zaire basin lies at the center of equatorial Africa (Figure 17.1) and is the second largest basin in the world. Its watershed (3.8×10^6 km^2) is mostly constituted by a large peneplain (altitudes lower than 400 m) surrounded by highlands to the north and the south and by the mountainous chain of the East African valley to the east. Lake Tanganyika and its drainage basin are also part of the Congo-Zaire basin. Its central region is covered by an evergreen forest (50% of the total area) surrounded by tree savannahs (De Namur, 1990). The Congo-Zaire basin is characterized by a wet tropical climate. The mean annual rainfall calculated for the 1980s (Mahe, 1993) is 1550 mm yr^{-1}, and the mean temperature is > 20°C. The hydrological regime of the Congo-Zaire River is mainly pluvial and discharge fluctuations are due to the distribution of its tributaries on both sides of the equator resulting in an annual hydrologic cycle with two maxima in December and May, and minimum flows in August and March. Long-term average discharge at the Kinshasa-

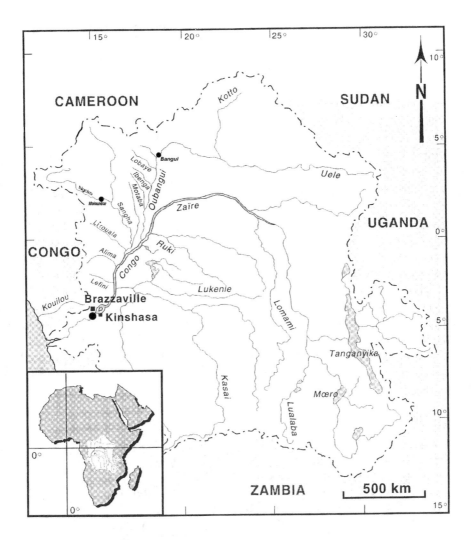

Figure 17.1 Map of the Congo-Zaire basin showing the location of collection sites.

Brazzaville hydrological station is about 40,600 m^3 s^{-1}, and average discharge during the study period was 37,700 m^3 s^{-1} or 11 l s^{-1} km^{-2} (Bricquet, 1995). The Congo-Zaire hydrological regime is one of the most steady in the world (irregular interannual ratio = 1.43).

Among the main tributaries of the Congo-Zaire River (Upper Zaire, Oubangui, Ngoko-Sangha, Likouala Mossaka, and Kasaï rivers), the Oubangui and the Ngoko-Sangha are the focuses of the present chapter.

With a drainage basin of about 489,000 km^2 and a mean flow of about 4,200 m^3 s^{-1} (5.8 l s^{-1} km^{-2}) at the Bangui gauge station, the Oubangui River is the second most important tributary of the Congo-Zaire River system. The mean annual rainfall of the Oubangui drainage basin is 1,540 mm yr^{-1} and its vegetation mainly comprises dry tree savannah (Boulvert, 1992).

The Ngoko-Sangha River constitutes the upper part of the Sangha River, a right bank tributary of the Congo-Zaire River. It drains an homogeneous forested basin which covers 67,000 km^2. The average rainfall extends upward of 1,700 mm yr^{-1}. The hydrological regime is mainly pluvial, with maximum discharge observed in October and minimum in March through April. A secondary discharge peak occurs in July. The mean annual discharge is 713 m^3 s^{-1} (11.3 l s^{-1} km^{-2}). Humid evergreen forest covers 95% of the basin.

Table 17.1 Physical Characteristics of the Congo–Zaire River and Its Tributaries

River	Station	Basin Area km²	Mean Annual Discharge m³ s⁻¹	Runoff l s⁻¹ km²	Forested Area in the Basin %
Oubangui	Bangui	489,000	3,750	7.7	22
Ngoko-Sangha	Moloundou	67,000	715	10.7	95
Congo-Zaire	Kinshasa-Brazzaville	3,500,000	40,600	11.6	50

Major features of the Congo and its tributaries are shown in Table 17.1. More complete information about morphology, lithology, and vegetation of the watersheds has been published by Bricquet (1995), Olivry (1986), Orange et al. (1999), Seyler et al. (1993), and Sigha Nkamdjou et al. (1995).

17.2.1.2 The Amazon Basin

The Amazon basin (Figure 17.2) covers 6.4×10^6 km² and has an average discharge of 209,000 m³ s⁻¹, supplying up to 20% of all the river water discharged into the ocean (Molinier et al., 1997). The basin is bordered by the Andes Cordillera and the sub-Andean region in the west, and by Guyana and Brazilian Shields to the north and south, respectively. The entire Amazon basin is covered by tropical rainforest (71%) and savannas (29%; Sioli, 1984). Native vegetation in the forested basins is classified as moist open tropical forest and consists of perennially evergreen broadleaf trees with a high number of Palms (Pires and Prance, 1986). An inundated forest predominates in the lowest part of Negro River basin.

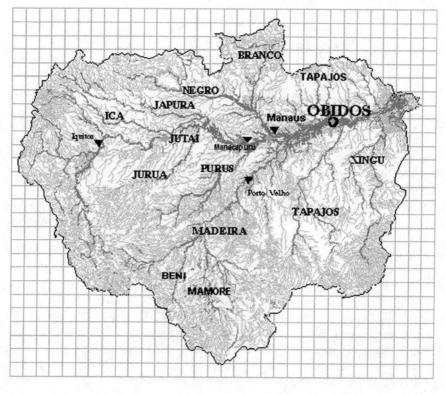

Figure 17.2 Map of the Amazon basin.

Table 17.2 Physical Characteristics of the Amazon River and Its Tributaries

Basin	River	Station	Basin Area km²	Mean Annual Discharge m³ s⁻¹	Runoff l s⁻¹ km²
Negro	Negro	Manaus	689,810	28,400	40.8
Solimoes	Solimoes	Manacapuru	2,147,740	103,000	48.0
Madeira	Beni	Villabela	282,500	8,920	32
	Mamore	Guajaramerin	599,400	8340	14
	Madeira	Foz	1,420,000	31,200	22.0
Amazon	Trombetas	Oriximina	128,000	2555	20.0
	Tapajos	Santarem	490,000	13,500	27.6
	Xingu	Porto de Moz	504,300	9,700	19.2
	Amazon	Obidos	4,618,750	168,000	36.5

In Brazil, the Amazon River refers to the mainstream channel downstream from the confluence of the Solimões and Rio Negro Rivers. Mean annual rainfall in the basin is about 2,000 mm yr⁻¹, and the water regime of the main channel is characterized by one high water stage and one low water stage, which occur between May and August and September and December, respectively. The main tributaries are the Solimões, the Negro, and the Madeira rivers. Its tributaries are classified according to their color, which varies depending on whether dissolved organic matter (black water tributaries) or suspended sediment (white water tributaries) predominates. The Solimões River is formed by the confluence of the Ucayali and Maranõn rivers, which originate from the Andes, and has a surface area of 2,240,000 km². It receives a mean rainfall of 2,900 mm yr⁻¹ and has a mean discharge of 103, 000 m³ s⁻¹. The Negro River, archetype of the blackwater rivers, is volumetrically the largest tributary of the Amazon with a surface area of 686, 810 km², a mean discharge of 28,400 m³ s⁻¹, and a mean rainfall of 2,566 mm yr⁻¹. From its left bank it receives the Branco River, a typical white water river, draining a dry savannah region situated in the north hemisphere, whereas the Negro River drains the densest part of the rain forest.

Two hundred kilometers downstream of the Solimões-Negro confluence, the Amazon River receives water from the Madeira River, which comes from the Bolivian Andes and passes through the central Amazon plain. The Madeira mainstream is formed by the confluence of the Mamore and Beni Rivers in Bolivia. The entire basin has a surface area of 1,420,000 km² and is characterized by a mean rainfall of 1,940 mm yr⁻¹, and a mean discharge of 31,200 m³ s⁻¹. The main tributaries of the lower course, the Trombetas, Tapajós, and Xingú rivers drain the Brazilian shield. Tapajós (area, 490, 000 km²; mean rainfall, 2,250 mm yr⁻¹; mean discharge, 13,500 m³ s⁻¹), Xingú (area, 504,300 km²; mean rainfall, 1,930 mm yr⁻¹; mean discharge, 9,700 m³ s⁻¹) and Trombetas (area, 128,000 km²; mean rainfall, 1,822 mm yr⁻¹; mean discharge, 2,555 m³ s⁻¹) are known as the clear water rivers of the Amazon basin (Sioli, 1984).

One of the largest riverine wetlands in the world is the floodplain of the Amazon River and its tributaries (Junk 1997). Estimates vary between 100,000 (Junk, 1985) to 300,000 km² (Junk, 1997). In the depressions of the terrain, covering a considerable area, the floodplain oxbow lakes called "várzeas" are formed, characterized by a high production of phytoplankton (Richey et al., 1990; Junk, 1997).

Major features of the Amazon basin are shown in Table 17.2. Additional details about main characteristics of the Amazon basin are reported by McClain et al. (2001).

17.3 SAMPLING AND ANALYTICAL METHODS

This chapter is based on an extensive dataset of analyses conducted for the Congo-Zaire basin between 1990 and 1996 by the PIRAT/PEGI program supported by INSU/CNRS, and for the Amazon basin between 1994 and 2000 by the HyBAm (Hydrology and geochemistry of the Amazon basin) project supported the IRD (French Research Institute for Development). The Following abbreviations are used throughout the text:

TSS: Total suspended solids (expressed in mg l^{-1})
POC: Particulate organic carbon (expressed in mg l^{-1})
POC%: Particulate organic carbon (expressed as a percentage of TSS)
DOC: Dissolved organic carbon (expressed in mg l^{-1})
TOC: Total organic carbon, i.e., sum of the DOC and the POC (expressed in mg l^{-1})

17.3.1 Sampling Frequency

With regard to Congo-Zaire River, samples were collected during various cruises carried out between Bangui (Central African Republic) and Brazzaville (Republic of Congo) during low and high water stages. Moreover, three "key stations" where monthly time series were carried out, have been selected on the basis of their particular characteristics (hydrology, vegetation type) in order to highlight the factors influencing the temporal distribution of organic C.

- Bangui station on the Oubangui River (savannah region) was sampled monthly between November 1990 and September 1996.
- Moloundou station on the Ngoko-Sangha River drains a basin, which is 95% tropical rain forest, and was sampled between January and December 1991.
- Brazzaville/Kinshasa station, which covers almost the entire Congo-Zaire basin, was sampled between November 1990 and October 1993.

In the Amazon basin, samples were collected during twelve cruises. During each cruise, one river was generally sampled preferentially, but all of the other key stations corresponding to the outlet of each sub-basin (Solimões, Negro, Branco, and Madeira rivers) were routinely sampled at least four times per year as well as the Óbidos station which controls 95% of water and 99% of the sediment discharge in the Amazon River (Callède et al., 2002; Filizola, in press). The key stations were hydrologically instrumented, and discharge data were collected on a daily basis.

17.3.2 Sampling and Analysis Procedures

A 1 liter sample was taken in the center of the river cross section in sterilized containers. After homogenization, precise volume of water was filtered through preheated and preweighed 0.70 μm Whatman GF/F fiberglass filters under reduced pressure to separate dissolved and particulate matters.

The filters were dried in an oven at 50°C for 24 h and weighed to determine TSS concentrations. Then filters were decarbonated with 2N HCl to eliminate carbonates, and dried at 60°C for 24 h. The POC contents were measured on a LECO CS 125 analyzer. Detailed description of POC analysis is reported by Etcheber (1986).

The water-fractions passing through the filter were acidified on a board with ultrapure H_3PO_4 and analyzed in the laboratory by high-temperature catalytic oxidation method (HTCO) using a Shimadzu TOC-5000 Instrument to determine DOC concentrations (Abril et al., 2002).

17.4 RESULTS

17.4.1 Spatial Variations of TSS, POC, and DOC Concentrations in the Congo-Zaire River and Tributaries

Analytical data for the Congo-Zaire basin are reported in Table 17.3. The mean concentration of TSS is relatively low, and ranges from 6 to 36 mg l^{-1} for three key stations. These concentrations

Table 17.3 Average Concentrations of TSS, POC, and DOC in the Oubangui, Ngoko–Sangha, and Congo–Zaire Basins

Rivers	Hydrological Stages	Water Discharges (m^3 s^{-1})	Specific Water Discharges (l s^{-1} km^{-2})	TSS (mg l^{-1})	POC (%)	POC (mg l^{-1})	DOC (mg l^{-1})	DOC/ TOC (%)
Oubangui at Bangui	Average	3,005	6.2	18.9	6.3	1.2	5.0	81.0
	High waters	4,556	9.3	28.1	5.7	1.6	6.1	79.0
	Low waters	834	1.7	6.1	11.5	0.7	3.5	83.0
Ngoko-Sangha at Moloundou	Average	862	12.9	28.8	6.9	2.0	10.5	84.0
	High waters	1,012	15.1	33.6	6.8	2.3	11.8	84.0
	Low waters	412	6.1	16.7	7.8	1.3	3.8	75.0
Congo-Zaïre at Brazzaville	Average	37,047	10.6	27.1	6.3	1.7	9.8	85.0
	High waters	41,232	11.8	24.6	6.1	1.5	11.0	88.0
	Low waters	32,861	9.4	29.5	6.4	1.9	8.6	82.0

are among the lowest reported in river water, and attributed to three factors: (1) flat terrain, (2) good vegetal cover, and (3) lack of highly erodible soils. The TSS values are higher in high water than low water stage, except for Congo-Zaire at the Brazzaville station where the difference is not significant due to the low seasonal variability of the hydrograph. The annual average concentration of POC varies from 1.2 mg l^{-1} (at Bangui station) to 2.0 mg l^{-1} (at Moloundou station) and follows the same pattern as the TSS with the maximum concentrations during the high water stages, except for the Brazzaville station as has been discussed before. The percentage of C contained in TSS (POC%) is relatively high as compared to temperate rivers (Meybeck, this volume), ranging from 5.7 to 11.5% in the Oubangui River, 6.8 to 7.8% in the Ngoko/Sangha River, and 6.1 to 6.3% in the Congo-Zaire River, for low and high water flows, respectively. Comparing these data obtained in the Congo-Zaire River with those already published and obtained in general from a small set of data, there exists a good agreement (6%, Cadet, 1984; 7%, Kinga Mouzeo, 1986). Comparing the mean POC values of the Oubangui River (1.2%) with others flowing in African savannah, an excellent agreement is observed with the Senegal (1.2%, Orange, 1990) and Gambia rivers (Lesack et al., 1984).

The concentration of dissolved organic carbon (DOC) differs between the Bangui station on the Oubangui River and Moloundou station on the Ngoko/Sangha River. At Bangui station, or the "Savannah observatory," the DOC concentrations ranged from 3.5 mg l^{-1} during the periods of low flow to 6.1 mg l^{-1} during the flood period with an annual average value of 5 mg l^{-1}. At Moloundou Station, or the "forest observatory," DOC concentrations ranged from 3.8 mg l^{-1} to 11.8 mg l^{-1}, with an annual average of 10.5 mg l^{-1}, i.e., three times higher than the former. Concerning the other main tributaries of the Congo-Zaire River, the highest DOC concentrations are found in the Upper Zaire at Mbandaka (18.1 mg l^{-1}, n = 5) and in the Ruki River (18.9 mg l^{-1}, n = 2), the latter draining the marshes and inundated forest zone situated in the center of the basin (Seyler et al., 1995). The right bank tributaries of the lower Congo-Zaire River upstream Kinshasa/Brazzaville, called Bateke Rivers (Djiri, Lefini, Nkeni, and Alima rivers), have an average DOC concentrations of 3.5 mg l^{-1} (Seyler et al., 1995).

It is also interesting to compare DOC/TOC ratio for each river, which is related with the specific phase (particulate or dissolved) on which organic C is primarily transported to the ocean. For the three key stations, DOC is apparently the dominant form with a mean concentration of 0.79 mg l^{-1} in the Oubangui River, 0.84 mg l^{-1} in the Ngoko-Sangha River, and 0.85 mg l^{-1} in the Congo-Zaire River.

17.4.2 Spatial Variations of TSS, POC, and DOC Concentrations in the Amazon River and Tributaries

Results for the Amazon basins at Óbidos station and for the sub-basins at the confluences with the main channel are shown in Table 17.4 and Figure 17.3. In the Negro basin, lower concentrations

Table 17.4 Average Concentrations of TSS, POC, and DOC in the Amazon River and Its Main Tributaries

Basin	River	TSS (mg l^{-1})	POC (mg l^{-1})	POC (%)	DOC (mg l^{-1})	DOC/TOC (%)
Negro	Negro	5	0.72	13.6	10.25	93%
Solimões	Solimões	81	1.3	1.6	4.5	78%
Madeira	Béni	451	2.9	0.6	4.11	59%
	Mamoré	109	1.12	1	7.68	87%
	Madeira	233	2.1	0.9	3.8	64%
Amazon	Trombetas	6	0.65	10.3	5.75	90%
	Tapajos	6	0.44	7.5	4.45	91%
	Xingu	9	0.95	10.7	3.4	78%
	Amazon	61	1.08	1.8	6.94	87%

of TSS (about 3.7 mg l^{-1}) were observed in low water period, whereas higher concentrations (up to 17 mg l^{-1}) were observed during the peak discharge. The same trend was observed for POC concentrations. The POC in the Negro represents only 9% of the TOC, and concentrations were relatively low (mean = 0.86 mg l^{-1}). However, the POC contents in terms of percentage of TSS are high, ranging from 5.4 to 31%, with a mean of 13%. The minimum POC corresponding to low waters is probably linked to the acidic and oligotrophic nature of the Negro River waters. This trend is one of the major differences with the POC pattern in the Congo-Zaire River. The organic C data obtained in the Negro River, confirmed that the entire Negro River basin contains large amounts of DOC and a low content of suspended matter. The concentrations of DOC ranges between 3 and 18 mg l^{-1} and the mean DOC concentration in the Negro basin is 12.7 mg l^{-1}. The lowest concentrations are observed during a rising water period. The DOC values represent an average of 93% of the TOC, whose contribution to Amazon does not vary much with different water levels. The DOC was always the principal component of TOC. Concerning the major tributary of the Negro River, the mean concentration of suspended solids in the Branco basin during peak discharge was about 22.7 mg l^{-1}, which is three times higher than that during the periods of low water level (7.5 mg l^{-1}). The Branco River contribution of organic C to Negro River is small, and represents the major contribution in TSS for the lower reach of the Negro basin. The POC concentrations are similar to those in other white rivers (0.31 to 2.67 mg l^{-1}) and seem to follow the same pattern as TSS. The POC does not vary significantly between different periods of low water and represented in average 5% of the TSS. As in the Negro river, the DOC fraction is also the main C fraction, representing around 80% of the TOC. The DOC concentrations are variable along the Branco River, with a mean of 3.5 mg l^{-1}, or four times the concentration of the Negro River. Already pointed out for the Congo basin, there is a strong influence of the vegetation cover on the POC concentration of riverine C.

In the Solimões River, the TSS concentrations obtained at the Manacapuru station for the entire hydrological cycle show that peak sediment discharge occurs when the water level begin to rise two months before the maximum flow (Filizola, 2003). The POC concentrations vary between 0.60 and 2.16 mg l^{-1} in July and April, respectively. With regard to the POC, it varies between 0.6 in March at the time of the high TSS discharge (141 mg l^{-1}) and 3.7 in July at the time when TSS concentrations are lowest (10 to 20 mg l^{-1}; Filizola, 2003). As in the Negro River, DOC was the dominant form of organic C, corresponding to about 76% of TOC. This percentage increases during low water periods when DOC concentrations are high. The mean DOC concentration was 5.88 mg l^{-1} for the whole of the Solimões basin, whereas the mean POC concentration was 1.19 mg l^{-1} and the mean POC% was 4.6%.

In the Madeira River, the concentration of TSS was correlated with river discharge. The highest concentrations (up to about 500 mg l^{-1}) were observed in April and the lowest concentrations (about 18 mg l^{-1}) in June. The POC concentration closely follows the same pattern. The POC concentration ranged between 0.17 and 4.46 mg l^{-1} in September and April, respectively. With regard to the POC,

ORGANIC CARBON TRANSPORTED BY THE EQUATORIAL RIVERS

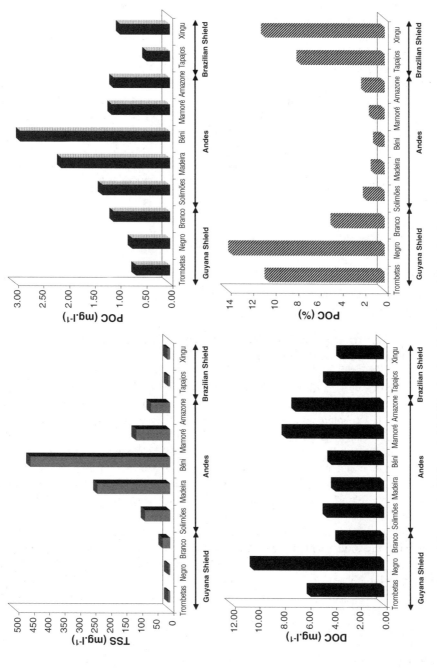

Figure 17.3 Average values of TSS, POC (mg l⁻¹), POC%, and DOC in the Amazon basin and its main tributaries.

it ranged between 3.3% in July during low waters and 0.8% in April at the time of the maximum TSS. The positive relations between TSS/POC/Water discharge and the opposite TSS/POC were also observed in the Congo-Zaire basin. During high flow period DOC concentrations ranged from 7.6 mg l^{-1} at the boundary between Bolivia and Brazil to 3.9 at its confluence with the Amazon River. The lowest concentrations were observed during the low stage (1.7 to 1.9 mg l^{-1}). The DOC/TOC ratio was generally close to 0.5, indicating that the C was transported as much in the dissolved form as in the particulate form.

Clearwater rivers (Tapajós, Xingú, and Trombetas) are characterized by intermediate concentrations of TSS and organic C compared to black and white rivers. The Tapajós River had a mean TSS concentration of 7.2 mg l^{-1}, a mean POC of 3.6 mg l^{-1} (POC% of 12.8%), and a mean DOC of 4.2 mg l^{-1} The concentrations of TSS, POC, and DOC are in the same range as those for the Xingú River (2.8 to 3.4 for the DOC mg l^{-1}, 10% for the POC). The ratio of DOC/TOC was similar in the Tapajós and the Xingú rivers (about 85%). In the Trombetas River the mean TSS concentration was 8 mg l^{-1}, the mean POC concentration was 0.69 mg l^{-1}, the mean POC% was 9.8%, and the mean DOC concentration was 5.5 mg l^{-1}.

Results obtained in these two large tropical basins indicate C dynamics. The main factors that explain the distribution of organic C in the different rivers of Congo-Zaire and Amazon basins are discussed below.

17.5 DISCUSSION

The results obtained show the impact of different parameters affecting the distribution of organic C concentrations, TSS, POC, and DOC in these rivers, and comparison of these data with other rivers of the world.

17.5.1 Factors Controlling the TSS, POC, and DOC Concentrations in the Congo-Zaire and Amazon Basins

The study of the Congo-Zaire tributaries and Amazon basin rivers show that the fluctuations in the TSS and POC concentrations are closely related to the hydrological factors. An increase in mechanical erosion is expected with increase in water discharge from the basin, leading to the mobilization by runoff of a huge POC stock associated with the mineral matrix of eroded soils and clay minerals but also contained in the litters of the topsoils of the basin and riparian zones. During the falling stage, POC concentrations decrease with the decrease in TSS, due to the progressive decline of the erosion processes. Such pattern is not typical of tropical rivers, but has been described for other major rivers (Telang et al., 1991; Kaplan et al., 1980; Colas, 1994; Maneux, 1998; Veyssy, 1998; Zhang et al., 1992).

For the entire data set, an inverse relation was observed between POC and TSS (Figure 17.4), which fits a logarithmic model with a correlation coefficient of 0.96. River erosion can be grouped into two categories: predominately chemical (weathering) or mechanical erosion processes. For the high turbid rivers, the mechanical erosion process dominates (TSS high and POC low) whereas for the low turbid rivers, the chemical weathering process is dominant. This distinction is independent of the type of basin since Ngoko/Sangha, Congo-Zaire, and Oubangui have the same characteristics as Tapajós (Amazon basin; Figure 17.4).

For the Amazon basin, the effect of altitude on organic C loads is shown in Figure 17.3, which compares concentrations of TSS, POC (mg l^{-1} and %), and DOC with the geographical parameters of the sub-basins. For instance, high concentrations of TSS and POC (mg l^{-1}) contents were observed in the rivers flowing from the Andean region, whereas high concentration in DOC and POC (%) were observed for lowland rivers draining the Shields. These observations are in agreement with the results obtained by other investigators (Milliman and Meade, 1983; Ludwig et al., 1996; Guyot

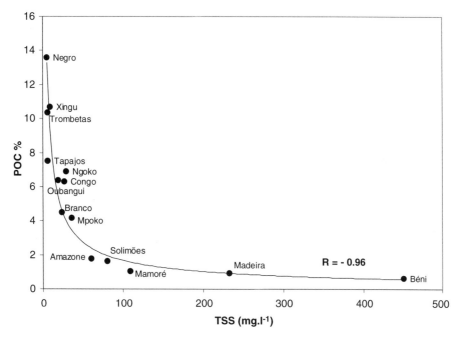

Figure 17.4 Variation of organic carbon content in river particulate matter (POC in %) for Congo-Zaire and Amazon sub-basins. Best fit is obtained with a logarithmic model (coefficient correlation = 0.92).

and Wasson, 1994), and show the importance of the contribution of the mountainous basins to the global fluxes of TSS and POC. The large and highly contrasting, and still relatively pristine basin of the Madeira River constitutes a suitable study area to test the impact of altitude on the riverine organic C concentration. The altitude decreases from 5,500 m at the summit of the basin to 120 m at the piedmont of the Andean Cordillera. The DOC concentrations vs. the altitude of sampling points are plotted in Figure 17.5. A marked enrichment in DOC was observed at 250 m altitude where the Madera River enters the flat plain. These results corroborate the observations made by Hedges et al. (2000) and Guyot and Wasson (1994), who also observed a strong contrast in DOC river concentrations between the Andean mountainous region (average of 2.2 mg l^{-1}) and the lowlands (average of 5.7 mg l^{-1}).

Whereas an "altitude effect" was observed in the distribution of organic loads of rivers, some variations in POC and DOC contents in lowlands rivers (e.g., Oubangui vs. Ngoko/Sangha rivers or Negro vs. Branco rivers) suggest that other key parameters such as the vegetative cover may explain these differences. The effect of the vegetation on the DOC river contents can be observed plotting the DOC concentrations and the percentage of forest cover in the basins of the Madeira River and the Congo-Zaire (Figure 17.6). This percentage was calculated for each sampling station watershed with the vegetation map of Africa and South America (Global Land Cover 2000 project EEC, 2003). The concentrations of DOC increased in both cases with an increase in the percentage of forested area in each basin. The comparison of the two basins shows two distinct trends with higher values observed for the Congo-Zaire River system. As previously mentioned, the higher altitude of the Madeira basin may explain lower DOC contents than those observed in the Congo-Zaire River for a similar vegetation cover.

17.5.2 Organic Carbon Flux from Congo-Zaire and Amazon Basins

To compute the organic C flux of the rivers, a monthly average was first calculated when more than one value was available by month (case of the Congo-Zaire basin). When just one value was

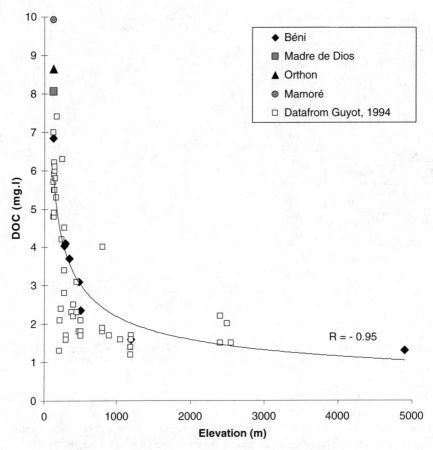

Figure 17.5 Mean DOC value vs. sampling point altitude (m above sea level); data from Guyot and Wasson, 1994 are also reported.

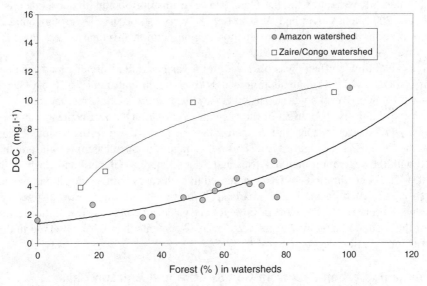

Figure 17.6 Mean DOC concentrations (mg l^{-1}) as a function of the forested area (%) in the Congo-Zaire and Madeira basins.

Table 17.5 Carbon Budget of the Congo–Zaire Basin

Basin	Annual Discharge[a] m³ s⁻¹	POC Flux Tg yr⁻¹	DOC Flux	TOC Flux	DOC Flux/ TOC Flux %
Oubangui	4,000	0.2	0.4	0.6	67
Ngoko/Sangha+Likouala	2,200	0.2	1.2	1.4	86
Kasai	8,500	0.4	3.1	3.5	89
Upper Zaire	16,960	1	6.5	7.5	87
Bateke Rivers	2,400	0.1	0.3	0.4	75
Congo-Zaire[b]	42,000	1.9	11.5	13.4	86

[a] Water discharge at the confluence with the main stream.
[b] Data for Kinshasa-Brazzaville Station.

available during a given month, this value was considered representative for the month. The mean annual flux of DOC, POC, and TSS were then computed by the summation of the 12 interannual monthly values.

17.5.2.1 Congo-Zaire Basin

Based on annual flux computed for the key stations and at the confluence of other tributaries (Upper Zaire, Kasai, Likouala, and Bateke rivers), a budget of the C species is presented for the Congo-Zaire basin (Table 17.5 and Figure 17.7). The data show the following: irrespective of the type of vegetation in the sub-basins, the dissolved C flux is a major component of the total flux. However, a marked difference is observed between the savannah and the forested basins. The ratio dissolved flux/particulate flux is 70% in the former and always up to 86% in the latter (Table 17.5). Since the savannah covers half of the basin (Global Land Cover 2000 project EEC, 2003), about 12.5% of TOC is produced by the savannah and 87.5% by the forested areas. Finally, considering the fluxes calculated at Kinshasa-Brazzaville station representative of the total flux

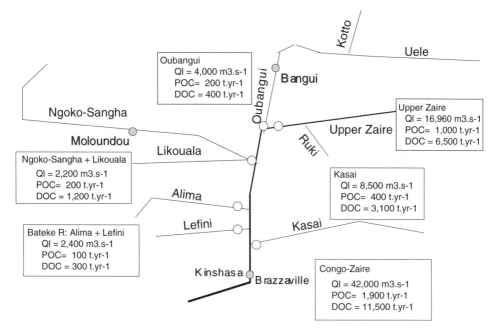

Figure 17.7 Fluvial budget of DOC and POC in the Congo-Zaire basin.

Table 17.6 Carbon Budget of the Amazon Basin

Basin	River	POC Flux Tg yr⁻¹	DOC Flux Tg yr⁻¹	TOC Flux Tg yr⁻¹	DOC Flux/ TOC Flux %
Negro	Negro	0.67	6.00	6.67	90
Solimões	Solimões	4.00	15.00	19.00	79
Madeira	Béni	0.80	1.10	1.90	58
	Mamoré	0.30	2.00	2.30	87
	Madeira	1.50	4.30	5.80	74
Amazon	Trombetas	0.05	0.50	0.55	91
	Tapajos	0.15	1.50	1.65	91
	Xingu	0.26	0.95	1.21	79
	Amazon	5.80	35.00	40.80	86

of the Congo-Zaire River to the Atlantic Ocean, this river contributes 1.9×10^6 t yr⁻¹ of POC and 11.5×10^6 t yr⁻¹ of DOC. Solid discharge makes up 30.6×10^6 t yr⁻¹. These values are similar to those reported by Probst et al. (1993) comprising 1.2×10^6 t yr⁻¹ of POC and 9.6×10^6 t yr⁻¹ of DOC, and Seyler et al. (1995) comprising 1.6×10^6 t yr⁻¹ and 11.4×10^6 t yr⁻¹ for the 1992 hydrological year, respectively.

17.5.2.2 Amazon Basin

The major contributors of organic C to the Amazon (Table 17.6, Figure 17.8) are the Solimões and Negro rivers. The Madeira River seems to contribute a relatively high amount of organic C to the Amazon River during high water periods. The present study shows that the Solimões River contributed about 40% of the TOC flux of the Amazon River. The Negro River also contributed 40%, the Madeira River ~14%, and the clear water rivers ~6%.

The Óbidos station is located upstream from the confluence of the Xingú and Tapajós rivers, but their water input (~3%) is low compared to other tributaries. The mean annual DOC flux was $35 \pm 4 \times 10^6$ t yr⁻¹ and the mean annual POC flux was $5.8 \pm 0.3 \times 10^6$ t yr⁻¹. POC flux was lower and DOC flux was higher than those reported by Richey et al. (1990). This discrepancy may be

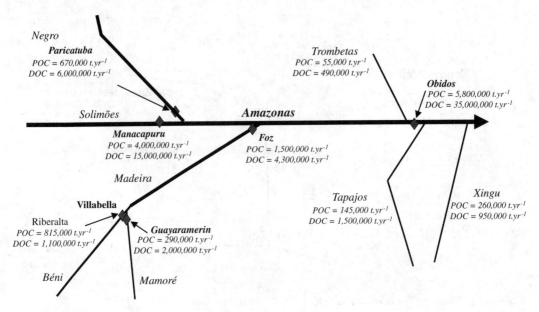

Figure 17.8 Fluvial budget of DOC and POC in the Amazon basin.

due to a difference in sampling separation procedure. Richey et al. (1990) estimated an annual TOC flux at the Óbidos station of 36.7×10^6 t yr^{-1}, which is similar to that of $40.8 \pm 4.3 \times 10^6$ t yr^{-1} reported in the present study. Nevertheless, the annual average TOC input from the principal Amazon tributaries (Negro, Solimões, Madeira, and Trombetas) was $36.9 \pm 3.3 \times 10^6$ t yr^{-1}. Comparing these inputs (36.9×10^6 t yr^{-1}), with the C flux in Óbidos (40.8×10^6 t yr^{-1}), a gain in organic C (about 4×10^6 t yr^{-1}) is observed which indicate, other important sources of organic C. These inputs are attributed to várzea systems. During the period in which the water level decreases, Richey et al. (1990) estimated that up to 400 kg s^{-1} of organic C came from the várzea lakes and between 60 and 120 kg s^{-1} of organic C came from other water stages associated with the Madeira River input. These additional inputs from floodplains to the mainstream are sufficient to account for the gain in C observed in Óbidos.

As mentioned above, two other large tributaries (Tapajós and Xingú rivers) flow through the Amazon River between Óbidos gauging station and the ocean. Taking into account the fluxes of these tributaries, up to 42×10^6 t of C is discharged each year into the Atlantic Ocean by the Amazon River.

17.5.3 Congo-Zaire and Amazon Organic Carbon Yields: Comparisons with World Ranges

Annual specific rates or yields of TSS (Figure 17.9) are of more than 100 g m^{-2} yr^{-1} for the basins in Andean ranges whereas these are about 10 g m^{-2} yr^{-1} for the stations corresponding to Congo-Zaire and Rio Negro basins. Specific rates of POC also show similar trends, i.e., a more significant flux for the Andean system than for Congo-Zaire and Negro basins. The values range between 0.2 and 2 g m^{-2} yr^{-1}. With regard to the DOC, there is a clear distinction between savannah and forest basins: the fluxes for Oubangui River draining savannah are 1 g m^{-2} yr^{-1} compared to 10 g m^{-2} yr^{-1} for the Amazon, Solimões, and Negro basins. Intermediate values of 4 g m^{-2} yr^{-1} were observed for Madeira, Congo-Zaire, and Ngoko.

The data in Figure 17.10 compare the computed yields of POC for Congo-Zaire and Amazon rivers along with those of others major world rivers. At the global scale, the relation between POC (%) concentration and TSS is not trivial. As discussed above, POC flux is directly linked to the sediment yield. The most turbid rivers are mountainous rivers such as the Beni, Mamore, and Madeira rivers. These rivers are in the same quadrant of Figure 17.10 as the temperate and semiarid rivers with a limited vegetation cover. Conversely the lowland rivers of the lowland regions of Congo-Zaire and Amazon basins fall among the "cold river" basins in which the large amount of POC may be eroded from organic-rich soil layer.

17.6 CONCLUSIONS

The extensive database of the organic C species in Congo-Zaire and Amazon rivers and its main tributaries allows us to determine factor controls and a quantitative estimate of TSS, DOC, and POC fluxes in the two major rivers of the world whose basin areas together cover ~70% of the world's humid tropics.

The common pattern observed in the rivers studied and other rivers of the world (Ittekkot et al., 1985; Meybeck, 1982; Spitzy and Ittekkot, 1991; Telang et al., 1991) shows that variation in POC is related to an increase in total suspended solids. This trend is attributed to the dilution of organic matter by mineral matter (Ittekkot et al., 1985; Meybeck, 1982). Reduced autochthonous production, due to a lack of light penetration at high sediment concentrations (Thurman, 1985), and differences in the sources and biogeochemical processes, affect the nature of organic matter at various stages of the hydrographic regime (Spitzy and Ittekkot, 1991; Wallace et al., 1982). While the highest contents of POC and DOC were observed during high water periods, the least

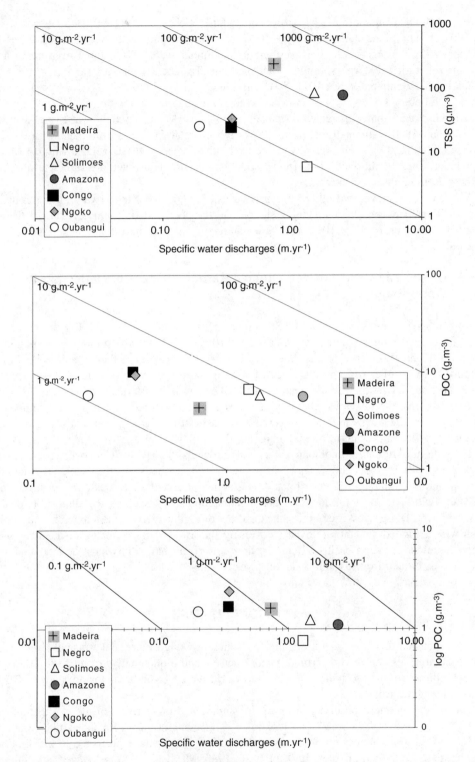

Figure 17.9 Annual specific rates of TSS, POC, and DOC in Congo-Zaire and Amazon basins (g m^{-2} yr^{-1}).

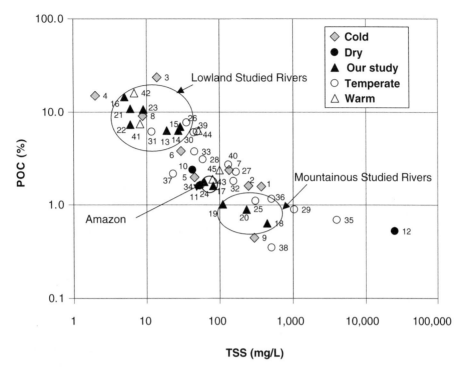

Figure 17.10 Relationship between annual average particulate organic carbon concentrations (POC in %) and total suspended solids (in mg/L) in rivers of Congo-Zaire and Amazon basins (our study) and major word river basins. **Cold** = (1) Mackenzie, (2) alpine Rhône, (3) N. Dvina, (4) Pechora, (5) Ob, (6) Lena, (7) Mackenzie, (8) Moose, (9) Yukon; **Dry** = (10) Gambia, (11) Orange, (12) Huang He; **Our study** = (13) Oubangui, (14) Ngoko, (15) Congo-Zaire, (16) Negro (17) Solimoes, (18) Beni, (19) Mamore, (20) Madeira, (21) Trombetas, (22) Tapajôs, (23) Xingu, (24) Amazone, **Temperate** = (25) Danube (26) Loire, (27) Mississippi, (28) Ohio, (29) Peace, (30) Rhine, (31) St. Lawrence, (32) Kuban, (33) Don, (34) Dniepr, (35) Choroch, (36) Changjiang, (37) Pee dee, (38) Rioni, (39) Seine, (40) Rhône; **Warm** = (41) Caroni, (42) Caura, (43) Orinoco, (44) Sanaga, (45) Tuy. (From Meybeck, M. 1993a. Riverine transport of atmospheric carbon: Sources, global typology and budget. *Water, Air, and Soil Pollution* 70:443; Meybeck, M. 1993b. C, N, P, and S in rivers: From sources to global inputs, In R. Wollast, F. T. Mackenzie, and L. Chou, eds., *Interactions of C, N, P, and S in Geochemical Cycles and Global Change*. Springer-Verlag, New York, p. 163. With permission.)

concentrated were observed during the low water periods. Thus, the main part of the organic C in these rivers is of terrestrial origin (allochtonous component) and in situ phytoplanctonic production (autochtonous component) is of limited significance in these rivers (Moreira-Turq et al., 2003). Both POC and DOC contents in the rivers studied were directly linked to weathering intensity, and their origin was associated with leaching of soils litter and humic layers (when DOC and POC% concentrations are high) and to the sedimentary rock containing fossil organic C (when POC% is low; Meybeck et al., this volume).

The distribution of organic C in different rivers of Congo-Zaire and Amazon basins depends on the mean altitude of the basins and the vegetation cover (%). With regard to the Congo-Zaire basin, Oubangui River draining mainly a savannah region, contained three times lower DOC than the Ngoko/Sangha River draining mainly a forested region. The trend is not clear for the POC. With regard to the Amazonian basin, the detailed study of the Madeira sub-basin indicated the role of vegetation type which itself depends on the mean altitude of the basin.

Computation of inputs to Atlantic Ocean show that the Congo-Zaire River contributes 13.4×10^6 t yr^{-1} of TOC of which 11.5×10^6 t yr^{-1} is DOC and 1.9×10^6 t yr^{-1} is POC. The Amazon contributed 42×10^6 t yr^{-1} of TOC of which 37×10^6 t yr^{-1} is DOC. Based on this study and the most recent estimate of the annual amount of organic C transported into oceans by rivers (500 × 10^6 t yr^{-1}; Spitzy and Ittekkot, 1991), it is concluded that the Congo-Zaire and Amazon rivers contribute a mean organic C flux of about 10 to 12% of the world's total.

REFERENCES

Abril, G. et al. 2002. Behaviour of organic carbon in nine contrasting European estuaries. *Est. Coast. and Shelf Sci.* 54:241.
Billen, G. et al. 1997. *Continental Aquatic Systems*, IGBP Water Group Report. D. Sahagian and L. Chou, eds. IGBP Water Meeting, Brussels, November 20–22.
Boulvert, Y. 1992. Carte phytogéographique au 1/1,000,000, République Centrafricaine. Notice explicative, No. 104, ORSTOM, Paris, 1992.
Bricquet, J. P. 1995. Les écoulements du Congo à Brazaville et la spatialisation des apports, in J. C. Olivry and J. Boulegue J., eds., *Grands Bassins Fluviaux Periatlantiques*. ORSTOM, Paris, 27 pp.
Cadet, G. C. 1984. Particulate and dissolved organic matter and chlorophyll-a in the Zaire River, estuary, and plume. *Neth. J. Sea Res.* 17 (2/4):429.
Callède, J., J. L. Guyot, J. Ronchail, M. Molinier, and E. de Oliveira E. 2002. L'Amazone à Obidos (Brésil). Etude statistique des débits et bilan hydrologique. *Hydrological Sci. J.* 47 (2):321.
Colas, C. 1994. Etude des flux de matière organique en milieu fluvial: Apports naturels et anthropiques. Rapport de DEA, Bordeaux.
Degens, E. T. 1982. Riverine carbon: An overview. In *Transport of Carbon and Minerals in Major World Rivers, Part 1*, Mitt. Geol.-Palaont. Inst. Univ. Hamburg. SCOPE/UNEP Sonderband Heft 52.
Degens, E. T., S. Kempe, and A. Spitzy. 1984. Carbon dioxide: A biogeochemical portrait, in C. O. Hutzinger, *The Handbook of Environmental Chemistry*, vol. 1. Springer-Verlag, Berlin.
Degens, E. T. and V. Ittekot. 1985. Particulate organic carbon: An overview. In *Transport of Carbon and Minerals in Major World Rivers, Part 3,* Mitt. Geol.-Palaont. Inst. Univ. Hamburg. SCOPE/UNEP Sonderband Heft 58.
Degens, E. T., S. Kempe, and J. E. Richey. 1991. *Biogeochemistry of Major World Rivers*. SCOPE Rep. 42, John Wiley & Sons, New York.
De Namur, C. 1990. Aperçu sur la végétation de l'Afrique Centrale Atlantique, in R. Lanfranchi and D. Schwartz, eds., *Paysages quaternaires de l'Afrique Centrale Atlantique*. ORSTOM, Paris.
Ertel, J. R. et al. 1986. Dissolved humic substances of the Amazon River system. *Limnol. Oceanogr.* 31(4):739–754.
Etcheber, H. 1986. Biogéochimie de la Matière Organique en milieu estuarien: Comportement, bilan, propriétés. Cas de la Gironde. *Mem. IGBA* 19.
Filizola N. 2003. *Transfert sédimentaire actuel par les fleuves amazoniens*. Thèse de l'Univ. Paul Sabatier Toulouse III, 292 pp.
Filizola N. and J. L. Guyot. Amazon river suspended sediment sampling and water discharge measurements at Obidos, Brazil, using an acoustic Doppler current profiler and traditional technologies. *Hydrological Sci. J.*, in press.
Global Land Cover 2000 project EEC. 2003. www:gvm.sai.it/glc2000.htm.
Guyot, J. L. and J. G. Wasson. 1994. Regional pattern of riverine dissolved organic carbon in the Amazon drainage basin of Bolivia. *Limnol. Oceanogr.* 39 2:452.
Hedges, J. I. et al. 1986. Composition and fluxes of particulate organic material in the Amazon River. *Limnol. Oceanogr.* 31 (4):717.
Hedges, J. I. et al. 1994. Origins and processing of organic matter in the Amazon River as indicated by carbohydrates and amino acids. *Limnol. Oceanog.* 39 (4):743.
Hegdes, J. I. et al. 2000. Organic matter in Bolivian tributaries of the Amazon River: A comparison to the lower mainstream. *Limnol. Oceanogr.* 45 (7):1449.
IGBP (International Geosphere Biosphere Program). 1995. *Land-Ocean interactions in the Coastal Zone*. IGBP Report, no. 33, LOICZ Implementation Plan, in J. C. Pernetta and J. D. Milliman, eds. Stockholm.

Ittekkot, V., S. Safiullah, B. Mycke, and R. Seifert. 1985. Seasonal variability and geochemical significance of organic matter in the River Ganges, Bangladesh. *Nature* 317:799–802.

Junk, W. J. 1985. The Amazon floodplain — a sink or source for organic carbon? *Mitt. Geol.-Palaont. Inst. Univ. Hamburg*. SCOPE/UNEP Sonderband Heft 58:267.

Junk, W. J. 1997. *The Central Amazon Floodplain: Ecology of a Pulsing System*. Ecological Studies 126, Springer-Verlag, Berlin.

Kaplan, L. A., R. A. Larson, and T. L. Bott. 1980. Patterns of dissolved organic carbon in transport. *Limnol. Oceanog.* 25:1034.

Kempe, S. 1979. Carbon in the rock cycle, in *The Global Carbon Cycle, SCOPE Rep 13*, B. Bolin, E. T. Degens, S. Kempe, and P. Ketner, eds. John Wiley & Sons, New York, 343 pp.

Kempe, S., D. Eisma, and E. T. Degens. 1993. *Transport of Carbon and Minerals in Major World Rivers*, vol. 6. Mitt. Geol.-Palaont. Inst. Univ. Hamburg. SCOPE/UNEP Sonderband 74, Universitat Hamburg, Hamburg.

Kinga-Mouzeo, M. 1986. Transport particulaire actuel du fleuve Congo et de quelques affluents; enregistrement quaternaire dans l'éventail détritique profond (sédimentologie, minéralogie et géochimie). Thèse Doct. Univ. Perpignan.

Lesack, L. F. W., R. E. Hecky, and J. Melack. 1984. Transport of carbon, nitrogen, phosphorus and major solutes in Gambia River, West Africa. *Limnol. Oceanogr.* 29:816.

Lewis, W. M. and J. F. Saunders, III. 1989. Concentration and Transport of dissolved and suspended substances in the Orinoco River, *Biogeochemistry* 7:203.

Ludwig, W., J. L. Probst, and S. Kempe. 1996. Predicting the oceanic input of organic carbon by continental erosion. *Global Biogeochem. Cycles* 10:23.

Mahé, G. 1993. Modulation annuelle et fluctuations interannuelles des précipitations sur le bassin versant du Congo, in J. C. Olivry and J. Boulegue, eds., *Grands Bassins Fluviaux Périatlantiques*. ORSTOM Editions, Paris.

Maneux, E. 1998. Erosion mécanique des sols et transports fluviaux de matières en suspension: Application des systèmes d'information géographique dans les bassins versants de l'Adour, de la Dordogne et de la Garonne. Thèse 3ème cycle, Univ. Bordeaux.

Martins, O. and J. L. Probst. 1991. Biogeochemistry of major African rivers: Carbon and mineral transport, in E. T. Degens, S. Kempe, and J. E. Richey, eds., *Biogeochemistry of Major World Rivers*. SCOPE/UNEP Sonderband, 243 pp.

McClain, M., R. L. Victoria, and J. E. Richey. 2001. *The Biogeochemistry of the Amazon Basin and Its Role in a Changing World*. Oxford University Press, Oxford.

Meybeck, M. 1982. Carbon, nitrogen and phosphorus transport by world rivers. *Am. J. Sci.* 282:401.

Meybeck, M. 1993a. Riverine transport of atmospheric carbon: Sources, global typology and budget. *Water, Air, and Soil Pollution* 70:443.

Meybeck, M. 1993b. C, N, P, and S in rivers: From sources to global inputs, In R. Wollast, F. T. Mackenzie, and L. Chou, eds., *Interactions of C, N, P, and S in Geochemical Cycles and Global Change*. Springer-Verlag, New York, p. 163.

Meybeck, M., A. Ragu, and L. Lachartre. 2006. Origins and behaviours of carbon species in world rivers, in *Soil Erosion and Carbon Dynamics*, Taylor & Francis, Boca Raton, FL.

Milliman, J. D. and Meade, R. H., World wide delivery of river sediment to the oceans. *J. Geology*, 91: 1, 1983.

Molinier, M. et al. 1997. Hydrologie du bassin amazonien, in H. Thery, ed., *Environnement et Développement en Amazonie Brésilienne*. Berlin, Paris, p. 24.

Moreira-Turcq, P., P. Seyler, J. L. Guyot, and H. Etcheber. 2003. Exportation of organic carbon in the Amazon Basin. *Hydrological Processes* 17 (7):1329.

Mounier, S., R. Braucher, and J. Y. Benaim. 1998. Differentiation of organic matter's properties of the Rio Negro basin by cross flow ultra-filtration and UV-spectrofluorescence. *Water Res.* 33:2363.

Olivry, J. C. 1986. *Fleuves et rivières du Cameroun*. Collection Monographies Hydrologiques, No. 9, ORSTOM, Paris.

Orange, D. 1990. *Hydroclimatologie du Fouta-Djalon et dynamique actuelle d'un vieux paysage latéritique*. Thèse Doct. Univ. Louis Pasteur, Strasbourg.

Orange, D., A. Laraque, and J. C. Olivry. 1999. Evolution des flux de matières le long de l'Oubangui et du fleuve Congo. Symposium International MANAUS'99, Manaus, November 16–19.

Paolini, J. 1995. Particulate organic carbon and nitrogen in the Orinoco river (Venezuela). *Biogeochemistry* 29:59.

Patel, N. et al. 1999. Fluxes of dissolved and colloidal organic carbon, along the Purus and Amazonas rivers (Brazil). *Sci. Total Environ.* 229:53.

Pires, J. M. and G. T. Prance. 1986. The vegetation types of the Brazilian Amazon, in G. T. Prance and T. M. Lovejoy, eds., *Key Environments: Amazonia*. Pergamon Press, Oxford, U.K.

Probst, J. L. 1990. Géochimie et Hydrologie de l'érosion continentale. Mécanismes, bilan global actuel et fluctuations au cours des 500 derniers millions d'années, vol. 1. *Sci. Geol. Mém.* 94, Strasbourg.

Probst, J. L., S. Mortati, and Y. Tardy. 1993. Carbon river fluxes and weathering CO_2 consumption in the Congo and Amazon River basins. *Applied Geo.* 7, 1–13.

Quay, P. D. et al. 1992. Carbon cycling in the Amazon River: Implications from the 13C compositions of particles and solutes. *Limnol. Oceanogr.* 37 (4):857.

Richey, J. E. et al. 1980. Organic carbon: Oxidation and transport in the Amazon River. *Science* 207:1348.

Richey, J. E., J. I. Hedges, A. H. Devol, and P. D. Quay. 1990. Biogeochemistry of carbon in the Amazon River. *Limnol. Oceanogr.* 35 (2):352.

Richey, J. E. et al. 1991. The biogeochemistry of a major river system: the Amazon case study, in E. T. Degens, S. Kempe, and J. E. Richey, eds. *Biogeochemistry of Major World Rivers*. John Wiley & Sons, New York, pp. 57–74.

Seyler, P., L. Sigha-Nkamdjou, and J. C. Olivry. 1993. Hydrogeochemistry of the Ngoko river, Cameroon: Chemical balances in a rain-forest equatorial basin, in IAHS Joint International Meeting, Yokohama, Japan, July 11–23.

Seyler, P. et al. 1995. Concentrations, fluctuations saisonnières et flux de carbone dans le bassin du Congo, in J. C. Olivry and J. Boulègue, eds., *Grands Bassins Fluviaux Peiatlantiques*. ORSTOM Editions, Paris, 217 pp.

Sigha-Nkamdjou, L., P. Carre, and P. Seyler. 1995. Bilans hydrologiques et géochimiques d'un écosystème forestier équatoraial de l'Afrique centrale: La Ngoko à Moloundou, in J. C. Olivry and J. Boulègue, eds., *Grands Bassins Fluviaux Periatlantiques*. ORSTOM Editions, Paris, 217 pp.

Sioli, H. 1984. *The Amazon: Limnology and Landscape Ecology of a Mighty Tropical River and Its Basin*, Dr. W. Junk, Publishers, Dordrecht, The Netherlands.

Spitzy, A. and V. Ittekkot. 1991. Dissolved organic carbon in rivers, in R. F. C. Mantoura, J. M. Martin, and R. Wollast, eds., *Ocean Margin in Global Change*. John Wiley & Sons Ltd.

Telang, S. A. et al. 1991. Carbon and mineral transport in major North American, Russian Arctic, and Siberian rivers: The St Lawrence, the Mackenzie, the Yukon, the Arctic Basin Rivers in the Soviet Union and the Yenisei, in E. T. Degens, S. Kempe, and J. E. Richey, eds., *Biogeochemistry of Major World Rivers*. John Wiley & Sons, New York, pp. 337–344.

Thurman, E. M. 1985. *Organic Geochemistry of Natural Waters*, N. Nijhoff and W. Junk, eds., Boston, MA.

Veyssy, E. 1998. *Transferts de matières organiques des bassins versants aux estuaires de la Gironde et de l'Adour (Sud-Ouest de la France)*. Thèse 3ème cycle, Univ. Bordeaux.

Wallace, J. B., G. W. Ross, and J. L. Meyer. 1982. Seston and dissolved organic carbon dynamics in a southern Appalachian stream. *Ecology* 53:824.

Zhang, S., W. B. Guan, and V. Ittekkot. 1992. Organic matter in large turbid rivers: The Huanghe and its estuary. *Mar. Chem.* 38:53.

CHAPTER **18**

Soil Carbon Stock and River Carbon Fluxes in Humid Tropical Environments: The Nyong River Basin (South Cameroon)

Jean-Loup Boeglin, Jean-Luc Probst, Jules-Rémy Ndam-Ngoupayou, Brunot Nyeck, Henri Etcheber, Jefferson Mortatti, and Jean-Jacques Braun

CONTENTS

18.1 Introduction 275
18.2 Characterization of the Nyong River Basin 276
18.3 Methodology 277
18.4 Results 278
    18.4.1 Soil Carbon 278
    18.4.2 Carbon in Surface Waters 279
        18.4.2.1 Dissolved Organic Carbon (DOC) 279
        18.4.2.2 Dissolved Inorganic Carbon (DIC) 279
        18.4.2.3 Total Suspended Solids (TSS) and Particulate Organic Carbon (POC) 280
18.5 Discussion 281
    18.5.1 The Major Role of Colored Waters 281
    18.5.2 The Important Contribution of DOC Fluxes 281
    18.5.3 Low CO_2 Uptake by Silicate Weathering 281
    18.5.4 Chemical Properties of the DOC and Transport Capacity of Trace Elements 282
    18.5.5 Global Carbon Budget at the Mengong Catchment Scale 283
18.6 Conclusion 284
Acknowledgments 285
References 285

18.1 INTRODUCTION

The role of organic and inorganic carbon (C) in the environmental equilibrium is now generally accepted. Therefore, on a global scale, the interactions between the C cycle and the weathering of rocks and soils or the continental water circulation have been frequently demonstrated (Berner et al., 1983; Meybeck, 1987; Amiotte-Suchet and Probst, 1993; Ludwig et al., 1998; Amiotte-Suchet et al., 2003). Hydrochemical studies of C transfers, realized on different-size catchments located in all climatic areas, are complex because this element is found under several forms (mineral or organic, dissolved or particulate). The C origin in the continental waters (atmospheric or soil CO_2, dissolution and erosion of rocks, soil organic matter, aquatic microorganisms), so that possible exchange or transformation reactions can be determined using isotopic tracing of C (Amiotte Suchet et al., 1999; Aucour et al., 1999; Barth and Veizer, 1999; Brunet et al., in press) and of other elements like strontium (Negrel et al., 1993; Gaillardet et al., 1999), or geochemical modeling (Probst et al., 1994; Amiotte-Suchet and Probst, 1995; Gaillardet et al., 1997, Mortatti and Probst, 2003). While data of C concentrations and fluxes are available for many rivers, including the Niger (Martins and Probst, 1991; Boeglin and Probst, 1996), the Congo (Probst et al., 1994; Seyler et al., 1995), and the Amazon (Richey et al., 1990), in the case of tropical basins, the C contents and stocks in the soil cover of corresponding watersheds have generally not been determined.

The objective of this chapter is to analyze the C concentrations in different water reservoirs in different forms in relation to the weathering rate and the C content in the soils of five nested catchments belonging to the Nyong basin in South Cameroon. This forested granitic watershed is representative of the humid tropical domain. Such a study has been rarely performed on this type of environment (in Puerto Rico, McDowell and Asbury, 1994; White et al., 1998; in the Ivory Coast, Stoorvogel et al., 1997a; Stoorvogel et al., 1997b), contrary to the temperate North American or European environments (see for example the compilations of Velbel, 1995; White and Blum, 1995; Drever and Clow, 1995).

18.2 CHARACTERIZATION OF THE NYONG RIVER BASIN

The Nyong basin, covering an area of 27,800 km² between 2°58' and 4°32' N latitude, is located in Cameroon. Five nested catchments have been sampled in the present study, i.e., from upstream to downstream: the experimental Mengong catchment near to Nsimi village, the Awout watershed (tributary of the So'o) at Messam, the So'o watershed (tributary of the Nyong) at Pont So'o, the upper Nyong basin at Mbalmayo, and at Olama station after the Nyong and So'o confluence (Figure 18.1).

The study took place on the southern Cameroon plateau, a smooth undulating area, with altitudes between 650 and 850 m. The morphology, presenting large depressed swampy zones (about 20% of the Nsimi catchment area) between eroded hills, is derived from an original half-orange landscape (Bilong et al., 1992).

The regional substratum is granite-type. The southern part of the study area is mainly constituted of Liberian granitoids of the Ntem group (2600 to 2900 M yr), and corresponds to the edge of the Congo craton, whereas the northern part, with gneisses and migmatites of the Yaoundé series metamorphized during the Panafrican orogenesis (~600 M yr) is the western continuation of the Oubanguide chain (Vicat, 1998).

A spatial distribution of pedologic horizons in the Nsimi catchment was done from pits and holes (Nyeck et al., 1993) and using resistivity measurements along several cross sections (Ritz et al., 1998). The lateritic profile developed on slopes and hills (up to 50 m thick) is characterized by a deferruginisation and a nodulation within the ferricrete, which is capped by a sandy-clayey homogeneous cover yellowish to reddish in color. In the depressed zone, hydromorphic soils (1 to 3 m thick), essentially constituted by grey colluvial sandy clays containing high organic matter content (> 10%) at their upper part, overlay the truncated mottled clay horizon.

Southern Cameroon is characterized by a four-season Guinean climate. The two rainy seasons (March to May, September to November) are separated by a short dry season, whereas the well-marked long dry season is from December to February. The annual rainfall is between 1500 and 2000 mm with a mean temperature of about 25°C presenting a 2 to 3°C annual amplitude, and a potential evapotranspiration of 1250 mm yr^{-1} (Olivry, 1986). For the hydrologic year 1998/99,

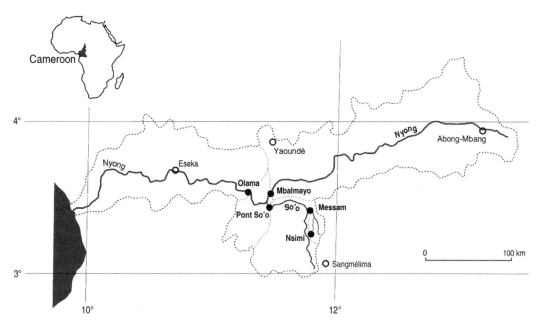

Figure 18.1 Location map of the Nyong basin (southern Cameroon); the studied stations are indicated by black circles.

rainfall (P) and runoff (R) averaged 1822 and 387 mm, respectively, for the Mengong catchment, and 1667 and 283 mm for on Nyong basin at Olama. These values correspond to a runoff ratio (R/P) of 21 and 17%, respectively.

The Yaoundé-Sangmélima-Abong Mbang triangular area (Figure 18.1) is mainly the domain of the semi-deciduous forest, characterized by Stertuliaceae and Ulmaceae. This type of vegetation is found in limited sectors surrounded with more degraded vegetation zones (Villiers in Santoir and Bopda 1995). The swampy depressions are generally colonized by raffias (*Raphia montbuttorum*). The anthrogenic effects on the environment are rather weak on the southern Cameroon plateau, because of the low population density (5 to 10 in km^{-2}, except in the urban zone of Yaoundé). Human activities include deforestation, shifting cultivation without fertilizer input, and growing food crops (tubers, banana, corn) or peanuts, tomatoes, cocoa bean, etc.

18.3 METHODOLOGY

Soils were sampled on Nsimi catchment, using profiles on summit and side slopes, and by auger in the valley. The organic C was analyzed in two ways: (1) wet oxidation method using potassium dichromate and titration with ammonium-iron sulfate (Anne's method, Rouiller et al., in Bonneau and Souchier, 1994), performed at the Hydrological Research Center in Yaoundé (Cameroon), and (2) measurement of the CO_2 volume evolved consecutively to the combustion of a soil aliquot, performed in the laboratories of the Centro de Energia Nuclear na Agricultura in Piracicaba (Brazil). The first method, appropriate for C contents > 1%, is not precise in the case of soils with low C content (C < 0.2%).

Waters samples were collected monthly at six stations (one on a spring, two on streams, three on rivers) of the Nyong network. After measurement of the physicochemical parameters (temperature, pH, electrical conductivity) and filtration in the field, samples were sent to the Centre de Géochimie de la Surface in Strasbourg (France) for determination of cations and anions, dissolved silica, alkalinity, and dissolved organic C (DOC). The alkalinity (= "acid neutralizing capacity") was determined by titration with H_2SO_4 using Gran method on a Mettler memotitrator DL 40RC,

Table 18.1 Organic Carbon Content and Stock in Different Soil Horizons of the Mengong Catchment

Horizon	A %	T (m)	V ($10^3 \cdot m^3$)	V %	AD	OC %	S_{OC} (Mg)	S_{OC} %
Lateritic Profile (with Overlay on Hills and Slopes) from Top to Depth								
Upper overlay	80	0.2	96	0.6	1.30	1.5	1,870	5.3
Overlay	80	2	960	5.6	1.28	0.30	3,690	10.5
Ferricrete/nodular	80	2	960	5.6	1.66	0.15	2,390	6.8
Mottled clays	99	5	2,970	17.4	1.29	0.15	5,750	16.3
Saprolite	99	20	11,900	69.7	1.50	0.10	17,850	50.7
Hydromorphic Soils of the Depression (Covering Truncated Mottled Clays) from Top to Depth								
Organo-mineral	15	0.5	45	0.3	1.44	4.0	2,590	7.3
Colluvial	15	1.5	135	0.8	1.62	0.50	1,090	3.1
Total catchment	—	—	17,066	100	—	—	35,230	100

Note: A % is the percentage of the catchment area (0.60 km²), T is the mean thickness (in m), V is the volume (in $10^6 \cdot m^3$), and AD the apparent density of an horizon; OC% is the percentage of organic carbon content; S_{OC} is the stock of organic carbon, S_{OC}% is the percentage of organic carbon stock in a given horizon compared the whole catchment; data in the last line correspond to the whole catchment.

with an analytical reproducibility better than 2%. The Shimadzu TOC 5000A analyzer was used to measure DOC content by catalytic combustion; the reproducibility is about 5%, the detection limit is 0.1 mg L^{-1}.

The determination of particulate organic C (POC) was done on the total suspended solids (TSS) collected on glass microfiber filters for the April 22 and 23, 2002 flood event. The concentrations were determined by catalytic combustion on a LECO CS125 carbon analyzer at the Centre de Recherche sur les Environnements Sédimentaires et Océaniques of Bordeaux I University (France).

18.4 RESULTS

18.4.1 Soil Carbon

Soils of the southern Cameroon plateau are laterites developed on hills and slopes, or hydromorphic soils in depressions and valleys, all formed on a granitoid-type bedrock. Carbonates have not been observed in the geological substratum and in the soil. Thus, total C present in the soil is contained in the organic matter. Soil samples were collected and analyzed only in the Mengong catchment for different horizons. The C contents exhibited a very heterogeneous distribution in the different soil profiles (Table 18.1). High concentrations were found in the first meter of the soil on hillslopes and hydromorphic soils (up to 6% in the sandy-clayey horizon) of the swampy depression, and the C concentration was very low in all subsoil horizons. These results are in accord with those reported by Humbel et al. (1977) for different ferrallitic soils of Cameroon.

Total soil C stock was estimated for the Mengong catchment, using the data of organic C analyses in the present study, pedological studies (description of toposequences, determination of apparent density) conducted by Nyeck et al. (1993), and geophysical measurements (surficial resistivity allowing an assessment of the soil profile depth along cross sections) conducted by Ritz et al. (1998). Different parameters used in the calculation for each horizon are summarized in Table 18.1. The organic C stock is estimated at 35,000 Mg in soils of the Mengong catchment. The surface horizons of hills, slopes, and especially of depression, which contain high organic matter content, represent only 13% of the total C stock, of which 51% is contributed by saprolite in which low C content is compensated by the thickness of this formation, which covers the entire catchment.

Table 18.2 Hydrological Characteristics of the Nyong Catchment Network and Average Values for Some Physicochemical Parameters Measured during the Year 1998 to 1999 in the Spring and River Waters

Station	A	Qm	DIC C	DIC F	TDS C	TDS F	DOC C	DOC F	TSS C	TSS F	FCO$_2$
				Mengong Experimental Catchment							
Spring	—	—	0.71	—	11.6	—	0.31	—	—	—	—
Stream	0.60	387	0.87	286	16.5	6000	14.3	5677	8.1	2873	23.8
				Nyong River Stations from Upstream to Downstream							
Messam	206	431	0.66	198	20.8	7087	22.9	9304	13.9	4112	16.5
Pont So'o	3,070	370	1.20	308	23.3	7175	15.1	6315	20.5	6966	25.7
Mbalmayo	13,555	257	1.51	264	24.3	4996	15.9	4670	12.8	3237	22.0
Olama	18,510	283	1.26	285	22.1	5609	15.6	4849	14.3	4488	23.7

Note: S is the catchment area in km^2; Qm is the annual mean runoff expressed in mm yr^{-1}; C is the mean annual concentration in mg L^{-1}; F, the annual specific flux in kg km^{-2} yr^{-1}; DIC is the dissolved inorganic carbon contained in bicarbonate; TDS is total inorganic dissolved solids; DOC is dissolved organic carbon; TSS is total suspended solids; FCO$_2$ is the atmospheric/soil CO$_2$ specific flux consumed by weathering (in 10^3 mol km^{-2} yr^{-1}).

18.4.2 Carbon in Surface Waters

18.4.2.1 Dissolved Organic Carbon (DOC)

In the Mengong catchment, an obvious distinction can be made between the clear spring waters and the colored outlet river waters. This color difference has previously been reported by Viers et al. (1997) and Oliva et al. (1999), and is attributed to differences in DOC content (Table 18.2). For the hydrologic year 1998/99, the mean annual DOC concentration was 0.31 mg.L^{-1} in the spring and 14.3 mg.L^{-1} in the stream at the outlet.

The mean DOC river fluxes were between 4700 and 9300 kg km^{-2} yr^{-1} (Table 18.2). The values corresponding to the southern part of the Nyong watershed are higher than those observed for the eastern part of the basin (Mbalmayo station). But this difference may be due to a higher drainage intensity in the southern part, rather than to differences in the vegetation and soil covers.

The organic-rich waters exhibit a very high ionic imbalance between the cationic sum (S_c) and the anionic sum (S_a), due to an important anionic deficit, as was reported by Probst et al. (1992) for the Congo River. The mean annual deficit averaged 51 to 98% depending on the stations. This imbalance is mainly due to organic anions that are not taken into account in the measurement of alkalinity (Fillion et al., 1998). Consequently, there is a good relationship between ionic deficit (in μeq L^{-1}) and DOC content (in mg L^{-1}) for the Mengong River (Figure 18.2) as well as for the other rivers of the Nyong basin. The slope of the regression lines obtained for different stations varies between 4.3 and 9.1 μeq mg^{-1} C; the correlation coefficients R calculated for all catchments are > 0.91.

18.4.2.2 Dissolved Inorganic Carbon (DIC)

Dissolved inorganic C (DIC) is present mainly as bicarbonates at the pH of most river waters. This ion is released consecutively to hydrolysis in which soil and rock minerals are weathered under the effect of the carbonic acid (Equation 18.3). This acid, originating from the atmospheric CO$_2$ (Equation 18.2), is abundant in soil solutions and is produced by the oxidation of soil organic matter (Equation 18.1):

$$CH_2O + O_2 \leftrightarrow CO_2 + H_2O \qquad (18.1)$$

$$CO_2 + H_2O \leftrightarrow H_2CO_3 \qquad (18.2)$$

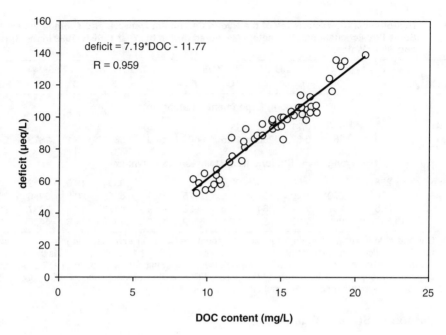

Figure 18.2 Relationship between the anionic deficit and the DOC content in the Mengong River during the year 1998/99.

For example, the weathering reaction of orthose (or microcline) into kaolinite can be written as follows (Garrels and Mackenzie, 1971):

$$2KAlSi_3O_8 + 3H_2O + 2CO_2 \leftrightarrow Si_2Al_2O_5(OH)_4 + 2K^+ + 4SiO_2 + 2HCO_3^- \qquad (18.3)$$

As shown in Equation 18.3, all bicarbonate ions released into solution originate from atmospheric/soil CO_2.

The mean annual bicarbonate concentrations are 3.6 mg L^{-1} in the spring water, and between 3.4 and 7.7 mg L^{-1} in different rivers. The corresponding concentrations of DIC for the Nyong River stations are summarized in Table 18.2.

Since the substratum of the Nyong drainage basin is exclusively composed of silicate rocks, the total bicarbonate fluxes measured in the river waters sampled at different stations can be directly related to the atmospheric/soil CO_2 (Equation 18.3). Consequently the calculation of the CO_2 fluxes used in chemical weathering is directly derived from the river's alkalinity fluxes. The values calculated for different stations range between 16.5 and 25.7 10^3 mol km^{-2} yr^{-1} (Table 18.2).

18.4.2.3 Total Suspended Solids (TSS) and Particulate Organic Carbon (POC)

Studies on the drainage waters of the Nyong basin showed that the suspended matter is enriched in organic matter; the mineral fraction is mainly composed of kaolinite, quartz, and goethite, and amorphous silica is present in diatom frustules or in phytoliths (Olivie-Lauquet et al., 2000). No measurement of POC concentration was made for the TSS during the hydrologic year 1998/99. However, POC was measured (17 samples) during the flood event of April 28 and 29, 2002 on the Mengong River. These measurements indicated that POC content represented on average 25.3% of the TSS content, with relatively low variations compared to water discharge fluctuations. This percentage is close to that (25.4%) obtained from the mean annual POC and TSS fluxes calculated by Ndam-Ngoupayou (1997) for the hydrologic year 1995/96 on the Mengong River. Whereas this percentage is higher than those obtained for the Nyong at Mbalmayo (17.3%) and Olama (13.8%), it shows that the suspended material decreases in organic C from upstream to downstream during the same time the TSS concentration increases due to the physical erosion of more inorganic materials. This pattern is comparable to those observed for most world rivers (Martins and Probst, 1991; Ludwig et al., 1996). Using these data, the POC specific fluxes were estimated between 600 and 1400 kg km^{-2} yr^{-1} for different stations of the Nyong River basin.

18.5 DISCUSSION

18.5.1 The Major Role of Colored Waters

There are two main types of water in the Nyong basin: clear waters and colored waters. The colored water, which remains after filtration and cannot be related to suspended matters, is attributed to a high content of dissolved organic matter (Ndam-Ngoupayou, 1997). This corroborates with our results, with mean DOC concentrations > 14 mg L^{-1} in all river waters, whereas these are about 0.3 mg L^{-1} in the Mengong spring waters. Moreover, Ndam-Ngoupayou (1997) shows that in the Mengong catchment clear waters come not only from springs, but also from hill or slope water tables and from the depressional deep water table, whereas colored waters, constituting the river waters, are also found in the surficial depression water table. The major role of dissolved organic matter in the chemical weathering was demonstrated by Viers et al. (1997) and Oliva et al. (1999) in this catchment, but also by Idir et al. (1999) in the Strengbach catchment in the Vosges Mountains (France). Organic acids, abundant in the swampy depression waters, are responsible for an enhanced dissolution of soil minerals and for the transport in a complexed form of insoluble metallic elements.

18.5.2 The Important Contribution of DOC Fluxes

In terms of specific fluxes in river waters, Table 18.2 shows that the DOC values are comparable to those of inorganic TDS (in which dissolved silica flux represents 40 to 55% according to the stations), whereas these DOC fluxes are one to two times higher than TSS fluxes. Considering that POC averages 20% of TSS, the mean annual flux of total organic carbon (DOC + POC) represents 38 to 49% of the total inorganic and organic materials (TDS + DOC + TSS) exported by the river at different stations; these percentages correspond to (DOC + POC) specific fluxes between 5300 and 10,100 kg km^{-2} yr^{-1}. Present results are comparable to those of Sigha-Nkamdjou et al. (1995) for other forested southern Cameroon catchments (Ntem, Kadéi, Boumba, Dja, Ngoko), where the calculated fluxes were between 3300 and 5300 kg km^{-2} yr^{-1} for DOC, and between 600 and 1200 kg km^{-2} yr^{-1} for POC. For savannah Cameroonian basins (Sanaga, Mbam), Ndam-Ngoupayou (1997) reported lower values for DOC (~1500 kg km^{-2} yr^{-1}) due to lower vegetation biomass and to lower drainage intensity. In contrast, the POC fluxes are rather high (1000 kg km^{-2} yr^{-1} for the Sanaga, 4200 kg km^{-2} yr^{-1} for the Mbam), due to a notable TSS transfer (18 and 98 Mg km^{-2} yr^{-1}, respectively). A comparison between the DOC fluxes exported on the Niger basin: 593 kg km^{-2} yr^{-1} at Lokodja (Martins and Probst, 1991), 455 kg km^{-2} yr^{-1} at Bamako (Boeglin and Probst, 1996), and on the Congo basin: 2.9 Mg km^{-2} yr^{-1} (Nkounkou and Probst, 1986) or 3.1 Mg km^{-2} yr^{-1} (Seyler et al., 1995), is a significant illustration of the influence of the vegetation type on the organic C solubilization. Ludwig et al. (1996) proposed a mean DOC flux of 1043 kg km^{-2} yr^{-1} in the dry tropical zone, and of 3818 kg km^{-2} yr^{-1} in the humid tropical zone.

18.5.3 Low CO_2 Uptake by Silicate Weathering

Concerning the dissolved inorganic C, the specific fluxes of CO_2 used in chemical weathering on the silicated Nyong watersheds are between 17 and 26*10^3 mol km^{-2} yr^{-1}, which are equivalent to 750 to 1140 kg CO_2 km^{-2} yr^{-1}. These values are low compared with those obtained on other lateritic

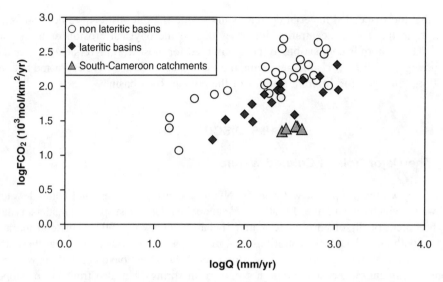

Figure 18.3 Relationship between annual CO_2 fluxes consumed by silicate weathering and mean annual runoff for rivers of different areas in the world (see Boeglin and Probst, 1998, for the list of river basins).

basins, 30 to $120*10^3$ mol km^{-2} yr^{-1} (Mortatti et al., 1992; Probst et al., 1994; Boeglin and Probst, 1998), except the one determined for the Senegal River whose watershed has a very low runoff Q (F_{CO2} = $17*10^3$ mol km^{-2} yr^{-1} for Q = 45 mm yr^{-1}). Considering only the part of F_{CO2} used in the silicate weathering, Boeglin and Probst (1998) demonstrated that, contrary to common belief, the F_{CO2} corresponding to lateritic basins — where the chemical weathering is supposedly very intense — are about two times lower than for the nonlateritic watersheds. In the case of the southern Cameroon catchments, the F_{CO2} values are three times lower than those previously calculated for the lateritic basins (Figure 18.3). In the case of the Nyong catchments, especially low F_{CO2} are attributed to the thickness of the lateritic cover on hills and slopes, which considerably slows the advance of percolating waters to the bedrock. This hypothesis is confirmed by the low mineral dissolved load in the clear waters and in the colored waters (Braun et al., 2002). The low chemical weathering intensity indicated by these weak F_{CO2} values is confirmed by the calculation done using the dissolved silica fluxes in the different upper Nyong catchments, which are low compared to the annual rainfall (see White and Blum, 1995). Supposing that 35% of the quartz of the bedrock remains in the saprolite, the weathering rate determined from the method proposed by Boeglin and Probst (1998) at Olama station averages 3.2 mm/1000 yr under present conditions, which is low compared to values generally obtained for tropical basins (6 to 15 mm/1000 yr; Nkoukou and Probst, 1987; Tardy, 1993; Mortatti and Probst, 2003). However, the weathering rate of 2.8 mm/1000 yr was reported by Seyler et al. (1993) for the equatorial Ngoko basin in southeastern Cameroon.

18.5.4 Chemical Properties of the DOC and Transport Capacity of Trace Elements

The high correlation obtained for all stations of the Nyong basin between the anionic deficit and the DOC content indicates that the presence of organic anions compensates for the lack of negative mineral charges as was previously reported by Sullivan et al. (1989) and Munson and Guerini (1993). The specific charge of the organic matter (in µeq mg^{-1} DOC), given by the slope of the regression line, differs among stations: 7.2 for the Mengong, 9.1 for the Awout at Messam, 5.3 for the So'o at Pont So'o, 4.6 for the upper Nyong at Mbalmayo, and 4.3 for Nyong at Olama. The decrease of this specific charge from upstream to downstream (with a particularly high value for the Awout stream) may be attributed to a progressive structural modification of the organic compounds in the river waters. For example, the length of the water course from Nsimi to Olama

SOIL CARBON STOCK AND RIVER CARBON FLUXES 283

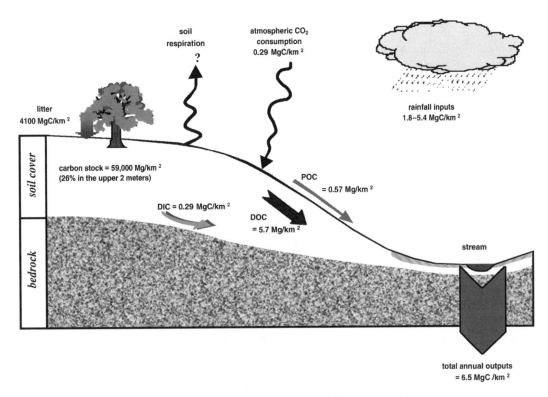

Figure 18.4 Carbon cycle and annual budget on the experimental Mengong catchment.

(~100 km) is about 3 to 6 days. That means that the capacity of the dissolved organic matter to complex the cations, particularly the heavy metals, decreases from upstream to downstream, thereby increasing the transport of heavy metals by the suspended matters.

18.5.5 Global Carbon Budget at the Mengong Catchment Scale

The amount of C transfer for the Mengong catchment is shown in Figure 18.4. Atmospheric inputs, essentially constituted by rainfalls, is a variable C content: mean DOC is 0.61 mg L^{-1} in open field precipitations and 3.6 mg L^{-1} in throughfalls (Ndam-Ngoupayou, 1997). In comparison, Freydier et al. (2002) determined a mean total organic C concentration (including fine suspended particles resulting from the combustion of biomass) of 2.5 mg L^{-1} in open field precipitations. However, these values were obtained from a low number of samples that did not cover the entire hydrologic year. Considering an annual rainfall of 1800 mm with a mean carbon content ranging from 1 to 3 mg L^{-1}, the corresponding input is 1.8 to 5.4 Mg C km^{-2} yr^{-1}.

The organic C stock in the soil of the Mengong catchment was estimated in this study at 59,000 Mg km^{-2}; however, only 26% of this amount of C is concentrated in the two upper meters of the profiles and can be drained by the surface and subsurface runoff.

Litter fall measurements made by Odigui Ahana (2000) in an experimental site at 100 km north of the Mengong catchment showed an annual value of 10.75 Mg of total wet organic matter with a mean C content of 38%, which is equivalent to 4100 Mg C km^{-2} yr^{-1}. For biomass, Likens et al. (1977) indicated a mean density of 450,000 Mg km^{-2} with an annual production of 22,000 t km^{-2} for the forested tropical area.

In such a lithological environment without any carbonate rocks, the riverine inorganic C is only found under dissolved inorganic C, there is no particulate inorganic C. In the Mengong catchment, the DIC riverine specific flux has been estimated from the alkalinity to 286 kg C km^{-2} yr^{-1}, corresponding to the same amount of atmospheric/soil CO_2 consumed by the chemical weathering of the silicate minerals.

Concerning the organic C, the riverine fluxes have been estimated for the Mengong catchment to 5700 and 575 kg C km^{-2} yr^{-1} respectively for the DOC and the POC. With the exception of Messam on the Awout brook, the DOC and DIC specific fluxes are comparable from one station to another over the Nyong River basin, indicating that the transfer dynamic of dissolved C does not change from upstream to downstream of the Nyong River basin, which presents the same type of lithology and vegetation covers, irrespective of the subcatchment size.

The residence time (R_t) of carbon in the Mengong catchment soils can be estimated from the total carbon stock (S_{OC}) in the soil cover (35,000 t), and from the annual fluxes of riverine carbon $\{F_{DOC}$ (3400 kg yr^{-1}) + F_{POC} (345 kg yr^{-1}) + F_{DIC} (172 kg yr^{-1})$\}$:

$$R_t = S_{OC}/(F_{DOC} + F_{POC} + F_{DIC}) \tag{18.4}$$

The residence time of 9000 yr indicates a rather slow turn over of the C in this catchment under present conditions. That means also that the riverine C fluxes represent only 0.1‰ of the soil C stock. This value is of the same order of magnitude as the global estimate (0.2‰) from the data of Meybeck (1982). However, if one considers that the reactive and mobilizable C is essentially contained in the upper part of the profiles (stock = 9200 Mg), or only in the hydromorphic depression soils (stock = 3700 Mg), the residence time notably decreases to 2350 yr and 950 yr, thereby increasing also the percentage of C exportation respectively to 0.4‰ and 1‰, respectively.

18.6 CONCLUSION

Analyses of the different forms of inorganic and organic C in the soils and in the waters for silicated catchments of the southern Cameroon forested area provide information on the transfer dynamic and on the mass balance for a tropical humid environment:

- The upper horizons of the soil and the corresponding draining waters exhibit rather high concentrations of organic C. Considering the soil cover containing a specific C stock of 59,000 Mg km^{-2}, about 26% of this amount is concentrated in the upper 2 m of the soil profiles (mean thickness = 30 m). In colored river and swamp waters, DOC fluxes are between 4700 and 9300 kg km^{-2} yr^{-1} (which is comparable to inorganic TDS fluxes), whereas POC fluxes (representing about 20% of TSS) are between 600 and 1400 kg km^{-2} yr^{-1}. In contrast, clear spring waters or ground waters contain low DOC contents.
- In such a noncarbonated environment, the DIC fluxes are normally low. Whereas for the Nyong River basin, these fluxes (about 300 kg C km^{-2} yr^{-1}) are lower than for other tropical silicated catchments. This can be related mainly to the low chemical weathering rate of silicate rocks estimated to be 3.2 mm/1000 yr for the upper Nyong basin at Olama. Such a weak weathering rate is explained by the thickness of the soil cover (until 50 m), insulating the bedrock from the interactions with the surficial draining waters.
- The turn over of the C (organic + inorganic) in the Mengong catchment is estimated to 9000 yr if one takes into account the whole C stock of the soil cover. However, considering that only the C contained in the hydromorphic soil horizons of the swampy zone is involved in the exchange and recycling processes, the residence time decreases to 950 yr.
- The high anionic deficit (51 to 98%) observed in all drainage waters is strongly correlated with the DOC content. The specific charge of the dissolved organic matter is estimated to 4.3 to 9.1 µeq mg^{-1} C. This specific charge decreases from upstream to downstream in the Nyong River basin, which indicates a progressive structural modification of the dissolved organic species during the transfer, reflecting a decreasing capacity of trace element complexation and transport.
- Values of C specific fluxes (DIC and DOC) obtained at different stations of the Nyong network are rather close. On the contrary, the POC specific fluxes are variable — depending on the TSS

flux — from one station to another. However, the POC/TSS ratio is slightly variable and is about 20%. These results are attributed to a comparable C dynamic at the scale of the Nyong River basin, independently of the surface area of the different subcatchments.

ACKNOWLEDGMENTS

The authors would like to thank D. Million and G. Krempp from the Centre de Géochimie de la Surface in Strasbourg for chemical analyses of river waters, J. P. Bedimo-Bedimo and J. Nlozoa from the Hydrological Research Center of Yaoundé for hydrological measurements and field assistance. This study has been realized with the financial support of a CAMPUS/CORUS cooperation project no. 99,412,102 between the University Paul Sabatier in Toulouse and the University of Yaoundé I. This work is also a contribution to the IGCP project no. 459 (UNESCO-IUGS) on "Carbon Cycle and Hydrology" and to DYLAT project of IRD.

REFERENCES

Amiotte-Suchet, P., D. Aubert, J. L. Probst, F. Gauthier-Lafaye, A. Probst, F. Andreux, and D. Viville. 1999. ^{13}C pattern of dissolved inorganic carbon in a small granitic catchment: The Strengbach case study (Vosges Mountains, France). *Chem. Geol* 159:87–105.

Amiotte-Suchet, P. and J. L. Probst. 1993. Modelling of atmospheric CO_2 consumption by chemical weathering of rocks: Application to the Garonne, Congo and Amazon basins. *Chem. Geol.* 107:205–210.

Amiotte-Suchet, P. and J. L. Probst. 1995. A global model for present-day atmospheric/soil CO_2 consumption by chemical erosion of continental rocks (GEM-CO_2). *Tellus* 47B:273–280.

Amiotte-Suchet, P., J. L. Probst, and W. Ludwig. 2003. Worldwide distribution of continental rock lithology: Implication for the atmospheric/soil CO_2 uptake by continental weathering and alkalinity river transport to the ocean. *Global Biogeochem. Cycles* 17(2), in press.

Aucour, A. M., S. Sheppard, O. Guyomar, and J. Wattelet. 1999. Use of ^{13}C to trace origin and cycling of inorganic carbon in the Rhône River system. *Chem. Geol.* 159:87–105.

Barth, J. and J. Veizer. 1999. Carbon cycle in the St Lawrence aquatic ecosystem at Cornwall (Ontario), Canada: Seasonal and spatial variations. *Chem. Geol.* 159:107–128.

Berner, R. A., A. C. Lasaga, and R. M. Garrels. 1983. The carbonate-silicate geochemical cycle and its effect on atmospheric carbon dioxide over the past 100 million years. *Am. J. Sci.* 283 (7):641–683.

Bilong, P., S. M. Eno Belinga, and B. Volkoff. 1992. Séquence d'évolution des paysages cuirassés et des sols ferrallitiques en zones forestières tropicales d'Afrique Centrale. Place des sols à horizons d'argiles tachetées. *C.R. Acad. Sci. Paris* 314 (série II):109–115.

Boeglin, J. L. and J. L. Probst. 1996. Transports fluviaux de matières dissoutes et particulaires sur un bassin versant en région tropicale: Le bassin amont du Niger au cours de la période 1990–1993. *Sci. Géol., Bull.* 49 (1–4) Strasbourg:25–45.

Boeglin, J. L. and J. L. Probst. 1998. Physical and chemical weathering rates and CO_2 consumption in a tropical lateritic environment: the upper Niger basin. *Chem. Geol.* 148:137–155.

Bonneau, M. and B. Souchier. 1994. *Pédologie 2. Constituants et propriétés des sols*, 2nd ed. Masson, 665 pp.

Braun, J. J., B. Dupré, J. Viers, J. R. Ndam-Ngoupayou, J. P. Bedimo-Bedimo, L. Sigha-Nkamdjou, R. Freydier, H. Robain, B. Nyeck, J. Bodin, P. Oliva, J. L. Boeglin, S. Stemmler, and J. Berthelin. 2002. Biogeohydrodynamic in the forested humid tropical environment: the case study of the Nsimi experimental watershed (South Cameroon). *Bull. Soc. Géol. France*, 173 (4):347–357.

Brunet, F., D. Gaiero, J. L. Probst, P. J. Depetris, F. Gauthier-Lafaye, and P. Stille ^{13}C tracing of dissolved inorganic carbon sources in Patagonian rivers. *Hydrol. Processes*, in press.

Drever, J. I. and D. W. Clow. 1995. Weathering rates of silicate minerals, In A. F. White and S. L. Brantley, eds., Chemical weathering rates of silicate minerals. *Miner. Soc. Amer. Rev. Mineral.* 31:463–483.

Fillion, N., A. Probst, and J. L. Probst. 1998. Natural organic matter contribution to throughfall acidity in French forests. *Environ. Internat.* 24:547–558.

Freydier, R., B. Dupré, J. L. Dandurand, J. P. Fortuné, and L. Sigha-Nkamdjou. 2002. Trace elements and major species in precipitations at African stations: Concentrations and sources. *Bull. Soc. Géol. France* 173 (2):129–146.

Garrels, R. M. and F. T. Mackenzie. 1971. *Evolution of Sedimentary Rocks.* W.W. Norton, New York, 397 pp.

Gaillardet, J., B. Dupré, C. J. Allègre, and P. Negrel. 1997. Chemical and chemical denudation in the Amazon river basin. *Chem. Geol.* 142 (3–4):141–173.

Gaillardet, J., B. Dupré, P. Louvat, and C. J. Allègre. 1999. Global silicate weathering and CO_2 consumption rates deduced from the chemistry of large rivers. *Chem. Geol.* 159 (1–4):3–30.

Humbel, F. X., J. P. Muller, and J. M. Rieffel. 1977. Quantité de matières organiques associées aux sols du domaine ferrallitique au Cameroun. *Cah. ORSTOM, sér. Pédol.* 15 (3):259–274.

Idir, S., A. Probst, D. Viville, and J. L. Probst. 1999. Contribution des surfaces saturées et des versants aux flux d'eau et d'éléments exportés en période de crue: Traçage à l'aide du carbone organique dissous et de la silice. Cas du petit bassin versant du Strengbach (Vosges, France). *C.R. Acad. Sci., Paris, Sc. Terre et planètes* 328:89–96.

Likens, G. E., F. H. Bormann, R. S. Pierce, J. S. Eaton, and N. M. Johnson. 1977. *Biochemistry of a Forested Ecosystem.* Springer-Verlag, Berlin.

Ludwig, W., P. Amiotte-Suchet, G. Munhoven, and J. L. Probst. 1998. Atmospheric CO_2 consumption by continental erosion: Present-day controls and implication for the last glacial maximum. *Glob. Planet. Changes* 16–17 (1–4):107–120.

Ludwig, W., J. L. Probst, and S. Kempe. 1996. Predicting the oceanic input of organic carbon by continental erosion. *Global Biogeochem. Cycles* 10(1):23–41.

Martins, O. and J. L. Probst. 1991. Biogeochemistry of major African rivers: Carbon and mineral transport, in E. T. Degens, S. Kempe, and J. E. Richey, eds., *Biogeochemistry of Major World Rivers.* SCOPE 42, John Wiley & Sons, chap. 6, pp. 129–155.

McDowell, W. H. and C. E. Asbury. 1994. Export of carbon, nitrogen and major ions from three tropical mountain watersheds. *Limnol. Oceanogr.* 39:111–125.

Meybeck, M. 1982. Carbon, nitrogen and phosphorus transport by world rivers. *Am. J. Sci.* 282:401–450.

Meybeck, M. 1987. Global geochemical weathering of surficial rocks estimated from river dissolved loads. *Am. J. Sci.* 287 (5):401–428.

Mortatti, J., J. L. Probst, and J. R. Ferreira. 1992. Hydrological and geochemical characteristics of the Jamari and Jiparana river basins (Rondonia, Brazil). *Geojournal* 26 (3):287–296.

Mortatti, J. and J. L. Probst. 2003. Silicate rock weathering and atmospheric/soil CO_2 uptake in the Amazon basin estimated from the river water geochemistry: Seasonal and spatial variations. *Chem. Geol.* 197:177–196.

Munson, R. K. and S. A. Gherini. 1993. Influence of organic acids on the pH and Acid Neutralizing Capacity of Adirondack lakes. *Water Resour. Res.* 29:891–899.

Ndam-Ngoupayou, J. R. 1997. *Bilans hydrogéochimiques sous forêt tropicale humide en Afrique: Du bassin expérimental de Nsimi-Zoételé aux réseaux hydrographiques du Nyong et de la Sanaga (Sud-Cameroun).* Thèse Doct. Univ. Paris VI, 214 p. 2 annexes.

Negrel, P., C. J. Allègre, B. Dupré, and E. Lewin. 1993. Erosion sources determined by inversion of major and trace element ratios and strontium isotopic ratios in the river waters: The Congo basin case. *Earth Planet. Sci. Letters* 120:59–76.

Nkounkou, R. R. and J. L. Probst. 1987. Hydrology and geochemistry of the Congo River system. *Mitt. Geol-Paläont. Inst. Univ. Hamburg*, SCOPE/UNEP 64:483–508.

Nyeck, B., P. Bilong, S. M. Eno Belinga, and B. Volkoff. 1993. Séquence d'évolution des sols sur granite dans le Sud du Cameroun. Cas des sols de Zoételé. *Ann. Fac. Sci. Univ. Yaoundé*, IHS, 1:254–277.

Odigui Ahana, D. H. 2000. Contribution à l'étude de la dynamique du carbone, de l'azote et des éléments minéraux dans un écosystème forestier: cas de la forêt secondaire de Nlobisson par Nkoabang (SE Cameroun). DEA Départ. *Sc. De la Terre, Univ. Yaoundé I* 93:4, annexes.

Oliva, P., J. Viers, B. Dupré, J. P. Fortuné, F. Martin, J. J. Braun, and H. Robain. 1999. The effect of organic matter on chemical weathering: Study of a small tropical watershed, Nsimi Zoetele, Cameroon. *Geoch. Cosmochim. Acta* 63:4013–4035.

Olivie-Lauquet, G., T. Allard, J. Bertaux, and J. P. Muller. 2000. Crystal chemistry of suspended matter in a tropical hydrosystem, Nyong basin (Cameroon, Africa). *Chem. Geol.* 170:113–131.

Olivry, J. C. 1986. Fleuves et rivières du Cameroun. Monogr. Hydro. *ORSTOM* 9:733, p. 2 cartes.

Probst, J. L., J. Mortatti, and Y. Tardy. 1994. Carbon river fluxes and global weathering CO_2 consumption in the Congo and Amazon river basins. *Applied Geochem.* 9:1–13.

Probst, J. L., R. R. Nkounkou, G. Krempp, J. P. Bricquet, J. P. Thiebaux, and J. C. Olivry. 1992. Dissolved major element exported by the Congo and the Ubangui rivers during the period 1987 to 1989. *J. Hydrol.* 135:237–257.

Richey, J. E., J. I. Hodges, A. H. Devol, P. D. Quay, R. L. Victoria, L. A. Martinelli, and B. R. Forsberg. 1990. Biogeochemistry of carbon in the Amazon river. *Limnol. Oceanogr.* 35 (2):352–371.

Ritz, M., H. Robain, E. Pervago, Y. Albouy, C. Camerlynck, M. Descloitres, and A. Mariko. 1998. Improvement to resistivity pseudosection modelling by removal of near surface inhomogeneity effects: Application to a soil system of Southern Cameroon. *Geophys. Research* 47:85–101.

Santoir, C. and A. Bopda. 1995. *Atlas regional Sud-Cameroun.* Ed. ORSTOM, Paris, 53 p., 21 pl.

Seyler, P., H. Etcheber, D. Orange, A. Laraque, L. Sigha-Nkamdjou, and J. C. Olivry. 1995. Concentrations, fluctuations saisonnières et flux de carbone dans le bassin du Congo. In J. C. Olivry and J. Boulègue, eds., *Grands Bassins Fluviaux Périatlantiques: Niger, Amazone, Congo.* Orstom Editions, Collect. Colloques et Séminaires, pp. 217–228.

Seyler, P., J. C. Olivry, L. Sigha-Nkamdjou. 1993. Hydrochemistry of the Ngoko River, Cameroon: Chemical balances in a rain forest equatorial basin, in *Hydrology of Warm Humid Regions*, Proceed. Symp. Yokohama, AIHS publ., 216, pp. 87–105.

Sigha-Nkamdjou, L., P. Carré, and P. Seyler. 1995. Bilans hydrologique et géochimique d'un écosystème forestier équatorial de l'Afrique Centrale: La Ngoko à Mouloundou, in J. C. Olivry and J. Boulègue, eds., *Grands Bassins Fluviaux Périatlantiques: Congo, Niger, Amazone.* ORSTOM Ed., Collect. Colloques Séminaires, pp. 199–216.

Stoorvogel, J.J., B. H. Jansen, and N. Van Bremen. 1997a. The nutrient budget of a watershed and its forest ecosystem in the Taï National Park in Côte d'Ivoire. *Biogeochem.* 37:159–172.

Stoorvogel, J. J., N. Van Bremen, and B. H. Jansen. 1997b. The material input by harmattan dust to a forest ecosystem in Côte d'Ivoire. *Biogeochem.* 37:145–157.

Sullivan, T. J., C. T. Driscoll, S. A. Gherini, R. K. Munson, R. B. Cook, D. F. Charles, and C. P. Yatsko. 1989. Influence of aqueous aluminium and organic acids on measurements of Acid Neutralizing Capacity in surface waters. *Nature* 338:408–410.

Tardy, Y. 1993. Petrologie des latérites et des sols tropicaux. Masson, Paris, 459 pp.

Velbel, M. A. 1995. Interaction of ecosystem processes and weathering processes, in S. T. Trudgill, ed., *Solute Modeling in Catchment Systems.* pp. 193–209.

Vicat, J. P. 1998. Esquisse géologique du Cameroun. Collect. Géocam 1/1998, J. P. Vicat and P. Bilong, eds. Presses Univ. Yaoundé, pp. 3–11.

Viers, J., B. Dupré, M. Polvé, J. Schott, J. P. Dandurand, and J. J. Braun. 1997. Chemical weathering in the drainage basin of a tropical watershed (Nsimi-Zoetele site, Cameroon): Comparison between organic-poor and organic-rich waters. *Chem. Geol.* 140:181–206.

White, A. F. and A. E. Blum. (1995) Effects of climate on chemical weathering in watersheds. *Geoch. Cosmochim. Acta* 59:1729–1747.

White, A. F., A. E. Blum, M. S. Schultz, D. V. Vivit, D. A. Stonestrom, M. Larsen, S. F. Murphy, and D. Eberl. 1998. Chemical weathering in a tropical watershed, Luquillo Mountains, Puerto Rico. I. Long-term versus short-term weathering fluxes. *Geoch. Cosmochim. Acta* 62:209–226.

CHAPTER **19**

Organic Carbon in the Sediments of Hill Dams in a Semiarid Mediterranean Area

Jean Albergel, Taoufik Mansouri, Patrick Zante, Abdellah Ben Mamou, and Saadi Abdeljaoued

CONTENTS

19.1 Introduction 289
19.2 Selection and Methods 290
    19.2.1 Choice of Lakes, Monitoring of Siltation 290
    19.2.2 Sampling Protocol 292
    19.2.3 Analytical Methods 292
19.3 Results 293
    19.3.1 Description of the Sedimentary Profiles 293
    19.3.2 Total Organic Carbon Contents of the Sediments 295
    19.3.3 Erosion and Loss of Organic Carbon in the Catchment 295
19.4 Conclusion 296
Acknowledgments 298
References 298

19.1 INTRODUCTION

Carbon is the principal element in organic matter and its presence is a decisive factor in determining soil quality. From the beginning of time, humankind has attempted to maintain or even increase the level of organic carbon (C) in fields to conserve soil fertility (Doran et al., 1996).

Soil erosion is selective in terms of preferentially removing the nutrients and colloids that are the essence of soil fertility. This preferential removal is clear for C, nitrogen (N), phosphorus (P), clay and loam up to 50 microns (Roose, 1984). Carbon losses from cultivated soils are a major agronomic issue, and are monitored on erosion plots of standardized dimensions (Wischmeier and Smith, 1978). The literature reports numerous values for losses of C and other elements found in the runoff water collected upstream of the erosion plots depending on soil type, plant cover, cropping practices, and intensity of erosion (Larson et al., 1983; Smith et al., 2001).

Over the last few years, soil C has become an environmental issue rather than just an agronomic one. The accumulation of C reserves in the soil can mitigate the accumulation of CO_2 in the atmosphere and limit the increase of greenhouse gases (Doran et al., 1998).

World soils, with 1,500 Gt of C to 1 m depth (more than twice that of vegetation-650 Gt), represent the greatest C reserve in the terrestrial ecosystems. Some authors consider that mineralization of soil C into CO_2 as land was brought under cultivation or as a result of erosion, has been one of the main causes of the increase in greenhouse gases over the last century. In retrospect, a parallel can be drawn between the increase in cultivated areas at the expense of forests and grasslands, losses of C and the increased sensitivity of the upper soil layer to degradation by erosion and desertification (Robert, 2000).

A runoff plot is not large enough to study the loss of soil C by water erosion in the context of climate change. In contrast, soil erosion measured at the catchment scale is more relevant. Assessment of temporal changes in soil C pool can be used to estimate the impact of changes in agricultural practices or land use (agroforestry, reforestation, erosion control) on C sequestration at the scale of a hydrological unit (Lal et al., 1995; 1997).

The experimental setup traditionally used to evaluate C losses at this scale is the catchment or watershed, where liquid flow, solid transports, and dissolved matter are measured at the outlet (Owens et al., 2002). This setup is very effective but is difficult to establish and is also very costly. It requires the presence of operators at each flood event or the availability of sampling devices and reliable automatic sensors.

Small dams in semiarid areas are very good sediment traps. Moreover, one of the aims of the 1,000 hill lakes currently being established in Tunisia is to trap erosional products from hill slopes and to protect large dams from rapid sedimentation. These small dams also enable an experimental setup for the measurement of hydrological balances and solid transport to be established at low cost (Albergel and Rejeb, 1997). A study conducted on a series of small dams in the Tunisian Dorsal showed that the reservoir water was low in P and N. (Rahaingomanana, 1998). There is no risk of eutrophication in these lakes due to sparse vegetation, therefore the water cannot contain much newly formed C. The dams are of recent construction and are subject to significant siltation. Some have discharged very little if at all.

These observations have led to the following question: Would analysis of the organic C in the dam sediments give correct information on C losses from the catchment by erosion? Thus, a sampling program was launched during the summer of 1997 and an analysis protocol introduced. This chapter describes the experiment and presents the first results.

19.2 SELECTION AND METHODS

19.2.1 Choice of Lakes, Monitoring of Siltation

Two hill lakes that were dry throughout the summer of 1997 were chosen for the experiment. They were in the same climate and had never discharged (in the first case) or had discharged very little (in the second). Figure 19.1 shows the situation of the lakes, the bathymetry at the time of the experiment, and the points at which the core samples were taken. Siltation was heavy in the first lake and more moderate in the second, it having been the subject to erosion control work over 60% of its area (Table 19.1).

The bathymetry of the reservoir was measured by localized sampling of the bottom of the reservoir, following transverse lines established by stretching a cable between the two shores. The ends of each line were leveled and positioned on the as-built drawing of the reservoir. A digital model of the terrain was constructed (Figure 19.1).

ORGANIC CARBON IN THE SEDIMENTS OF HILL DAMS IN A SEMIARID MEDITERRANEAN AREA

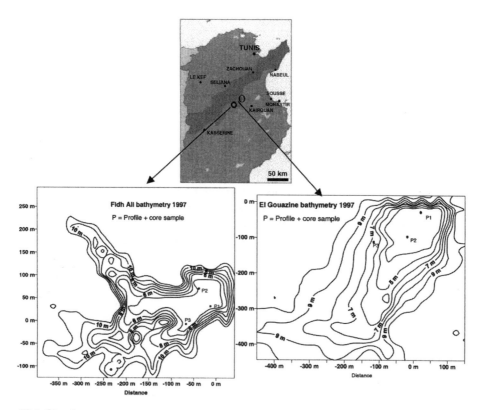

Figure 19.1 Situation map.

Table 19.1 Dams Used in the Experiment (1999 Data)

Dam	Location	Catchment Area (ha)	Year of Constr./Vol. of Dam (m³)	Siltation Volume (m³)	Sediment Flux (t)	Specific Erosion (t/ha/year)
Fidh Ali	Central Tunisia, anticlinal — Eocene. Gypseous marls and shelly limestones (400 mm isohyet)	238	1991/134,710	49,840	0	24.2
El Gouazine	Central Tunisia, valley in old cemented alluvial deposits — Pleistocene. (400 mm isohyet)	1,810	1990/237,030	16,030	1,300	1.8

Comparison between the reservoir volumes at discharge level between one measurement and another was used to estimate the amount of material retained. A mean concentration of suspended matter, obtained by sampling, was attributed to the discharged volumes. Solid transport between two bathymetry measurements was obtained by adding the mass of soil exported by the discharged liquid flows to the volume of silt retained in the dam multiplied by its density:

$$T = V_s \times d + \sum_{i=1}^{n} S_i C_i \tag{19.1}$$

where T is the total solid transport between two bathymetric measurements (t); V_s the measured volume of silt (in m³); d the density of the silt; n the number of floods that have caused discharges between two measurements; S_i the volume discharged during flood i (in m³); and C_i the mean measured concentration of suspended matter during flood i (t/m³).

The solid transports were reconstituted for each flood using Williams's hypothesis (1982). At the scale of an elementary catchment, solid transport is determined by the volume of the flood and the shape of its hydrograph. It can be expressed by the relation shown:

$$T_i = \lambda (Q_i \times V_i)^\beta \tag{19.2}$$

where T_i is the sediment input of flood i (in metric tons); V_i, the volume of the natural flood entering the reservoir (in m³); Q_i, the peak flood flow (in m³ s⁻¹); λ, a parameter representing the soils of the slopes and their use; and β a parameter characterizing the river system. For each catchment, both parameters were calculated by optimization, comparing the sum of the reconstituted solid transports between two bathymetry measurements and the measured sum from Equation 19.1 (Albergel et al., 1999).

19.2.2 Sampling Protocol

In each dam, three points were selected for sediment sampling (Figure 19.1) and a soil pit was dug to the level of liquid mud. A PVC tube was then driven in with a sledgehammer until it could not be driven any farther (bottom of dam before flooding). During penetration, moist sediments were forced into the tube. The height of the sediment was measured on the outside and the inside of the tube to establish the real depth of sampling. The core was removed and taken to the laboratory. Each core was cut longitudinally with an electric saw. One half was kept intact for a fine description, while the other half was used for the analyses.

The description of the lithofacies observed in the shafts was based on lithological properties, particle size distribution and color variation using the Munsell code. It served to match up the sedimentation levels between the three cores. The hydrological events responsible for the sedimentary deposits during the period of hydrological measurements were identified. For instance, a fine layer of very fine dark bluish-grey clay (2FOR GLEY 3/10B) was considered an indication of a flood-free period with a full lake, and a grey-beige sandy-clayey layer (5Y 5/2) of a major flood event. Samples were taken for analysis every 5 cm along each profile. In other words, not all measurements were made on the same profile. The parameters defined for each sediment layer were moisture content, bulk density, particle size distribution, clay type, and organic matter content.

19.2.3 Analytical Methods

All analyses were conducted at the sedimentology laboratory of the Geology Department of the Faculty of Sciences, University of Tunis II. The techniques employed to analyze the physical and mineralogical properties of the sediments were those traditionally used to study soils (Chamayou and Legros, 1989): bulk density by the core method, laser granulometry measurements, and mineralogical identification by differential thermal analysis. Organic carbon determination was done by Rock-Eval pyrolysis, a method commonly used in oil prospecting (Tissot and Welte, 1984). Small samples (100 mg) were placed in a pyrolysis oven. The products were detected with a flame ionization detector (FID). The parameters listed below were collected for eight samples at El Gouazine and thirteen samples at Fidh Ali. Both catchments had similar soil types. Three soil samples were taken from the surface layer of the El Gouazine catchment corresponding to three different land uses (pine forests, grasslands, and cultivated land). The parameters measured were:

- TOC: the amount of total organic C, expressed in mg of CO_2 produced during pyrolysis per mg of sample or as a percentage
- S1: the amount of free hydrocarbons present in the sediments and released at 300°C (volatile organic matter: gas and oil)
- S2: the amount of hydrocarbons produced from cracking of nonvolatile organic matter
- Tmax: the temperature at which maximum release of hydrocarbons from cracking of kerogen during pyrolysis occurs, indicating the "maturation" stage of organic matter
- HI: the hydrogen index expressed in mg of hydrocarbons/g of TOC, representing the amount of potential S2 hydrocarbons compared to that of TOC. This index is used to determine the origin of the organic matter. Aquatic organisms and algae, rich in proteins and lipids, have a higher H:C ratio than land plants, which are richer in sugars.

19.3 RESULTS

19.3.1 Description of the Sedimentary Profiles

Figure 19.2 shows the sedimentary profiles of the dams for the horizons described in the field, excluding the core samples. Table 19.3 presents the sediment input volumes for each flood in application of the Williams model and the thickness of the attributed sediments. Only the heaviest floods for the two catchments (peak flow > 1 m^3/s) are shown in this table, although the data were calculated for all floods, i.e., 34 at El Gouazine and 28 at Fidh Ali. The elevation of the top of the core was established relative to the lake's water level.

The sediments deposited in the two lakes have very high fine fractions. The predominant fraction is silty-clayey, and other fractions are incidental (Mansouri, 2001). At Fidh Ali, the < 63 μm fraction was over 90% at all depths. At El Gouazine, the presence of sand was more marked and the < 63 μm fraction ranged from 55 to 95% depending on the sample. Microgranulometric analysis of the fine fraction showed a deposit per loss of load, followed by settling.

The densiometric analyses of the same depositional cycle showed a gradual decrease in bulk density. The granulometric analyses confirmed this type of deposit, in fact each cycle was formed of clayey-silty strata toward the top and coarser-grained strata at the base. The dry bulk density of the sediments ranged from 1 to 1.5 Mg/m^3, showing a pattern of deposits. The mean bulk density for a full cycle was 1.3 Mg/m^3 (Mansouri, 2001).

Variations in moisture content were observed along the length of the sedimentary columns, with a maximum below the cracks and a decrease toward the base, close to the substrate (from 60 to 30%). A significant decrease in the volume of the dry sediments was noted in the surface layer.

The sediments of both lakes contained approximately the same minerals. A large quantity of quartz and clay and a significant quantity of calcite were observed in both lakes, and a few gypsum crystals and small quantities of dolomite were present at Fidh Ali. The clay fraction consisted of smectite, kaolinite, and small quantities of illite. Comparison with the clay minerals of the catchment soils showed an increase in smectite and illite and a decrease in kaolinite (Mansouri, 2001). The

294 SOIL EROSION AND CARBON DYNAMICS

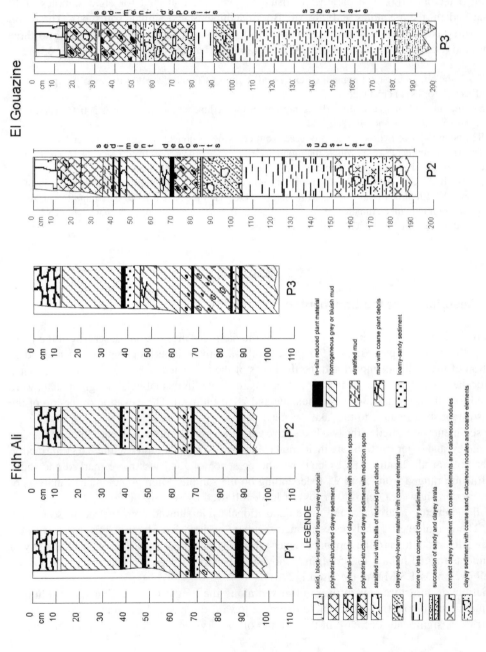

Figure 19.2 Description of the sedimentary profiles.

Table 19.2 Parameters of the Analyses by Rock-Eval Pyrolysis

	Fidh Ali					El Gouazine				
Sample	S1	S2	Tmax	TOC%	HI	Sample	S1	S2	Tmax	TOC %	HI
H10	0.03	0.13	405	0.57	22.80	C1-1	0.14	0.73	412	1.41	52
H12	0	—	—	0.56	—	C1-2	0.14	106	425	0.49	216
H13	0.08	0.24	401	0.59	40.67	C1-3	0.13	0.91	432	0.56	163
P3-10	0.04	0.24	416	1.45	16.55	C1-4	0.18	1.32	423	1.48	89
P3-11	0.07	0.16	403	0.29	55.17	C1-5	—	—	—	0.37	—
P3-20	0.05	0.23	402	0.46	50.00	C1-6	0.17	1.24	423	1.25	99
P3-21	0.07	0.18	394	0.40	45.00	C1-7	0.17	1.05	429	1.25	84
P3-30	0.05	0.29	402	0.84	34.52	C1-9	0.2	1.56	413	1.50	104
P3-40	0.03	0.14	404	1.62	8.64						
P3-50	0.04	0.21	414	1.43	14.68						
P3-60	0.04	0.14	406	1.63	8.58						
P3-71	0.02	0.07	403	1.24	5.64						
P3-80	0.03	0.06	414	0.73	8.21						

properties of these sedimentary profiles established in small dams were similar to those described by Ben Mamou for larger dams in central Tunisia (1998).

19.3.2 Total Organic Carbon Contents of the Sediments

The results of the organic matter analysis by Rock-Eval pyrolysis for the Fidh Ali and El Gouazine lakes are shown in Table 19.2. In core P1, taken 1 m from the El Gouazine embankment, the TOC contents range between 0.37 and 1.5% ($1.03 \pm 0.47\%$). For shaft P3 at Fidh Ali, the TOC contents range between 0.4 and 1.63% ($0.91 \pm 0.49\%$). The variation in TOC in the sedimentary profiles is very irregular and reflects the heterogeneous nature of the lithological composition, the low values are observed for the coarser-grained beige horizons and the highest values in the very fine-grained bluish horizons. The TOC contents of the sediments tend to be higher than those of the catchment soils (0.66% in the surface horizons of soils under cultivation, representing 88% of the area of the Fidh Ali catchment and 55% of the El Gouazine catchment; 0.88% in El Gouazine pine forest soils; and 1% in the maquis and grassland soils). There was no decrease in organic matter with depth of the sediments. The lakes are recent (< 10 years old), so the organic matter is preserved in the sediments without change.

The HI values are shown in Figure 19.3 as a function of Tmax. All analyses fall within the category of type III organic matter: continental ligneous of external origin, i.e., particulate and amorphous material originating from the soil horizons. These products are introduced into the lake by erosion. Only one point falls outside this category but is close to its boundary.

There is a significant positive correlation between the amounts of free hydrocarbons (S1) and those of hydrocarbons produced from the cracking of nonvolatile organic matter (S2). The correlation coefficient r is 0.85 for 21 analyses. The correlation between the TOC content and each of these fractions (S1 and S2) is positive but not significant (r = 0.5).

19.3.3 Erosion and Loss of Organic Carbon in the Catchment

The results of the analyses by pyrolysis confirm that most of the organic C found in the dams comes from soil erosion and that this organic matter has not been transformed in the recently accumulated sediments. The question asked in the introduction can thus be answered in the affirmative: The organic matter found in the hill lake sediments is that lost from the catchment as a result of water erosion.

Using mean values, rough estimates can be given of the interannual loss of C for each catchment. There is a mean specific loss of organic C of 218 kg/ha/year at Fidh Ali and of 19 kg/ha/year at

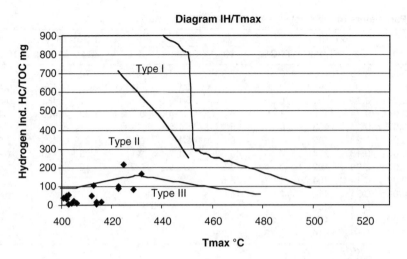

Figure 19.3 HI as a function of Tmax and organic matter type.

El Gouazine, where conservation measures were installed in 1996. Reconstitution of the amounts of organic C collected in the dam for each flood over the period of the hydrological observations (1993 to 1997) shows significant fluxes corresponding with the high flood events (Table 19.3).

Figure 19.4 shows a significant relation between the peak flow and the amount of C transported to the dam, and that the heaviest floods are usually accompanied by high TOC contents, except for the exceptional flood at Fidh Ali on October 3, 1994 (specific peak flow: 14.6 $m^3/s/km^2$), which caused spectacular erosion with the collapse of banks and the appearance of large gullies. This anomaly is attributed to the fact that a large proportion of the solid transport resulting from the flood corresponded to materials washed from the banks of the river system and from the bottom of the gullies. This involved gypseous marls containing a low organic matter content. The scouring of the surface layer by this flood was such that no high TOCs are found in the following floods.

19.4 CONCLUSION

The results observed at the scale of the erosion plot are confirmed at the scale of the small catchment: the concentrations of TOC in the erosion products downstream of the catchment are higher than in the soils they originate from. The strongest floods carry large amounts of organic C. The exceptional floods have a scouring effect that impoverishes the soils of organic C over a long period of time.

The experiment presented herein from two catchments in the Tunisian Dorsal is conclusive, and has given a rough estimate of the total organic C losses for two different catchments. In the first catchment, 80% under cultivation and consisting primarily of soil derived from argilites and gypseous marls, the losses are estimated at 200 kg/ha/year, i.e., ten times higher than from a wooded catchment (40% of the area) where erosion control measures had been installed. Soil losses from the cultivated catchment are 15 times higher than from the wooded catchment, but the sediments from the latter are richer in TOC.

These results need to be confirmed by new samples from the same dams in 2003 and 2004 (the 1997 samples corresponded to 6 or 7 years of accumulation), and from other dams. In Tunisia, 30 artificial reservoirs have been selected in the Atlas Mountains, between Cape Bon and the Algerian border, to form a network for the observation and monitoring of the hill lakes and their catchments. They are diverse in catchment areas, ranging from an anthropic semiforest to entirely agricultural environments. Excluding the dams where there are major discharges, the experiment

ORGANIC CARBON IN THE SEDIMENTS OF HILL DAMS IN A SEMIARID MEDITERRANEAN AREA 297

Table 19.3 Transported Carbon for the Three Floods with Flows above 1 m³/s

Catchment	Date	Rainfall (mm)	Flood Volume (m³)	Flood Q max (m³/s)	Deposition (m³)	Siltation Level (m)	TOC%	Carbon (kg)
El Gouazine	02-Sept-94	8.5	13,300	6.23	290	1.95	1.25	4,714
El Gouazine	30-Sept-94	14.0	12,140	4.86	233	2.03	1.25	3,780
El Gouazine	03-Oct-94	38.0	126,800	28.8	2,549	2.56	1.25	41,414
El Gouazine	11-Oct-94	8.0	26,000	6.67	435	2.62	0.37	2,091
El Gouazine	22-Oct-94	8.0	4,000	1.33	58	2.63	0.37	278
El Gouazine	04-Nov-94	21.5	27,000	8.33	506	2.7	0.37	2,433
El Gouazine	01-Aug-95	3.0	5,200	2.50	97	2.71	0.37	466
El Gouazine	11-Aug-95	2.5	18,700	7.50	385	2.76	1.45	7,399
El Gouazine	13-Sept-95	13.5	37,500	9.33	654	2.82	1.46	12,575
El Gouazine	17-Sept-95	14.5	12,800	1.19	106	2.83	0.48	772
El Gouazine	20-Sept-95	27.0	147,280	35.00	3,109	3.08	1.23	49,703
El Gouazine	24-Sept-95	21.5	15,218	6.67	314	3.12	1.41	5,750
El Gouazine	18-Oct-95	29.9	19,200	3.33	244	3.14	1.41	4,472
El Gouazine	10-Dec-95	63.5	44,000	3.33	395	3.17	1.41	7,234
El Gouazine	13-Jan-96	25.0	21,000	2.22	203	3.19	1.41	3,723
El Gouazine	15-Feb-96	11.0	10,000	3.33	167	3.21	1.41	3,063
El Gouazine	14-Mar-96	25.0	22,000	11.67	546	3.22	1.41	10,008
El Gouazine	12-May-96	18.0	16,000	1.33	129	3.24	1.41	2,365
El Gouazine	09-Sept-96	50.5	32,800	3.83	28	3.24	1.41	516
Fidh Ali	04-Sept-93	29.5	8,180	3.80	839	0.66	0.73	7,963
Fidh Ali	14-Sept-93	18.5	2,590	2.17	311	0.84	0.73	2,950
Fidh Ali	27-Sept-94	39.5	14,550	6.88	2,959	1.92	1.62	62,312
Fidh Ali	03-Oct-94	51.0	63,800	34.30	9,900	3.4	0.61	77,222
Fidh Ali	11-Jun-95	25.5	12,000	12.80	2,123	3.61	0.59	16,281
Fidh Ali	04-Sept-95	4.0	6,932	1.33	415	3.66	0.59	3,184
Fidh Ali	16-Sept-95	17.0	12,500	7.17	1,550	3.8	0.59	11,891
Fidh Ali	19-Sept-95	31.0	28,219	12.67	3,459	4.09	0.46	20,687
Fidh Ali	22-Sept-95	22.5	30,396	7.33	2,631	4.3	0.47	16,072
Fidh Ali	23-Sept-95	16.0	1,100	1.50	153	4.31	0.57	1,133
Fidh Ali	15-Oct-95	36.5	20,901	6.83	2,032	4.46	0.57	15,057
Fidh Ali	17-Oct-95	14.0	4,254	3.33	532	4.5	0.57	3,944
Fidh Ali	09-Sept-96	23.5	3,700	2.83	447	4.54	0.57	3,310

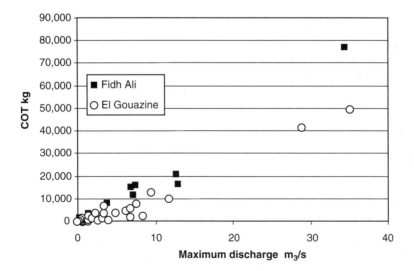

Figure 19.4 TOC flux as a function of peak flow (all are measured data points).

described in this chapter can be repeated and an estimate obtained of the erosion of organic matter under different environmental conditions. Soil conservation measures are being installed on several of these catchments. The effects of these measures on runoff, soil erosion, and C flux need to be assessed.

ACKNOWLEDGMENTS

This research was conducted as part of the European HYDROMED program and was supported by the ACTA Directorate-General of the Tunisian Ministry of Agriculture. The authors thank Director-General Si Habib Fahrat. They also thank Garth Evans for the translation into English.

REFERENCES

Albergel J. and N. Rejeb. 1997. Les lacs collinaires en Tunisie: Enjeux, contraintes et perspectives. *Comptes Rendus à l'Académie d'Agriculture de France* 83:77–88, 101–104.

Albergel J., Y. Pépin, S. Nasri, and M. Boufaroua. 1999. *Modeling small dams siltation with MUSLE*. Dept. of water resources engineering, Lund Institute of Technology, June 29–July 3, 1998. Report 3222, LUND, Suede. pp. 195–204.

Ben Mamou, A. 1998. *Barrages Nabeur, Sidi Salem, Sidi Saad et Sidi Boubaker. Quantification, étude sédimentologique et géotechnique des sédiments piégés. Apports des images satellitaires*. Thèse de Doctorat en Géologie. Tunis II.

Chamayou H. and J. P. Legros. 1989. *Les bases physiques, chimiques et minéralogiques de la science du sol*, Edition ACCT-CILF-PUF, 593 p.

Doran, J. W., E. T. Elliott, and K. Paustian. 1998. Soil microbial activity, nitrogen cycling, and long term changes in organic carbon pools as related to fallow tillage management. *Soil Tillage Research* 49:3–18.

Doran, J. W., M. Sarrantonio, and M. A. Liebig. 1996. Soil health and sustainability, in: D. L. Sparks, ed., *Advances in Agronomy*, vol. 56. Academic Press San Diego, CA, pp. 1–54.

Lal, R., J. M. Kimble, and R. F. Follett. eds. 1997. *Soil properties and their management for carbon sequestration*. USDA, National Resources Conservation Service, National Soil Survey Center, Lincoln, NE, 150 pp.

Lal, R., J. M. Kimble, E. Levine, and B. A. Stewart. eds. 1995. *Soil and Global Change: Advances in Soil Science*, Lewis publishers, Chelsea, MI, 440 pp.

Larson, W. E., F. J. Pierce, and R. H. Dowdy. 1983. Loss in long-term productivity from soil erosion in the U.S., in El-Swaîfy, Moldenhauer, and Lô, eds., *Soil Erosion and Conservation*. Soil Cons. Soc. Am., Ankeny, USA:262–271.

Mansouri, T. 2001. *Modélisation spatialisée des écoulements et du transport solide des bassins versants des lacs collinaires de la Dorsale tunisienne et du Cap Bon*. Thèse de doctorat en Géologie. Tunis, El Manar, 286 pp.

Owens, L. B., R. W. Malone, D. L. Hothem, G. C. Starr, and R. Lal. 2002. Sediment carbon concentration and transport from small watersheds under various conservation tillage practices. *Soil & Tillage Research* 67:65–73.

Rahaingomanana, N. 1998. *Caractérisation géochimique des lacs collinaires de la Tunisie semi-aride et régulation géochimique du phosphore*. Thése de doctorat, Géosci Université de Montpellier 1, Montpellier, France, 313 pp.

Robert, M. 2000. En arrière plan du protocole de Kyoto, des enjeux qui dépassent la lutte contre le changement climatique. *Le courrier de l'environnement de l'INRA*. No. 41 INRA, www.inra.fr/dpenv/roberc41.htm.

Roose, E.1984. *Introduction à la gestion conservatoire de l'eau, de la biomasse et de la fertilité des sols*. (GCES) FAO, Rome (ITA), Bulletin Pédologique de la FAO (ITA), No 70, 438 pp.

Smith, W. N., G. Wall, R. Desjardins, and B. Grant. 2001. *Le carbone organique des sols. C. Qualité du sol*. www.agr.gc.ca/policy/environment/eb/public_html/pdfs/aei/fchap09.pdf, pp. 85–93.

Tissot, B. P. and D. H. Welte. 1984. *Petroleum Formation and Occurrence*. Springer-Verlag, Berlin, 699 pp.
Williams, J. R. 1982. Testing the Modified Universal Soil Loss Equation, in Estimating erosion and sediment yield on rangelands, USDA.ARM-W-26, pp.157–164.
Wischmeier, W. H. and D. D. Smith. 1978. *Predicting Rainfall Erosion Losses: A Guide to Conservation Planning*. USDA Handbook, U.S. GPO, Washington, D.C., 537 pp.

CHAPTER 20

Monitoring Soil Organic Carbon Erosion with Isotopic Tracers: Two Case Studies on Cultivated Tropical Catchments with Steep Slopes (Laos, Venezuela)

Sylvain Huon, Boris Bellanger, Philippe Bonté, Stéphane Sogon, Pascal Podwojewski,
Cyril Girardin, Christian Valentin, Anneke de Rouw, Fernando Velasquez,
Jean-Pierre Bricquet, and André Mariotti

CONTENTS

20.1 Introduction 302
20.2 Physiographic Settings and Material Sampling 303
    20.2.1 The Houay Pano Catchment in Northern Laos 303
    20.2.2 The Rio Boconó Watershed in Northwest Venezuela 304
20.3 Analytical Methods 305
20.4 Results and Discussion 306
    20.4.1 Monitoring SOC Erosion with ^{137}Cs for Cultivated Soils of the
           Houay Pano Catchment in Laos 306
        20.4.1.1 Application of ^{137}Cs Measurements in Soil Erosion Studies 306
        20.4.1.2 Estimates of Soil Erosion Rates on the Houay Pano Catchment 307
    20.4.2 Monitoring Erosion of Organic Carbon Derived from Soil and Rock Sources
           in Runoff and Stream Flows with Stable C and N Isotopes for the Boconó
           Watershed in Venezuela 311
        20.4.2.1 Principles of Natural $^{13}C/^{12}C$ and $^{15}N/^{14}N$ Labeling of Soils and Suspended
               Sediments 311
        20.4.2.2 Monitoring Erosion of Fine-Size (< 50 μm) SOC with Field Plot
               Experiments 312
        20.4.2.3 Monitoring Sources of Fine-Size (< 50 μm, < 200 μm) Suspended
               Organic Carbon in Stream Flows during Flood Events 314
        20.4.2.4 Monitoring the Evolution of Suspended Organic Matter in a Water
               Reservoir at the Outlet of the Watershed 320
20.5 Conclusion 321
Acknowledgments 322
References 322

20.1 INTRODUCTION

With an estimated content of 1600 Pg carbon (C) in the first m, soil organic matter (SOM) is the largest terrestrial reservoir of organic C, exceeding the terrestrial biosphere (560 Pg C) and atmosphere (750 Pg C) storage capacities (Post et al., 1982; Eswaran et al., 1993; Sundquist, 1993). Tropical regions account for nearly 30% of the soil organic C (SOC) reservoir (Milliman et al., 1987; Ross, 1993; Dixon et al., 1994; Lal, 1995; Zech et al., 1997). SOM plays a key role in soil physical and chemical properties such as structural stability, porosity, nutrient availability, and ion-exchange capacity (Oades, 1986; Le Bissonnais, 1990; Tiessen et al., 1994; Lal et al., 1999; Roscoe et al., 2001; Puget et al., 2001).

Rising atmospheric CO_2 levels coupled with global warming may stimulate terrestrial photosynthesis (fertilization effect) and enhance SOC sequestration (e.g., Houghton et al., 1993; Kirschbaum, 2000; Schlesinger and Andrews, 2000). However, present-day deforestation and land-use change question the efficiency of C storage in soils because enhanced runoff and tillage erosion may induce opposite trends. Erosion leads to physical removal of SOC through soil particle redistribution along slopes and enhances mineralization of SOM, increasing as much the C flux to the atmosphere. Since the behavior of soils as a sink or a source for atmospheric CO_2 is still debated (Ciais et al., 1995; Houghton et al., 1998), monitoring SOC erosion in cultivated tropical environments may provide significant information on the C flux to the atmosphere and contribute to a better management of soil resources for sustainable development purposes (e.g., Lal, 1990; Houghton, 1991; Houghton et al., 1993).

The transfer by erosion of SOC to the hydrographic network may be addressed using relationships between total organic C concentrations in top soil horizons and in suspended loads of river discharges (Meybeck, 1982; 1993; Probst, 1992; Ludwig et al., 1996) or hydrological and GIS-based models that link organic C outputs to climatic and physiographic settings on the catchments such as precipitation, slope, vegetation cover, and soil properties (Post et al., 1982; Esser and Kohlmaier, 1991; Probst, 1992; De Roo, 1993; Browne, 1995). However, these approaches generally provide little information on the nature and the source of organic matter exported from the catchments and its further evolution in the hydrographic network. Moreover, providing a direct link between SOC erosion and suspended organic matter loads is a difficult assignment. Soil erosion rates monitored with field plot experiments (Morgan, 1986; Lang, 1992; Loughran and Campbell, 1995) often do not match the sediment delivery ratio of stream flows, reflecting sorting, deposition, or remobilization processes that may take place along slopes and river banks on the watersheds (Meade, 1988; Dedkov and Mozzherin, 1992; Milliman and Syvitski, 1992; Trimble and Crosson, 2000). Accordingly, scaling up field plot measurements to catchments for SOC erosion budgets and models involves large and unexplained variability with respect to soil and sediment composition. Another major uncertainty on the impact of soil erosion on the C cycle is the mineralization rate of suspended organic matter in the hydrological network (range: 0 to 100%, see references in Lal, 1995). It is likely that the extent of organic carbon degradation is controlled by the nature of organic matter (i.e., labile vs. resistant or protected compounds, Hedges et al., 2001). The behavior in surface environments of refractory organic matter derived from geological basement sources (such as shales or other organic-matter bearing rocks) contrasts sharply with that of vegetation debris, charcoals, or humic substances generated by soil erosion. Discrimination between lithic and soil sources of organic matter in stream flows is important to constrain both soil erosion and global C cycle budgets (Meybeck, 1993; Kao and Liu, 1996; 2000; Raymond and Bauer, 2001a; 2001b; Megens et al., 2002). The mineralization rate of suspended organic matter is also controlled by redox conditions, in particular for water reservoirs located at the outlet of the watersheds, which may constitute ultimate receptacles for runoff and undergo severe oxygen depletion with high organic matter supply from the drainage areas (Likens, 1972; Chapra and Dobson, 1981; Stumm and Morgan, 1996). With lower mineralization rates than in soils and high residence times, riparian zones, wetlands, and lacustrine environments may possibly behave as C sinks with respect to atmospheric CO_2 (e.g., Dean and Gorham, 1998) and bias organic C budgets based on direct link between soil erosion on catchments and riverine transport.

Isotope tracers such as radionuclides or stable isotopes may support identification and quantification of the different pools of organic matter generated by erosion on watersheds and the sources of suspended organic matter in stream flows and river discharges (Mariotti et al., 1980; 1984; 1991; Hedges et al., 1986; Wada et al., 1987; Cai et al., 1988; Ittekot, 1988; Bird et al., 1992; 1994; Walling et al., 1993; Thornton and McManus, 1994; Onstad et al., 2000; Masiello and Druffel, 2001) or provide information to refine SOC budgets (Arrouays et al., 1995; Balesdent et al., 1988; Balesdent and Mariotti, 1996; Ritchie and McCarty, 2003). This chapter presents the results of several isotopic studies carried out on two cultivated watersheds with steep slopes located in tropical regions. The objectives were to (1) link the erosion status and the organic C content of cultivated soils using ^{137}Cs and total organic carbon measurements for the Houay Pano catchment (Laos), (2) better constrain the source of suspended organic matter during flood events by monitoring the composition of suspended loads in runoff and stream flows with stable N and C isotope measurements for the Rio Boconó watershed (Venezuela), and (3) evaluate the impact of a water reservoir located at the outlet of the Rio Boconó watershed (Venezuela) on suspended organic C exportation to the floodplain. The principles of the isotopic methods used in this study are summarized in the text along with each case study.

20.2 PHYSIOGRAPHIC SETTINGS AND MATERIAL SAMPLING

20.2.1 The Houay Pano Catchment in Northern Laos

Soil sampling for ^{137}Cs and organic C measurements was carried out for selected soils of the Houay Pano catchment (67 ha), located near Luang Prabang in northern Laos (19°51'00"–19°51'45"N, 102°09'50"–102°10'20"E). Soils were collected for 10 cm depth increments, using a corer with an internal 4.3×4.7 cm area, especially designed by the LSCE (laboratoire des Sciences du Climat et de l'Environnement). Catenas mainly cultivated with Job's tears (*Coix lacryma Jobi*) were sampled in October 2000 along two toposequences located on both sides of the main thalweg (Figure 20.1). Soils of the catchment are Entisols (18.5%; clay soils with medium

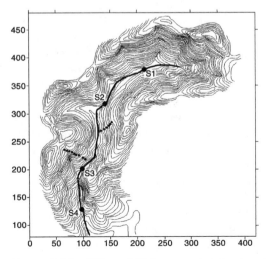

Figure 20.1 Location of soils sampled for ^{137}Cs and TOC measurements along two toposequences on the Houay Pano catchment (Laos). S1 to S4 refer to the main weirs. Grey circles correspond to soil locations.

fertility, pH = 6.4), Ultisols (33.1%; clay soils with medium fertility, pH = 5.5), and Alfisols (48.5%; heavy clay soils with medium fertility, pH = 6.2), settled on a geological basement, composed of Permian to Upper Carboniferous argillite series (shales, mudstones, and fine-grained sandstones) overlaid by limestone cliffs (NAFRI, 2001). After removal of organic matter, soils sampled for this study contain 43 to 50% of clays (< 2 μm), 34 to 42% of "fine silts" (2 to 50 μm) and 14 to 18% "coarse silts" and sands (50 to 2000 μm). The catchment is characterized by steep slopes. Slope gradients range 3 to 150% and average 60% for altitudes between 425 to 718 m above sea level. Mixed-deciduous and dry forests, mainly located on hilltops, account for less than 26% of land cover. Farmers mainly grow upland rice (*Oryza savita*), maize (*Zea mays*), and Job's tears, using slash-and-burn shifting cultivation with rotating fallows. Detailed surveys and interviews with farmers, cross checked by aerial photographs, showed that during the past four decades, catenas were generally cultivated for 7 to 8 years and maintained under fallow the remaining time. Mean annual temperature, precipitation, and evapo-transpiration (1986 to 1996 record) were 25.3°C, 1403 mm, and 1022 mm, respectively (NAFRI, 2001). Between 81% and 99% of total rainfall currently occurs during the rainy season, from April to October, with a maximum supply in July and August. Average precipitation records for Luang Prabang (LP), where precipitation data are available before 2000, can be used for Houay Pano (HP) where this study was conducted ($P_{LP} = 1.07\ P_{HP}$, r = 0.94, n = 39 with P = monthly precipitations in mm). Between 1954 and 1976, during the period of maximum ^{137}Cs fallout in Southeast Asia (see further in the text), precipitation averaged 1139 ± 225 mm in Luang Prabang, with an excursion to 1564 mm in 1963 (Bricquet et al., 2001). In 2000, the mainstream discharge of the Houay Pano catchment averaged 9.4 l s^{-1} with a total sediment yield of 7 t ha^{-1} (NAFRI, 2001).

20.2.2 The Rio Boconó Watershed in Northwest Venezuela

Erosion of SOC was monitored under natural rainfall conditions at various space and time scales by collecting runoff samples generated on experimental field plots and in stream flows during flood events. Located in northwestern Venezuela (Figure 20.2a, 08°57'–09°31'N, 70°02'–70°34'W), this watershed covers about 8% of the Venezuelan Andes (1620 km^2). Its outlet in the Llanos floodplain is closed by the Peña Larga dam built in 1983 (area, 122 km^2, volume, 2850 10^6 m^3, maximum depth, 72 m, Figure 20.2b). This large reservoir, designed for water and electricity production, also provided the opportunity to monitor offsite effects of SOC erosion on water quality (Bellanger et al., 2004b). Runoff experiments were performed on 30 m^2 experimental field plots (Felipe-Morales et al., 1977; Lal, 1990) on a bare field plot, formerly under maize and maintained manually free of vegetation, a maize field plot, and a coffee field plot (*Coffea arabica*). After removal of organic matter, soils are composed of 24 to 28% clays (< 2 μm), 18 to 26% silts (2 to 63 μm) and 46 to 56% sands (63 to 2000 μm). Coffee is currently cultivated on hill slopes at intermediate altitudes (500 to 1500 m), under 4 to 5 m high tree covers (Ataroff and Monasterio 1997), whereas maize fields are often established near the main streams. Runoff and suspended sediment discharge were monitored between April 15 and September 15 for each field plot with weekly measurements of water volume and sediment yield in the collection tanks. In addition, a single flood event (June 14, 1998) was covered in detail with a sampling frequency of 5 to 30 min. All suspended sediments were collected in polyethylene bottles in runoff and stream flows of the watershed during flood events and at several depths in the water column of the Peña Larga reservoir.

The tropical climate conditions prevailing in the region are highly influenced by altitude, ranging 200 to 4000 m above sea level within the watershed. The average temperature decreases with altitude by 0.6°C per 100 m. Precipitation records also display decreasing trends, of 1500 to 2500 mm in the lower parts of the watershed to less than 1000 mm for the higher altitude Paramó grasslands (2700 to 4000 m; Cornièles, 1998). About 87% of total rainfall takes place during the rainy season, between April and October with a maximum in June and July (MARNR, 1991 in Rodríguez, 1999). Mean annual temperature, precipitation, and evapo-transpiration at the sampling

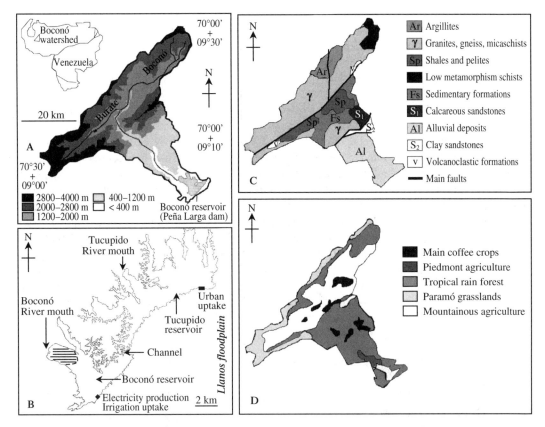

Figure 20.2 Location of the Rio Boconó watershed in Venezuela with main physiographic characteristics (a) altitudes, (b) physiographic map of the Peña Larga water dam at the outlet of the Rio Boconó watershed with open and black circles refer to sampling locations in September 1999. (Modified from Bellanger, B., S. Huon, P. Steinmann, F. Chabaux, F. Velasquez, V. Vallès, K. Arn, N. Clauer, and A. Mariotti. 2004a. *Applied Geochemistry*, 19, 1295–1314. With permission.) (c) Simplified lithological map, and (d) simplified vegetation cover. (Modified from Bellanger, B., S. Huon, F. Velasquez, V. Vallès, C. Girardin, and A. Mariotti. 2004b. *Catena*, 58, 125–150. With permission.)

area location (Corojó subcatchment, 1200 m above sea level) are 21°C, 1600 mm, and 980 mm, respectively (Rómulo-Quintero, 1999). Ultisols are the most common soil types found on the Boconó watershed (Pérez, 1997). They are derived from a geological basement composed of Paleozoic meta-sedimentary rocks and Quaternary alluvial deposits for the southern part of the watershed whereas, crystalline rock formations mainly outcrop on the northern part (Figure 20.2c). Approximately 42% of the watershed is occupied by rain forest (500 to 1700 m) and dry mountainous forest (1500 to 2700 m). Paramó-type vegetation and fallows make up 17.5% of the area. The remaining land is used for agriculture (i.e., maize, coffee, and living productions) often established on steep slopes (Figure 20.2d). The main stream of the watershed is the Boconó River whose discharge averages 77.7 m^3 s^{-1} (1952 to 1973 record; MARNR, 1991 in Rodríguez, 1999) with a mean sediment yield of 7 10^6 t yr^{-1} (Cornièles, 1998).

20.3 ANALYTICAL METHODS

Total organic carbon (TOC), total nitrogen (TN), $\delta^{13}C$ and $\delta^{15}N$ isotope ratios were measured, after carbonate removal, on the same sample aliquot by EA-IRMS (Carlo-Erba NA-1500 NC Elemental Analyser on line with a Fisons Optima Isotope Ratio Mass Spectrometer; Girardin and

Mariotti, 1991). Carbonate removal for TOC analyses was performed using 1N HCl under pH control (above 4.0), in order to reduce possible leaching effects (Huon et al., 2002). Results for isotope abundance are reported in per mil (‰) relative to Pee Dee Belemnite (PDB) standard and relative to air N_2, for $\delta^{13}C$ and $\delta^{15}N$, respectively (Coplen et al., 1983). TOC and TN concentrations are reported in mg g^{-1} of dry sample (equivalent to weight ‰). During the course of this study, analytical precision was better than the means ($\pm 1\sigma$): $\pm 0.1\%$ for $\delta^{13}C$, $\pm 0.3\%$ for $\delta^{15}N$, 0.1 mg g^{-1} for TOC and 0.05 mg g^{-1} for TN. Data reproducibility was controlled by replicated analysis of samples (50%) and by a tyrosine standard that yielded $-23.2 \pm 0.1\%$ and $10.1 \pm 0.30\%$ for $\delta^{13}C$ and $\delta^{15}N$, respectively (mean values for tyrosine: $\delta^{13}C = -23.2\ \%$, $\delta^{15}N = 10.0\ \%$).

All soil samples were ground using a hand mortar and sieved at 2 mm in order to remove coarse vegetation debris, stones, and other coarse lithic aggregates. Fine-size fractions of soils (i.e., < 50 μm or < 200 μm) were separated by wet sieving from grounded soil sample aliquots. For suspended sediments no grinding was necessary and fine fractions were directly recovered from sieving of samples collected in streamflows and runoff flows and stored in polyethylene bottles. In both cases the fractions are representative of the total fine fraction of soils or sediments. Dry bulk soil "densities" (in g cm^{-3}) were calculated using the ratio of lyophilized soil weight to sampled volume. ^{137}Cs activities were measured from subsamples (60 to 80 g) that were put in tightly closed plastic boxes and submitted to 24 hours γ-counting. Coaxial HP Ge N-type detectors were used for γ-spectrometry (8000 channels, low background). Efficiencies and backgrounds were periodically controlled with sediment and soil standards (Soil-6, IAEA-135, IAEA-375, and KCl).

20.4 RESULTS AND DISCUSSION

20.4.1 Monitoring SOC Erosion with 137Cs for Cultivated Soils of the Houay Pano Catchment in Laos

20.4.1.1 Application of ^{137}Cs Measurements in Soil Erosion Studies

The erosion status of soils can be assessed using fallout ^{137}Cs labeling techniques. ^{137}Cs is an artificial radionuclide (half-life of 30.17 years) produced by atmospheric testing of thermonuclear weapons during the 1950s and 1960s in the Northern Hemisphere (UNSCEAR, 1969; 1993; 2000) and after the Chernobyl reactor accident in 1986 (Cambray et al., 1989; SPARTACUS, 2000). During these events, large amounts of ^{137}Cs were released and dispersed in the stratosphere. Most of the fallout occurred between 1954 and 1976 with a maximum supply in 1963. Once ^{137}Cs reaches land surface, it is strongly adsorbed by fine-grained soil particles and no longer exchangeable (Rogowski and Tamura, 1965; Ritchie and McHenry, 1973; Cremers et al., 1988). Therefore, the redistribution of ^{137}Cs is likely to be controlled only by erosion, transport, and sedimentation of soil particles with limited migration with soil depth, usually not exceeding around 30 cm, due to local bioturbation and convective-diffusive processes (Ritchie et al., 1974; Brown et al., 1981a; 1981b; Jong et al., 1983; Ritchie and McHenry, 1990; Walling and Quine, 1992; Loughran et al., 1987; Higgitt and Walling, 1993; He and Owens, 1995).

In cultivated soils, ^{137}Cs is mixed to the plow depth and surface concentrations are lower than for noncultivated soils (Quine et al., 1999). Undisturbed sites should have ^{137}Cs inventories that reflect the amount of ^{137}Cs fallout minus the loss due to radioactive decay. They provide reference values used as baselines for assessing the local erosion and deposition status of soils (Sutherland, 1994). Estimates of soil redistribution rates on cultivated and uncultivated soils may be derived from simple proportional (de Jong et al., 1983) to more complex mass balance models that link the amount of remaining soil ^{137}Cs activity with the assumed reference inventory, time, tillage dilution, particle size, and time-variant fallout ^{137}Cs input (Katchanovski and de Jong, 1984; Zhang et al., 1990; Quine, 1995; Walling and He, 1999; 2001). For each site, the calculated erosion rates account for both tillage erosion (downslope redistribution of soil particles due to tillage, mainly weeding operations, e.g., for tropical soils, Turkelboom et al., 1999; Dupin et al., 2002) and erosion induced by water runoff (overland flow and rill erosion, e.g., Lal, 1990).

20.4.1.2 Estimates of Soil Erosion Rates on the Houay Pano Catchment

Soils of the Houay Pano catchment are theoretically suitable for ^{137}Cs investigations because most of the fields were under forests or long-term fallows when ^{137}Cs fallout took place in the Southeast Asian regions. Cultivation only started at the beginning of the 1970s with a maximum intensification during the 1990s. One of the major uncertainties for the assessment of soil erosion rate is the reference ^{137}Cs inventory used to discriminate undisturbed from eroded soils. Three different methods were used to constrain this reference value: (1) the average ^{137}Cs inventory of "assumed" undisturbed soils, located on long-term fallow fields on the catchment; (2) the "global" ^{137}Cs fallout value provided by the Csmodel1 developed by Walling and He (2001) and; (3) the ^{137}Cs deposition inventory derived from average precipitation records in Vietnam (Hien et al., 2002). The ^{137}Cs inventory of three assumed undisturbed soils provided an average value (± 1 σ) of 613 ± 48 Bq m^{-2} (range: 585 to 671 Bq m^{-2}). Based on the average annual precipitation of 1564 mm for the Houay Pano catchment in 1963 (Bricquet et al., 2001), the year of maximum fallout, the ^{137}Cs deposition inventory should be approximately 596 Bq m^{-2} (Hien et al., 2002) but decreases to 474 ± 88 Bq m^{-2} if the overall 1954 to 1976 precipitation record is taken into account (average ± 1σ: 1168 ± 263 mm, Bricquet et al., 2001). The ^{137}Cs reference inventories calculated with Csmodel1 (Walling and He, 2001) are slightly higher, 635 Bq m^{-2} and 559 Bq m^{-2} for 1963 and 1954 to 1976 precipitation records, respectively. Both methods assume 100% homogenous deposition of ^{137}Cs fallout. However, the extent of radionuclide deposition bears some uncertainties with respect to local rainfall intensity (i.e., for ^{7}Be and ^{210}Pb; Caillet et al., 2001). Since the average ^{137}Cs value measured for undisturbed soils and the rainfall-derived ^{137}Cs reference inventories for 1963 are consistent (596 Bq m^{-2}, 613 Bq m^{-2}, and 635 Bq m^{-2}), the average of the three values (615 Bq m^{-2}) was assumed to provide a first order estimate of the ^{137}Cs fallout reference for the Houay Pano catchment (with respect to year 1963). The ^{137}Cs inventories determined with the 1954 to 1976 precipitation records (474 and 559 Bq m^{-2}; average, 517 Bq m^{-2}) are lower than the levels derived from undisturbed soils (range: 585 to 671 Bq m^{-2}). However, this value cannot be rejected because it accounts for the overall period of ^{137}Cs fallout in Southeast Asia. The results obtained with the low reference estimate (517 Bq m^{-2}) will also be discussed as an alternative solution for soil redistribution rate estimates.

Plots of ^{137}Cs vs. TOC inventories for the first 30 cm of 15 Ultisol–Alfisol soils sampled along two toposequences indicate that the distribution of ^{137}Cs inventories is significantly linked to the total SOC content. Equivalent patterns are displayed for each soil horizon (Figure 20.3a–c). A common process apparently relates the soil erosion status with the amount of SOM in the top 30 cm of soils (Mabit and Bernard, 1998; Ritchie and McCarty, 2003; Figure 20.3c–d). The K content of these 15 soils is rather low (range, 0.15 to 0.37%; average ± 1σ: 0.24 ± 0.06%; Table 20.1) with respect to other soils of the watershed (range, 0.15 to 1.43%, n = 49; average ± 1σ: 0.60 ± 0.36%, data not shown) and, thus, provides constrain for "homogenous" K-bearing clay content, a first-order proxy for fine-size particles content in soils. Therefore, it is likely that the distribution of ^{137}Cs in the selected soils is apparently not linked to differences in soil retention properties. No relationship between ^{137}Cs inventory and local slope gradient could be put forward, which indicates that the redistribution of soil particles is not directly linked to slope. Surprisingly, TOC (Figure 20.4a) and ^{137}Cs (Figure 20.4b) inventories decrease with the amount of fine-size (< 50 µm) particles in the soils. Given that (1) ^{137}Cs is preferentially bound to fine-size soil particles (i.e., clay particles) and (2) TOC concentration and ^{137}Cs activity both decrease with soil depth (Table 20.1), these trends are best explained by selective removal of topsoil layers by runoff and tillage erosion. Local outcrop of deeper soil layers with lower SOC and ^{137}Cs levels but higher fine-size particle content explain

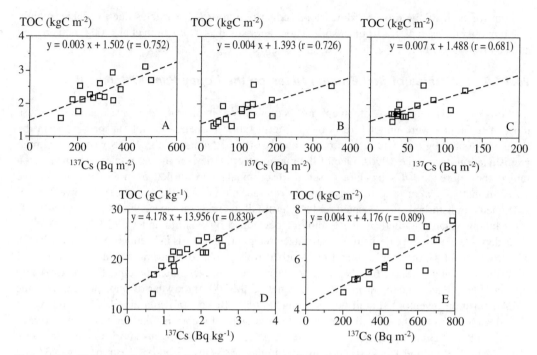

Figure 20.3 Plots of TOC (< 200 μm) vs. ^{137}Cs inventories of 15 cultivated soils of the Houay Pano catchment (Laos) for several soil depths (a) 0 to 10 cm, (b) 10 to 20 cm, (c) 20 to 30 cm, (d) and (e) integrated 0 to 30 cm. Regression coefficients are displayed between brackets for each plot.

Table 20.1 Average Data for the Top 30 cm of 15 Soils Sampled on the Houay Pano Catchment

Soil depth (cm)	[< 50 μm] (wt.%)	K (wt%)	Soil density (g cm^{-3})	^{137}Cs (Bq kg^{-1})	TOC (mgC g^{-1})	^{137}Cs (Bq m^{-2})	TOC (kgC m^{-2})
0–10	78.5 ± 7.2	0.26 ± 0.06	0.84 ± 0.09	3.5 ± 1.3	28.4 ± 3.9	284 ± 105	2.3 ± 0.4
10–20	84.2 ± 3.2	0.24 ± 0.06	0.90 ± 0.13	1.4 ± 0.9	20.7 ± 3.1	121 ± 81	1.8 ± 0.4
20–30	83.1 ± 5.4	0.22 ± 0.06	1.15 ± 0.09	0.5 ± 0.3	16.4 ± 2.6	55 ± 30	1.9 ± 0.3

Note: ± 1σ, standard deviation.

the observed pattern. Traditional farming practices do not involve plowing but two to four hoeing and weeding operations to a maximum depth of 2.5 cm (range, 2 to 3 cm) for each year of cultivation. The resulting tillage erosion, i.e., superficial redistribution of soil particles with down slope soil movements, may account for the large variability of ^{137}Cs inventory observed along the slopes and explain local outcrop of deeper soil horizons (Turkelboom et al., 1999; Dupin et al., 2002).

The ^{137}Cs inventories indicate that most of the soils display lower ^{137}Cs specific activities than the assumed reference (615 Bq m^{-2}). Based on the first-order relationship that links TOC and ^{137}Cs inventories, a reference TOC inventory of 6.64 kgC m^{-2} for 30 cm of soil (maximum penetration depth of ^{137}Cs, Figure 20.3e) was derived. Using the average soil density of the 15 soils (± 1σ : 0.98 ± 0.08 g cm^{-3}; range: 0.84 to 1.15 g cm^{-3}, Table 20.1), the TOC inventory corresponds to an average TOC concentration of 21.93 mgC g^{-1}. This value is in the range of expected TOC concentrations for the undisturbed soils that decrease with depth from ca. 20 to 30 mgC g^{-1} (0 to 10 cm) to ca. 16 to 20 mgC g^{-1} (20 to 30 cm). Assuming a TOC concentration of 21.93 mgC g^{-1} in 1963 when these soils were still under fallow, the redistribution of soil particles along slopes provided, in 2000, average (± 1σ) soil and SOC net losses and gains of −1.28 ± 0.66 kg m^{-2} yr^{-1} with −28.1 ± 14.5 gC m^{-2} yr^{-1} and 0.66 ± 0.46 kg m^{-2} yr^{-1} with 14.4 ± 10.1 gC m^{-2} yr^{-1}, respectively (Table 20.2). Although these averages bear high uncertainties (expressed as high standard deviations), soil

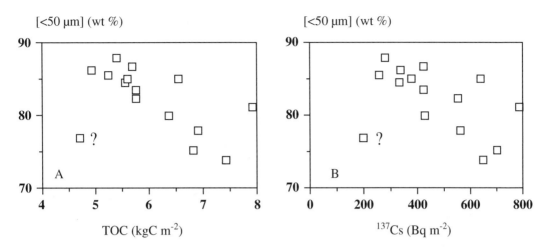

Figure 20.4 Plots of: (a) TOC and (b) ^{137}Cs inventories vs. fine-size (< 50 µm) fraction content of 15 cultivated soils (0 to 30 cm) of the Houay Pano catchment (Laos).

Table 20.2 Soil and TOC Redistribution Rate Estimates for 15 Soils Sampled on the Houay Pano Catchment

Soil Sample Number	TOC Inventory (kgC m^{-2})	TOC Change since 1963[1] (gC m^{-2} yr^{-1})	Predicted Soil Redistribution[2] (kg m^{-2} yr^{-1})	Soil Redistribution Rate Using ^{137}Cs[3] (kg m^{-2} yr^{-1})	Predicted TOC Redistribution[4] (gC m^{-2} yr^{-1})	Δ TOC[5] (gC m^{-2} yr^{-1})
HP33	5.26	−37.29	−1.70	−0.64	−13.96	−23.33
HP34	6.55	−2.47	−0.11	−0.40	−8.72	6.24
HP35	5.72	−24.92	−1.14	−0.31	−6.75	−18.17
HP36	6.36	−7.63	−0.35	−0.30	−6.51	−1.11
HP37	5.07	−42.49	−1.94	−0.48	−10.53	−31.96
HP38	5.76	−23.83	−1.09	−0.31	−6.69	−17.14
HP39	5.55	−29.47	−1.34	−0.49	−10.79	−18.67
HP40	6.91	7.39	0.34	−0.07	−1.60	8.98
HP41	7.61	26.27	1.20	0.29	6.34	19.93
HP42	7.35	19.26	0.88	0.05	1.16	18.10
HP43	6.81	4.66	0.21	0.15	3.17	1.49
HP44	4.71	−52.18	−2.38	−0.91	−19.88	−32.30
HP45	5.61	−27.89	−1.27	0.04	0.92	−28.80
HP46	5.77	−23.47	−1.07	−0.09	−1.92	−21.55
HP47	5.24	−37.93	−1.73	−0.70	−15.34	−22.60
Mean	6.02	−16.30	−0.77	−0.28	−6.07	10.73
± 1σ	0.87	23.47	1.07	0.34	7.40	17.94

Note: Negative and positive signs refer to soil or TOC losses and gains, respectively. Due to low uncertainties on ^{137}Cs activity and TOC concentration measurements, all redistribution rates are displayed with less than 5% error.

[1] Ratio of the difference between reference (6.64 kgC m^{-2}) and measured TOC inventories to time in years since 1963 (1963 to 2000 = 37 years).
[2] Ratio of TOC change to reference TOC concentration (21.93 mgC g^{-1}).
[3] Estimated using mass balance model 2 of csmodel1 (Walling and He, 2001) with a ^{137}Cs reference inventory of 615 Bq m^{-2}, an average plough depth of 24.7 kg m^{-2} cm for 30 cm of soil and using year 1970 for the start of cultivation.
[4] Ratio of soil redistribution rate to reference TOC concentration (21.93 mgC g^{-1}).
[5] Difference between TOC change (column 3) and predicted TOC redistribution (column 6).

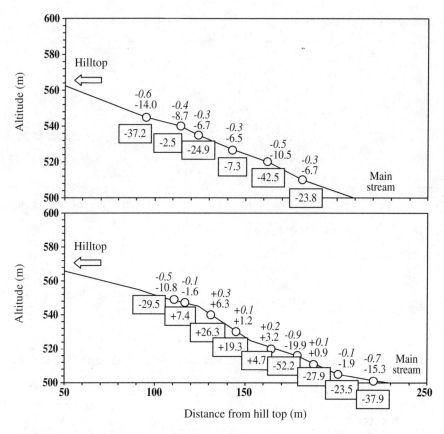

Figure 20.5 Distribution of soils along the two toposequences of the Houay Pano catchment (Laos). TOC changes (in gC m^{-2} yr^{-1}) are included in rectangles. Soil redistribution rates (in kg m^{-2} yr^{-1}) in italic and predicted TOC redistribution (in gC m^{-2} yr^{-1}) are reported above the slope line.

gains and losses are not randomly distributed along slopes. For one of the two toposequences (Figure 20.5b), soil and TOC depletions are found on hilltops and in bottom parts of the slopes whereas accumulations are observed for high to midslope soils. This unexpected pattern may be explained by the chronology of soil occupation along the slopes of the catchment. In 1963, all fields were under fallow and, cultivation started in the early 1970s for catenas located near the main stream, with progressive extension to hilltops during the 1980s and 1990s. Therefore, the duration of cultivation apparently controls the extent of soil particles redistribution calculated with the TOC inventory. Direct monitoring of soil erosion on selected subcatchments of the Houay Pano watershed provided exportation budgets of –0.573 kg m^{-2} and –0.058 kg m^{-2} for soils under traditional cultivation and fallow grounds, respectively (NAFRI, 2001). Making the simple assumption that soil erosion rates can be estimated on the basis of 7 years of cultivation with 30 years of fallow during the past 37 years (time elapsed between 2000 and 1963), an overall soil erosion rate estimate of –0.16 kg m^{-2} yr^{-1} was calculated. This value is in the same order of magnitude as the soil loss rate estimates made with mass balance model 2 (range: –0.07 kg m^{-2} yr^{-1} to –0.91 kg m^{-2} yr^{-1}; Table 10.2; Csmodel1, Walling, and He, 2001).

However, modeled redistribution rates are much lower than those predicted with soil TOC depletion. With mass balance model 2, removal and accumulation rates averaged (± 1σ) –0.43 ± 0.25 kg m^{-2} yr^{-1} and 0.13 ± 0.11 kg m^{-2} yr^{-1}, respectively (Table 20.2), vs. –1.28 ± 0.66 kg m^{-2} yr^{-1} and 0.66 ± 0.46 kg m^{-2} yr^{-1}, respectively, using TOC-^{137}Cs inventories. Lower redistribution rates indicate additional losses of SOC with respect to ^{137}Cs-bound particle redistribution. This deviation most likely accounts for other processes controlling the SOC content such as mineralization of SOM (i.e., direct conversion to CO_2 through soil respiration) and soil leaching (i.e., dissolved organic carbon [DOC] exportation by runoff). But this interpretation is more puzzling for accumulation sites, for which TOC contents rise with respect to the assumed initial inventory (6.64 kgC m^{-2}) and to hypothesized CO_2 and DOC releases. The soil redistribution rates calculated with the low ^{137}Cs reference inventory previously outlined (517 Bq m^{-2}) averaged (± 1σ): –0.36 ± 0.21 kg m^{-2} yr^{-1} and +0.25 ± 0.16 kg m^{-2} yr^{-1} for erosion and accumulation, respectively. The soil redistribution budget that can be derived from these values (using 2.5 cm depth as plowing depth, Walling and He, 2001) indicate that local soil losses may be compensated by local soil accumulations along slopes and is not corroborated by direct soil erosion estimates at the scale of the catchment (–0.16 kg m^{-2} yr^{-1}).

The results obtained for the Houay Pano catchment suggest that the redistribution pattern of ^{137}Cs and the SOC content in the first 30 cm of cultivated soils are significantly linked (i.e., Mabit and Bernard, 1998; Ritchie and McCarty, 2003). However, SOC erosion rates estimated using TOC depletion since 1963 are higher than those calculated with mass balance models based on ^{137}Cs inventories. This deviation is best explained by decoupling processes. Soil particles redistribution along slopes is not the only pathway for SOC depletion that is also controlled by mineralization of SOM through respiration and DOC releases by runoff. Although TOC losses attributed to soil erosion are highly variable from site to site (ca. 8 to 85% of total loss, average ± 1σ: 36 ± 21%), coupling ^{137}Cs measurements with SOC concentration measurements reduces uncertainty on soil redistribution rates derived from fallout ^{137}Cs models. However, major uncertainties on the assessment of SOC erosion (and accumulation) remain: (1) the reference ^{137}Cs value used to discriminate erosion and deposition sites by comparison with undisturbed soils and (2) the depth of the cultivation layer required to compute soil erosion rates (Csmodel1 of Walling and He, 2001), which was assumed to be low (ca. 2.5 cm) according to the traditional farming methods used on the Houay Pano catchment. Better constrain on these two parameters may allow quantitative determination of SOC budgets that account for both physical and biogeochemical processes, i.e., soil stability, SOC mineralization, and SOC storage capacities.

20.4.2 Monitoring Erosion of Organic Carbon Derived from Soil and Rock Sources in Runoff and Stream Flows with Stable C and N Isotopes for the Boconó Watershed in Venezuela

20.4.2.1 Principles of Natural $^{13}C/^{12}C$ and $^{15}N/^{14}N$ Labeling of Soils and Suspended Sediments

The $\delta^{13}C$ composition of SOM is directly inherited (within 1‰) from plant cover $\delta^{13}C$ in surface horizons (Balesdent and Mariotti, 1996). The ^{13}C natural abundance in plants is mainly controlled by the $\delta^{13}C$ of the inorganic C source (subject to variations in time and space), the photosynthetic pathway utilized (C_3, C_4, CAM) and, to a lesser extent, by environmental conditions (temperature, humidity, pCO_2). Aerial plants with a C_3 photosynthetic cycle (85% of plant species) are strongly ^{13}C-depleted with a mean $\delta^{13}C$ of –26‰ due to large isotopic fractionation between CO_2 ($\delta^{13}C$ = –7.8 ‰, present-day value) and plant organic carbon (Deines, 1980; O'Leary, 1988; Farquhar et al., 1989). In contrast, plants with C_4 photosynthesis discriminate less against $^{13}CO_2$ and display a mean $\delta^{13}C$ of –12‰ (Farquhar, 1983; O'Leary, 1988). The $\delta^{13}C$ of SOM in tropical regions is primarily influenced by the relative contribution of C_3 vs. C_4 plants, but is also affected by isotopic fractionation that occurs during the decomposition of plant tissues. This latter effect results from the metabolic activity of decomposer organisms or from differential decomposition of biochemical fractions of the plants that are isotopically distinct (Galimov, 1985; Fogel and Cifuentes, 1993). Accordingly, the composition of SOM in undisturbed soils generally displays a ^{13}C-enrichment pattern with depth (up to 1 to 3‰) associated with SOC mineralization (Melillo et al., 1989). Similar trends with soil depth are also displayed for $^{15}N/^{14}N$ ratios (Mariotti et al., 1980, Tiessen et al., 1984; Yoneyama,

1996). However, changes in the concentration and the $\delta^{13}C$ of atmospheric CO_2 due to fossil fuel combustion may also partly explain the ^{13}C-enrichment of stabilized SOM (e.g., Marino and McElroy, 1991). In the past 20 years, $\delta^{13}C$ values have been successfully used as tracers for SOM dynamics studies in particular with respect to vegetation changes (i.e., C_3-C_4 plants or vice versa; Cerri et al., 1985; Balesdent et al., 1987; Cerling et al., 1989; Martin et al., 1990; Mariotti and Peterschmitt, 1994; Desjardins et al., 1994; Arrouays et al., 1995; Schwartz et al., 1996; Balesdent, 1996; Balesdent and Mariotti, 1996). However, caution must be used when applying decompositional models to simulate $\delta^{13}C$ and $\delta^{15}N$ changes because the chemical and mineralogical variability of soils also possibly controls the rate of SOM decomposition and the extent of isotopic fractionation with soil depth (e.g., Ehleringer et al., 2000; Buchmann and Kaplan, 2001; Krull et al., 2003).

Numerous studies have also used the potential of ^{13}C and ^{15}N natural abundance measurements in suspended sediments to derive information on the origin of particulate organic carbon (POC) in rivers (e.g., Hedges et al., 1986; Mook and Tan, 1991; Mariotti et al., 1991; Bird et al., 1992; 1994; Martinelli et al., 1999). Suspended sediment $\delta^{13}C$ are generally used to estimate the relative contribution of land-derived (allochthonous) vs. phytoplanktonic-derived (autochthonous) organic C sources. However, degradation of organic matter or interactions with dissolved phases during transport (e.g., Richey et al., 1980; 1988) may alter the $^{13}C/^{12}C$ content of suspended material and complicate the interpretation based on an ideal two components mixing. Since organic matter in most plants fractionate against air $^{15}N/^{14}N$, fresh or "fossil" vegetation debris transported in suspended sediments may be discriminated from SOM on the basis of their $\delta^{15}N$ values (e.g., Mariotti et al., 1984; Kao and Liu, 1996; Kao and Liu, 2000; Huon et al., 2002). Due to limited $^{15}N/^{14}N$ fractionation during burial of rapidly accumulating organic matter in sedimentary basins, refractory organic C derived from geological sources should have $\delta^{15}N$ (and $\delta^{13}C$) inherited from plant compositions, contrasting with that of mineralized SOM more enriched in ^{15}N and ^{13}C (Precambrian-Jurassic sedimentary records, in Tyson, 1995; Hayes et al., 1999). Additional constraints on the origin of riverine particulate organic carbon (POC) have been obtained by coupling $\delta^{13}C$ values with C/N ratio, $\Delta^{14}C$, POC/Chlorophyll-a ratio or lipid characterization measurements (Kennicut et al., 1987; Hedges et al., 1994; Thornton and Mc Manus, 1994; Barth et al., 1998; Onstad et al., 2000; Masiello and Druffel, 2001; Meyers, 1997; Raymond and Bauer, 2001; Krusche et al., 2002).

20.4.2.2 Monitoring Erosion of Fine-Size (< 50 µm) SOC with Field Plot Experiments

During the rainy season (April to September 1998), mean soil loss and mean runoff can only be correlated for the bare field plot ($r = 0.83$, Figure 20.6). Cumulative losses of organic C in the fine-size (< 50 µm) fraction amounted 29.5 gC m^{-2} and 2.3 gC m^{-2} for the bare field plot and the coffee field plot, respectively (Figure 20.7, corresponding to soil losses of 1.8 kg m^{-2} and 0.05 kg m^{-2}, respectively). During the same period, cumulative runoff was 60 times higher for the bare field plot than for the cultivated field plots. Maximal runoff and in SOC loss occurred in June and July. Differences in the magnitude of soil loss and runoff between cultivated and bare field plots reflect the classical protective effect of the vegetation cover that reduces raindrop impact and runoff generation (Roose, 1977; Wischmeier and Smith, 1978). The vegetation cover apparently also controlled the grain size distribution of suspended sediments (Wan and El Swaify, 1997; Hairshine et al., 1999) and induced a sorting effect for cultivated soils. During the single rainfall event experiment, soils from the coffee field plot released almost exclusively fine-size sediments (mean weighted < 50 µm size fraction contribution: 93%, Figure 20.8) originating from the breakdown of surface soil aggregates (sheet erosion, Lal, 1990) and transported by limited runoff. In contrast, suspended sediments exported from the bare field plot by higher runoff had a grain size distribution that matches that of topsoil horizons (mean weighted < 50 µm size fraction contribution: 45% for suspended sediments vs. 47% for 0 to 20 cm soil horizons (Figure 20.8) and accounted for additional rill erosion processes (Lal, 1990). During rainfall, the overall fine (< 50 µm) SOC exportation amounted to 1.3 gC m^{-2} and 0.03 gC m^{-2} for the bare and the coffee field plots, respectively.

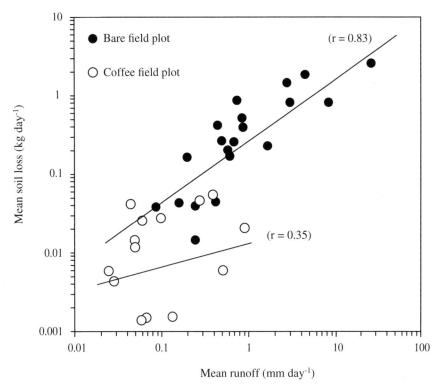

Figure 20.6 Plot of mean soil loss vs. mean runoff for the bare and the coffee field plots on the Rio Boconó watershed (Venezuela) during the rainy season. (From Bellanger, B., S. Huon, F. Velasquez, V. Vallès, C. Girardin, and A. Mariotti. 2004a. Experimental $\delta^{13}C$ and $\delta^{15}N$ study of soil organic carbon loss by erosion in the Venezuelan Andes. *Catena*, 58, 125–150. With permission.)

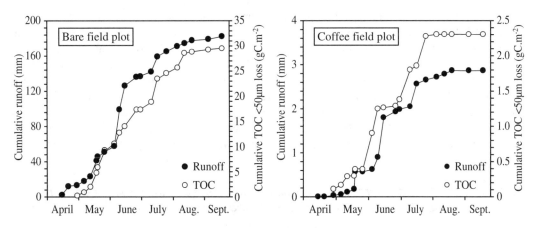

Figure 20.7 Plots of cumulative runoff and fine size TOC loss for the two experimental field plots of the Rio Boconó watershed (Venezuela) during the rainy season.

The composition ($\delta^{13}C$, $\delta^{15}N$, TOC, TOC/TN) of fine-size (< 50 μm) suspended organic matter is displayed as a function of suspended sediment concentration in Figure 20.9 for the bare field plot. For low sediment concentrations, the composition of fine organic matter reflected the transport of vegetation debris mobilized from top soil horizons and characterized by high organic C contents, high TOC:TN ratios and $^{13}C-$ and $^{15}N-$ depleted compositions with respect to SOM (Meyers, 1997). An equivalent pattern is displayed for the coffee field plot (not shown here, see in Bellanger

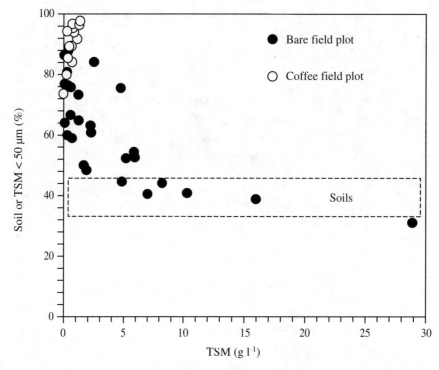

Figure 20.8 Plots of fine-size particles in suspended sediments vs. TSM delivery by soils of two experimental field plots of the Rio Boconó watershed (Venezuela) during a single rainfall event (June 14, 1998). Soil composition is indicated by the dashed line. TSM = total suspended matter concentration. (From Bellanger, B., S. Huon, F. Velasquez, V. Vallès, C. Girardin, and A. Mariotti. 2004a. Experimental $\delta^{13}C$ and $\delta^{15}N$ study of soil organic carbon loss by erosion in the Venezuelan Andes. *Catena*, 58, 125–150. With permission.)

et al., 2004b). For higher runoff and suspended sediment concentrations, approximately above 30 ml s^{-1} and 0.5 g l^{-1} (95% of suspended matter yield for the bare field plot), respectively, the composition of fine-size (< 50 μm) organic matter evolves toward constant values that closely reflect the composition of SOM, fingerprinting isotopic contrasts inherited from the vegetation cover. A constant composition for fine-size (< 50 μm) suspended particles can be assumed for most of the sediment yield, which provides quantitative information on SOC erosion. Monitoring the release of POC from coarser size fractions will only be meaningful for steady state conditions and during periods of high runoff. Sorting effects related to erosion intensity suggest that only fine-size organic fractions transported in stream flows will accurately reflect the contribution of soil fine fractions.

20.4.2.3 Monitoring Sources of Fine-Size (< 50 μm, < 200 μm) Suspended Organic Carbon in Stream Flows during Flood Events

Monitoring changes in the source of organic matter in suspended sediments of stream flows was carried out for two flood episodes (Cornièles, 1998; Bellanger, 2003). The first event (June 14, 1998) of medium intensity had a return period of 1 year and generated for the main stream, the Boconó River before the confluence with the Burate (Figure 20.2a), a maximum water discharge of 110 m^3 s^{-1} (Figure 20.10a). During the second event (July 9 to 10, 1995), water discharges rose to 240 m^3 s^{-1} and 290 m^3 s^{-1} for the Rio Boconó and the Rio Burate (Fig. 20.11a), respectively, with a return period of 10 years. Total suspended sediment concentrations (TSM) and water discharges (Q) are correlated through power functions with no hysteresis (TSM = 5 10^{-4} × Q$^{1.9}$, r^2

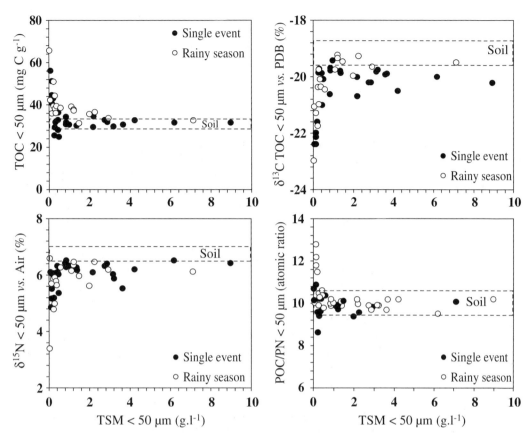

Figure 20.9 Plots of changes in TOC, $\delta^{13}C$, $\delta^{15}N$ and TOC/TN in fine-size suspended sediments released by soils of the bare field plot on the Rio Boconó watershed (Venezuela). Soil composition is indicated by the dashed line. TSM = total suspended matter concentration. (From Bellanger, B., S. Huon, F. Velasquez, V. Vallès, C. Girardin, and A. Mariotti. 2004a. Experimental $\delta^{13}C$ and $\delta^{15}N$ study of soil organic carbon loss by erosion in the Venezuelan Andes. *Catena*, 58, 125–150. With permission.)

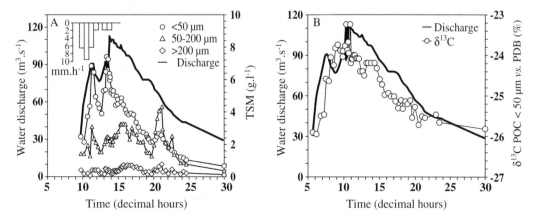

Figure 20.10 Plots of water discharge and (a) suspended matter grain size compositions and (b) $\delta^{13}C$ of fine-size POC for the Rio Boconó (Venezuela) during the June 14, 1998 flood event. TSM = Total suspended matter concentration. Precipitations are reported in mm h^{-1}.

= 0.97 for the Rio Burate in 1995; TSM = 3.5 $10^{-3} \times Q^{1.35}$, r^2 = 0.77, and TSM = 4 $10^{-2} \times Q^{1.14}$, r^2 = 0.69, for the Rio Boconó in 1995 and 1998, respectively). These relationships indicate that erosion along slopes of the watershed and remobilisation of riverbank sediments was synchronous to water discharge (Nordin, 1985). Therefore, monitoring the composition of POC in suspended sediments may provide both information on erosion intensity and on the contributing sources of organic carbon if they have sufficiently contrasted isotopic signatures.

During the June 1998 flood event of the Rio Boconó, fine- (< 50 µm) and coarse- (> 200 µm) size fractions account for 60 and 6% of total suspended sediment load, respectively. The $\delta^{13}C$ composition of fine (< 50 µm) suspended organic matter varies linearly with suspended sediment concentration (Figure 20.10b and Figure 20.12a). This relationship reflects a correlative change with erosion intensity of the type of suspended organic matter. For high suspended matter concentrations (high erosion levels) the source of organic C is most likely derived from SOM, characterized by enriched ^{13}C compositions with respect to depleted ^{13}C compositions for vegetation debris (Meyers, 1997). This interpretation is also supported by $\delta^{15}N$ changes in the composition of fine-size (< 50 µm) suspended sediments. When plotted in a $\delta^{15}N$-$\delta^{13}C$ mixing diagram (Fig. 20.12c; see discussion in Ganeshram et al., 2000), suspended POC compositions are scattered between a vegetation-derived organic matter pool with $\delta^{15}N$ values close to atmospheric N_2 composition (0‰) and soil-derived organic matter pools enriched in ^{15}N with respect to the composition of vegetation debris (Figure 20.12c). Additional information may also be derived from POC concentrations in fine-size suspended matter. Low POC contents reflect enhanced supply of SOM during peak flow and falling water stages whereas; high POC concentrations involve more vegetation debris during rising and falling water stages (Figure 20.12b). However, precise monitoring of the contribution of each SOM pool may still remain difficult because the $\delta^{13}C$ values (range, –25.5‰ < $\delta^{13}C$ < –23.5‰) that can be assigned to soil and vegetation debris end-members may overlap. For this flood event that lasted 24 hours, we derived SOC exportation budgets of 166 10^6 gC and 139 10^6 gC for < 200 µm and < 50 µm size fractions of suspended sediments, respectively. Scaling up these values to the upper Boconó catchment's (574 km^2) provides specific erosions of 0.29 gC m^{-2} for < 200 µm size fractions and 0.24 gC m^{-2} for < 50 µm size fractions. These budgets are in the range of the organic C budgets derived from the field plots experiments (1.3 gC m^{-2} and 0.03 gC m^{-2} for the bare and the coffee field plots, respectively; see above) carried out during the same flood event (June 14, 1998). Given that (1) less than 20% of the catchment is occupied by bare fields (excluding outcrops of geological basement rocks) and (2) the experiments carried out for bare and coffee field plots account for high and low erosion conditions with respect to cultivated and fallow lands, monitoring fine-size suspended fractions in stream flows provides organic C budgets that are consistent with the results obtained with the field plot experiments. It is therefore possible to derive SOC erosion budgets from fine-size suspended POC concentrations in stream flows as far as only top SOM is involved by erosion processes and transported by runoff to stream flows during the flood event.

During the July 1995 flood event, the discharge of the Rio Burate rose by a factor 10 whereas the suspended sediment concentration increased by a factor 150 (Figure 20.11a). This event is more intense than the previous one. The fine-size (< 200 µm) fraction accounts for 93% of total suspended sediment yield. Monitoring ^{15}N abundances in these fine sediments with $\delta^{15}N$ measurements provides minimum values when water discharge is maximal, indicating a concomitant change in the composition of suspended sediments (Figure 20.11b). When plotted against suspended matter concentration, TOC, TOC/TN, and $\delta^{15}N$ values follow decreasing hyperbolic trends with increasing water discharge and suspended sediment load (Figure 20.13a–b). The decrease of TOC concentrations in suspended sediments is usually interpreted as a dilution trend toward a minimum value (ca. 4 to 5 mgC g^{-1} in our study) that either reflects: (1) enhanced contribution of mineral particles (i.e., quartz, micas, etc.) with respect to organic matter, derived from erosion of the geological basement on the watershed (Meybeck, 1982); or (2) supply of SOM from deeper soil horizons, more depleted in organic C than top soil horizons (Ludwig et al., 1996). In our study, the TOC/TN

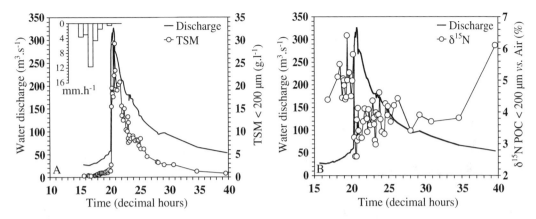

Figure 20.11 Plots of water discharge and (a) TSM delivery and (b) $\delta^{15}N$ of suspended fine-size (< 200 µm) fractions for the Rio Burate (Rio Boconó watershed, Venezuela) during the July 9 and 10, 1995 flood event. TSM = total suspended matter concentration.

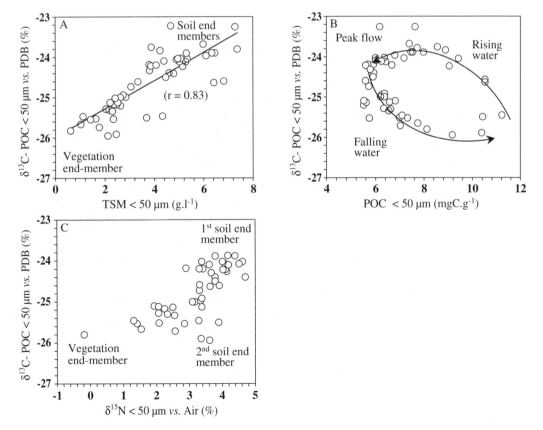

Figure 20.12 Plots of $\delta^{13}C$ values vs. (a) TSM, (b) POC, and (c) $\delta^{15}N$ for fine (< 50 µm) suspended fractions during the June 14, 1998 flood event of the Rio Boconó (Venezuela). TSM = total suspended matter concentration. Estimated soil and vegetation end-members compositions are reported.

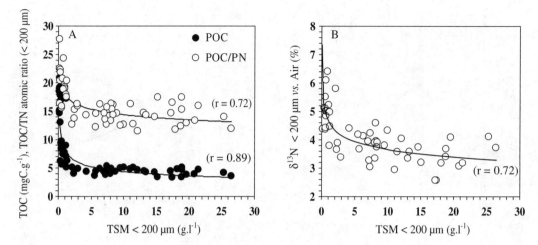

Figure 20.13 Plots of: (a) TOC and TOC/TN vs. TSM and B) d15N vs. TSM for fine-size (< 200 μm) suspended fractions during the June 14, 1998 flood event of the Rio Burate (Boconó watershed, Venezuela). TSM = total suspended matter concentration.

Table 20.3 Average Composition of Fine Organic Matter in Soils, Vegetation, and Rocks of the Rio Burate Watershed

< 200 μm	TOC (mgC g^{-1})	TN (mgN g^{-1})	TOC/TN	δ^{15}N vs. AIR (%)	δ^{13}C vs. PDB (%)
Soils (0 to 20 cm)	35 ± 10	3 ± 1	22 ± 7	5.5 ± 0.5	Variable*
Vegetation debris	460 ± 50	30 ± 10	17 ± 7	0 ± 2	Variable*
Shales–lutites	15 ± 3.8	1.0 ± 0.2	14 ± 5	2.5 ± 0.5	−25.5 ± 0.3

* Due to occurrence of both C$_3$ and C$_4$ plant communities on the watershed.

and δ^{15}N of fine (< 200 μm) organic matter of carboniferous shales (organic-matter bearing rocks; Table 20.3) located along the banks of the Rio Burate (Figure 20.2c) match those obtained during the period of high suspended sediment transport corresponding to high erosion stages. SOM cannot be the exclusive supplier of organic C because TOC/TN and δ^{15}N values in suspended sediments are higher and lower, respectively, than for soil surface horizons. The contribution of the different pools of organic matter can be quantified using isotopic mass balance equations if suspended sediments that compose the mixtures derive from end-members with contrasted compositions (Balesdent and Mariotti, 1996). The contribution of particulate nitrogen sources of vegetation debris may be assumed to be negligible with respect to the high turbidity of stream flow waters. Meaningful discrimination between lithic and soil sources of suspended organic matter is only obtained with TOC/TN and δ^{15}N values. The use of ^{13}C abundance measurements might be confusing because SOM might be inherited from either C$_3$ or C$_4$ plants and, because long-term cultivation also involves rotating fallow periods and local replacement of C$_3$ by C$_4$ plant covers and vice versa.

In an ideal mixture between soil- and rock-derived organic particles, the instantaneous N and δ^{15}N contributions of the two end-members in suspended (< 200 μm) sediments can be expressed by the following equations:

$$N_{sample} = N_{soils} + N_{rocks} \text{ with } fN_{rocks} = N_{rocks}/N_{sample} = 1 - fN_{soils}$$

$$N_{sample} \times \delta^{15}N_{sample} = (N_{soils} \times \delta^{15}N_{soils}) + (N_{rocks} \times \delta^{15}N_{rocks})$$

$$\delta^{15}N_{sample} = [(1 - fN_{rocks}) \times \delta^{15}N_{soils}] + (fN_{rocks} \times \delta^{15}N_{rocks})$$

$$fN_{rocks} = (\delta^{15}N_{sample} - \delta^{15}N_{soils})/(\delta^{15}N_{rocks} - \delta^{15}N_{soils})$$

The fraction of organic carbon derived from rock erosion (fC_{rocks}) can be calculated using the C:N specific ratios (equivalent to TOC/TN in Table 20.3) of each end-member in the mixture:

$$fC_{rocks} = C_{rocks}/C_{sample} = [N_{rocks} \times (C/N)_{rocks}]/[N_{sample} \times (C/N)_{sample}]$$

$$fC_{rocks} = fN_{rocks} \times [(C/N)_{rocks}/(C/N)_{sample}]$$

According to these calculations, the contribution of rock-derived particles (fC_{rocks}) in suspended sediments accounts for 0 to 80% of the (< 200 µm) POC transported during the flood event (Figure 20.14). Erosion of the geological basement is maximal during the period of high water discharge of the Rio Burate, reflecting enhanced rock and sediment contribution to the suspended load. However, these estimates bear a high uncertainty, mainly due to a rather low discrimination between the average $\delta^{15}N$ values of the two end-members ($\Delta\delta^{15}N$ = 3‰, Table 20.3), with respect to the $\delta^{15}N$ variability in each pool (σ = 0.5‰, Table 20.3). The overall average weighted contribution of fine organic C derived from rocks accounts for 50% of the suspended (< 200 µm) POC load. During the flood event that lasted 24 hours, 254 10^6 gC were exported in the < 200 µm size fraction of suspended sediments on the Rio Burate watershed (364 km²). Specific organic C erosion was ca. 0.70 gC m⁻², approximately half of it (0.35 gC m⁻²) originating from organic matter bearing rocks. As already outlined by Meybeck (1993) on a more global scale, mountainous catchments with large outcrops of sedimentary to low-grade metamorphic rocks may provide a significant fraction of POC to river discharge and bias organic carbon budgets solely based on SOM erosion. Moreover, the behavior in surface environments of more refractory organic matter released from geological basement sources (such as shales or other organic-matter bearing rocks) will contrast sharply with that of vegetation debris, charcoals or humic substances generated by soil erosion.

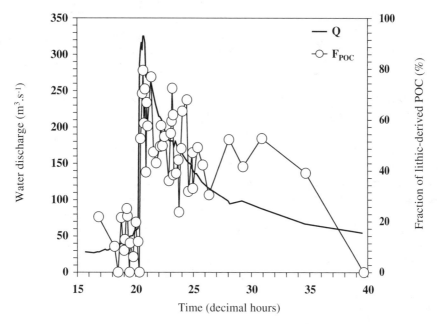

Figure 20.14 Plot of water discharge and fraction of lithic-derived POC for fine-size (< 200 µm) suspended fractions during the June 14, 1998 flood event of the Rio Burate (Boconó watershed, Venezuela). TSM = total suspended matter concentration.

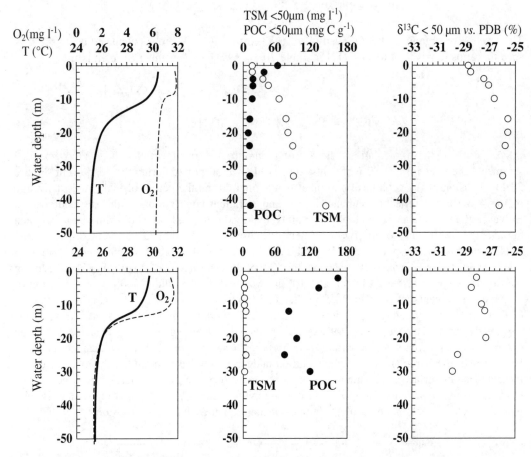

Figure 20.15 Plot of dissolved oxygen (O_2), temperature (T), total suspended matter and POC concentrations (TSM and POC), $\delta^{13}C$ for fine size fractions in the water column of the Boconó reservoir (upper graphs) and the Tucupido reservoir (lower graphs) in September 1999. TSM = total suspended matter concentration.

20.4.2.4 Monitoring the Evolution of Suspended Organic Matter in a Water Reservoir at the Outlet of the Watershed

The Peña Larga water dam can be divided into two parts, the Boconó and the Tucupido reservoirs that communicate through a narrow and shallow channel with a maximum depth of 20 m (Figure 20.2b). Thermal stratification in both reservoirs was stable throughout the year with, oxic conditions (dissolved $O_2 > 4.5$ mg.l^{-1}, water residence time: $\tau = 0.36$ years, Bellanger et al., 2004b) and hypoxic to anoxic conditions (dissolved $O_2 < 1$ mg l^{-1}, water residence time $\tau = 6.2$ years; Bellanger et al., 2004b) below the thermocline, for the Boconó and Tucupido reservoirs, respectively. Suspended sediment loads are mainly controlled by Boconó River discharge with high and depth increasing concentrations for Boconó Reservoir and low and constant concentrations for Tucupido Reservoir. The composition of fine-size (< 50 μm) suspended organic matter in the oxic reservoir reflects that of the sources of suspended sediment in the water column: i.e., surface productivity above the thermocline (more depleted in ^{13}C, enriched in organic C, Figure 20.15) and suspended organic matter derived from erosion on the watershed below the thermocline (Figure 20.15; average composition for the Boconó river: $\delta^{13}C = -25.8‰$, [TOC] = 10.3 mgC g^{-1}). In contrast, fine suspended sediments in the water column of the hypoxic to anoxic reservoir are more enriched in organic C and depleted in ^{13}C, indicating additional contribution of chemoautotrophic or methanotrophic biomass to suspended material load (Harvey and Macko, 1997; Hollander and Smith, 2001; Lehmann et al., 2002). POC budgets indicate that during the past two decades erosion on the Boconó watershed involved the transfer of 32 10^9 gC yr^{-1} to the Peña Larga water dam (ca. 19.8 gC m^{-2} yr^{-1}; Cornièles 1998). Of this, approximately 21 ± 5 10^9 gC yr^{-1} (66%) was buried in bottom lake and alluvial fan sediments and only ca. 2 10^9 gC yr^{-1} (6%) was exported from the water dam to the Llanos floodplain. A net amount of 8 ± 5 10^9 gC yr^{-1} (ca. 25%, not including ca. 1 10^9 gC involved in the suspended load) has been mineralized in the Boconó basin (Bellanger et al., 2004b). These organic C burial rates are one order of magnitude higher than those usually reported for small lakes (Einsele et al., 2001) due to relatively high erosion rates on the watershed. Although providing a significant flux of dissolved C produced by in situ oxidation of suspended organic matter, the Peña Larga water reservoir reduced by nearly 94% the POC flux generated by erosion on the Rio Boconó watershed.

20.5 CONCLUSION

Monitoring the composition of organic C in soils and suspended sediments of stream flows with isotopic tracers provides significant information that may help to better constrain SOC erosion budgets. The redistribution pattern of fallout ^{137}Cs and the SOC content in the first 30 cm of cultivated soils of the Houay Pano catchment (Laos) are significantly linked. A common process apparently relates the soil erosion status with the amount of organic matter in the topsoil horizons. However, SOC erosion rates, estimated using total SOC depletion since 1963, are higher than those calculated with mass balance models based on ^{137}Cs inventories. With mass balance model 2 (Walling and He, 2001), erosion and accumulation rates along two toposequences averaged (± 1σ): –0.43 ± 0.25 kg m^{-2} yr^{-1} and 0.13 ± 0.11 kg m^{-2} yr^{-1}, respectively, vs. –1.28 ± 0.66 kg m^{-2} yr^{-1} and 0.66 ± 0.46 kg m^{-2} yr^{-1}, respectively, using TOC – ^{137}Cs inventories. The difference reflects additional organic C losses controlled by mineralization of SOM through respiration and dissolved organic C releases by runoff. Monitoring SOC erosion with ^{137}Cs may help to better constrain soil leaching and mineralization rates by subtracting from SOC inventories the effect of particles redistribution along slopes due to runoff and tillage erosion. The reliability of these estimates is mainly linked to the reference ^{137}Cs inventory used to discriminate erosion and deposition sites by comparison with undisturbed soils and to the cultivation depth used to compute soil erosion rates with mass balance models.

Monitoring fine-size (< 50 μm or < 200 μm) suspended organic matter transported by runoff and stream flows with δ^{13}C and δ^{15}N measurements may allow quantification of the contributing pools of organic matter if the isotopic signature of each end-member in the mixture is sufficiently discriminating. During an important flood event, the contribution of particulate organic C derived from incisive erosion of organic matter bearing rocks of the geological basement of the Rio Boconó watershed (Venezuela) could be distinguished from that of topsoil horizons using δ^{15}N measurements on fine suspended sediments. The contribution from lithic sources was approximately equivalent to that of soils, which indicates that soil erosion budgets and models that only account for the contribution of SOC might be partly biased. The transport of fine-size (< 50 μm or < 200 μm) suspended organic matter in stream flows during flood events was apparently conservative for different scales (30 m^2, 364 km^2, and 574 km^2) with soil and SOC specific erosions in the same order of magnitude, between those of bare fields (high soil erosion conditions) and coffee fields (low soil erosion conditions) end-members (Table 20.4).

Since fine-size fractions account for most of the organic C stored in soil surface horizons, monitoring fine organic matter in runoff and stream flows may provide significant constraint on SOC erosion, possibly allowing to scale up field plot measurements to catchments with appropriate grain size correction factors, which still need to be determined. Although water quality might be affected by high-suspended matter loads derived from erosion on the watershed, the extent of organic C released in the hydrographical network of the floodplain can be thoroughly reduced by

Table 20.4 Selected Soil and SOC Erosion Estimates for the Rio Boconó Watershed (Venezuela)

	Bare Soil (30 m²)		Soil under Coffee (30 m²)		Rio Boconó (574 km²)	Rio Burate (364 km²)
	Season	Flood Event	Season	Flood Event	Flood Event	Flood Event
Soil loss (kg m^{-2})	1.8	0.09	0.05	0.0007	0.06	0.17
[< 200 μm] (wt%)	—	62	—	96	90	93
[< 50 μm] (wt%)	—	45	—	93	56	—
POC loss (gC m^{-2})	—	—	—	—	—	—
< 200 μm (gC m^{-2})	—	1.3	—	0.03	0.29	0.35
< 50 μm (gC m^{-2})	29.5	1.2	2.3	0.03	0.24	—

water reservoirs located at the outlet of the watersheds. The Peña Larga water reservoir reduced by nearly 94% the suspended organic matter load discharged by the Rio Boconó through sediment burial and in situ mineralization processes and behaved as a major sink for C with respect to suspended POC supply from the watershed.

ACKNOWLEDGMENTS

This work was supported by the French INSU (PROSE 53 and PNSE 12 programs), the Venezuelan MARNR (Ministerio del Ambiente et de Los Recursos Naturales Renovables, Caracas) and the Laotian NAFRI (National Agricultural and forestry Research Institute, Vientiane). The authors are grateful to V. Vallès (Université de Provence, Marseille, France), M. Cornièles-Vallès, A. Moreau, and Colonel R. Mena Nava (Instituto Geografico de Venezuela Simón Bolivar) for their hospitality and support during field studies in Venezuela. Bathymetric and hypsometric data for the Peña Larga water dam were kindly provided by A. Montilva, M. J. Guerrero, H. Briceño, and M. Alvarado (DESURCA-CADAFE, Venezuela). We are also very grateful to Ty Phommasack and to late Somphanh Thonglasamee (NAFRI, Laos) for their help, assistance, and marked interest for our study in northern Laos. M. Grably and G. Bardoux (UMR BioEMCo) are also thanked for their help during field and laboratory sample processing.

REFERENCES

Arrouays, D., J. Balesdent, A. Mariotti, and C. Girardin. 1995. Modelling organic carbon turnover in cleared temperate forest soils converted to maize cropping by using ^{13}C natural abundance measurements. *Plant and Soil* 173:191–196.

Ataroff, M. and M. Monasterio M. 1997. Soil erosion under different management of coffee plantations in the Venezuelan Andes. *Soil Technol.* 11:95–108.

Balesdent. J., A. Mariotti, and B. Guillet. 1987. Natural ^{13}C abundance as a tracer for studies of soil organic matter dynamics. *Soil Biol. Biochem.* 19:25–30.

Balesdent, J., G. H. Wagner, and A. Mariotti. 1988. Soil organic matter turnover in long-term experiments as revealed by carbon-13 natural abundance. *Soil. Sci. Am. J.* 52:118–124.

Balesdent, J. (1996). The significance of organic separates to carbon dynamics and its modelling in some cultivated soils. *Eur. J. Soil Sci.* 47: 485–493.

Balesdent, J. and A. Mariotti. 1996. Measurement of soil organic matter turnover using ^{13}C natural abundance, in T. W. Boutton and S. Yamasaki, eds., *Mass Spectrometry of Soils*. Marcel Dekker Pub. New York, pp. 83–111.

Barth, J. A. C., J. Veizer, and B. Mayer. 1998. Origin of particulate carbon in the upper St. Lawrence: Isotopic constraints. *Earth and Planet. Sci. Lett.* 162:111–121.

Bellanger, B. 2003. *Transfert de carbone organique dans le réseau hydrographique par érosion hydrique. Application à un bassin versant en zone tropicale humide (Rio Boconó, Andes vénézueliennes)*. Ph.D. diss., Univ. Paris VII (France). 236 p.

Bellanger, B., S. Huon, P. Steinmann, F. Chabaux, F. Velasquez, V. Vallès, K. Arn, N. Clauer, and A. Mariotti. 2004a. Oxic — anoxic conditions in the water column of a tropical freshwater reservoir (Peña-Larga, NW Venezuela). *Applied Geochemistry*, 19, 1295–1314.

Bellanger, B., S. Huon, F. Velasquez, V. Vallès, C. Girardin, and A. Mariotti. 2004b. Experimental $\delta^{13}C$ and $\delta^{15}N$ study of soil organic carbon loss by erosion in the Venezuelan Andes. *Catena*, 58, 125–150.

Bird, M. I., W. S. Fyfe, D. Pinheiro-Dick, and A. R. Chivas. 1992. Carbon isotope indicators of catchment vegetation in the Brazilian Amazon. *Global Biogeochem. Cycles* 6, 3:293–306.

Bird, M. I., P. Giresse, and A. R. Chivas. 1994. Effect of forest and savannah vegetation on the carbon-isotope composition of sediments from the Sanaga River, Cameroon. *Limnol. Oceanogr.* 39, 8: 1845–1854.

Bricquet, J. P., A. Boonsaner, T. Phommassack, and T. D. Toan. 2001. Statistical analysis of long series rainfall data: A regional study in South-East Asia, in Integrated catchment management for land and water conservation and sustainable production in Asia. IWMI-ICRISAT-ADB Joint Annual Review and Planning. Dec. 10–14, 2001. Hanoi, Vietnam.

Brown, R. B., N. H. Cutshall, and G. F. Kling. 1981a. Agricultural erosion indicated by ^{137}Cs redistribution: I. Levels and distribution of ^{137}Cs activity in soils. *Soil Sci. Soc. Am. J.* 45:1184–1190.

Brown, R. B., G. F. Kling, and N. H. Cutshall. 1981b. Agricultural erosion indicated by ^{137}Cs redistribution: II. Estimates of erosion rates. *Soil Sci. Soc. Am. J.* 45:1191–1197.

Browne, R. B. 1995. The role of Geographical Information Systems in hydrology, in I. D. L. Foster, A. M. Gurnell, and B. W. Webb, eds., *Sediment and Water Quality in River Catchments*. John Wiley & Sons Ltd., pp. 33–48.

Buchmann, N. and J. O. Kaplan. 2001. Carbon isotope discrimination of terrestrial ecosystems — how well do observed modeled results match? in E. D. Schulze, M. Heimann, S. Harisson, E. Holland, J. Lloyd, I. C. Prentice, and D. Schimel, eds., *Global Biogeochemical Cycles in the Climate System*. Acad. Press., San Diego, pp. 253–266.

Cai, D. L., F. C. Tan, and J. M. Edmond. 1988. Sources and transport of particulate organic carbon in the Amazon River and estuary. *Estuary Coastal Shelf Res.* 26:1–14.

Cambray, R. S., K. Playford, G. N. J. Lewis, and R. C. Carpenter. 1989. *Radioactive fallout in air and rain: Results to the end of 1988*. United Kingdom Atomic Energy Authority Report AERE-R 13575. HMSO. London.

Caillet, S., P. Arpagaus, F. Monna, and J. Dominik. 2001. Factors controlling ^7Be and ^{210}Pb atmospheric deposition as revealed by sampling individual rain events in the region of Geneva, Switzerland. *J. Environ. Radioactivity* 53:241–256.

Cerling, T. E., J. Quade, Y. Wang, and J. R. Bowman. 1989. Carbon isotopes in soils and paleosols as ecology and palaeoecology indicators. *Nature* 341:138–139.

Cerri, C., C. Feller, J. Balesdent, R. Victoria, and A. Plenecassagne. 1985. Application du traçage isotopique naturel en ^{13}C à l'étude de la dynamique de la matière organique dans les sols. *CRAS. Paris* 300:423–428.

Chapra, S. C. and H. F. H. Dobson. 1981. Quantification of the lake trophic typologies of Naumann (surface quality) and the Thienemann (oxygen) with special reference to Great Lakes. *J. Great Lakes Res.* 7, 2:182–193.

Ciais, P., P. P. Tans, M. Trolier, J. W. C. White, and R. J. Francey. 1995. A large northern hemisphere terrestrial CO_2 sink indicated by the $^{13}C/^{12}C$ ratio of atmospheric CO_2. *Science* 269:1098–1102.

Coplen, T. B., C. Kendall, and J. Hopple. 1983. Comparison of stable isotope reference samples. *Nature* 302:236–238.

Corniéles, M. 1998. *Etude et modélisation des transferts d'eau, d'éléments dissous et particulaires dans un bassin versant torrentiel. Cas du Rio Boconó dans les Andes vénézuéliennes*. Ph.D. diss., Univ. Avignon, France. 183 pp.

Cremers, A., A. Elsen, P. De Preter, and A. Maes. 1988. Quantitative analyses of radiocaesium retention in soils. *Nature* 335:247–249.

de Jong, E., C. B. M. Begg, and R. G. Kachanoski. 1983. Estimates of soil erosion and deposition from Saskatchewan soils. *Can. J. Soil Sci.* 63:607–617.

de Roo, A. P. J. 1993. Validation of the ANSWERS catchment model for runoff and soil erosion simulation in catchments in the Netherlands and the United Kingdom, in K. Kovar and H. P Nachtnebel, eds., *Application of Geographic Information Systems in Hydrology and Water Resources*. Proceedings of Vienna Conference, April 1993, IAHS Pub. No. 211, pp. 465–474.

Dean, W. E. and E. Gorham. 1998. Magnitude and significance of carbon burial in lakes, reservoirs and peat lands. *Geology* 26:535–538.

Dedkov, A. P. and V. T. Mozzherin. 1992. Erosion and sediment yield in mountain areas of the world, in *Erosion, Debris Flows and Environment in Mountain Regions*. Proceedings Chengdu symposium, July 1992, IAHS publ. 209:29–36.

Deines, P. 1980. The isotopic composition of reduced organic carbon, in A. F. Fritz and J. C. Fontes. eds. Elsevier, Amsterdam, pp. 329–406.

Desjardins, T., F. Andreux, B. Volkoff, and C. C. Cerri. 1994. Organic carbon and ^{13}C contents in soils and soil size-fractions, and their changes due to deforestation and pasture installation in eastern Amazonia. *Geoderma* 61:103–118.

Dixon, R. K., S. Brown, R. A. Houghton, A. M. Solomon, M. C. Trexler, and J. Wisniewski. 1994. Carbon pools and flux of global forest ecosystems. *Science* 263:185–190.

Dupin, B., K. B. Phantahvong, A. Chanthavongsa, and C. Valentin. 2002. Assessment of tillage erosion rates on steep slopes in the Northern Lao PDR. *Lao J. Agriculture & Forestry* 4:52–59.

Ehleringer, J. R., N. Buchmann, and L. B. Flanagan. 2000. Carbon isotope ratios in below ground carbon cycle processes. *Ecological Applications* 10:412–422.

Einsele, G., J. Yan, and M. Hinderer. 2001. Atmospheric carbon burial in modern lake basins and its significance for the global carbon budget. *Global Planet. Change* 30:167–195.

Esser, G. and G. H. Kohlmaier. 1991. Modeling terrestrial sources of nitrogen, phosphorus, sulphur and organic carbon to rivers, in E. T. Degens, S. Kempe, and J. E. Richey, eds., *Scope UNEP 42*. John Wiley & Sons, New York, pp. 297–320.

Eswaran, H., E. Van Den Berg, and P. Reich. 1993. Organic carbon in soils of the world. *Soil Sci. Soc. Am. J.* 57:192–194.

Farquhar, G. D., J. R. Ehleringer, and K. T. Hubick. 1989. Carbon isotope discrimination and photosynthesis. *Annual Rev. Plant Phys. Plant Mol. Biol.* 40:503–537.

Farquhar, G. D. 1983. On the nature of carbon isotope discrimination in C_4 species. *Aust. J. Plant. Phys.* 19:205–226.

Felipe-Morales, C., R. Meyer, C. Alegre, and C. Vitorelli. 1977. Determination of erosion and runoff under various cultivation systems in the Santa Amahuancayo region. I. Preliminary results of the 1974 to 1975 and 1975 to 1976 seasons. *An. Cient. Univ. Nac.* 15, 1–4: 75–84.

Fogel, M. and L. Cifuentes. 1993. Isotope fractionation during primary production, in H. Engel and S. Macko, eds. Plenum Press, New York, pp. 73–97.

Galimov, E. 1985. *The Biological Fractionation of Isotopes*. Academic Press, New York.

Ganeshram, R. S., T. F. Pedersen, S. E. Calvert, G. W. McNeill, and M. R. Fontugne. 2000. Glacial-interglacial variability in denitrification in the world's oceans: causes and consequences. *Paleoceanography* 15, 4:361–376.

Girardin, C. and A. Mariotti. 1991. Analyse isotopique du ^{13}C en abondance naturelle dans le carbone organique: un système automatique avec robot préparateur. *Cah. Orstom sér. Pédol.* 26, 4:371–380.

Hairshine, P. B., G. C. Sander, C. W. Rose, J. Y. Parlange, W. L. Hogarth, I. Lisle, and H. Rouhipour. 1999. Unsteady soil erosion due to rainfall impact: A model of sediment sorting on the hill slope. *J. Hydrol.* 220:115–128.

Harvey, H. R. and S. A. Macko. 1997. Kinetics of phytoplankton decay during simulated sedimentation: Changes in lipids under oxic and anoxic conditions. *Org. Geochem.* 27, 3/4:129–140.

Hayes, J. M., H. Strauss, and A. J. Kaufman. 1999. The abundance of ^{13}C in marine organic matter and isotopic fractionation in the global biogeochemical cycle of carbon during the past 800 Ma. *Chemical Geol.* 161:103–125.

He, Q. and P. Owens. 1995. Determination of suspended sediment provenance using caesium-137, unsupported lead-210 and radium-226: A numerical mixing model approach, in I. D. L. Foster, A. M. Gurnell, and B. W. Webb, eds., *Sediment and Water Quality in River Catchments*. John Wiley & Sons Ltd., New York, pp. 207–227.

Hedges, J. I., W. A. Clark, P. D. Quay, J. E. Richey, A. H. Devol, and U. M. Santos. 1986. Composition and fluxes of particulate organic material in the Amazon River. *Limnol. Oceanogr.* 31, 4:717–738.

Hedges, J. I., G. L. Cowie, J. E. Richey, and P. D. Quay. 1994. Origins and processing of organic matter in the Amazon River as indicated by carbohydrates and amino acids. *Limnol. Oceanogr.* 39, 4:743–761.

Hedges, J. I., J. A. Baldock, Y. Gélinas, C. Lee, M. Peterson, and S. G. Wakeham. 2001. Evidence for nonselective preservation of organic matter in sinking marine particles. *Nature* 409:801–804.

Hien, P. D., H. T. Hiep, N. H. Quang, N. Q. Huy, N. T. Binh, P. S. Hai, N. Q. Long, and V. T. Bac. 2002. Derivation of ^{137}Cs deposition density from measurements of ^{137}Cs inventories in undisturbed soils. *J. Environ. Radioactivity* 62:295–303.

Higgitt, D. L. and D. E. Walling. 1993. The value of caesium-137 measurements for estimating soil erosion and sediment delivery in an agricultural catchment, Avon, U.K., in S. Wicherek, ed., *Farm Land Erosion in Temperate Plains Environment and Hills*. Elsevier, Amsterdam, pp. 301–315.

Hollander, D. J. and M. A. Smith. 2001. Microbially mediated carbon cycling as a control on the $\delta^{13}C$ of sedimentary carbon in eutrophic Lake Mendota (U.S.): New models for interpreting isotopic excursions in the sedimentary record. *Geochim. Cosmochim. Acta* 65, 23:4321–4337.

Houghton, R. A. 1991. Tropical deforestation and atmospheric carbon dioxide. *Climatic Change* 19:99–118.

Houghton, R. A., J. D. Unruh, P. A. Lefebvre. 1993. Current land cover in the tropics and its potential for sequestering carbon. *Global Biogeochem. Cycles* 7, 2:305–320.

Houghton, R. A., E. A. Davidson, and G. M. Woodwell. 1998. Missing sinks, feedbacks, and understanding the role of terrestrial ecosystems in the global carbon balance. *Global Biogeochem. Cycles* 12, 1:25–34.

Huon, S., F. E. Grousset, D. Burdloff, G. Bardoux, and A. Mariotti. 2002. Sources of fine-sized organic matter in North Atlantic Heinrich layers: $\delta^{13}C$ and $\delta^{15}N$ tracers. *Geochim. Cosmochim. Acta* 66, 2:223–239.

Ittekot, V. 1988. Global trends in the nature of organic matter in suspensions. *Nature* 332:436–438.

Kao, S. J. and K. K. Liu. 1996. Particulate organic carbon export from a subtropical mountainous river (Lanyang His) in Taiwan. *Limnol. Oceanogr.* 41, 8:1749–1757.

Kao, S. J. and K. K. Liu. 2000. Stable carbon and nitrogen isotope systematics in a human-disturbed watershed (Lanyang-I) in Taiwan and the estimation of biogenic particulate organic carbon and nitrogen fluxes. *Global Biochemical Cycles* 14, 1:189–198.

Katchanovski, R. G. and E. de Jong. 1984. Predicting the temporal relationship between cesium-137 and erosion rate. *J. Environ. Qual.* 13, 2:301–304.

Kennicut, M. C., C. Barker, J. M. Brooks, D. A. DeFreitas, and G. H. Zhu. 1987. Selected organic matter source indicators in the Orinoco, Nile and Changjiang deltas. *Org. Geochem.* 11, 1:41–51.

Kirschbaum, M. U. F. 2000. Will changes in soil organic carbon act as a positive or negative feedback on global warming? *Biogeochemistry* 48, 1:21–51.

Krull, E. S. and J. O. Skjemstad. 2003. $\delta^{13}C$ and $\delta^{15}N$ profiles in ^{14}C-dated oxisol and vertisol as a function of soil chemistry and mineralogy. *Geoderma* 112:1–29.

Krusche, A. V., L. A. Martinelli, R. L. Victoria, M. Bernardes, P. B. Camargo, M. V. Ballester, and S. E. Trumbore. 2002. Composition of particulate and dissolved organic matter in a disturbed watershed of southeast Brazil (Piracicaba River basin). *Water Res. Res.* 36:2743–2752.

Lal, R. 1990. *Soil Erosion in the Tropics: Principles and Management*. McGraw Hill, New York, 580 pp.

Lal, R. 1995. Global soil erosion by water and carbon dynamics, in R. Lal, J. Kimble, E. Levine, and B. A. Stewart, eds., *Soils and Global Change*. Lewis Publishers, Boca Raton, FL, pp. 131–142.

Lal, R., D. Mokma, and B. Lowery. 1999. Relation between soil quality and soil erosion, in R. Lal, J. Kimble, E. Levine, and B. A. Stewart, eds., *Soil Quality and Soil Erosion,* CRC Press, Boca Raton, FL, pp. 237–259.

Lang, R. D. 1992. Accuracy of two sampling methods used to estimate sediment concentrations in runoff from soil-loss plots. *Earth Surf. Process. Landforms* 11:307–319.

Le Bissonnais, Y. 1990. Experimental study and modelling of soil surface crusting processes, in R. B. Bryan, ed., Soil erosion, experiments and models. *Catena suppl.* 17:13–28.

Lehmann, M. F., S. M. Bernasconi, A. Barbieri, and J. A. McKenzie. 2002. Preservation of organic matter and alteration of its carbon and nitrogen isotope composition during simulated and in-situ early diagenesis. *Geochim. Cosmochim. Acta* 66, 20:3573–3584.

Likens, G. E. 1972. Eutrophication and aquatic ecosystems, in G. E. Likens, ed., *Nutrients and Eutrophication*. Am. Soc. Limnol. and Oceanogr. Spec. Symp. 1, pp. 3–13.

Loughran, R. J., B. L. Campbell, and D. E. Walling. 1987. Soil erosion and sedimentation indicated by caesium-137, Jackmoor Brook catchment, Devon, England. *Catena* 14:201–212.

Loughran, R. J. and B. L. Campbell. 1995. The identification of catchment sediment sources, in I. D. L. Foster, A. M. Gurnell, and B. W. Webb, eds., *Sediment and Water Quality in River Catchments*. John Wiley & Sons Ltd., New York, pp. 189–205.

Ludwig, W., J- L. Probst, and S. Kempe. 1996. Predicting the oceanic input of organic carbon by continental erosion. *Global Biogeochem. Cycles* 10, 1:23–41.

Mabit, L. and C. Bernard. 1998. Relationship between soil ^{137}Cs inventories and chemical properties in a small intensively cropped watershed. *C.R.A.S. Paris* 327:527–532.

Marino, B. D. and M. B. McElroy. 1991. Isotopic composition of atmospheric CO_2 inferred from carbon in C_4 plant cellulose. *Nature* 349:127–131.

Mariotti, A., D. Pierre, J. C. Vedy, and S. Bruckert. 1980. The abundance of natural nitrogen-15 in the organic matter of soils along an altitudinal gradient (Chablais, Haute-Savoie, France). *Catena* 7:293–300.

Mariotti, A., C. Lancelot, and G. Billen. 1984. Natural isotopic composition of nitrogen as a tracer of origin for suspended organic matter in the Scheldt estuary. *Geochim. Cosmochim. Acta* 48:549–555.

Mariotti, A., F. Gadel, P. Giresse, and K. Mouzeo. 1991. Carbon isotope composition and geochemistry of particulate organic matter in the Congo River (Central Africa): Application to the study of Quaternary sediments off the mouth of the river. *Chemical Geology* 86:345–357.

Mariotti, A. and E. Peterschmitt. 1994. Forest savanna ecotone dynamics in India as revealed by carbon isotope ratios of soil organic matter. *Oecologia* 97:475–480.

Martin, A., A. Mariotti, and J. Balesdent. 1990. Estimate of organic matter turnover rate in a savanna soil by ^{13}C natural abundance measurements. *Soil Biol. Biochem.* 22, 4:517–523.

Martinelli, L. A., M. V. Ballester, A. V. Krusche, R. L. Victoria, P. B. Camargo, M. Bernardes, and J. P. H. B. Ometto. 1999. Land-cover changes and $\delta^{13}C$ composition of riverine particulate organic matter in the Piracicaba River basin (southeast region of Brazil). *Limnol. Oceanogr.* 44:1827–1833.

Masiello, C. A. and E. R. M. Druffel. 2001. Carbon isotope geochemistry of the Santa Clara River. *Global Biogeochem. Cycles* 15, 2:407–416.

Meade, R. H. 1988. Movement and storage of sediment in river systems, in A. Lerman and M. Meybeck, eds., *Physical and Chemical Weathering in Geochemical Cycles*. Kluwer Academic Pub., Boston, pp. 165–179.

Megens, L.,J. van der Plicht, J. W. de Leeuw, and F. Smedes. 2002. Stable carbon and radiocarbon isotope compositions of particle size fractions to determine origins of sedimentary organic matter in an estuary. *Org. Geochem.* 33:945–952.

Melillo, J. M., J. D. Aber, A. E. Linkins, A. R. Turner, B. Fry, and K. J. Nadelhoffer. 1989. Carbon and nitrogen dynamics along the decay continuum: Plant litter to soil organic matter. *Plant and Soil* 115:189–198.

Meybeck, M. 1982. Carbon, nitrogen, phosphorous transport by world rivers. *Am. J. Sci.* 282:401–450.

Meybeck, M. 1993. Riverine transport of atmospheric carbon: sources, global typology and budget. *Water, Air and Soil Pollution* 70:443–463.

Meyers, P. A. 1997. Organic geochemical proxies of palaeoceanographic, palaeolimnologic and paleoclimatic processes. *Org. Geochem.* 27, 5/6:213–250.

Milliman, J. D., Y. S. Quin, M. E. Ren, and Y. Saito. 1987. Man's influence on the erosion and transport of sediments by Asian rivers: The Yellow River (Huanghe) example. *J. Geol.* 95:751–762.

Milliman, J. D. and J. P. M. Syvitski. 1992. Geomorphic/tectonic control of sediment discharge to the ocean: The importance of small mountainous rivers. *J. Geol.* 100:325–344.

Mook, W. C. and F. C. Tan. 1991. Stable carbon isotopes in rivers and estuaries, in E. T. Degens, S. Kempe, and J. E. Richey, eds., *Scope UNEP 42*, John Wiley & Sons, New York, pp. 245–264.

Morgan, R. P. C. 1986. *Soil Erosion and Conservation*. Longmans. London.

National Agricultural and Forestry Research Institute (NAFRI). 2001. An innovative approach to sustainable land management in Lao PDR. MSEC Report. 39 pp.

Nordin, C. F., Jr. 1985. The sediment discharge of rivers — a review, in Erosion and transport measurement. *IAHS Pub.* 133:3–47.

Oades, J. M. 1986. The retention of organic matter in soils. *Biogeochemistry* 5:35–70.

O'Leary, M. H. 1988. Carbon isotopes in photosynthesis. *Bioscience* 38:328–336.

Onstad, G. D., D. E. Canfield, P. D. Quay, and J. I. Hedges. 2000. Source of particulate organic matter in rivers from the continental U.S.: Lignin phenol and stable carbon isotope compositions. *Geochim. Cosmochim. Acta* 64, 20:3539–3546.

Pérez, G. J. 1997. *Validación de métodos de muestreo no convencionales de sedimentos en suspensión. Caso cuenca alta Río Boconó*. Unpublished Thesis Ing. Forestal. Universidad de Los Andes, Mérida, Venezuela, 96 pp.

Post, W. M., W. R. Emanuel, P. J. Zinke, and A. G. Stangenberger. 1982. Soil organic carbon pools and world life zones. *Nature* 298:156–159.

Probst, J. L. 1992. Géochimie et hydrologie de l'érosion continentale. Mécanismes, bilan global actuel et fluctuations au cours des 500 derniers millions d'années. *Mém. Sci. Géologiques* 94, 161p.

Puget, P., C. Chenu, and J. Balesdent 2000. Dynamics of soil organic matter associated with particle-size fractions of water-stable aggregates. *Eur. J. Soil Sci.* 51:595–605.

Quine, T. A., D. E. Walling, Q. K. Chakela, O. T. Mandiringana, and X. Zhang. 1999. Rates and patterns of tillage and water erosion on terraces and contour strips: Evidence from caesium-137 measurements. *Catena* 36:115–142.

Quine, T. A. 1995. Estimation of erosion rates from caesium-137 data: The calibration question, in I. D. L. Foster, A. M. Gurnell, and B. W. Webb, eds., *Sediment and Water Quality in River Catchments.* John Wiley & Sons Ltd., New York, pp. 307–329.

Raymond, P. A. and J. E. Bauer. 2001a. Use of ^{14}C and ^{13}C natural abundances for evaluating riverine, estuarine, and coastal DOC and POC sources and cycling: A review and synthesis. *Organic Geochem.* 32:469–485.

Raymond, P.A. and J. E. Bauer. 2001b. Riverine export of aged terrestrial organic matter to the North Atlantic Ocean. *Nature* 409:497–500.

Richey, J. E., J. T. Brock, R. J. Naiman, R. C. Wissmar, and R. F. Stallard. 1980. Organic carbon: Oxidation and transport in the Amazon River. *Science* 207:1348–1350.

Richey, J. E., A. H. Devol, S. C. Wofsy, R. Victoria, and M. N. G. Riberio. 1988. Biogenic gas and the oxidation and reduction of carbon in Amazon River and floodplain waters. *Limnol. Oceanogr.* 33:551–561.

Ritchie, J. C. and J. R. McHenry. 1973. Determination of fallout Cs-137 and natural gamma-ray emitting radionuclides in sediments. *Int. J. Applied Radiation Isotopes* 24:575–578.

Ritchie, J.C., J. R. McHenry, A. C. Gill, and P. H. Hawkes. 1974. Fallout ^{137}Cs in the soils and sediments of three small watersheds. *Ecology* 55:887–890.

Ritchie, J. C. and J. R. McHenry. 1990. Application of radionuclide fallout cesium-137 for measuring soil erosion and sediment accumulation rates and patterns. *J. Environ. Quality* 19:215–233.

Ritchie, J. C. and G. W. McCarty. 2003. ^{137}Cesium and soil carbon in a small agricultural watershed. *Soil and Tillage Res.* 69:45–51.

Rodríguez, A. A. 1999. Aplicación preliminar de técnicas de diseño hidrológico a parcelas de erosión del municipio Boconó, Estado Trujillo. Unpublished report, Universidad de los Andes, Boconó, 62 pp.

Rogowski, A. S. and T. Tamura T. 1965. Movement of ^{137}Cs by runoff, erosion and infiltration on the alluvial captina silt loam. *Health Phys.* 11:1333–1340.

Rómulo-Quintero, J. 1999. Determinación y análisis del coeficiente de escorrentía en parcelas de erosión del sector la Corojó. Boconó, estado Trujillo. Unpublished report, Universidad de los Andes, Boconó, 100 pp.

Roose, E. 1977. Erosion et ruissellement en Afrique de l'Ouest: vingt années de mesures en petites parcelles expérimentales. Paris ORSTOM coll. Trav. et Doc. 78, 108 pp.

Roscoe, R., P. Buurman, E. J. Velthorst, and C. A. Vasconcellos. 2001. Soil organic matter dynamics in density and particle size fractions as revealed by the $^{13}C/^{12}C$ isotopic ratio in a Cerrado's oxisol. *Geoderma* 104:185–202.

Ross, S. M. 1993. Organic matter in tropical soils: current conditions, concerns and prospects for conservation. *Progress in Physical Geography* 17, 3:265–305.

Schwartz, D., H. de Foresta, A. Mariotti, J. Balesdent, J. P. Massimba, and C. Girardin. 1996. Present dynamics of the savanna-forest boundary in the Congolese Mayombe: A pedological, botanical and isotopic (^{13}C and ^{14}C) study. *Oecologia* 106:516–524.

Schlesinger, W. H. and J. A. Andrews. 2000. Soil respiration and the global carbon cycle. Controls on soil respiration: implications for climate change. *Biogeochemistry* 48, 1:7–20.

SPARTACUS. 2000. Spatial redistribution of radionuclides within catchments: Development of GIS-based models for decision support systems. Final report. M. Van der Perk, A. A. Svetlitchnyi, J. W. den Besten, and A. Wielinga, eds. Utrecht University, 165 pp.

Stumm, W. and J. J. Morgan. 1996. *Aquatic Chemistry: Chemical Equilibria and Rates in Natural Waters*, 3rd ed. John Wiley & Sons, New York, 1022 pp.

Sundquist, E. T. 1993. The global carbon budget. *Science* 259:934–941.

Sutherland, R. A. 1994. Spatial variability of ^{137}Cs and the influence of sampling on estimates of sediment redistribution. *Catena* 21:57–71.

Tiessen, H., R. E. Karamanos, J. W. B. Stewart, and F. Selles. 1984. Natural nitrogen-15 abundance as an indicator of soil organic matter transformations in native and cultivated soils. *Soil Sci. Soc. Am. J.* 55:312–315.

Tiessen, H., E. Cuevas, and P. Chacon. 1994. The role of organic matter in sustaining soil fertility. *Nature* 371:783–785.

Thornton, S. F. and J. McManus. 1994. Application of organic carbon and nitrogen stable isotope and C/N ratios as source indicators of organic matter provenance in estuaries systems: Evidence from the Tay estuary, Scotland. *Estuarine, Coastal and Shelf Science* 38:219–233.

Trimble, S. W. and P. Crosson. 2000. U.S. soil erosion rates — Myths and reality. *Science* 289:248–250.

Turkelboom, F., J. Poesen, I. Ohler, and S. Ongprasert. 1999. Reassessment of tillage erosion rates by manual tillage on steep slopes in northern Thailand. *Soil and Tillage Res.* 51:245–259.

Tyson, R. V. 1995. *Sedimentary Organic Matter: Organic Facies and Palynofacies.* Chapman and Hall, London, 615 pp.

United Nations Scientific Committee on the effects of Atomic radiation (UNSCEAR). 1969. 24th session, Supp. No. 13 (A/7613). New York. United Nations.

United Nations Scientific Committee on the effects of Atomic radiation (UNSCEAR). 1993. 24th session, Supp. No. 13 (A/7613). New York. United Nations.

United Nations Scientific Committee on the effects of Atomic radiation (UNSCEAR). 2000. 55th session, Supp. No. 46 (A/55/46). New York. United Nations.

Wada, E., M. Minagawa, H. Mizutani, T. Tsuji, R. Imaizuni, and K. Karazawa. 1987. Biogeochemical studies on the transport of organic matter along the Otsuchi river watershed Japan. *Estuarine, Coastal and Shelf Science* 25,321–336.

Walling, D. E. and T. A. Quine. 1992. The use of caesium-137 measurements to provide quantitative erosion rate data. *Land. Degrad. Rehabil.* 2:161–175.

Walling, D. E., J. C. Woodward, and A. P. Nicholas. 1993. A multi-parameter approach to fingerprinting suspended-sediment sources, in N. E. Peters, E. Hoehn, C. Leibundgut, N. Tase, and D. E. Walling, eds., *Tracers in Hydrology.* IAHS Pub. No 215, IAHS Press. Wallingford, pp. 329–338.

Walling, D. E. and Q. He. 1999. Improved models for estimating soil erosion rates from cesium-137 measurements. *J. Environ. Qual.* 28:611–622.

Walling, D. E. and Q. He. 2001. Models for converting ^{137}Cs measurements to estimates of soil redistribution rates on cultivated and uncultivated soils (including software for model implementation). A contribution to the IAEA coordinated research programs on soil erosion and sedimentation.

Wan, Y. and S. A. El-Swaify. 1997. Flow-induced transport and enrichment of erosional sediment from a well-aggregated and uniformly textured Oxisol. *Geoderma* 75:251–265.

Wischmeier, W. H. and D. D. Smith. 1978. *Predicting Rainfall Erosion Losses: A Guide to Erosion Planning.* Agriculture Handbook 537, U.S. Department of Agriculture, Washington D.C.

Yoneyama, T. 1996. Characterization of natural ^{15}N abundance of soils, in T. W. Boutton and S. -I. Yamasaki, eds., *Mass Spectrometry of Soils.* Marcel Dekker Inc., New York, pp. 205–223.

Zech, W., N. Senesi, G. Guggenberger, K. Kaiser, J. Lehmann, T. M. Miano, A. Miltner, and G. Schroth. 1997. Factors controlling humification and mineralization of soil organic matter in the tropics. *Geoderma* 79:117–161.

Zhang, X. B., D. L. Higgitt, and D. E. Walling. 1990. A preliminary assessment of the potential for using caesium-137 to estimate rates of soil erosion in the Loess Plateau of China. *Hydrol. Sci. J.* 35:267–276.

SECTION 4

Conclusions

CHAPTER 21

Erosion and Carbon Dynamics: Conclusions and Perspectives

Eric Roose, Michel Meybeck, Rattan Lal, and Christian Feller

CONTENTS

21.1 Basic Concepts 331
21.2 Carbon Erosion at the Plot Scale (1 to 100 m^2) 332
21.3 Carbon Transport in Rivers 334
21.4 Perspectives 336
    21.4.1 Plot Scale 336
    21.4.2 Hillside 336
    21.4.3 Large Watershed Scale 336
    21.4.4 Global Scale 336

21.1 BASIC CONCEPTS

Exponential population growth, intensification of agricultural production at the expense of forests, and rapid urbanization have increased problems of soil degradation, floods, and erosion in areas considered relatively stable until the end of the twentieth century. These changes in land uses have repercussions not only on local soil erosion and agricultural productivity but also on the global carbon budget and the accelerated greenhouse gas (GHG) effects.

The main focus of this book is to examine the role of biomass management on soil C dynamics and how selective erosion can impoverish soil organic matter (SOM) and transfer dissolved and particular soil organic carbon (SOC). SOC, locked within soil aggregates, may mineralize at an accelerated rate or concentrate in colluvial, alluvial deposits, or buried with marine sediments.

The chapter by Robert states that the climatic change is the most serious problem for the twenty-first century. Even if there are many uncertainties, preventive actions are necessary to minimize the risks and reduce net anthropogenic emissions. Soil C sequestration can have a significant contribution but requires appropriate policies to implement changes in forestry and agricultural land uses. These changes seem to be beneficial to soil protection and environmental quality required for a sustainable agriculture.

Bernoux et al. discuss soil carbon sequestration and state that because of its numerous ancillary benefits (economic, environmental, agronomic), it is an opportune solution to mitigating the atmospheric GHG emissions. Soils constitute approximately 1500 Gt C, which is about three times the terrestrial C pool and twice the atmospheric pool. Land-use conversion, change in agricultural system or soil and water management can strongly alter the soil C pool. For global warming caused by increase in atmospheric GHG concentration, it is necessary not only to increase the soil carbon stocks but also to decrease the losses of dissolved OC in runoff and drainage, and reduce the emission of gases like CO_2, CH_4, and N_2O. While the use of mineral fertilizers or legume cover crops often has a very positive impact on biomass production and SOC stocks, they may induce emissions of N_2O or CH_4 and reduce the positive impact on net carbon sequestration.

The chapter by Lal highlights the significance of erosion by water and wind. Erosion preferentially removes SOM, a light fraction concentrated in the vicinity of the soil surface. Yet the fate of eroded soil carbon is not well understood. Sedimentologists argue that eroded C transported in the rivers is buried and sequestered in alluviums and marine sediments. Some soil scientists believe that 20 to 30% of the C transported over the landscape is mineralized and released in the atmosphere. The total amount of SOC displaced by erosion annually is estimated at 4 to 6 Pg of which 0.8 to 1.2 Pg is emitted into the atmosphere. The debate is accentuated by numerous uncertainties: (1) breakdown of aggregates by erosion leading to more exposure of SOC to microbes enhancing mineralization, (2) alterations in soil moisture and temperate regimes of eroded soils enhancing oxidation of SOM, (3) redistribution of the SOC-enriched sediments over the landscape accentuating the oxidation, (4) the aggregation of the SOC buried in the depressions and protected sites, and (5) mineralization of SOC transported in the rivers depending on the climatic conditions. Assuming a delivery ratio of 13 to 20%, SOC concentration in sediment of 2%, and oxidation rate of 20% of the C displaced, erosion induced C emission is estimated at 0.8 to 1.2 Pg C yr^{-1}. Of this, 56% is from Asia, 26% from America, 8% from Europe, 6% from Africa, and 4% from Oceania.

Changes in SOC stocks caused by water erosion have been evaluated but the dynamic responses of C fluxes to redistribution of eroded C and terrestrial sedimentation are still poorly understood. Van Oost et al. describe the SPEROS-C model, which combines spatially distributed models of water and tillage erosion with a model of carbon dynamics. The main advantage of this approach is the explicit consideration of spatial C transfers between landscape elements, facilitates calculation of the overall C budget at the landscape scale. A sensitivity analysis clearly showed that soil erosion may considerably affect the C balance of agricultural fields but that additional research is needed for an accurate quantification of erosion-induced C fluxes between soil and atmosphere.

21.2 CARBON EROSION AT THE PLOT SCALE (1 TO 100 m²)

Numerous data have been presented about the C losses by various erosion processes at the scale of conventional runoff plots (100 m²) in tropical and Mediterranean areas (Roose and Barthès; Barthès et al.; Blanchart et al.; Morsli et al.; Bilgo et al.; Rodriguez et al.). Compared with biomass production (1 to 20 t ha^{-1} yr^{-1}), losses of particulate organic carbon (POC) by hydric erosion are moderate: 1 to 50 kg C ha^{-1} yr^{-1} when the soil is effectively protected by vegetative cover of residue, 50 to 500 kg C ha^{-1} yr^{-1} under clean weeded crops, burned or overgrazed grasslands, and > 1000 kg C ha^{-1} yr^{-1} on bare plots under erosion-prone conditions (very steep slopes, intensive and abundant rainstorms). The POC losses mainly depend on the amount of soil eroded and on SOC concentration of the soil surface. Land-use changes cause more drastic changes in soil erosion and SOC stocks than POC erosion because when erosion increases, the C content of sediments decreases, through reduction in interrill erosion and increase in rill erosion.

Processes governing losses of dissolved organic carbon (DOC) in runoff or drainage water are not well known. These losses range from 1 to 600 kg C ha^{-1} yr^{-1} depending on the volume of runoff or percolating water (Roose and Barthès; Blanchart et al.; Bep et al., 2004).

Carbon losses by erosion and drainage are of the same order of magnitude as the SOC sequestration (0.1 to 3 t C ha^{-1} yr^{-1}) rate with improved management (Barthès et al.; Blanchart et al.; Rodriguez et al.; Morsli et al.). Consequently, the biologic soil conservation systems (like direct drilling under mulch, mulching or covercrops, ley-farming, agro-forestry, etc.) and better covering the soil surface, reduce losses of C by erosion and drainage, and produce more C in the biomass. This is a win-win situation because it increases the SOC sequestration while restoring soil quality and increasing the resistance to erosional forces.

Carbon enrichment ratio (CER) compares the C content of sediments to the C content of soils for 2.5 or 10 cm depth. To make an easier comparison among authors and taking into account that most of the soil analyses are made on 0 to 10 cm horizon, which represents a part of plow layer, Roose suggested comparing the C content of sediments (fine particles in suspension + coarse sediments) to the SOC content of the top 10 cm layer. The sheet erosion preferentially removes POC: linear erosion scours all the upper horizons (CER = 1), gullies cut also deep horizons low in SOM (CER < 1), and mass movements like tillage erosion, scour the whole profile over the landslide horizon (CER < 1). Analyzing 56 runoff plots data, Roose and Barthès identified five classes of CER: bare soils (CER = 1.3 ± 0.2), cropped fields and overgrazed grasslands (2.0 ± 0.5), grassed fallows (2.7 ± 1.0), protected bush savannas (4.2 ± 3.2), and dense forests (7.5 ± 4.5). Low CER level are mostly related to high clay content, high SOC in deep horizons, Andosols, Nitosols or Vertisols, swelling clays (more linear erosion), steep slopes, bare soils, and high intensity rainstorms and runoff. The value of CER are generally higher from soils with sandy clay surface horizon than from those with loamy or clay deep soils, on long and gentle slopes than on steep hillslopes (Bilgo et al.; Boye and Albrecht; Roose and Barthès).

In Andosols of Canary Islands, Rodriguez et al. reported that C erosion varied between 1.2 and 251 kg C ha^{-1} yr^{-1} with a runoff rate of 19%. More than 90% of eroded SOC is comprised of POC and very little, if any, of the DOC. In sediments, aggregates < 0.5 mm predominated although a slight predomination of larger aggregates is observed in the topsoil. All aggregates are very stable to dispersion in water. It seems that in erosion, fine aggregates from the soil surface are lost due to impact of raindrops. The SOC mainly associated with the fine aggregates (POC) is washed from the hilltop and redistributed downslope in small terraces used for crop production where the conditions favor the rupture of stable organo-mineral bindings that protect SOC from oxidation. In this way, erosion of Andosols organic horizon leads to dispersion and the net emission of CO_2. The selective processes of sheet erosion depends on scouring the top surface soil that is enriched in organic carbon than the deeper horizons. It also depends on the selective transport and deposit of lighter fraction along the hillslope when the topsoil is sandy or gravelly, or contains very stable aggregates like Andosols, Nitosols, and calcic Vertisols that are very rich in clay and loam (> 70%).

Four chapters focus on data involving eroded carbon on natural or simulated rainfall in 1m^2 micro-plots. They report on the initial stage of runoff, splash dispersion, and sheet erosion, but do not address the issue of the variability of the soil surface irregularities and physical properties associated with the roughness, nor the risk of rill erosion in conventional runoff plot of 100 m^2.

Brunet et al. showed that the restoration of pastures in Brazil decreased the risk of runoff and erosion through effective soil cover. The loss of C by sheet erosion was limited to 1 to 3 gC m^{-2} yr^{-1} under pasture and 62 gC m^{-2} yr^{-1} for the bare soil (pasture denuded with roots left in the soil). The CER was 1.1 to 1.3 for pasture and 1.4 for degraded pasture (bare soil). These low rates are explained by the high percentage of clay in the surface layer of this Ferrallitic soil.

Chaplot et al. showed that soil erosion was only 11.2 gC m^{-2} yr^{-1} on a 0.6 ha watershed compared to 31.2 and 75.2 gC m^{-2} yr^{-1} on 1 m^2 plots. Soil losses where twice as much with continued rice cropping compared to prior 4 years of natural fallow. The difference in erosion was mainly due to one large rainfall event and to a difference in soil resistance to detachment by the raindrop impact. The preceding fallow period reduced mean runoff and mean C content by 10 % but total soil and C losses by 60%. The CER ranged from 1.7 to 1.5 with the fallow on this 46% slope gradient and 55% of clay content in the topsoil. The differences in erosion between small watershed (0.6 ha) and the 1 m^2 plots suggest that erosion can be drastically increased in small plots (border effect and less roughness) and that these agro-pedologic problems of carbon erosion must be studied at the scale of at least 100 m² to correctly integrate the soil variability and risks of rill erosion. Similarly, Nyamadzawo et al. used a rainfall simulator to demonstrate that runoff and soil losses in three diverse fallow treatments (*Sesbania sesban*, *Tephrosia vogeli*, and natural fallow) generally were lower relative to continuous maize in two sites in Zambia. After one year of cropping, however, there were no differences among treatments. Boye and Albrecht also reported through rainfall simulation studies that improved fallow (with *Crotalaria grahamiana* or *Tephrosia candida*) reduced runoff, soil and C losses under subsequent maize crop, but the reduction was higher for the clay soil than for the sandy loam soil of Western Kenya. The proportion of topsoil C lost in sediments was more in the sandy loam than in the clay soil. After one cropping season, C losses on the sandy loam soil were much higher under no-till than under conventional tillage. Long-term experiments on large fields are needed to confirm these results.

21.3 CARBON TRANSPORT IN RIVERS

Riverine carbon is transported as dissolved inorganic (DIC = CO_2, HCO_3, CO_3) or particulate forms (mostly calcite) or organic (DOC and POC). The DIC originates from the weathering of noncarbonated rocks (aluminum silicate minerals), where 100% of it is of atmospheric origin (carbonic acid and soil organic acids) and from the dissolution of carbonate minerals where only half of it originate from the parent rock, and the other half comes from the atmosphere. The DOC and POC originate mostly from soil leaching and mechanical erosion and PIC from the erosion of carbonaceous rocks. Carbon species are characterized by a wide range of age from a few days (autochthonous PIC and algal POC) to 100 of millions years (dissolution of carbonate rocks). The mean concentration ranges by over 2 or 3 orders of magnitude depending on the river basins.

In river systems the exportation of Total Organic Carbon (TOC) is generally limited, about 1% of the biomass produced on the continent. Thus, it is often neglected in the carbon balance studies and usually corresponds to a constant flux of 500 gC ha^{-1}/yr^{-1}. Anthropogenic sources of DIC and PIC can be neglected but direct sources of POC and even DOC from major cities or food industries are common. The transfer of carbon in river systems is determined by numerous physico- or biochemical processes which function as successive filters: humid areas, lakes, reservoirs, alluvial plains, estuaries, and deltas. Soils function as the first filter. Specific studies conducted at a range of temporal and spatial scales have helped identify underlying mechanisms of these C transfers. The chapter by Boeglin et al., for the humid tropical plains such as the NSIMI watershed (60 ha in Western Cameroon) shows that forest biomass is large and did influence the fluxes of ions and carbon released from the rock weathering and a low rate of particulate erosion (9 mg l^{-1} of suspended sediments in average). The soils of the ferrallitic toposequence are deep (10 to 30 m) and OC is mostly contained in the low lands of these landscapes. The leaching is permanent and the riverine exportation of materials is mainly organic. The DOC represents 57% of the dissolved load at the exit and POC is only 22% of the exported carbon at the river exit. Depending on the landscape position, river waters are either clear (DOC = 1 mg l^{-1}) or brown (DOC = 15 mg l^{-1}) along the Amazon River (Seyler et al.)

In Laos and Venezuela watersheds located on steep mountainous hillslopes, mechanical erosion of sedimentary rocks are about twice as much as from the Nsimi basin (Huon and Valentin). For these watersheds, it has been possible to distinguish other sources of carbon from mechanical erosion using various isotopes (^{13}C, ^{14}C, ^{18}O) allowing us to separate the part in the schist from that in the SOC (sometimes > 10%).

Anthropogenic actions have numerous impacts on transfer and retention of carbon in the river systems. The net balance of human perturbation on the carbon cycle in fluvial waters must be addressed under site-specific conditions. The Piracicaba basin in Sao Paulo state, Brazil (12400 km², 3 millions of inhabitants with high level of industrialization) is a good example of diverse influences (Martinelli). The initial forest has been entirely cleared in 50 years between 1950 and

2000 and replaced by tree plantations, pastures, and industrial crops like sugarcane. The summit slopes of the Amazon basin (the JI-Parana, Rondonia) were also extensively deforested during 1980s, but the foot slopes are still under forest. A comparison of the two basins show the influences of deforestation, diverse land uses and inadequately treated urban waste.

Small stream dams are numerous in semiarid areas of Africa and America. Such dams protect water reservoirs and alter the transfer of carbon. The data from Tunisia reported by Albergel et al. show that these reservoirs trap sediments and POC. Some chemical markers indicate that the POC stored in the lake sediments (between 0.4 and 1.6% of the sediments) originate from change in soils and vegetations. This type of C storage continues as long as the reservoir is heavily silted (between 10 and 100 years).

In large river basins, other natural filters are functioning like in the interior delta of the Niger (30,000 km^2 of humid areas; Orange et al.). The seasonal cycles of DOC and POC are complex. The first peak in POC concentration occurs in August when the water stage is low and a second in October when the water stage is at the highest level. A part of this POC is related to the chlorophyll produced by algae that grow in the inundated plains and ponds. At the end of the dry season when the water level is the lowest, macrophytes debris produced in the lake of the delta are mixed with suspended sediments. In this complex cycle, the organic carbon content of water transport in and out of the lake is not strongly altered (DOC = 1.7 mg l^{-1}; POC = 0.6 to 0.8 mg l^{-1}). Studies based on isotopic or geochemical markers are needed to understand these cycles.

The two biggest rivers of the world, the Congo-Zaire and Amazon, have been studied for several years under the auspices of PEGI, PIRAT, and HIBAM. In the case of the Congo-Zaire River, concentration and carbon fluxes are naturally moderated by the hydrologic cycle of this river (Orange et al.). Because of the absence of mountains and the very good protection of the soils by the humid tropical rainforest, the Congo-Zaire River has low suspended load of 20 to 30 mg l^{-1}. Similar to the Amazon River, the runoff is distinguished into brown acid (Black or Coca-Cola rivers), with very low rate of minerals but rich in DOC, and clear waters neutral and rich in minerals. In the Congo-Zaire River, the contents of organic carbon (DOC = 10 mg l^{-1} at Brazzaville and 5.7 mg l^{-1} at Bangui for the Oubangui) are higher than in the Niger, but the POC content is moderate (POC = 1.7 and 1.3 mg l^{-1}). Finally the fluxes of carbon exported by the Congo-Zaire River are ten times higher than those of Total Organic carbon in the Niger River.

In the Amazon River system, Papajos, Xingu, and Prombetas have clear waters while the Negro river is a typical example of black waters. For the first time, the carbon fluxes transported by all the main tributaries of this large network (6,400,000 km^2) have been accurately measured during one whole hydrologic cycle. The carbon balance thus obtained shows that TOC transported at Obidos is 4 TgC yr^{-1}, in relation to the sum of the three main rivers constituting the Amazon River (Negro, Solimoes, Madeira). The authors attribute this excess to the organic matter produced in the flood plains called "várzeas." The Amazonian system contributes 8 to 10% to the TOC transported in rivers to the oceans of the world.

The fate of the different carbon forms (organic and inorganic) can be studied during the hydrologic cycle at a specific gauging station or compared among major rivers (Meybeck, 2002). Some properties are quasi-universal such as the decrease in POC content with increase in sediment flux, the increase in POC and DOC transport with the increase in river flow, and the increase of the fossil carbon flux (DIC, PIC, POC, C-fossil) with increase in sediment flux of C. The increase in DOC concentration is often observed with the increase in total flux However, there are some exceptions to this general rule. In rivers where anthropogenic eutrophication occurs, autochton carbon (from algae or calcite precipitation) can dominate in the summer. The influence of global warming on the C transfer in the rivers at the global scale remains to be evaluated because it depends on many competing factors.

The fluvial carbon transported into the coastal ecosystems is subject to numerous factors (Ludwig). More than 90% of the particles are stored on the land. The POC is subject first to the estuarian filter prior to being mineralized very early during the "early diagenesis" in the coastal zone. The DIC is gradually precipitated in carbonate sediments, ensuring "the loop of the river carbon." The DOC stored in the marine sediments constitutes an enormous reservoir (30,000 Gt C) of carbon in comparison with the terrestrial vegetation (650 Gt) or the atmosphere (750 Gt). Analyses of the ^{13}C show that this pool is of marine origin. The river DOC is often very old, up to a few thousands of years as determined by the ^{14}C analyses. The loop of the carbon may partially obliterate the differences in CO_2 emissions between the northern and the southern hemispheres since 1950.

The role of the fluvial carbon, generated by chemical or mechanical erosion, at the global and geological time scale has been studied since 1980s (Probst). The atmospheric CO_2 absorbed during the weathering of the silicate rocks constitutes an important carbon sink.

Detailed studies remain to be done on the weathering of silicate rocks, either on monolithologic basins, or on large river basins (Amazon, Congo-Zaire, Brahmapoutra, Mackenzie).

21.4 PERSPECTIVES

21.4.1 Plot Scale

Several chapters have been presented about the losses of particulate carbon by sheet erosion in tropical and Mediterranean conditions. The enrichment ratio for carbon has been measured, which makes it possible to evaluate the carbon losses with data on soil erosion, and carbon content of the top 10 cm layer. However, few, if any data are available on the losses of dissolved carbon in the runoff and drainage waters. In humid tropics, DOC contents are not negligible (1 to 600 kg ha^{-1} yr^{-1}) and the composition of runoff is similar to the DOC in the rivers. Few distinction has been made between POC eroded in fine suspensions (which is carried into the rivers) and coarse sediments (sands, aggregates, litter mostly retained behind the roughness of the soil surface, or at the base of the hills).

There is a need for more data on the percentage of aggregates remaining in the sediments at the exit of the fields, the river and the ocean, and also about their susceptibility to be mineralized during transport and after deposition.

21.4.2 Hillside

The erosion processes are complex and the methodologies to separate them are not well known. Of course ^{137}Cesium and other tracers are very useful to show the spatial variation in flux, but refinements are needed to confirm that the initial deposition is really homogenous on the entire landscape even when the rainfalls is wind driven with varying direction.

21.4.3 Large Watershed Scale

A good synthesis has been made for rivers with regards to physiography, climates, rocks, slopes, and vegetation. The origin of various carbon forms has been described using biotracers.

The analysis of past carbon fluxes, such as during the glaciations (18,000 BP), involves a fine knowledge of the climate change and the exposure of surface rocks to erosion. The evolution of the DOC in coastal zones and in oceans remains to be assessed to quantify different components of the C cycle.

21.4.4 Global Scale

Field scale assessments of the particulate carbon losses by erosion are characterized by numerous uncertainties with regards to the impact of the kinetic energy of raindrops and the runoff on aggregates sequestering the POC, the delivery ratio of the sediments from the fields at the base of hill slopes to the nearest river, the fate of the carbon particles in the rivers and, the deposition rate at the foot of slopes, in valleys, ponds, lakes, estuaries, etc. Few experimental data are available that assess the fate of aggregates when transported over the field. Some experiments have shown that about 30% of macro aggregates remain intact at the end of 1 m² plots (Nitosols of Martinique, Khamsouk and Roose, Vertisols in Morocco, Roose, Simonneaux, Sabir, Blavet et al., unpublished data), and 1% in the Wadi during low waters or 10% during the flood events. The important issue to be addressed is with regard to the proportion of organic matter being mineralized during its transfer from the eroded field to the nearest river and in the river to the large rivers and the estuaries.

Soil and water conservation measures (particularly biologic ones) have a positive impact on SOC stocks because they reduce the TOC losses and increase the biomass deposited over (litter) or into the soil (roots), while also improving aggregate stability (erosion resistance) and water infiltration. Additional research is also needed to assess the impact of covercrops (grasses, legumes, or bushes) and the quality of the SOC in relation to gaseous emissions (CO_2, CH_4, N_2O) from the plots to the rivers and estuaries.

Index

A

Acid oxalate extractable minerals, 78
Afforestation, 6, 15
Agriculture
  conservation, 8, 10, 32
  management of SOM in, 8
  pesticides in, 9
  resource-based, 30
  subsistence, 25
  sustainable, 9, 10, 11, 15
  tillage in, 9
Agroforestry, 8, 9, 11, 15, 120, 290
  with leguminous plant integration, 20
  as management option, 182
  pasture *vs.*, 17
Agronomic systems, 8. *See also* Cropping systems
Alfisol(s), 67, 150, 304
  in catchments, 169
  description, 170
  fallow and, 176
  infiltration rate, 177
  origins, 57
  Utisols and, 307
  with Vertic properties, 57
Algeria, 104–119. *See also* Beni-Chougran Mountains, Algeria
Allophane, 79, 82
Aluminum, 78
Amazon River, 211, 213, 226
  annual carbon yields, 269
  basin, 258–259
    DOC, 262, 263
    DOC/TOC, 262
    factors controlling organic carbon in, 264–265
    POC, 262, 263
    TSS, 262, 263
  organic carbon flux, 268
  physical characteristics, 259
  tributaries, 259
  watershed, 241, 243
Andosol(s), 74
  allophanic, 76, 77, 80
  characteristics, 74, 77
  genesis, 75
  laboratory analysis, 77–80
    acid oxalate extractable iron, aluminum, and silicon in, 78
    allophane content, 79
    dissolved SOC, 78
    potassium sulfate extractable SOC, 78
    pyrophosphate extractable iron and aluminum, 78
    soil moisture, 79
  properties, 75
  resistance to water erosion, 74
  runoff plots, 76–77
    samples from, 79
  sediment samples, 79
  as sink for organic carbon, 81
  SOC content, 74
  soil samples, 78–79

B

Banana cropping
  bulk density of soil and, 93, 94
  C stock changes, 95
  carbon content of soil and, 93, 94
  rotating systems, erosion and, 87–99
  soil C budgets, 99
  soil C stock, 93, 94
  soil losses, 95
Beni-Chougran Mountains, Algeria, 104–119. *See also* Mascara, Algiers; Medea, Algiers; Tlemcen, Algiers
  crops, 105
  erosion, 108–110
  geological description, 105
  rainfall, 108, 116
  run-off plots, 105
  runoff, 108, 116–117
    plots, 106
    rates, 108
  soils, 105
    brown calcareous, 105, 109, 112 (*See also* Brown calcareous soil)
    erosion, 107, 109
      measurement of, 107, 109
      sheet, 108–110
    experimental treatments, 106

339

Fersiallitic, 105, 109 (*See also* Fersiallitic soils)
  properties, 106
  Vertic, 105, 109 (*See also* Vertic soils)
Benin, 143–153
Bicarbonate, as dominating anion for river waters, 226
Biodiversity, 10, 11, 25
  international convention, 9
  soil C stock and, 88
Biogeochemical models, 38
Brahmaputra River, 228
Brazil
  central plateau, 157–165
  rivers, 239–252
    watersheds, 241
Brown calcareous soils, 105, 109
  bulk density, 111
  treatment
    C stock changes and, 110–113
    CER and, 111
    eroded organic C in, 111, 118–119
    rainfall during, 107–108, 116
    runoff and, 108, 109, 116–117
    sheet erosion and, 108–110
    SOC and, 110–113, 118
Budgets, 31, 37, 97, 99, 229–231, 283–284, 331
Bulk density, 92, 94
Burkina Faso, 127–140
  carbon, 135, 136
    in dissolved rains, 134–135
    losses, 135, 136, 138
  cropping systems, 126, 127
  erosion, 131–132
    carbon losses and, 135, 136
    causes of, 138
    fine particle loss by, 135, 136
    measurement of, 132–135
    nitrogen losses and, 135, 136
  losses
    carbon, 135, 136, 138
    fine particle, 135, 136
    nitrogen, 135, 136
    rainfall and, 130–131
  rainfall
    runoff and, 130–131
    soil losses and, 130–131
  runoff
    complete cropping and, 131
    rainfall and, 130–131
    soil treatment and, 131–132
  soil
    analysis of, 129
    carbon, 132, 133
    carbon losses, 135, 136
    erosion, 131–135
    nitrogen losses, 135, 136
    physical properties of, 130

C

C stock changes, 194. *See also* Soil C stock, changes
  in banana cropping, 95
  in Fersiallitic soils, 110–113
  intensive tillage and, 95
  with no tillage, 95, 150
  in pineapple cropping
    intensive-till, 95
    no-till, 95
  in sugar cane cropping, 95
Cabras River, 241, 243
Carbon, 132, 133
  dissolved, 134–135, 209, 212, 217
    inorganic, 209, 211
      atmospheric *vs.* total river, 231
      carbonate weathering, 212, 227
      concentrations, 226, 235
      export rates, 226
      flux increases, 221, 235
      global distribution, 227
      occurrence, 212
      origins, 210, 211, 212, 213, 231, 232
      particulate carbon and, 215
      ratio to DOC, 213, 214
      sensitivity to anthropic influence, 220
      silicate weathering, 212
      stability, 215
      in surface runoff water, 215
      TAC and, 223, 228
      trends, 221
      weathering, 212
    organic, 78, 217, 221, 225
      analyses, 209, 210
      ^{14}C dating, 211
      chemical properties, 282–283
      concentrations, 216, 218
      DIC *vs.*, 213, 214
      dominance in TAC, 212
      evolution of, 216
      export, 218, 225, 226
      fluxes, importance of, 281
      increase in degradable, 222
      leaching, 28, 211
      in Mengong catchment, 279, 280
      origins, 211, 212, 213
      POC and, 217
      prime sources, 217
      ratio to DIC, 213, 214
      regional discrepancies, 229
      river specific discharge and, 217
      sinks, 217
      TOC and, 261, 262
      total, 209, 213
      trends, 221
      TSS and, 217
  erosion-induced global emission, 37
  exportation, 224
  fluvial (*See also* Riverine carbon)
    anthropogenic impacts on, 220–221
    autochthonous, 220

inorganic
    dissolved, 209, 211
        atmospheric *vs.* total river, 231
        carbonate weathering, 212, 227
        concentrations, 226, 235
        export rates, 226
        flux increases, 221, 235
        global distribution, 227
        occurrence, 212
        origins, 210, 211, 212, 213, 231, 232
        particulate carbon and, 215
        ratio to DOC, 213, 214
        sensitivity to anthropic influence, 220
        silicate weathering, 212
        stability, 215
        in surface runoff water, 215
        TAC and, 223, 228
        trends, 221
        weathering, 212
    particulate, 209, 210
        autochthonous, 213, 217, 220, 228, 233
        export of, 227
        origins, 211, 212, 213
        production, 217
        sinks, 217
        TSS and, 211
    total, 213
isotropic composition, 240
organic
    dissolved, 78, 217, 221, 225
        analyses, 209, 210
        ^{14}C dating, 211
        chemical properties, 282–283
        concentrations, 216, 218
        DIC *vs.*, 213, 214
        dominance in TAC, 212
        evolution of, 216
        export, 218, 225, 226
        fluxes, importance of, 281
        increase in degradable, 222
        leaching, 28, 211
        in Mengong catchment, 279, 280
        origins, 211, 212, 213
        POC and, 217
        prime sources, 217
        ratio to DIC, 213, 214
        regional discrepancies, 229
        river specific discharge and, 217
        sinks, 217
        TOC and, 261, 262
        total, 209, 213
        trends, 221
        TSS and, 217
    particulate, 213
        algal, 220
        analysis, 210
        autochthonous, 210, 211, 220
        average age, 211
        ^{14}C dating, 211
        concentrations, 218
        detrital, 220
        DOC and, 217
        export, 218, 226, 234
        floods and, 216
        fluxes, 230, 234
        fossil, 211
        fossil controversy, 211, 212, 213, 222–224
        nonalgal, 220
        origins of, 222–224
        sediment yield and export of, 218
        sedimentary, 211
        soil, 230
        TOC and, 218, 219
        trapping efficiency, 217
        TSS and, 211, 217–219, 230, 231
    total, 228
particulate
    inorganic, 209
        autochthonous, 213, 217, 220, 228, 233
        export of, 227
        origins, 211, 212, 213
        production, 217
        sinks, 217
        TSS and, 211
    organic, 213
        algal, 220
        analysis, 210
        autochthonous, 210, 211, 220
        average age, 211
        ^{14}C dating, 211
        concentrations, 218
        detrital, 220
        DOC and, 217
        export, 218, 226, 234
        floods and, 216
        fluxes, 230, 234
        fossil, 211
        fossil controversy, 211, 212, 213, 222–224
        nonalgal, 220
        origins of, 222–224
        sediment yield and export of, 218
        sedimentary, 211
        soil, 230
        TOC and, 218, 219
        trapping efficiency, 217
        TSS and, 211, 217–219, 230, 231
    in rivers (*See* Riverine carbon)
    selectivity, 138
    sinks, forests as, 15
    storage, 143–154, 302, 335
        erosion and, 38
    total atmospheric
        calculation of, 212
        dominant species in limestone regions, 212
        export, 228
        global budget of, 229
    transfer among adjacent ecosystems, 19
Carbon budgets, 31, 37, 97, 99, 229–231, 283–284, 331
Carbon cycle, 5, 23
    in continental biosphere, 5–8
    fossil fuel combustion and, 23
    global climate change and, 3

land use change and, 23
soil cultivation and, 23
Carbon dioxide, 4
    emission
        fossil carbon and, 5
        quantification of, 49
        fertilizer effect, 5
            in cereals, 5
            in grasslands, 5
            in trees, 5
    release, quantification of, 49
    sequestration, 16
    sinks, 25
Carbon dynamic models, 39
Carbon enrichment ratio (CER), 105, 112, 201, 203
    defined, 56, 164
    effect of eroded SOC and erosion on, 62
    eroded SOC and, 66–68
    as expression of relationship between eroded SOC, erosion and topsoil SOC, 64, 68–69
    land use and, 62–63
Carbon losses, 138, 193
    agricultural systems and, 96
    by erosion, 91–92, 95, 135
    by leaching, 91, 95
    by runoff, 95, 96
    sugarcane cropping and, 95, 96
    in suspended sediments, 95
Carbon redistribution, 45–46
Carbon sequestration, 13–21, 234, 331
    challenges to, 9–11
    defined, 14–15
    erosion and, 74
    international conventions, 9–10
    mean residence time, 17
    potential, 17
    processes, 25
    in soils, 8
    space scales, 17
    time scale, 17–19
    in trees, 8
Carbonaceous pollution, 221–222
Carbonates, secondary, 25
Carbonic acid, 25
"Carriage," 132
Catchment(s), 56, 167–178
    alfisols in, 169
    Cameroon, 281
    cropping systems
        natural wooded/grassy fallow vs. rice cultivation, 170–172
    description, 168–170
    land use and management, 170
    Laos, 303–304
    natural wooded/grassy fallow vs. rice cultivation, 170–176
        in bulk density, 172
        in CER, 174
        in eroded organic C, 174
        in organic C content, 172
        in organic C stock, 172
        in runoff, sediment, and carbon losses, 173–176
        in soil characteristics, 172–173
        in soil structural stability, 176
        in soil surface coverage by vegetation, 176
        in soil texture, 176
    Nsimi, 177
        as representative of slash and burn systems, 170
    rice cultivation vs. natural wooded/grassy fallow
        in bulk density, 172
        in CER, 172
        in eroded organic C, 174
        in organic C content, 172
        in organic C stock, 172
        in runoff, sediment, and carbon losses, 173–176
        in soil characteristics, 172–173
        in soil structural stability, 176
        in soil surface coverage by vegetation, 176
        in soil texture, 172
    sediment analysis, 171
    soil sampling, 171
    soil structural stability, 171–172
CER. See Carbon enrichment ratio (CER)
Cereal(s)
    fertilizer effect of carbon dioxide in, 5
    production
        decrease in, 104
        soil nutrients required for, 105
    rotation with legumes, 106, 111
    SOC and, 110
^{137}Cesium analysis, 56, 306
CIRAD, 8, 88
Clay soil, 89, 185, 187, 188
Climate change
    global
        carbon cycle and, 3
        causes, 4–5
        consequences, 5
        remediation, 3
    soil degradation and, 26
Colored waters, 281
Congo-Zaire River
    annual carbon yields, 269
    basin, 256–258
        DOC, 261
        DOC/TOC, 261
        factors controlling organic carbon in, 264–265
        POC, 261
        TSS, 260, 261
    organic carbon flux, 267–268
    tributaries, 257
Contaminant(s), 25
Corn Belt, U.S., 6
Cropping systems, 93, 94, 126, 127
    continuous cultivation, 185–191
        bulk density of soil in, 185, 186
        C/N ratio in, 185, 186
        CER in, 189
        OC and, 185, 201, 204
        runoff in, 188, 189
        sediment carbon in, 188, 189
        sediment OC in, 201, 204

SOC stock in, 185, 186
soil C losses in, 189
soil loss in, 188, 189, 201, 204
soil strength in, 187
steady state infiltration rates and, 202
water stable aggregates in, 187
erosion and, 87–99
improved fallow
bulk density of soil in, 185, 186
C/N ratio in, 185, 186
CER in, 189
OC loss, 201
runoff in, 188, 189, 192
sediment carbon, 188, 189
sediment OC in, 201, 203, 204
SOC stock in, 185, 186
soil C content in, 185, 186
soil C losses in, 189
soil loss in, 188, 189, 193, 201, 203, 204
soil strength in, 187
steady state infiltration rates and, 202
water stable aggregates in, 187
intensive-till
bulk density of soil in, 93, 94
C stock changes in, 95
carbon content of soil in, 93, 94
soil C stock in, 93, 94
soil losses in, 95
maize, 144
acidity and, 146
C stock changes in, 147, 150, 152
clay content in, 146
eroded carbon in, 149, 151, 152
residue biomass in, 147, 148, 149–150
runoff in, 149, 151, 152
soil carbon in, 146–147, 149
soil losses in, 149, 151, 152
maize and fertilizers, 144
acidity and, 146
C stock changes in, 147, 150, 152
clay content in, 146
eroded carbon in, 149, 151, 152
residue biomass in, 147, 148, 149–150
runoff in, 149, 151, 152
soil carbon in, 146–147
soil losses in, 149, 151, 152
no-till
bulk density of soil in, 93, 94, 185, 186
C/N ratio in, 185, 186
C stock changes in, 95, 150
carbon content of soil in, 93, 94
CER in, 189
runoff in, 188, 189
sediment C losses, 189
sediment concentration in, 188, 189
SOC stock in, 185, 186
soil C losses in, 189
soil C stock in, 93, 94
soil losses in, 95, 188, 189, 193
soil strength in, 187
water stable aggregates in, 187

relay maize and legume crop cover, 144
acidity and, 146
C stock changes in, 147, 150, 152
clay content in, 146
eroded carbon in, 149, 151, 152
residue biomass in, 147, 148, 149–150
runoff in, 149, 151, 152
soil carbon in, 146–147
soil losses in, 149, 151, 152

D

Dams, 291
sedimentary profiles, 293, 294
Danube River, 213
Deforestation, 5, 6, 167
alternative to, 8
as land use change, 25, 27
mineralization after, 83
SOC losses by water erosion after, 75, 76
Desertification, 9–10, 26
international convention, 9–10
DIC. *See* Dissolved inorganic carbon (DIC)
Dissolved inorganic carbon (DIC)
atmospheric *vs.* total river, 231
carbonate weathering, 212, 227
concentrations, 226, 235
export rates, 226
factors affecting, 226–227
flux increases, 221, 235
global distribution, 227
occurrence, 212
origins, 210, 211, 212, 213, 231, 232
particulate carbon and, 215
ratio to DOC, 213, 214
sensitivity to anthropic influence, 220
silicate weathering, 212
stability, 215
in surface runoff water, 215
TAC and, 223, 228
trends, 221
weathering, 212
Dissolved organic carbon (DOC), 78, 217, 221, 225
analyses, 209, 210
^{14}C dating, 211
chemical properties, 282–283
concentrations, 216, 218
DIC *vs.*, 213, 214
dominance in TAC, 212
evolution of, 216
export, 218, 225, 226
fluxes, importance of, 281
increase in degradable, 222
leaching, 28, 211
in Mengong catchment, 279, 280
origins, 211, 212, 213
POC and, 217
prime sources, 217
ratio to DIC, 213, 214

regional discrepancies, 229
river specific discharge and, 217
sinks, 217
TOC and, 261, 262
total, 209, 213
trends, 221
TSS and, 217
DOC. *See* Dissolved organic carbon (DOC)
Dolores River, 215
Dordogne River, 217
Dranse River, 213, 227

E

Ecoenergy International Corporation, 15
"Edge-effects," 127
Eroded organic carbon (EOC), 32, 111, 113
 fate of, 28–31
  sedimentologist view, 29–30
  soil scientist view, 29–30
 following treatment of
  brown calcareous soils, 111, 118–119
  fersiallitic soils, 111, 118–119
 measurement, 107
 monitoring, 311–321
  fine size, 312–320
  with isotropic labeling of soils and suspended sediments, 311–312
  of suspended organic matter, 320–321
 seasonal variations and, 113–116
 as sink, 74
 SOC and, 113
 soil losses and, 165
 in various cropping systems
  maize, 149, 151, 152
  maize and fertilizers, 149, 151, 152
  natural wooded/grassy fallow *vs.* rice cultivation, 174
  relay maize and legume crop cover, 149, 151, 152
  rice cultivation *vs.* natural wooded/grassy fallow, 174
Erodibility, 204
 index, 89
Erosion, 64–66
 carbon losses and, 91–92, 95, 135, 136 (*See also* Eroded organic carbon (EOC))
 carbon sequestration and, 74
 carbon storage and, 38
 causes, 10, 138
 CER and, 62, 66–68
 control, 104
 cropping systems and, 87–99
 evaluation, 170
 fine particle loss by, 135, 136
 impact of, 139
 interrill, 74, 82
 measurement, 56, 106–107
  qualitative approach, 132–135
 nitrogen losses and, 135, 136

quantification, 56
relationship between topsoil content, SOC erosion and, 61–62
seasonal variations and, 113–116
selectivity, 68
sheet, 108–110, 117, 119
 significance, 105
 slope gradient and, 117
 SOC and, 118
of SOC, 24, 61, 80–81, 304–306
soil
 in Beni-Chougran Mountains, Algeria, 107, 109
  measurement of, 107, 109
  sheet, 108–110
 carbon dynamics and, 26–28
 crop production and, 50
 defined, 56
 effect on carbon storage, 38
 measurement, 57–60
 mineralization, 24
 as multistage process, 37
 on site effects, 24
water, 56, 64, 332
 andosol resistance to, 74
 fallowing and, 171, 177–178
 importance of rainfall and slope in, 64
 mechanisms, 56
 OC losses by, 56, 75, 80–81
 scale effect, 176–177
 SOC losses by, 56, 75
Eutrophication, 210, 211, 222, 233

F

Fallow, 126, 138–139
 grazed, 106, 109, 111, 113
 improved, 183–194, 198–205
  OC and, 185, 203, 204
  OC enrichment ratio and, 203, 204
  soil loss and, 188, 189, 193, 201, 203, 204
 plowed, 106, 113
 protected, 106, 108, 109, 110, 111, 112, 113
 runoff, 117
 soil losses, 117
 reduction, 168
 vegetated, 106, 113
 water erosion and, 168
Fast-wetting test, 171
Ferralsols, 158–165
 bulk density, 158
 particle size distribution, 158
 SOC, 158
 treatment, 159–164
  effects, 160–164
   on CER, 164
   on OC in sediment, 161–163
   on runoff coefficient, 161
   on soil losses, 160
  rainfall during, 160, 161

Ferrihydride, 82
Fersiallitic soils, 105, 109
    bulk density, 111
    treatment
        C stock changes and, 110–113
        CER and, 111
        eroded organic C in, 111, 118–119
        rainfall during, 107–108, 116
        runoff and, 108, 109, 116–117
        sheet erosion and, 108–110
        SOC and, 110–113, 118
Fertility, 88, 104
Fertilizer(s), 144
    leaching, 25
    nitrogen, 240
Floods, 168, 296
Forest(s)
    as carbon sinks, 15
    evergreen vegetation, 75
    humid soils, 6
    rain, 168
        soil, 251
    soil C content, 7
Fossil fuel combustion, 23
French Institute CIRAD, 8
Fungicides, 88

G

Gambia River, 212, 213, 227, 228
Garonne River, 210
Geomorphological models, 3
GHG. *See* Greenhouse gas(es)
Gironde River, 210
Global Assessment of Soil Degradation, 10
Global budgets, 229–231, 331
Global climate change
    carbon cycle and, 3
    causes, 4–5
    consequences, 5
    remediation, 3
Global models, 229–231
Global warming potential, 16
Grasslands, 5
Greenhouse effect, 3
Greenhouse gas(es), 23, 44, 331, 332
    efflux, 26
    emission, 5, 150–151
        forestry and, 56
        land use and, 56
    main sources of, 5
    runoff and, 24
Groundwater recharge, 168

H

Heanes technique, 171

Herbicides, 88
Hill lakes, 289–296
Hillside, 336
Horton-type model, for estimating infiltration rate, 199
Houay Pano catchment, 307–311
Huang Re. *See* Yellow River
Hudson River, 211

I

Imogolite, 82
Inceptisols, 57
Infiltration rate, Horton-type model for estimating, 199
Inorganic carbon
    dissolved
        atmospheric *vs.* total river, 231
        carbonate weathering, 212, 227
        concentrations, 226, 235
        export rates, 226
            factors affecting, 226–227
        flux increases, 221, 235
        global distribution, 227
        occurrence, 212
        origins, 210, 211, 212, 213, 231, 232
        particulate carbon and, 215
        ratio to DOC, 213, 214
        sensitivity to anthropic influence, 220
        silicate weathering, 212
        stability, 215
        in surface runoff water, 215
        TAC and, 223, 228
        trends, 221
        weathering, 212
    particulate, 209
        autochthonous, 213, 217, 220, 228, 233
        export of, 227
        origins, 211, 212, 213
        production, 217
        sinks, 217
        TSS and, 211
Intergovernmental Panel on Climate Change (IPPC), 4
Interill erosion, 74, 82
International conventions, 9–10
    on biodiversity, 9
    on climatic change, 9
    on desertification, 9–10
International Panel on Climate Change, 15
IPPC. *See* Intergovernmental Panel on Climate Change (IPPC)
Iron
    acid oxalate extractable, 78
    pyrophosphate extractable, 78

J

Ji-Paraná River, 241, 243, 251

K

Kenya, 181–194
  clay soil, 185, 187, 188
  cropping systems
    continuous cultivation, 185–191
      bulk density of soil in, 185, 186
      C/N ratio in, 185, 186
      CER in, 189
      runoff in, 188, 189
      sediment C loss, 189
      sediment concentration in, 188, 189
      SOC stock in, 185, 186
      soil C losses in, 189
      soil loss in, 188, 189
      soil strength in, 187
      water stable aggregates in, 187
    improved fallow
      C/N ratio in, 185, 186
      CER in, 189
      runoff in, 188, 189, 192
      sediment C loss, 189
      sediment concentration in, 188, 189
      SOC stock in, 185–191, 186
      soil C losses in, 189
      soil loss in, 188, 189, 193
      soil strength in, 187
      water stable aggregates in, 187
    no-till
      bulk density of soil in, 185, 186
      C/N ratio in, 185, 186
      CER in, 189
      runoff in, 188, 189
      sediment C losses, 189
      sediment concentration in, 188, 189
      SOC stock in, 185, 186
      soil C losses in, 189
      soil losses in, 95, 188, 189, 193
      soil strength in, 187
      water stable aggregates in, 187
Kyoto Protocol, 6, 15

L

Lake milking, 217
Lakes, 217
Land cover, 6, 143–153
Land cultivation, 5
Land management, 3, 170
Land use
  CER and, 62–63
  change, 23, 25, 167
  GHG emission, 56
  heterogeneous, 127
  homogenous, 126–127
  soil organic carbon and, 6
Laos, 167–178. *See also* Catchment(s)
Le Bissonnais laboratory test, 171

Leaching, 211
  carbon losses by, 91, 95
  fertilizer, 25
Legume cover crop, 143–153
Loire River, 213
Lysimeter, 91

M

Mackenzie River, 213
Martinique, 87–99
  banana cropping in
    bulk density of soil and, 93, 94
    C stock changes, 95
    carbon content of soil and, 93, 94
    rotating systems, erosion and, 87–99
    soil C budgets, 99
    soil C stock, 93, 94
    soil losses, 95
  pineapple cropping
    intensive-till
      bulk density of soil and, 93, 94
      C stock changes in, 95
      carbon content of soil and, 93, 94
      soil C stock and, 93, 94
      soil losses in, 95
    no-till
      bulk density of soil and, 93, 94
      C stock changes in, 95
      carbon content of soil and, 93, 94
      soil C stock and, 93, 94
      soil losses in, 95
    soil C budgets, 97
Mascara, Algiers, 105
  CER, 112
  maximum runoff rates, 108
  rainfall, 107
  runoff plots, 105
  soil, 105
Mean residence time, 20
Medea, Algiers, 105
  CER, 112
  maximum runoff rates, 108
  rainfall, 105
  runoff plots, 105
  soil, 107
Mediterranean regions, 105. *See also* Beni-Chougran Mountains, Algeria
Mengong catchment, 278
  DIC in, 279–280
  DOC in, 279, 280
  global C budget at, 283–284
  total soil C content, 278
Methane, 4, 16
  radiative forcing, 16
Mineralization, 24, 38, 120
  after deforestation, 83
  tillage erosion and, 49
Mississippi River, 228

Mogi River, 241, 243
Mollisols, 57
Multilayer landscape model structure, 40

N

Nitisol, 89
Nitrogen, 132, 133, 305
    in dissolved rains, 134–135
    losses by erosion, 135
Nitrogen fertilizers, 240
Nitrous oxide, 4, 16
    emission, 150–151
Nyong River basin, 275–284
    characterization, 276–277
    TSS in, 280

O

Oak Ridge National Laboratory, 15
OC. *See* Organic carbon (OC)
Organic carbon (OC), 209, 212, 214. *See also* Soil organic carbon (SOC)
    annual transport of particulate, 30
    dissolved, 78, 217, 221, 225
        analyses, 209, 210
        ^{14}C dating, 211
        chemical properties, 282–283
        concentrations, 216, 218
        DIC *vs.*, 213, 214
        dominance in TAC, 212
        evolution of, 216
        export, 218, 225, 226
        fluxes, importance of, 281
        increase in degradable, 222
        leaching, 28, 211
        in Mengong catchment, 279, 280
        origins, 211, 212, 213
        POC and, 217
        prime sources, 217
        ratio to DIC, 213, 214
        regional discrepancies, 229
        river specific discharge and, 217
        sinks, 217
        TOC and, 261, 262
        total, 209, 213
        trends, 221
        TSS and, 217
    eroded, 32, 111, 113
        fate of, 28–31
            sedimentologist view, 29–30
            soil scientist view, 29–30
        following treatment of
            brown calcareous soils, 111, 118–119
            fersiallitic soils, 111, 118–119
        measurement, 107
    monitoring, 311–321
        fine size, 312–320
        with isotropic labeling of soils and suspended sediments, 311–312
        of suspended organic matter, 320–321
    seasonal variations and, 113–116
    as sink, 74
    SOC and, 113
    soil losses and, 165
    in various cropping systems
        maize, 149, 151, 152
        maize and fertilizers, 149, 151, 152
        natural wooded/grassy fallow *vs.* rice cultivation, 174
        relay maize and legume crop cover, 149, 151, 152
        rice cultivation *vs.* natural wooded/grassy fallow, 174
    fallow treatment and, 185, 201, 204
    loss, in catchment, 295–296
    particulate, 209, 210, 213, 280–281
        algal, 220
        analysis, 210
        autochthonous, 210, 211, 220
        average age, 211
        ^{14}C dating, 211
        concentrations, 218
        detrital, 220
        DOC and, 217
        export, 218, 226, 234
        floods and, 216
        fluxes, 230, 234
        fossil, 211
        fossil controversy, 211, 212, 213, 222–224
        nonalgal, 220
        origins of, 222–224
        recycled sedimentary, 211
        sediment yield and export of, 218
        sedimentary, 211
        soil, 230
        TOC and, 218, 219
        trapping efficiency, 217
        TSS and, 211, 217–219, 230, 231
    runoff, 82
    sedimentary, 30, 82
    sink, 81
    topsoil, 82, 302

P

Particulate inorganic carbon (PIC), 209
    autochthonous, 213, 217, 220, 228, 233
    export of, 227
    origins, 211, 212, 213
    production, 217
    sinks, 217
    TSS and, 211
Particulate organic carbon (POC), 213
    algal, 220

analysis, 210
autochthonous, 210, 211, 220
average age, 211
^{14}C dating, 211
concentrations, 218
detrital, 220
DOC and, 217
export, 218, 226, 234
floods and, 216
fluxes, 230, 234
fossil, 211
fossil controversy, 211, 212, 213, 222–224
nonalgal, 220
origins of, 222–224
sediment yield and export of, 218
sedimentary, 211
soil, 230
TOC and, 218, 219
trapping efficiency, 217
TSS and, 211, 217–219, 230, 231
Particulate organic matter (POM), 239
analysis, 243
isotropic carbon composition, 240, 250
phytoplankton as source, 251
sources, 248–249
Pesticides, 9, 88
PIC. See Particulate inorganic carbon (PIC)
Pineapple cropping
intensive-till
bulk density of soil and, 93, 94
C stock changes in, 95
carbon content of soil and, 93, 94
soil C stock and, 93, 94
soil losses in, 95
no-till
bulk density of soil and, 93, 94
C stock changes in, 95
carbon content of soil and, 93, 94
soil C stock and, 93, 94
soil losses in, 95
soil C budgets, 97
Piracicaba River, 241
carbon-nitrogen ratio, 251
watershed analysis, 243
Pisca River, 241, 243
POC. See Particulate organic carbon (POC)
POM. See Particulate organic matter (POM)
Principle component analysis, 190–191
Pyrophosphate extractable minerals, 78

R

Radiative forcing, 16
Rain forest, 168, 251
Rainfall simulation, 69, 127, 199
Reforestation, 6, 15
Relay-cropping maize, 144
Reservoirs, 217
Residue biomass, 147–148

Rhone River, 210, 217
Rio Boconó, 304, 305
Rio Negro, 213, 226
basins, 269
carbon export, 228
River(s)
acidification, 234
bicarbonate as dominating anion for, 226
black, 264
carbon in (See Riverine carbon)
carbon transport in, 334–336
chemical contamination, 233
clearwater, 264
course segmentation, 233
flow, POM and, 239
flow regulation, 233
holocene, 231–233
major syndromes, 233
neoarheism, 233
noncarbonated basins, 212
organic carbon, 243
POM, 239
salinization, 234
stilting, 233
systems, evolution of, 233–235
total suspended solids, 211
watersheds, 241, 243, 251, 304, 305, 336
white, 264
Riverine carbon
annual yields, 269
discharge relations, 215–216
dissolved, 209, 212, 217
analyses, 209, 210
inorganic, 209 (See also Dissolved inorganic carbon (DIC))
carbonate weathering, 212, 227
export rates, 226
origins, 210, 211, 232
ratio to DOC, 213, 214
silicate weathering, 212
organic, 217, 221, 225 (See also Dissolved organic carbon))
chemical properties, 282–283
DIC vs., 213, 214
export, 226
fluxes, importance of, 281
total, 209, 213
effect of waterbodies, 217
eutrophication and, 222
export rates, 226
basin lithology and, 226
runoff and, 226
exportation, 224–231
future evolution, 233–235
global flux to oceans, 229
holocene, 231–233
inorganic, 209, 212, 214
atmospheric vs. total river, 231
carbonate weathering, 212, 227
concentrations, 226, 235

dissolved, 209 (*See also* Dissolved inorganic carbon (DIC))
   carbonate weathering, 212, 227
   export rates, 226
   origins, 210, 211, 232
   ratio to DOC, 213, 214
   silicate weathering, 212
equilibrium of, 222
export rates, 226
   factors affecting, 226–227
flux increases, 221, 235
global distribution, 227
occurrence, 212
organic *vs.*, 213, 214
origins, 210, 211, 212, 213, 231, 232
particulate, 209, 215 (*See also* Particulate inorganic carbon (PIC))
   autochthonous, 213, 217, 220, 228, 233
   export of, 227
   origins, 211, 212, 213
   production, 217
   sinks, 217
   TSS and, 211
sensitivity to anthropic influence, 220
silicate weathering, 212
stability, 215
in surface runoff water, 215
TAC and, 223, 228
total, 209, 213, 214
trends, 221
weathering, 212
megacity impact on, 221–222
organic, 78, 209, 212, 214, 217, 221, 225
   analyses, 209, 210
   ^{14}C dating, 211
   chemical properties, 282–283
   concentrations, 216, 218
   DIC *vs.*, 213, 214
   dissolved, 217, 221, 225 (*See also* Dissolved organic carbon (DOC))
      chemical properties, 282–283
      DIC *vs.*, 213, 214
      export, 226
      fluxes, importance of, 281
      particulate, 209, 210, 213, 280–281
      total, 209, 213
   dominance in TAC, 212
   evolution of, 216
   export, 218, 225, 226
   fluxes, importance of, 281
   increase in degradable, 222
   leaching, 28, 211
   in Mengong catchment, 279, 280
   origins, 211, 212, 213
   particulate, 213, 217–219
      algal, 220
      analysis, 210
      autochthonous, 210, 211, 220
      average age, 211
      ^{14}C dating, 211
      concentrations, 218
      detrital, 220
      DOC and, 217
      export, 218, 226, 234
      export rates, 226
      floods and, 216
      fluxes, 230, 234
      fossil, 211
      fossil controversy, 211, 212, 213, 222–224
      nonalgal, 220
      origins of, 222–224, 232
      sediment yield and export of, 218
      sedimentary, 211
      soil, 230
      TOC and, 218, 219
      trapping efficiency, 217
      TSS and, 211, 217–219, 230, 231, 233
   POC and, 217
   prime sources, 217
   regional discrepancies, 229
   river specific discharge and, 217
   sinks, 217
   TOC and, 261, 262
   total, 209, 213
   trends, 221
   TSS and, 217
origins, 210–214, 240
   carbonaceous pollution and, 212
   from rock erosion and weathering, 212
   from soil and atmosphere, 212
particulate, 209, 210, 211, 217
   inorganic, 209
      autochthonous, 213, 217, 220, 228, 233
      export of, 227
      origins, 211, 212, 213
      production, 217
      sinks, 217
      TSS and, 211
   organic, 213, 217–219
      algal, 220
      analysis, 210
      autochthonous, 210, 211, 220
      average age, 211
      ^{14}C dating, 211
      concentrations, 218
      detrital, 220
      DOC and, 217
      export, 218, 226, 234
      export rates, 226
      floods and, 216
      fluxes, 230, 234
      fossil, 211
      fossil controversy, 211, 212, 213, 222–224
      nonalgal, 220
      origins of, 222–224, 232
      sediment yield and export of, 218
      sedimentary, 211
      soil, 230
      TOC and, 218, 219
      trapping efficiency, 217
      TSS and, 211, 217–219, 230, 231, 233
terrestrial sources, 240

total atmospheric, global flux to oceans, 229
total suspended solid, 213, 217–219, 230
transport, 334–336
variability, 212
Runoff
channeling, 127
depth, 188
DIC in, 215
GHG and, 24
measurement, 106–107
plot (*See* Runoff plot(s))
from protected fallow, 117
rate, 188
river carbon export rates and, 226
seasonal variations and, 113–116
shearing and transport capacities, 56
Runoff plot(s), 88, 89
size, 290

S

Sandy loam, 185, 187, 188
Savannas, 126–140
SCOPE/CARBON program, 256
Seine River, 214
Sheet erosion, 108–110, 117, 119, 218
significance, 105
slope gradient and, 117
SOC and, 118
Siberian rivers, 211
Silicate weathering, 212
carbon dioxide uptake by, 281–282
Silicon, acid oxalate extractable, 78
Sinks, 15, 25, 74, 81, 217
Slash and burn system, 170
Slow-wetting test, 171
SOC. *See* Soil organic carbon (SOC)
Soil(s)
eroded (*See also* Soil erosion)
physical composition of, 137–138
resistance to penetration, 187
resistance to shear, 187
Soil C budgets, 99
in banana treatment, 97
in intensive-till pineapple treatment, 97
in sugarcane treatment, 97
Soil C content, 92, 94
Soil C sequestration, 13–21, 117, 331
challenges to, 9–11
defined, 14–15
international conventions, 9–10
mean residence time, 17
potential, 17
processes, 25
in soils, 8
space scales, 17
time scale, 17–19
in trees, 8

Soil C stock, 82, 92, 94, 147
changes, 91, 99, 112, 332
entropic factors affecting, 6
increase, 8
as indicator of agrosystem sustainability, 88
measurements, 90
soil fertility and, 88
variations resulting from lateral transfers, 19
Soil carbon, 92, 146–147. *See also* Soil organic carbon (SOC)
decomposition, geomorphological model, 44–47
fate of eroded, 28–31
sedimentologist view, 29–30
soil scientist view, 29–30
outputs, 98, 99
Soil carbon stock. *See* Soil C stock
Soil cultivation, 23
Soil degradation
climate change and, 26
human-induced, 167
Soil erosion
carbon dynamics and, 26–28
characteristics, 56
crop production and, 50, 60
effect on carbon storage, 38
impact, 139
measurement, 57–60
mineralization, 24
as multistage process, 37
selectivity, 289
on site effects, 24
slope gradient and, 60
Soil fertility, 88
Soil inorganic carbon (SIC), 27. *See also* Inorganic carbon
Soil losses
in banana cropping, 95
by erosion, 80–81, 91
in no-till cropping systems, 95, 193
in pineapple cropping
intensive-till, 95
no-till, 95
from protected fallow, 117
rainfall and, 130–131
in sugar cane cropping, 95
in various cropping systems, 149, 151, 152
Soil organic carbon (SOC), 110–111, 331. *See also* Organic carbon (OC)
agricultural practices and, 6
analysis, 60, 77–80
Walkley-Black procedure in, 78
atmospheric carbon dioxide and, 38
beneficial impacts, 25
budget, 31, 37
density, 56
displaced, 25–26, 30, 74
dissolved, 78
dynamics
in deeper layers of soil profile, 49
soil erosion and, 26–28
enrichment ratio, 80, 81

eroded, 66
   CER and, 62, 64, 66–68
   defined, 64
   erosion and, 61, 68–69
   land management and, 61
   quantification of, 56
   rainfall and, 61
   slope gradient and, 61
   topsoil SOC and, 62
   vegetation and, 61
 erosion and, 24, 56, 60–61, 74, 104
   losses by, 80–81
   relationship between erosion, topsoil content and, 61–62
   water, 56
 fertility and, 104
 forms, 49
 global budget, 31
 global variations in, 118
 land cover and, 6
 land use management and, 6, 118
 losses, 80–81
   after deforestation, 75, 76
   rate of, 61
 measurement, 107
 mineralization, 38, 120
 partitioning, 6
 pool, 25–26
   depletion of, 38
   dissolved, 74
   labile, 74
   passive, 74
   slowly oxidizable, 74
   stability, 74
 potassium sulfate extractable, 78
 pyrophosphate extractable, 78
 redistribution, 24, 26
 in sediments, 56, 60, 80, 81
 sequestration, 117
 soil erosion and dynamics of, 26–28
 soil quality and, 37
 storage, 106, 311
 temperature and, 6
 in topsoil, 57, 61, 106, 311
 total stock, 6
 transported into rivers, 24
 water erosion and, 80–81
Soil organic matter (SOM), 24, 331
 composition of, 25
 management, 8
Soil Science Society of America, 15
Soil strength, 186–188
Solimões River, 213, 259
 carbon export, 228
 TSS concentrations, 262
Spatially distributed soil erosion model (SPEROS). See SPEROS; SPEROS-C
SPEROS, 39
SPEROS-C model
 application, 47–49
 description, 39–40
 implementation, 46–47
 structure, 44
 tillage erosion, 40–42
 water erosion, 40–42
St. Lawrence River, 250
Sudanese savannas, 126–140. *See also* Burkina Faso
Sugarcane cropping
 bulk density of soil and, 93, 94
 C stock changes, 95
 carbon content of soil and, 93, 94
 soil C budgets, 97
 soil C stock, 93, 94
 soil losses, 95
Suspended solids
 carbon-nitrogen ratio, 245
 coarse, 245–248
 fine, 245–248
 total, 213, 217–219, 224, 230, 280

T

Temperature, 6
 change, 3, 16
   effect of, 5
   projected, 4
Tillage, 9, 95, 150
 erosion, 40–42, 50
   mineralization and, 49
 intensive, 95
 soil displacement distributions, 43
Time scale
 diachronic approach, 18–19
 synchronic approach, 19
Tlemcen, Algiers, 105
 maximum runoff rates, 108
 rainfall, 107
 runoff plots, 105
 soil, 105
 soil C stock changes, 112
TOC. *See* Total organic carbon (TOC)
Topsoil, 46, 68, 313, 316
 carbon content, 7, 82, 302
 erosion, 165
 relationship between erosion, SOC erosion, and, 61–62
Total atmospheric carbon (TAC), 223, 224, 228
Total organic carbon (TOC), 305
 analyses, 306
 in sediments, 295
Total suspended solids (TSS), 213, 217–219, 224, 230, 280
Tropical rain forests, 168
TSS. *See* Total suspended solids (TSS)

U

United Nations Framework Convention on Climate Change, 16
U.S. Department of Agriculture, 15

U.S. Department of Energy, 14
"USLE" scale, 127
Utisols, 57, 150
    Alfisols and, 307

V

Verisols, 57
Vertic soils, 105, 109
    bulk density, 111
    treatment
        C stock changes and, 110–113
        CER and, 111
        eroded organic C in, 111, 118–119
        rainfall during, 107–108, 116
        runoff and, 108, 109, 116–117
        sheet erosion and, 108–110
        SOC and, 110–113, 118
Volcanic ash soils, 57, 74, Se also Andosol(s)

W

Walkley-Black procedure, 78, 171
Water erosion, 56, 332
    andosol resistance to, 74
    fallowing and, 171, 177–178
    importance of rainfall and slope in, 64
    mechanisms, 56
    OC losses by, 56, 75, 80–81
    scale effect, 176–177
    SOC losses by, 56
        following deforestation, 75
Water runoff samples, 79
Water stable aggregates, 89, 186–188
Watersheds, 241, 243, 251, 304, 305, 336
Weathering, 212, 227, 281–282
Wet oxidation technique, 171
Wetlands, 217
Wind erosion, 56
World Agroforestry Centre, 194

Y

Yellow River, 213, 217
    carbon export, 228

Z

Zaire River, 228
Zambia, 197–205
    soil properties, 198